Advances in Science, Technology & Innovation

IEREK Interdisciplinary Series for Sustainable Development

Advances in Science, Technology & Innovation (ASTI) is a series of peer-reviewed books based on important emerging research that redefines the current disciplinary boundaries in science, technology and innovation (STI) in order to develop integrated concepts for sustainable development. It not only discusses the progress made towards securing more resources, allocating smarter solutions, and rebalancing the relationship between nature and people, but also provides in-depth insights from comprehensive research that addresses the **17 sustainable development goals (SDGs)** as set out by the UN for 2030.

The series draws on the best research papers from various IEREK and other international conferences to promote the creation and development of viable solutions for a **sustainable future and a positive societal** transformation with the help of integrated and innovative science-based approaches. Including interdisciplinary contributions, it presents innovative approaches and highlights how they can best support both economic and sustainable development, through better use of data, more effective institutions, and global, local and individual action, for the welfare of all societies.

The series particularly features conceptual and empirical contributions from various interrelated fields of science, technology and innovation, with an emphasis on digital transformation, that focus on providing practical solutions to **ensure food, water and energy security to achieve the SDGs.** It also presents new case studies offering concrete examples of how to resolve sustainable urbanization and environmental issues in different regions of the world.

The series is intended for professionals in research and teaching, consultancies and industry, and government and international organizations. Published in collaboration with IEREK, the Springer ASTI series will acquaint readers with essential new studies in STI for sustainable development.

ASTI series has now been accepted for Scopus (September 2020). All content published in this series will start appearing on the Scopus site in early 2021.

Stella Kostopoulou •
Gricelda Herrera-Franco •
Jacob Wood • Kheir Al-Kodmany
Editors

Cities' Vocabularies and the Sustainable Development of the Silkroads

Springer

Editors
Stella Kostopoulou
Regional Economic and Tourism Development
Faculty of Economics and Political Science
Aristotle University of Thessaloniki
Thessaloniki, Greece

Gricelda Herrera-Franco
Faculty of Engineering Sciences
Universidad Estatal Península de Santa Elena (UPSE)
La Libertad, Ecuador

Jacob Wood
JCU Singapore Business School
James Cook University
Singapore, Singapore

Kheir Al-Kodmany
University of Illinois Chicago
Chicago, IL, USA

ISSN 2522-8714 ISSN 2522-8722 (electronic)
Advances in Science, Technology & Innovation
IEREK Interdisciplinary Series for Sustainable Development
ISBN 978-3-031-31026-3 ISBN 978-3-031-31027-0 (eBook)
https://doi.org/10.1007/978-3-031-31027-0

The title of the photo: Rotunda (or Rotonda) monument in Thessaloniki, Region of Central Macedonia, Greece by Miltiadis Maleas. Brief info: An early 4th-century AD Roman monument, next to the Arch of Galerius in the city center, inscribed on the UNESCO World Heritage List in 1988 as part of the Paleochristian and Byzantine monuments of Thessaloniki.

This Springer imprint is published by the registered company Springer Nature Switzerland AG
The registered company address is: Gewerbestrasse 11, 6330 Cham, Switzerland

Scientific Committee

Andreas Papatheodorou, Department of Tourism Economics and Management, University of the Aegean
Andrea Giampiccoli, Department of Hospitality and Tourism, Durban University of Technology
Andrzej Tucki, Maria Curie-Skłodowska University
Angeliki Ziaka, Department of Theology, Aristotle University of Thessaloniki
Carlos Mora-Frank, Centro de Investigaciones y Proyectos Aplicados a las Ciencias de la Tierra (CIPAT), ESPOL Polytechnic University, Guayaquil, Ecuador
Efthymia Sarantakou, Department of Tourism Management, University of West Attica
Fernando Morante-Carballo, Facultad de Ciencias Naturales y Matemáticas, Escuela Superior Politécnica del Litoral, Guayaquil, Ecuador
Giorgos Gritzas, Department of Spatial Planning and Development, Aristotle University of Thessaloniki
Grzegorz Iwanicki, Maria Curie-Skłodowska University in Lublin, Poland
Haejin Jang, James Cook University
Jhon Caicedo Potosi, Centro de Investigaciones y Proyectos Aplicados a las Ciencias de la Tierra (CIPAT), ESPOL Polytechnic University, Guayaquil, Ecuador
Joselyne Bárbara Solórzano Chauca, Centro de Investigaciones y Proyectos Aplicados a las Ciencias de la Tierra (CIPAT), ESPOL Polytechnic University, Guayaquil, Ecuador
K. Thirumaran, James Cook University, Singapore
Karen Yambay de Armijos, ESPOL Polytechnic University
Katarzyna Michalczuk, Department of Socio-Economic Geography, Faculty of Earth Sciences and Spatial Management, Maria Curie-Sklodowska University, Lublin, Poland
María Fernanda Jaya Montalvo, Centro de Investigaciones y Proyectos Aplicados a las Ciencias de la Tierra (CIPAT), ESPOL Polytechnic University, Guayaquil, Ecuador
Maribel Aguilar-Aguilar, Centro de Investigaciones y Proyectos Aplicados a las Ciencias de la Tierra (CIPAT), ESPOL Polytechnic University, Guayaquil, Ecuador
Marilena Papageorgiou, Department of Spatial Planning and Development, Aristotle University of Thessaloniki
Mir Ali, University of Illinois at Urbana-Champaign, USA
Niki Papageorgiou, Department of Theology, Aristotle University of Thessaloniki
Pantoleon Skayannis, Professor Emeritus (Infrastructure Policy), Department of Planning and Regional Development, University of Thessaly
Paúl Carrión-Mero, Centro de Investigaciones y Proyectos Aplicados a las Ciencias de la Tierra (CIPAT), ESPOL Polytechnic University, Guayaquil, Ecuador
Paulo César Escandón Panchana, Escuela Superior Politécnica del Litoral, ESPOL, Geo-Recursos y Aplicaciones GIGA, ESPOL Polytechnic University, Guayaquil, Ecuador
Peng Du, Thomas Jefferson University
Simona Azzali, James Cook University
Styliani Lefaki, Department of Architecture, Aristotle University of Thessaloniki

Acknowledgments

We want to thank the authors of the research papers that were selected for addition to this book. We would also like to thank the reviewers who contributed with their knowledge and constructive feedback to ensure the manuscript is of the best quality possible. Special thanks goes to the editors of this book for their foresight in organizing this volume and diligence in doing a professional job editing it. Finally, we would like to thank the IEREK team for supporting the publication of the best research papers submitted to the conference.

Introduction

The Silk Road, a historic network of trade routes that connected the East and the West, played an important role in the exchange of goods, ideas, and cultural practices across civilizations. Along this extensive network of routes, cities emerged into significant trade and culture centers, evolving distinctive identities that were reflected in their architecture and arts as a result of their diverse populations and cultural influences.

This conference proceedings book is a selection of research papers from the first edition of the International Conference on "Silk Road Sustainable Tourism Development and Cultural Heritage (SRSTDCH)" held in 2021 in collaboration with Aristotle University of Thessaloniki, the European Interdisciplinary Silk Road Tourism Centre, Greece, and the fifth edition of the International Conference on "Cities' Identity through Architecture and Arts (CITAA)" held in 2021 in collaboration with University of Pisa, Italy. The book brings together scholars from various disciplines, economists, architects, planners, tourism experts, and decision-makers to explore how the Silk Road sparked not only trade to thrive, but was also a catalyst for cultural exchange, tourism, and entrepreneurship. Through a range of case studies and interdisciplinary approaches, this book aims to provide a nuanced and comprehensive understanding of the cultural interaction and sustainable development along the Silk Roads, as well as the formation of cities' identities through architecture and arts.

The chapters in the book examine how the Silk Road influenced trade, architecture and the arts of cities along its routes. From the caravanserais and temples of Central Asia and China, to the bustling marketplaces of the Middle East and the Mediterrannean, the Silk Roads connected a vast and diverse region that was rich in cultural exchange and innovation.

In addition to its emphasis on the Silk Road legacy, this book also highlights how local cultures and traditions shaped the architecture and arts of cities along the routes. From the indigenous styles and motifs found in the art and architecture of Central Asia and China, to the blending of local and foreign influences in the arts of the Middle East and the Mediterrannean, this volume showcases the diversity and complexity of civilizations along the Silk Roads.

Overall, this conference proceedings book offers a thorough and nuanced examination of how the Silk Road impacted travel, trade, entrepreneurship, tourism, cultural exchange and artisanal crafts, and the development of cities' identities through their built environments. Through its interdisciplinary approach and diverse range of case studies, the book provides a valuable resource for scholars and researchers interested in the cultural exchange and sustainable tourism development along this ancient network of trade routes, as well as the formation of distinctive identifiers of the cities' architecture and the arts throughout the centuries and over modern times.

Contents

Urbanism, Place, and Design

Skyphilia: The Role of High Structures in the Formation of Space Identity

Haifei Zhou, Jiawei Leng, Yu Xu, and Lijun Sun

Abstract

This study explores how high structures shape the space identity through a vertical form of self-alienation and transcendence, which is proposed as skyphilia to be the counterpart of topophilia in this study. Three hierarchical concepts, namely, topophilia, topophobia, and skyphilia are elaborated to understand the relationship between man and the environment. Accordingly, different motivations to form the space identity are discovered based on various psychological cognitions. Typical high structures characterized by distinctive identities, ranging from the primitive societies to modern metropolia, such as the Babel Tower, the Egyptian Pyramids and Obelisks, the Eiffel Tower, the Sagrada Familia, the Burj Khalifa Tower, and the Golden Gate Bridge, are studied comparatively to discover their common base and differences in forming the space identity. The constituent motivations of skyphilia are specifically discussed, i.e., (1) the functional requirements of various high structures, (2) the materialistic symbolism to represent material sufficiency or technical excellence by alienating itself through height, and (3) the spiritual symbolism to reflect people's desire to transcend the world skyward. At last, the Zifeng Tower of Nanjing will be taken as an example with four modes of horizontality to build the equilibrium between topophilia and skyphilia. The results show that skyphilia and topophilia are two parallel ways to overcome the primitive topophobia through self-affirmation and self-alienation, but they may also lead to the other two kinds of topophobia characterized by over-affirmation and over-alienation. Skyphilia could be achieved by balancing its constituent motivations, and equilibrium between topophilia and skyphilia is essential for the space identity in modern cities.

Keywords

Skyphilia • Topophilia • Topophobia • High structures • Space identity

1 Introduction

Space identity involves a process of seeing and defining self in the given environment. For a city with spatial sophistication, its physical space is a reservoir of places with various meanings projected onto them. A city's identity, even physically seen the same, is then extremely different for individuals with various backgrounds (i.e., age, gender, education, profession, religion, culture, experience, and so on). Identification is like a mirror, in which those who have identified with it will see themselves reflected (Leach 2007). Given that people inhabit the same physical environment, how their perceptions origin, develop and vary among different groups, and in which level they share a common base and differentiate themselves from the other, becomes a valuable issue in dealing with the rapidly developing and increasingly homogeneous city space characterized by high structures, especially in China today.

A lot of prominent humanistic geographers have made great contributions to the understanding of human identification with the environment since the 1970s. Among them, Yifu Tuan is held up as a key voice in the popularization of Topophilia, a concept which literally means love of place and was first elaborated in his influential book *Topophilia: A Study of Environmental Perception, Attitudes and Values*. Topophilia involves the positive affective bond between

H. Zhou · J. Leng (✉) · L. Sun
School of Architecture, Southeast University, Nanjing, China
e-mail: jw_leng@seu.edu.cn

H. Zhou
e-mail: zhf_harris@126.com

L. Sun
e-mail: s-lijun@163.com

Y. Xu
School of Architecture, Sapienza University of Rome, Rome, Italy

S. Kostopoulou et al. (eds.), *Cities' Vocabularies and the Sustainable Development of the Silkroads*, Advances in Science, Technology & Innovation, https://doi.org/10.1007/978-3-031-31027-0_1

people and place or setting (Tuan 1974), it encapsulates any pleasant relationship we have with particular bits of the world we experience. This emotion calls for a delicate balance between the environmental characteristics and the aforementioned individual backgrounds, both bodily and mentally. For comparison, the feeling of aversion towards the environment was also studied by Tuan in his later books such as *Landscapes of Fear* and *Space and Place: The Perspectives of Experience*, in which the topophobic relation between the inhabitants and the built world are also sensitively investigated. Topophilia and topophobia, are then considered as two antipodal but interacted psychological patterns when describing our emotional relationship with the world, especially the modern cities. The enrichment and extension of the notion of topophilia have been ongoing for decades, another essential scholar who must be mentioned is Relph (1976, 1981, 1993), his theory of Place and Placelessness has some similar connotations to that of topophilia and topophobia, by criticizing the "eradication of distinctive places and the making of standardized landscapes that result from an insensitivity to the significance of place" (Relph 1976). Besides, the book *Topophilia and Topophobia: reflections on twentieth-century human habitat* edited by Xing Ruan and Paul Hogben, is a crystallization of the symposium with the same title held in Beijing in 2005 to reflect our emotional bonds to architecture and place in the new technology-driven century (Ruan and Hogben 2007). It contains several essential articles brilliantly addressing the diversity of topophilia, including that of Yifu Tuan, Joseph Rykwert, and Neil Leach. In addition, some quantitative methods were also developed based on topophilia and topophobia to better understand the affective bond between people and the environment, and thus stimulate local and regional development (Oliveira 2010; Šimáček 2020).

Topophilia and topophobia are philosophical musings of the modern city, from the 1970s to the 2020s, and from the West to Asia. The metropolises formed by widespread networks and mega-structures are extremely different from the ancient cities, in which human's biological and spiritual perceptions were closely connected to the cosmos, nature, and architecture. Topophilia and topophobia are then proposed timely to remind the disassociation of man and the built environment, to recall our positive affective bonds to places, and to find a new relationship between people and place of the modern city. However, in this introspection process, topophilia and topophobia are roughly considered as two opposite poles, i.e., the sensitive, nostalgic, and self-affirmed topophilia always conflicts with the insensitive, rootless, and self-alienated topophobia. The applicability and limitation of the antipodes, which were ordinarily cited in a way that favors topophilia over topophobia, have not been fully realized from both a theoretical and practical view. In addition, in investigating the regression of topophilia, the towering glass-and-steel buildings, which are considered as the materialized symbols of the modern city, are frequently questioned or criticized for their indifferent attitudes towards the city space (Kingwell 2005; Tuan 2007). Now that a majority of human beings will live in cities in the process of global urbanization, how to re-interpret the meaning of the omnipresent high structures in our inhabited city space, and how could they form a positive space identity for various citizens, becomes a tantalizing and valuable issue.

This paper will reconstruct the relationship between topophilia and topophobia, the concept skyphilia is coined to stand as the new counterpart of topophilia, and both of them will be evaluated rationally to recognize their positive aspects to form the space identity, and their adverse aspects leading to topophobia. Specifically, skyphilia is used to illustrate how it could help the formation of city identity, and the re-connection of citizens and the city space they inhabit. Following that, the Zifeng Tower of Nanjing is exemplified to show how skyphilia plays a unique role in the formation of city identity with diversity. At last, the extreme situation derived from skyphilia which may lead to topophobia is also discussed to further prescribe the scope of skyphilia.

2 Topophilia and Skyphilia: Two Modes to Overcome Topophobia

The scope of topophilia and topophobia is wide. The seemingly all-embracing concepts that guided Tuan's efforts to structure and enrich their intensions give a straightforward understanding of our affections to the environment, i.e., like and dislike, love and aversion from a dialectical perspective. In Tuan's two major books (Tuan 1974, 1979), he gave a sophisticated description of our perceptual system, individual human differences, and diverse environments characterized by geographical, political, social, and cultural features. One of the essential contributions by Tuan could be his elaborations on the mechanism forming what he encapsulated as topophilia, and his meticulous listing of all kinds of fears that humans may suffer, such as nature, disease, ghosts, and a lot of urban maladies (noise, traffic, fire, violence, crime, and discrimination, etc.). Tuan's work is more like revealing the phenomenon of topophilia and topophobia through biological, historic, and social investigations, and he didn't set topophilia and topophobia as opposite poles deliberately. Tuan never advocated for the prosperity of topophilia by dispraising topophobia in a radical way that many successors used to preach their vacuous conception of "topophilia".

In comparing the traditional societies and the modern world, Tuan pointed out that the primitive and traditional peoples lived in a vertical, rotary, and richly symbolical world, whereas modern man's world tends to be broad of surface, low of ceiling, nonrotary, aesthetic, and profane (Tuan 1974). This point was made based on the transformation of our relationship with nature, i.e., in the pre-scientific age, people's living vision is to adapt to nature whereas now we tend to dominate it. However, this shifting is not about the replacement of the primary and secondary relationship between humans and nature, but the scope of our understanding of nature and how we identity ourselves within it. In ancient times, people are more confined in their horizontal living space due to their lack of geographical knowledge, the topographic limitations are then compensated by vertical spaciousness to overcome people's fear of the undeveloped land and build human self-affirmation vertically in both constructive and spiritual ways. While nature is more stratified for our ancestors who gained their world view more from heaven than land, people living in modern societies are prone to consider nature as the ground beneath the feet, a constellation of landscapes or sceneries, the space that encompasses our settling, entertainment, activities, and memories, etc. Though Tuan's primitive proposing of topophilia seems to cover the two different attitudes towards nature, i.e., transcending nature and immersing in nature, the subsequent topophils [lovers of place (Betjeman 1947)] living in modern societies tend to understand topophilia as its literal meaning: the love of the very topography that we attach to. The other attitude towards nature, mostly held by our ancestors through a vertical form of transcendence, seems to be buried in oblivion or applied inappropriately. It deserves an excluding naming as a recurrent reminder that is distinctive from topophilia, and it could serve as a counterpart of topophilia to make up its unacknowledged shortcomings when in dealing with the enormously complex space of modern cities. That is what we expect for the newly coined conception in this study: skyphilia.

Besides the meaning of the space above ground, "sky" also indicates nothingness in the Chinese language. Skyphilia, as endowed in its literal meaning, refers to a kind of love driven by the urge to transcend the complex of the profane world, and in most cases, it facilitates our self-satisfaction with a vertical form towards the sky characterized by determination, purity, inanimateness, infinity, and imagination (Zhou et al. 2022).

Skyphilia is not the opposite of topophilia, but its parallel which reflects another kind of human affection for the world. As shown in Table 1, a lot of similar theories could be found to have a common base with topophilia and skyphilia. In illustrating empathy and abstraction, Worringer concluded the two poles of human artistic volition: whereas the precondition for the urge to empathy is a happy pantheistic relationship of confidence between man and the phenomena of the external world, the urge to abstraction is the outcome of a great inner unrest inspired in man by the phenomena of the outside world (Worringer 1997). Concerning Relph's theory of place and placelessness, he identifies two modes of spatial experience: the first one is what he called pragmatic space, perceptual space and existential space, which are instinctive, bodily, and immediate; the second that he identified is planning space, cognitive space, and abstract space, which are more cerebral, ideal, and intangible (Relph 1976). Relph used several dialectical opposites to conceptualize place experience, which is often misunderstood by critics, who claim that he favors places over placelessness, insideness over outsideness, authentic over inauthentic places, and rootedness over mobility. However, one could also find the extraordinary coverage and flexibility of Relph's place and placelessness, which may cover the connotations of topophilia and skyphilia to some extent. Besides, some similar dialectical oppositions embedded in the other theories, such as Neil Leach's *Topophilia and Topophobia*, and Georg Simmel's *Bridge and Door*, also subconsciously indicate the two parallel modes that we adapt to settle in the world: immersion and transcendence, although sometimes one pole of the antipodes is mistaken as the negative.

A rational re-evaluation of topophilia could then be conducted with its positive counterpart skyphilia discovered, and their relationship with topophobia requires an authentic reset. As illustrated in Fig. 1, two circles are defined separately to represent the world consisting of the two fundamental emotions: phobia and philia. The inner one is made up of topophilia and skyphilia, which jointly contribute to a pleasant world that is not only empathetic but also transcendent. The outer one is surrounded by various types of phobias which could be distinguished as:

- **Topophobia (Type 1)**: the horror of non-self. It is a bit similar to Agoraphobia, an abnormal fear of being helpless in a situation one couldn't build a tangible connection between the self and the environment. In the ancient world characterized by unformed nature, people are threatened by the wildness full of dangers and the uncertainty of the elusive climate. Efforts have been made by generations to escape from the phobia of dangerous and unfathomed nature, in which one couldn't gain his or her self-affirmation when dominated by feelings of insecurity. Two separate ways are then formed: the first is to explore and adapt to the land we live in horizontally (topophilia), and the other is to build a vertical anthropomorphic form alienated from the living plane (skyphilia).
- **Topophobia (Type 2)**: the horror of undifferentiated self. It exposes the danger of the loss of self that might be caused by over-affirmation with the environment. Based on two seminal texts (Roger Caillois's "Mimicry and Legendary Psychasthenia" and Jacques Lacan's "The Mirror Stage"), Leach (2007) provided a unique insight

Table 1 Binary oppositions for illustration of topophilia and skyphilia

Relative theories	Topophilia	Skyphilia
Yifu Tuan's *Topophilia*	Surface	Vertical
	Nonrotary	Rotary
	Profane	Symbolical
	Security	Grandeur
	Shade under the baobab	Circle under the heaven
	Worldliness	Wilderness
Wilhelm Worringer's *Abstraction and Empathy*	Empathy	Abstraction
	Beauty of the organic	Beauty of the life-denying inorganic
	Self-affirmation	Self-alienation
	Representation of space	Suppression of space
Edward Relph's *Place and Placelessness*	Place	Placeless
	Insideness	Outsideness
	Authenticity	Inauthenticity
	Rootedness	Mobility
	A static, bounded site	A dynamic, globally-connected process
Neil Leach's *Topophilia/Topophobia*	Attraction	Repulsion
	Identify the self	Distinguish the self
	Return home	Flee the nest
Georg Simmel's *Bridge and Door*	Connection	Separation
	Finitude	Infinitude
	Limitation	Limitlessness

into the formation of identity, which involves a continual shuttling between self-affirmation and self-alienation. In his opinion, the process to adapt to the environment is the search for the similar, whose ultimate objective is assimilation to the surroundings. We often ignore the extreme case of the empathetic assimilation, in which one may find himself dissolving into space. Devoured by the environment, the living creature is no longer the origin of the coordinates, but one point among others; it is dispossessed of its privilege and literally no longer knows where to place itself (Caillois and Shepley 1984). The typical horror of the undifferentiated self could be caused by the standardized places in cities and the extremely populist landscapes such as fascism.

- **Topophobia (Type 3)**: the horror of alienated self. It refers to the over-alienation that may lead to one's failure to identify himself or herself with the surroundings. Leach mentioned two conditions of the mirror stage: homomorphic identification and heteromorphic identification, i.e., identification with what is similar and identification with something that is the "other" (Leach 2007). The primitive urge of skyphilia is to distinguish oneself from the environment, a form of alienating identification. But if this process is reinforced actively or passively, one may fall into the phobia of isolation, discrimination, and rootlessness. The cloud-piercing skyscrapers that are indifferent to the outside world, and the gentrification of places developed for a specific group or class, are convincing demonstrations of this kind of topophobia.

Envisaged in the novel framework reconstituting topophobia, topophilia, and skyphilia, the traditional binary oppositions, in which one always governs the other or one pole is regarded as the negative, could reestablish itself to contribute to a new balanced, dynamic and introspective world of philias. Identity could then be formed between self-affirmation and -alienation, between absorption into the environment and transcendence of it, and between the extreme states of the potentially fascistic loss of self and the melancholic withdrawal into the self.

3 Skyphilia: The Motivations of High Structures

High Structures are not unique to the modern cities full of high-rises, they have covered almost the entire history of human development, ranging from the ancient Egyptian pyramids to the Gothic cathedrals, from the Eiffel Tower to

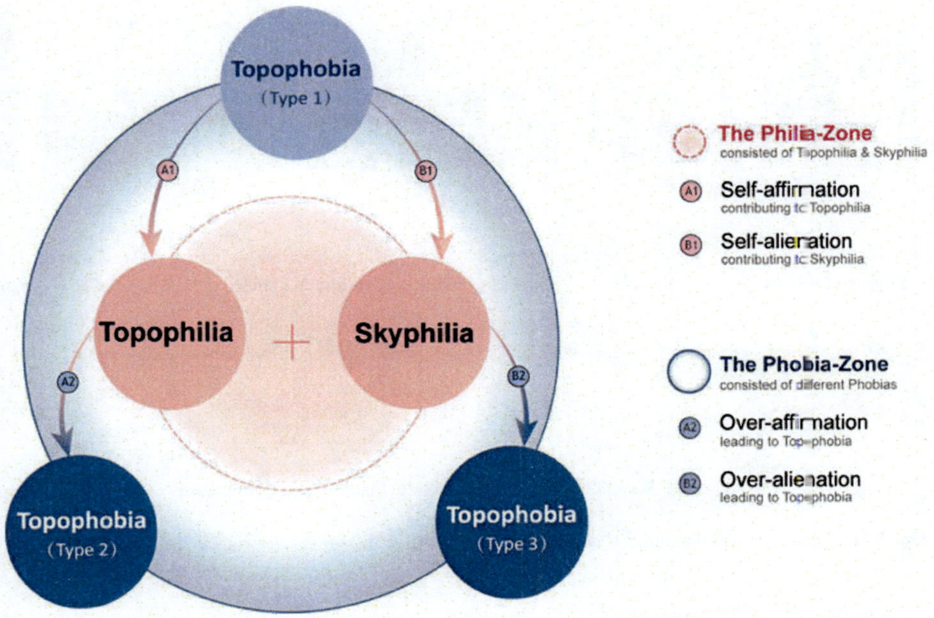

Fig. 1 The framework to identify the environment based on Philia and Phobia

the Burj Khalifa Tower. Skyphilia seems to be the nature of man, who not only tries to stay in touch with the earth but also obsesses about transcending it. The cosmic city is itself an outstanding example of a creation that transcends the routines of biological and social life, that directs people's attention heavenward (Tuan 2007). Alienating from our living plane, exploring the mystery of the sky, and searching for a pleasant balance between topophilia and skyphilia, are proven to be a ceaseless theme of human development.

The reasons for our pursuit of high structures are various, and it should not simply be attributed to the insatiable greed for material sufficiency, which is frequently blamed for the over-development of high buildings. It is irrational to belittle or speak highly of the high structures in their extreme situations, a thorough investigation of all the motivations of high structures may help to form the positive and impressing space identities characterized by a pleasant atmosphere of skyphilia and topophilia.

Three motivations of high structures could be roughly classified as (1) the functional role to satisfy architectural and structural requirements; (2) the materialistic symbolism to represent material sufficiency or Technical excellence by alienating itself through height; and (3) the spiritual symbolism to reflect people's desire to transcend the world skyward. As shown in Fig. 2, a comparison of the eminent high-rises throughout human development is beneficial to a comprehensive understanding of them. By distinguishing the differences on the aforementioned three motivations, they could be elaborated hereunder in search of the ideal state of skyphilia:

- ***Pyramid***
  The ancient pyramids were not only pharaohs' tombs and ritual sites, but also the symbolism reflecting people's view of life and death, earth and heaven. As Giedion addressed, the ziggurat had multiple symbolic meanings: it was the solid rock that emerged out of the chaos; the mountain which represented the center of the universe; the earthly throne of the gods; the monumental site for the offering of sacrifices; and the latter to the heaven (Giedion 1962). In addition, the extraordinary technologies building them are impressive even for contemporary people.
- ***Babel Tower***
  The Babel Tower was built to have an explicitly religious purpose, i.e., to link to heaven and seek the protection of the Patronus, it was a sacred gift symbolizing the royal sovereign and people's spiritual sustenance. Regular events were held in the tower every year, and the faithful would come from all over the country to worship. The revolving trail leading to the Marduk Temple at the top of this 90 m high tower, highlights the majesty and honor of that time with political and technological power.
- ***Obelisk***
  The Egyptian obelisks played a vital role in the religion of ancient Egyptians, placed in pairs at the entrance of the temples. In Egyptian mythology, the obelisks symbolized the sun god Ra, and it is hypothesized that their shapes were derived from natural phenomena associated with the sunset. There were few practical functions of the Egyptian obelisks apart from their monumental significance. Besides, the ancient obelisks are usually

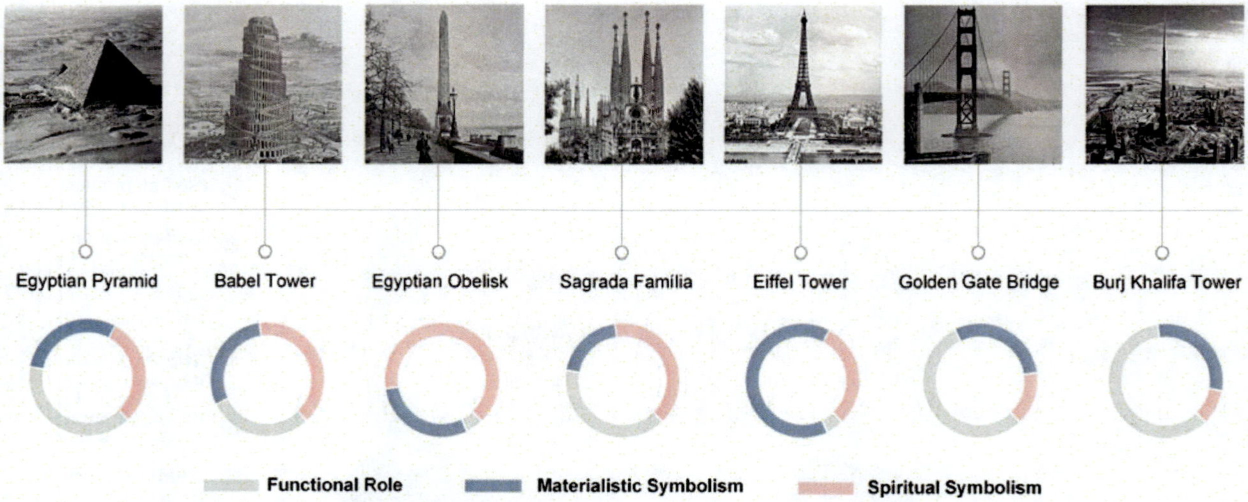

Fig. 2 Comparative motivation analysis of high structures

monolithically consisting of a single stone, which represents the superb stone craft of that time.

- ***Gothic Church***
  In comparing Gothic churches with the skyscrapers in a modern city, Tuan pointed out that: Church spires point to the sky, reminders that our destiny is not confined to this world, that we follow a path—a directional time—that begins with the earth but ends in heaven. Inside the church, however, services continue to cater to life's biological and social stages, and rituals are conducted to conform with nature's cycles (Tuan 2007). There seems to be an equilibrium between the inside topophilia and the outside skyphilia, which could be uniformly driven by the functional, material, and spiritual requirements of a Gothic church.
- ***Eiffel Tower***
  Finished in 1889 as the entrance to the 1889 World's Fair, the Eiffel Tower is still one of the most recognizable structures in the world. Standing 324 m high, the Eiffel Tower surpassed the Washington Monument to become the tallest man-made structure in the world after its completion. And before that, no structure had ever been constructed to a height of 300 m, or even 200 m. It is a remarkable piece of structural art due to the great technological progress, despite its lack of architectural function.
- ***Golden Gate Bridge***
  From the 1930s, the suspension bridge and the cable-stayed bridge which adopted high towers structurally to support the long spans became impressing landscape icons in the environment. Though the towers of bridges are mainly constructed for the structural requirement, the difficulty of their completion also reflects a country's highest technical strength in construction, and the cultural meanings are frequently given in the shape of the towers implying a solidified fusion of topophilia and skyphilia.
- ***Burj Khalifa***
  As a representative of the skyscrapers, the Burj Khalifa has been the tallest structure and building in the world since its topping out in 2009 with a total height of 829.8 m, which may represent Dubai's economic level in a way. Despite a feast for our eyes and their centralized functions that cater to business and finance, skyscrapers like Burj Khalifa are also questioned for their indifference to the environment and the self-serving images skyward. And in Tuan's opinion, the skyscrapers are subject to technological advances and fashions, which may diminish the eternality that we humans expect to hang on (Tuan 2007). The spiritual connection to the outsideness of skyscrapers calls for mending to define a new kind of skyphilia in modern societies.

4 Space Identity Formed by High Structures: Zifeng Tower as an Example

In a modern city characterized by diversification and complexity, many contradictions are symbiotic, such as the urban convenience and maladies, the comfort and chaos, the patrician and the indigent, the magnate and the self-employed, etc. Identity, as Leach addressed, is born of an interaction between the self and the other, between the horror of the undifferentiated self and the horror of the alienated self, between becoming one with the world and becoming distinct from that world (Leach 2007), or from a perspective of directionality in this study, between topophilia and skyphilia, which are reciprocal in the formation of space identity, especially that of the high-rises.

The Zifeng Tower is a 450 m high skyscraper located in Nanjing, Jiangsu Province of China, and it remains as the highest building in Nanjing since its completion in 2010. The tower's stepping is functional; office, hotel, and retail are located above grade, while restaurants and a public observatory are housed at the top of it. Despite its usual spatial functions as a typical skyscraper, Zifeng Tower stands in the old town district of Nanjing, which is known as the ancient capital of six dynasties of ancient China. The vertical image of Zifeng Tower symbolizing modernity and its surrounding urban texture and space formed through dynasties, could then demonstrate a diversified and encompassing space identity that could be reinforced by the fusion of topophilia and skyphilia.

To better understand the reciprocal relationship between the Zifeng Tower and its surroundings, four graphical modes of horizontality (Fig. 3) are herein used to investigate how an impressing space identity could be formed by the conjunction of topophilia and skyphilia, i.e. (1) the linear water axis; (2) the linear road axis; (3) the open ground; and (4) the enframed scenery. As shown in Fig. 4a, b, the water between the Sakura Island and the Ring Island of Xuanwu Lake generates two water axis bringing our eyes straight up to the Zifeng Tower. The horizontal scenery which is free to the public extends to a pure form of a skyphilia without the urban chaos. In Fig. 4c, d, two streets (Xiqiao Road and GaoYunLing Road) whose directions could be extended to the Zifeng Tower, encompass the profane world full of restaurants, shops, and various passers-by. Scenes in the streets are true reflections of the duality in a modern city, giving people implications to live in both reality and dreams. In Fig. 4e, f, two playgrounds are extremely active with distinctive identities under the silhouette of the Zifeng Tower, while the one located at Southeast University is well maintained, but the other is obsolete in an old community. In Fig. 4g, the Zifeng Tower rises as a symbolic landmark enframed by the gate of the ancient Nanjing city wall, forming a unique space identity composed of history and currency, and of topophilia and skyphilia.

Given that the existence of high structures is essential and controllable in a modern city, their images skywards deserve a rational aesthetic evaluation to be woven into the outside city space by strategically connecting their verticality with the horizontality of the profane world, by pursuing the equilibrium between topophilia and skyphilia.

5 Conclusion

This study proposes a novel concept "skyphilia" as the counterpart of topophilia. Based on an investigation of the previous theories concerning topophilia and topophobia, we

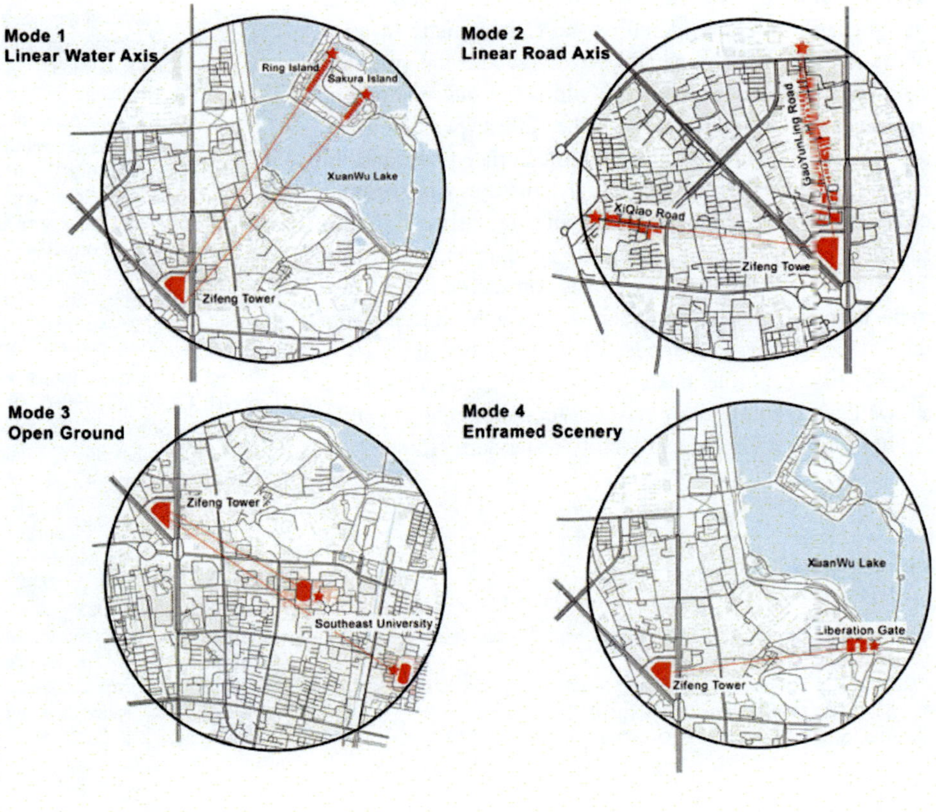

Fig. 3 Four modes of horizontality to interact with the verticality of Zifeng Tower

Fig. 4 Zifeng Tower observed from the four kinds of observation points

discovered the dual aspects of topophilia, which is often blindly praised for its virtue. Self-affirmation and self-alienation are considered as the characteristic of topophilia and skyphilia separately to overcome the primitive kind of topophobia of non-self, while over-affirmation and over-alienation could lead to the other two kinds of topophobia, i.e., the horror of undifferentiated self and the horror of alienated self. Typical high structures in all ages are compared for their motivations from three aspects: (1) the functional role; (2) the materialistic symbolism; and (3) the spiritual symbolism, and skyphilia is considered to be rooted in the harmony of these motivations. Through the example of the Zifeng Tower, four modes of horizontality are used to build a reciprocal relationship with the high-rises, the result shows that an equilibrium between topophilia and skyphilia is vital for the diversity and vitality of modern cities.

References

Betjeman, J.: Slick But Not Streamlined: Poems & Short Pieces. Doubleday (1947)

Caillois, R., Shepley, J.: Mimicry and Legendary Psychasthenia, pp. 17–32 (1984). https://doi.org/10.2307/778354

Giedion, S.: The Eternal present. Pagheon Books (1962)

Kingwell, M.: The city of tomorrow: searching for the future of architecture in Shanghai. Harpers Mag. **310**(1857), 62–71 (2005)

Leach, N.: Topophilia/topophobia: the role of the environment in the formation of identity. In: Ruan, X., Paul, H. (eds.) Topophilia and Topophobia: Reflections on Twentieth-Century Human Habitat, pp. 31–43. Routledge (2007). https://doi.org/10.4324/9781003060758

Oliveira, J., Roca, Z., Leitão, N.: Territorial identity and development: from topophilia to terraphilia. Land Use Policy **27**(3), 801–814 (2010). https://doi.org/10.1016/j.landusepol.2009.10.014

Relph, E.: Place and Placelessness. Pion (1976)

Relph, E.: Rational Landscapes and Humanistic Geography. Routledge (1981). https://doi.org/10.4324/9781315672601

Relph, E.: Modernity and the reclamation of place. In: Seamon, D. (ed.) Dwelling, Seeing, and Designing: Toward a Phenomenological Ecology, pp. 25–40. SUNY Press (1993)

Ruan, X., Hogben, P.: Topophilia and Topophobia: Reflections on Twentieth-Century Human Habitat. Routledge (2007). https://doi.org/10.4324/9781003060758

Seamon, D., Sowers, J.: Place and placelessness: Edward Relph. In: Hubbard, P., Kitchin, R., Valentine, G. (eds.) Key Texts in Human Geography, pp. 43–52. Sage (2008). https://doi.org/10.4135/9781446213742

Šimáček, P., Šerý, M., Fiedor, D., Brisudová, L.: To fear or not to fear? Exploring the temporality of topophobia in urban environments. Morav. Geogr. Rep. **28**(4), 308–321 (2020). https://doi.org/10.2478/mgr-2020-0023

Tuan, Y.F.: Topophilia: A Study of Environmental Perceptions, Attitudes and Values. Prentice-Hall (1974)

Tuan, Y.F.: Landscapes of Fear. Pantheon Books (1979)

Tuan, Y.F.: Time, space, and architecture: some philosophical musings. In: Ruan, X., Paul, H. (eds.) Topophilia and Topophobia:

Reflections on Twentieth-Century Human Habitat, pp. 22–30. Routledge (2007). https://doi.org/10.4324/9781003060758

Worringer, W.: Abstraction and Empathy: A Contribution to the Psychology of Style. Ivan R. Dee, Inc. (1997)

Zhou, H., Currà, E., Leng, J., Xu, Y., Hu, W.: Contradiction and consistency: deconstruction of landscape bridges based on multiple temporal-spatial scales. Front. Arch. Res. **11**(1), 53–72 (2022). https://doi.org/10.1016/j.foar.2021.08.002

The Place Identity After Urban Transformation in China—A Case Study of Nanjing

Ting Zhang

Abstract

This study explores the place identity of different urban areas in Nanjing by examining the built environment. Focusing on the built environment of the *Laochengnan* area and *Hexi* area, this study investigates the different identities of Nanjingnese and new Nanjingnese. The *Laochengnan* is the main historical area that has a large local memory of Nanjingnese. The historical image of this area is rooted in architecture, newspapers, photos and movies which document the local culture before urban transformation. The Hexi area is a new urban space produced by the Youth Olympic Games in 2014 that has attracted new Nanjingnese to live there. By interviewing respondents of two different groups, this study analyzes how their place identities are formed through their urban daily life in built environment, and the role of time in understanding the place identity. Finally, place identity is interpreted as a dynamic relationship between built environment, social group and living experience, making the city gradually form a sense of division between non-administrative areas.

Keywords

Place identity • Built environment • Urban transformation • Urban China

T. Zhang (✉)
Wuxi Taihu University, Wuxi, 214064, China
e-mail: ting.zhang@polito.it; zhangt_wxu_edu@163.com

T. Zhang
Department of Architecture and Design, Politecnico di Torino, Turin, Italy

1 Introduction

Nanjing is the capital of Jiangsu province, it is also one of the major historical cities in China. During the urbanization process after 1980s, many urban areas have been redeveloped in various ways, forming highly hybrid urban images of contemporary Nanjing. These hybrid urban landscapes reflect the diverse culture, history and social-economic, especially the urban daily life. In decades of urban development, a large number of native households have been displaced and moved away from their original home. At the same time, millions of migrants have entered and settled in Nanjing. In contemporary Nanjing, how much do we know about the place identity? Do locals and migrations have the same place identity?

The new urban landscape of Nanjing has formed various patterns, and the strong contrast between the new and the old complicates the sense of identity. Although Nanjing is a famous historical and cultural city, the local government is trying to build a globalization Nanjing in terms of urban development goals. Due to the preservation policies, many historical areas cannot be demolished. Therefore driven by the Youth Olympic Games in 2014, the Nanjing government has made great efforts to develop the Hexi New town. Differ from the historical image of the Old City South (Laochengnan area), the modern landscape of the Hexi area is unfamiliar to many locals.

Many urban scholars examined the quality and quantity of housing for migrants, which has produced positive results (Wu 2004; Li and Yu 2009; Day 2013). There are also some studies conducted on local displaced persons that pointed out their sense of homelessness (Shao 2013; Zhang 2017). Based on the previous emotional survey of urban identity, this study focuses on two areas of Nanjing: The Old City South (Laochengnan area) and Hexi area to answer the question of how did the native Nanjingnese and new Nanjingnese form their different place identity.

S. Kostopoulou et al. (eds.), *Cities' Vocabularies and the Sustainable Development of the Silkroads*, Advances in Science, Technology & Innovation, https://doi.org/10.1007/978-3-031-31027-0_2

The next section starts by reviewing previous studies on identity. In the second section, the reasons for selecting the cases are explained and the framework for analysis is presented. Next, the case study provides the basis for the discussion section. Finally, the conclusion section is drawn through a comparative discussion.

2 Literature Review

Many urban scholars examined urban identity from various perspectives, focusing on historical resources, urban social changes and the actors in urban transformation process. The connection between people and important places is deeply needed. If people select to ignore this need, it will create an irrelevant environment. On the contrary, if people choose to respond to this demand, it is possible to develop an environment suitable for human living, thereby enhancing people's sense of experience (Relph 1976). Lynch (1961) believes that identity is the degree to which a person can recognize or recall a place that is different from other places. Cities, like individuals, should have individuality and distinction.

Many studies argued that the identity of an urban settlement depends on a certain density, it seems to satisfy the basic needs of mankind. However, this viewpoint does not make much sense unless space is viewed as an entity in a constantly changing environment (Oktay 2002). An urban settlement is not only composed of the physical environment but also the construction ways of people in different periods. Lefebvre (1911) proposed that space is the product of social relations, prompting people to pay attention to the spatial dimension of the city, not just the time scale. The various activities and changes in cities are expressed through the spatial dimension. Therefore, cities are constantly changing due to many factors, such as social, economic and political changes. During this process, some elements are destroyed or replaced. It caused the necessity to consider the urban space from historical perspective, not only the physical space, but also the people who live in them. These are very important for investigating the urban identity (Relph 1976; Punter 1991).

When considering the urban environment, the area is a symbol of urban assessment and urban expansion, as well as a basis for sustainable development (Oktay 1998). The architectural entities as physical manifestations of urban areas have been examined from various perspectives. Previous studies have found that one traditional urban area is a basic urban settlement type combining place and activity. This is related to the unique identity of the city, which differs from a single building or an entire city (Benevolo 1968). The environmental psychologists proposed that place identity is related to the relationships between individuals, social groups and families or neighbors. This idea is useful to interpret the daily activities of people in their physical environment.

Due to rapid globalization and continuous advancement in research approaches, it has become critical to define human and their environments in various ways (Cheshmehzangi 2020). Previous studies on Chinese urban study have focused on social-economic and political system (Xu 2004; He and Wu 2005; Zhang 2017; Wu 2022), few studies have concerned on how to form place identity of locals and new locals under urban transformation. Therefore, this research examines the relationship between people and their physical environment in attempt to understand not only the formation process of local place identity in their built environment, but also to explain the place identity of urban migrants.

3 Methodology

In order to examine the place identity of different groups during urbanization process, this study selects two case areas in Nanjing: the Old City South (Laochengnan area) and Hexi area (Fig. 1). Nanjing is a famous historical city in China, which has experienced many historical dynasties. In 1982, the State Council released the first batch of famous historical and cultural cities (Li Shi Wen Hua Min Cheng), Nanjing is one of them. It can be said that the contemporary Nanjing is a typical city with history and modernity.

Case area 1-Laochengnan area is located in the inner city of Nanjing, it has many historical buildings and is the birthplace of local culture. In the process of rapid urbanization, this area has undergone extensive demolition and renovation, causing many residents to be moved to other urban areas. Case area 2 is the Hexi area, a new urban area with modern atmosphere, which was rapidly formed under the impetus of the Olympic Games.

This study is based on structured interviews with two groups, including Nanjingnese in the Laochengnan area and new Nanjingnese in the Hexi area (Fig. 2). Nanjingnese refers to the native residents who were born and lived in the Laochengnan area. Those who came from other cities and settled in Nanjing are identified as New Nanjingnese. In this study, most of the people in this group live in the Hexi area. All respondents of this study were born from 1980 to 1990 and were college-educated. In the Laochengnan area, they are the last generation group that has lived in the old built environment characterized by lane (Linong) life. In the Hexi area, they were the early generation of highly educated migrants who settled in Nanjing. Focusing on these two

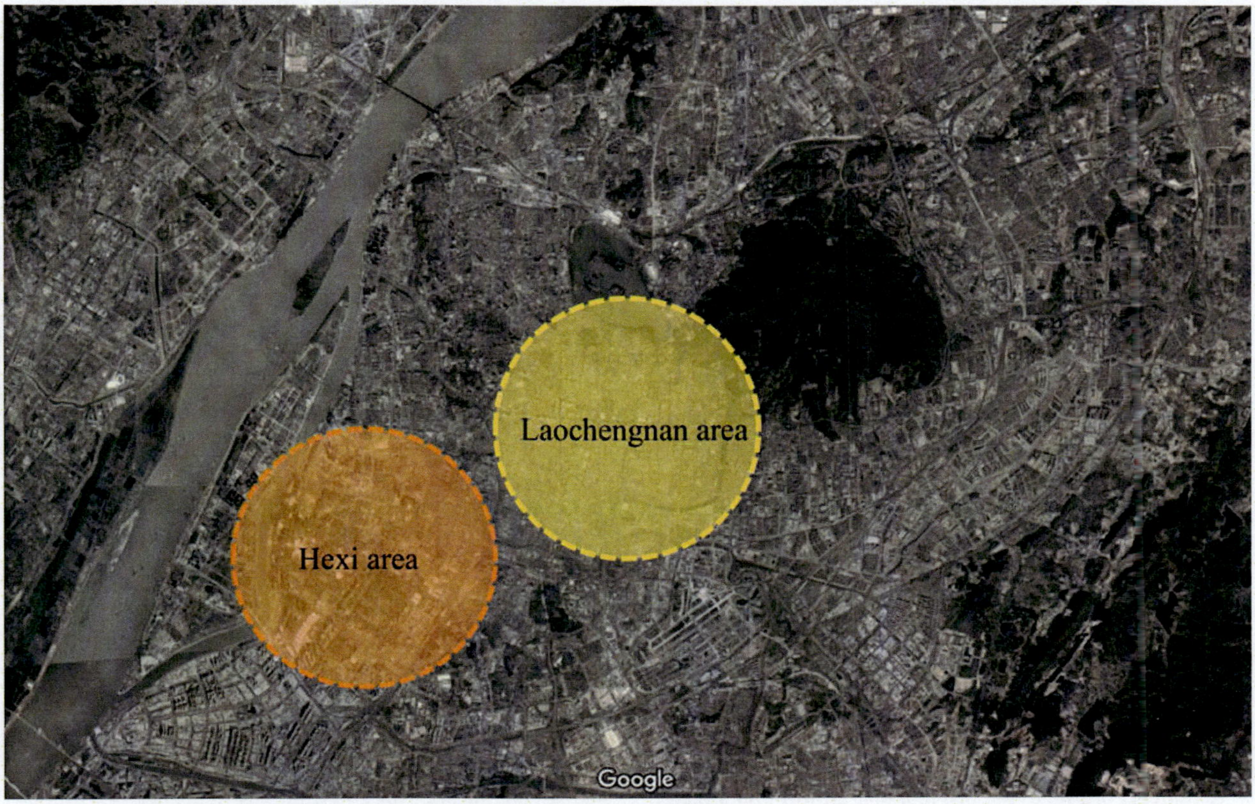

Fig. 1 The Location of two case areas. *Source* by author (base map from Google)

groups, this study investigates their identification with the built environment, historical memory, and urban daily life in Nanjing. The architectures of physical settings can be understood as the representation of their living activities, which can tell the story of the city and form the place identity of people and the built environment. The main attention of this study is not function and technologies but to people and the environment, to living approaches that encompass their emotions and sensations. This study mainly focuses on the relationship associated with place identity as a place where stories are formed and developed. In this study, the identities of two groups reflect different understanding of place identity, depending on their life experiences and memory in the different built environments.

In order to control the influence of variables, both groups of respondents are similar in terms of age, family structure, education and social economy. By analyzing and comparing two groups, this study argues that the symbolic meaning of Laochengnan is not necessarily sought by non-natives, which reveals the role of time and memory in understanding individual and place identity. These differences have gradually formed the sense of regional difference in contemporary Nanjing.

4 Case Study

4.1 The Identity of Nanjingnese

At the city scale, Nanjing's unique east–west character is related to its history. From the traditional house type of the Ming and Qing dynasties (1368–1912) to the fusion of east–west buildings of the Republic of China (1912–1949), these physical characteristics are preserved in the built environment of Laochengnan. For native Nanjingnese, these special physical environments are their living environment since birth and are recognized as authenticity. During the mass demolition of Nanjing's urban construction, many historical buildings were replaced by modern buildings, forming a dialogue between the old and the new. Following the changing physical environment, the individual identity of one place is always shaping and reshaping (Massey 1995).

Historically, Nanjing was the capital of the Ming dynasty (1368–1644) and the Republic of China (1912–1949). In the Ming Dynasty, Zhu Yuanzhang set the capital of Nanjing and divided the city into three parts. The Confucius Temple was the core of the southern part of the city and was the most densely populated area of Nanjing. During the Republic of

Fig. 2 The analysis framework. *Source* by author

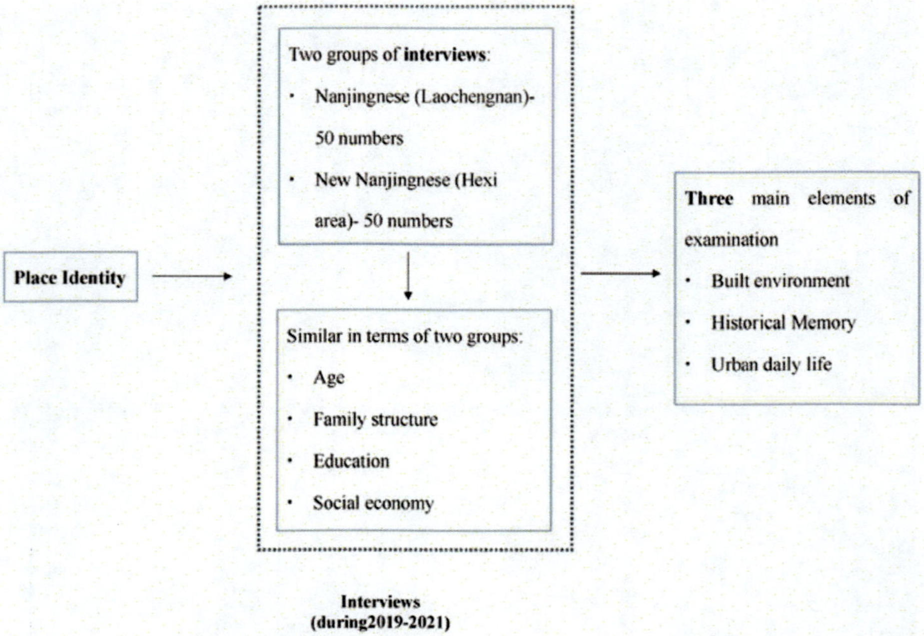

China period, a large population gathered here. This area suffered from large-scale wars during this period, resulting in the destruction of many buildings. After 1949, row houses were built according to the original layout, forming a pattern of lane living. Until today, this physical pattern continues and is remembered by locals. Nanjing's unique urban identity was embedded in the urban landscape before urban regeneration, especially in the Laochengnan area.

The respondents are the last generation who grew up in this area, most of them have moved to the other area after urban transformation. Although their new living environment is more spacious and comfortable than the original, they still miss their previous life in Laochengnan.

> The Laochengnan is our Nanjing. The low houses, the long and narrow lanes, and the mobile vendors, the playful children and the relationship with the neighborhood, this is our living environment.

During the interview, they all agreed that the Laochengnan was the real Nanjing in their memory. The live experience of built environment is concrete rather than images or records. The modern Shuiyou shopping mall, the reconstructed Laomendong are usually regarded as contemporary Nanjing, but the physical environment that represents the cultural and soul of Nanjing is terrace house with courtyard and narrow lanes. This physical form was influenced by the evolution of time, which has historical memory (Fig. 3).

The courtyard house is a traditional residential building type in South China, almost every house includes a courtyard and living function. Usually, the courtyard is connected to the entrance gate and the main building, forming an enclosure pattern. Each courtyard unit is juxtaposed to form the terraced house. Unlike the courtyard houses in north China, Nanjing is located in a more population and rainy climate area, resulting in the smaller scale of the courtyard. The streets are long and narrow, also called lane (Linong). The units facing streets are usually set up shops on the first floor, selling general goods for residents (Fig. 4). This physical structure provides semi-public space between streets and private houses within units, which shapes a unique urban space with a strong social community. A similar perspective was also proposed on the Shikumen study in Shanghai (Lu 1999; Wasserstrom 2009). The neighborhood in the Laochengnan area gradually developed a unique pattern of lane life (linongshenghuo) through the built environment.

Despite courtyard buildings, there are also a large number of residential areas built in the 1980s, most of them are more than 4 stories. Because of their acceptable construction quality, these buildings were not listed as Weifang (danger house) in the urban regeneration period. Until today, this type is the main residential building in Laochengnan (Fig. 5). Before the regeneration, this type of buildings and courtyard buildings mixed together to form a unique urban landscape feature.

More than two generations of the respondents' families were born and grew up in this physical environment. Many locals spent their entire lives in this area until they were forced to relocate. Through the interviews, many respondents expressed their attachment to the Laochengnan area, not only because of the traces of individuals in the physical environment, but also a strong sense of familiarity in life. The large-scale demolition and relocation have made them

Fig. 3 The Linong life model. *Source* from Sina Blog

Fig. 4 The courtyard of building. *Source* http://blog.sina.com.cn/s/blog_5139b8ee0101fbmq.html

lose the physical space of their place identity, and the original mode of their own social interaction of the lane-life model has gone. Facing the modern environment, it is difficult for them to find a sense of place identity for generations of living and memory.

It is the physical environment and life experience that form their place identity. The built environment takes a long time to form patterns of social interaction. According to the information provided by the interviewees, the living experience is often formed by the memories of several generations, and also shape the unique place identity of locals in Laochengnan area.

4.2 The Identity of New Nanjingnese

In 2010, Nanjing successfully applied for the right to host the 2014 Youth Olympic Games, which brought opportunities for the development of the Hexi area. The local government has positioned Hexi area as a commercial (CBD), business, culture and sports center, a new area that encompasses both residence and employment. In the Hexi area, architectural spectacles and modern high-rise apartments can be found everywhere, in contrast to the Laochengnan.

In order to achieve the goal of international development of Nanjing, the Hexi area has been built in a way that

Fig. 5 The traditional built environment of Laochengnan. *Source* by author

ignored the local urban history and the living experience of the local Nanjingnese. Landmark buildings designed by international architects and modern enclosed residential quarters have become the dominant physical environment in Hexi area (Fig. 6). Here, there is no trace of Nanjing's history and past life.

Since 2010, respondents of this study have gradually bought houses and settled in the Hexi area. As Miss. Zhao, who was still a university student that year, now is already working at a university in Nanjing, and her husband is also nonnative in our 2019 interview, they said that:

> We like to live in the Hexi area, where there are modern buildings, communities and many high-income people. We don't like the crowded streets and poorly-developed neighborhoods in the Laochengnan. We bought the house ten years ago, it was very cheap at the time, and now it has developed into the most expensive area in Nanjing.

In the follow-up study, another respondent Ms. Ye and her family, were also visited. They moved from Nantong, and they have rented and lived in Laochengnan before buying their current house. When they were asked about their reason for moving to Hexi area, they expressed interest

Fig. 6 The modern resident community in Hexi area. *Source* by author

in the modern facilities, clean community and spacious roads for cars.

Neither respondents mentioned the memory or identity of Laochengnan area. Their choices of housing location prefer the newly built environment with modern landscape and accessibility. The representative of local history in the built environment of Laochengnan is less attraction to New Nanjingnese, who are interested in construct their own place identity with the new neighborhood in the modern community of Hexi area.

Duyvendak (2011) proposed that the newcomers lack attachment to their new environment. They are not interested in the original features of the new environment because they can't find a sense of home. Thus, the respondents' responses demonstrate that place identity formation requires an accumulation process between people and environment (Xu 2020).

5 Discussion

In contemporary Nanjing, the difference between Laochengnan and Hexi area has a profound meaning rather than urban or suburban, or class preference. For native Nanjingnese, they recognize the Laochengnan as authentic Nanjing. This is related to "refers to the look and feel of a place as well as the social connectedness that place inspires" (Zukin 2010). These natives have not experienced certain history, but more the physical environment they grew up in, such as old row houses and lanes. They have gradually constructed their unique identity through the built environment and living experience. In the interview with them, it can be felt that the Hexi area is a very far place for them. Although there are modern and clean environments in Hexi area, the lesser urban memory makes it impossible for native Nanjingnese to have a place identity.

In contrast, the new Nanjingnese prefers Hexi area, because they do not have deep memories of living in Laochengna. Although Ms. Ye and her family have the experience of renting house in Laochengnan area, it was difficult for them to find a sense of identity among the native Nanjingnese. In Hexi area, there are not many historical memories. The new Nanjingnese have formed a new social structure through their own life modes in the built environment, such as gated communities, shopping malls and modern theatre. Over time, they have gradually formed their place identity, which is between them and their built environment. However, it is unfamiliar to native Nanjingnese.

By comparing Laochengnan and Hexi, the historical memory and modern development are juxtaposed and represented through the built environment. They describe two completely different stories, reflecting different social-culture, especially different lifestyles. The comparison reveals that although both areas are urban landscapes of Nanjing, different social groups have different understandings of them. Native Nanjingnese believes that the place identity is related to their own life experience in built environment. It is an emotion and habit that is rooted in their memory, this sense that is also imprinted in the built environment and is difficult to replace. For the new Nanjingnese, the modernization of Hexi area is more suitable for them.

6 Conclusion

Based on many fieldworks and interviews conducted in Nanjing during 2019–2021, this work reveals that the place identity of different built environments depends on the living experiences of the residents. By comparing their perceptions of built environments, it can be argued that different groups have significant differences in place identities formed by the built environment. Two groups of respondents construct their place identity by defining their relationship to the urban built environment. As response to urban development, this relationship advocates the importance of historical memory and living experience for the construction of place identity.

By examining two areas in Nanjing, this study demonstrates that place identity is a dynamic relationship among built environment, social group and living experience, rather than a fixed concept. This relationship is materialized in urban space and reflects the interrelationship between people and the built environment. In this study, the Nanjingnese and the new Nanjingnese have gradually formed a sense of regional difference in non-administrative areas through the physical environment of their residential areas. This sense of difference makes the place identity become an invisible boundary between Laochengnan area and Hexi area.

The city is not only the built environment, the more important is the social activities of space. After the rapid urbanization of Chinese cities, there are many urban problems to be solved in further urban development. In further urban strategy and management, it is necessary to take into account the role of people in the built environment, and adjust urban strategies in terms of different groups and scales.

Funding This study is supported by Philosophy and Social Science of Jiangsu Higher Education (2021SJA0913). This study is also supported by Soft Science Research Project of Wuxi Association for Science and Technology, 2023 (Research on the Implementation Mechanism of Urban Old Community Renewal under the Multiple partnership in Wuxi, China).

References

Benevolo, K.: The History of the City. Scolar Press, London (1968)

Cheshmehzangi, A.: Identity of Cities and City of Identities, pp. 116–117. Springer, New York (2020). https://doi.org/10.1007/978-981-15-3963-3

Day, J.: Effects of involuntary residential relocation on household satisfaction in Shanghai, China. Urban Policy Res. **31**(1), 93–117 (2013)

Duyvendak, J.W.: The Politics of Home: Belonging and Nostalgia in Europe and the United States. Palgrave Macmillan, Hampshire (2011)

He, S.J., Wu, F.L.: Property-led redevelopment in postreform China: a case study of Xintiandi redevelopment project in Shanghai. J. Urban Aff. **27**, 1–23 (2005)

Lefebvre, H.: The Production of Space (trans: Nicholson-Smith, D.). Blackwell, Oxford (1911)

Li, S.M., Yu, L.S.: Redevelopment, displacement, housing conditions, and residential satisfaction: a study of Shanghai. Environ. Plan. A **41**(5), 1090–1108 (2009)

Lu, H.C.: Beyond the Neon Lights: Everyday Shanghai in the Early Twentieth Century. University of California Press, Berkeley, CL (1999)

Lynch, K.: The Image of the City. MIT Press, Cambridge, MA (1961)

Massey, D.: Places and Their Pasts. Hist. Work. J. **39**, 182–192 (1995)

Oktay, D.: Urban spatial patterns and local identity: evaluations in a Cypriot town. Open House Int. **23**(3), 17–23 (1998)

Oktay, D.: The quest for urban identity in the changing context of the city. Cities **19**, 261–271 (2002)

Punter, J.: Participation in the design of urban space. Landsc. Des. **200**, 24–27 (1991)

Relph, E.C.: Place and Placelessness. Pion, London, UK (1976)

Shao, Q.: Shanghai Gone: Domicide and Defiance in a Chinese Megacity. Rowman & Littlefield, Lanham, MD (2013)

Wasserstrom, J.N.: Global Shanghai, 1850–2010: A History in Fragments. Routledge, Abingdon, UK (2009)

Wu, F.L.: Residential relocation under market-oriented redevelopment: the process and outcomes in Urban China. Geoforum **35**(4), 453–470 (2004)

Wu, F.L., Zhang, F.Z., Liu, Y.Q.: Beyond growth machine politics: understanding state politics and national political mandates in China's urban redevelopment. Antipode **54**, 608–628 (2022)

Xu, M.Q.: Redevelopment of Old Neighborhoods in Inner City Shanghai. Xuelin Chubanshe, Shanghai, China (2004) (in Chinese)

Xu, F.: Pudong is not my Shanghai: displacement, place-identity, and right to the "City" in Urban China, City Comm. **19**, 330–351 (2020)

Zhang, Y.P.: Domicide, social suffering and symbolic violence in contemporary Shanghai, China. Urban Geogr. **39**(2), 190–213 (2017)

Zukin, S.: Naked City the Death and Life of Authentic Urban Places. Oxford University Press, Oxford, UK (2010)

Waterfront Urban Regeneration and Its Impact on Socio-cultural Activities (The Case of Jeddah Waterfront)

Lara E. Al-Saud and Jawdat S. Goussous

Abstract

Waterfronts as public spaces have always been the most vibrant part of the city. Their association within the city is a unique product of environmental, economic, and social factors. A successful waterfront will reflect the quality of living, enhance the image of the city and create strong identity within the city inhabitants and visitors. Many coastal cities thrive to remain competitive, that's why they aspire to regenerate their waterfronts in an attempt to implement design measures that lead to increased tourism, stimulate the growth of the economy, boost the vitality of the city, and create a unique cultural experience for the users. The research addresses the lack of proper knowledge that the urban designers and local officials often have about integrating physical design features with the local social dynamics during the design process of new urban public spaces. The research aim is to study and evaluate the latest physical features and urban patterns of Jeddah waterfront. The purpose being is to investigate the major impacts of the waterfront urban regeneration on the socio-cultural activities happening at the waterfront, and whether this regeneration enhanced both the quality of life and the users' perception of the city. The study explored the literature to develop the evaluation framework process, and later created the public survey to measure users' perception. The methodology used though the proposed framework aims to evaluate the success of urban waterfronts regeneration from a socio-cultural perspective. At the end, the study reveals the regeneration provided major improvements and positive impacts on the socio-cultural activities.

Keywords

Urban regeneration • Waterfronts • Public spaces • Socio-cultural activities

1 Introduction

Waterfronts have long been considered an livly, active social hub urban space. Their association within the city is a unique product of environmental, economic, and social factors (Bruttomesso 1993). A waterfront is considered as a cultural hub for the residents, providing a sense of place where the symbolic meanings of settings influence the social context of human interactions (Tibbalds 2000). Thus, the physical morphology results in particular meanings to the people, generating a significant impact on their perceptions and socio-cultural activities (Canter 1977).

A successful waterfront will reflect the quality of living in the city hosting it. It also enhances the image of the city and creates strong identity within the city inhabitants and visitors. Many coastal cities thrive to remain competitive, that's why they aspire to regenerate their waterfronts in an attempt to implement design measures that lead to increase tourism, stimulate the growth of the economy, boost the vitality of the city, and create a unique cultural experience for the users (Shang 2011).

In the context of the Kingdom of Saudi Arabia, the country has a long history and culture that establishes the foundation for this endeavor. In recent years, Saudi Arabia introduced the vision of 2030. This vision aims to undertake major transformations to change image of Saudi cities and to create a world tourist destination. On the longer term, that aim is for all of Saudi cities to achieve a better quality of life. This included the initiation of social and cultural transformation and liberation for citizens. The major urban regeneration projects happening at Jeddah waterfront measure the enhancement of the city image, raise the quality of living,

L. E. Al-Saud (✉) · J. S. Goussous
Department of Architecture Engineering, University of Jordan, Amman, Jordan
e-mail: larzalsaud@gmail.com

S. Kostopoulou et al. (eds.), *Cities' Vocabularies and the Sustainable Development of the Silkroads*, Advances in Science, Technology & Innovation, https://doi.org/10.1007/978-3-031-31027-0_3

and attract more international tourists to explore the Saudi culture and life.

This paper discusses the research problem, aims and objectives of the study, a literature review exploring the theoretical concepts of place, urban regeneration and the results of the main case study of the paper which is latest urban regeneration project of Jeddah waterfront.

1.1 Research Problem

When it comes to waterfront regeneration projects, American approaches have become a model guideline in developing urban waterfronts (Wrenn et al. 1983). The successful redevelopment of Baltimore Inn Harbor was idolized by urban designers and public officials. It became a dominant prototype to be followed when recreating a successful public space in old waterfronts (Tibbalds 2000). The biggest drawback of implementing an international style is the loss of the cultural identity of the city. It can affect people's relationships to the place, contribute to identity formation, and impact social behavior (Marshall 2004). In the case to be reviewed, Jeddah waterfront has been going through rapid changes due to the recent regeneration projects, which resulted in introducing new users, functions, scale, and sensibility; outlandish to what used to exist. Therefore, the lack of integration of the waterfront design with the local culture will be reflected on the use and activity of the public spaces and the social behaviors and activities (Gehl 1987). This creates the need to evaluate the newly designed public space from a social and cultural standing point, and whether it succeeded in achieving better quality of life, and promoting socio-cultural activities for its users.

1.2 Research Aim and Objectives

The aim of this research is to study and evaluate the new physical features and urban patterns of Jeddah waterfront. The purpose being is to investigate the major impacts of the waterfront urban regeneration on the socio-cultural activities happening at the waterfront, and whether this regeneration enhanced both the quality of life and the users' perception of the city. In addition, to attempt to formulate an evaluation framework that could be used for similar waterfront regeneration projects in other cities.

2 Theoretical Concept of Place

Under the field of urban design literature, there are many theoretical concepts of place and public space design theories. By reviewing these theories, a clear understanding will highlight the importance of public spaces in urban design. The review includes a discussion on people's physical activities, place connections, the relationship between users and their physical setting, the psychological progressions that leads to understanding the urban public spaces and how they are essential in place making processes (Canter 1977).

2.1 Morphological Dimension

From a morphological perspective, urban areas can be analyzed by certain elements. Those elements define the configurations of the urban form and space, the infrastructure supporting the physical and spatial patterns. Elements such as: Streets, Parks, Garden, and buildings are mere ideas that help in shaping the city. They are considered part of the dynamic organism that keeps transforming the interrelationship with urban area over time (Whitehand et al. 2009) (Carmona 2010) (Moudon 1997). Some of these elements are: Street patterns, Urban and plot patterns, Buildings footprint and Land uses. The way that these elements overlap is that the street patterns/networks are the foundation for the urban plot, where users' movement and activity patterns are determined and characterized in each plot. Smaller blocks tend to provide for better movement pattern than larger blocks. Reason being is that they offer more vitality legibility, walkability and social interaction of public space. Being the foundation or urban network, streets are also the foundation of social networking, allowing users to participate in various socio-cultural and economic activities (Carmona 2010).

2.2 Perceptional Dimension

Since the 1960s, much research has been conducted on the way people perceive their urban environments. The research developed significant studies on architectural symbolism and place meaning in the built environment. The perceptional dimension of urban design can be described as environmental awareness, perception, experience and appreciation of place (Carmona 2010). Waterfronts have unique characteristics that portrait visual depth, movement, sound and color, which provide containment for users. The different types of waterfronts (sea, river, lake) convey multiple images and perception on the users because they also vary in characteristics. The natural elements trigger peoples' senses through restorative values that they possess and offer distinctive experience (Carr et al. 1992). As a result, the physical form of urban spaces has a direct effect on people's social and cultural activities, and the way people perceive and create place meaning emphasizes the value of the perceptual dimension of urban design (Lynch and Rodwin 1958).

2.3 Place-Making in Public Spaces

Place-making as an approach was introduced to inspire urban designers, planners and people to reimagine and reinvent places. It's a cooperative process that can help in shaping the urban community by strengthening the connections between people and places. It also, promotes better planning and design for urban places though facilitating a creative patter in use (activity) in the space by taking into consideration the physical, cultural and social attributes (Wyckoof 2014).

During the process of place-making, the first important step for urban designers is; identifying who's using the space, and what sort of activing pattern is happening in the space. Only after identifying the previous steps, the physical settings can start to be formed (Whyte 1982). Therefore, urban designers and planners must have strong analytical skills to understand the various effects that the physical features have on human activities, and comprehend the potential relationship between them. Such analysis can be achieved by studying the physical features in the space and the activity pattern distribution at an urban scale (Lynch and Rodwin 1958).

According to Canter (1977), another perspective to measure the success of place is understanding the purpose of this place, in relation to the activities and experiences happening at the place. This way, it can help understand more why the activity is happening instead of analyzing what the activity is (Canter 1977).

Later on, urban theorists (Relph 1976; 1977; Montgomery 1998) expanded the knowledge on behavioral setting to include personal experience, which they referred to as "place identity". They introduced and presented place components such as; meaning, activity, conceptions and physical attributes, that contribute to human experience in the place. As a result, places or environments can be understood by the physical characteristics of the place, in relation to the activates happening in them and the meaning people hold for them (Canter 1977).

In essence, (Relph 1976) argues that places provide unique address though identifying three basic elements that constitute the place: the physical setting, the activity and the meaning. Building on his theory, (Canter 1977)designed the following diagram to reintroduce place components as a product of: activity, physical attributes and conceptions.

3 Urban Regeneration

Many authors define urban regeneration based on the major physical, social and economic challenges and issues that lead to urban change. This is what (Reberts and Skyes 2000) tried to construct in his definition of urban regeneration. He defined it as "comprehensive and integrated vision and action which leads to the resolution of urban problems and which seeks to bring about a lasting improvement in the economic, physical, social and environmental condition of an area that has been subject to change".

3.1 Necessity of Urban Regeneration

The needs for urban change lead to the emergence of the current modern practice of urban regeneration. This change is a demanded response to attend to the urban problems and opportunities in the area (Hussien 2015). From Roberts's previous definition, we can break down the urban regeneration process (Fig. 1):

3.2 Benefits of Successful of Waterfront Regeneration

In a review of waterfront renewal schemes in the US, a number of positive outcomes were the deriving principle in pushing more waterfront regeneration projects (Jones 1998; Wakefield 2007). Some of the benefits are listed as the following:

- An increase in the economic values of properties, as well as investments in the area.
- The attention towards the waterfront helped in raising environmental awareness, which lead to improving the quality of the water.
- Increasing tourism by reviving the image of the city and place promotion.

3.3 Waterfront Urban Regeneration Examples

In the past, port cities were mainly occupied by ports, factories, warehouses, and transportation-related facilities. Due to deindustrialization, there has been a decline in the

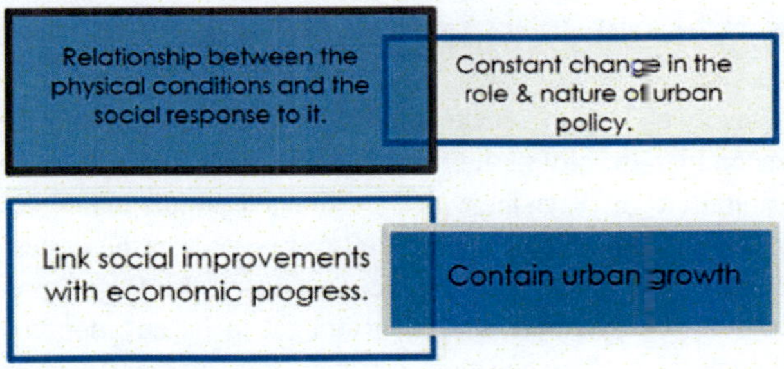

Fig. 1 Necessity of urban regeneration. *Source* Author, 2021

shipping and industrial sectors. This lead to the emergence of special rebranding and transformation of waterfronts environment (Hutton 2009). This change led the city planners to seek redevelopment of the urban waterfront in the interest of economic, touristic and place promotion competitiveness. In recent years, many waterfronts shifted the orientation from brownfields to more commercial, residential, and green recreational areas. The success in waterfront rebranding boosted the creation of themes landscapes, making it a global trend (Xie 2015). Major examples of this trend includes, the Baltimore waterfront in Maryland, USA (Vallega 2001), the waterfront regeneration in the Mersey Basin in Northwest England (Wood and Handley 1999), The Odaiba Waterfront Regeneration, Tokyo (Murayama and Parker 2006), and the waterfront development in Singapore (Chang and Huang 2011).

Even though the popularity of waterfront regeneration started in post-industrial cities, that doesn't mean the phenomenon didn't spread to all types of water settlements. In developing countries, urban communities focused on waterfront renewal in terms of theoretical perception for greater outcome in social, economic and political aspects (Hoyle 2002). The variance of aims from one waterfront regeneration to another made it difficult for many researchers to properly classify the types of regenerations. However, several literatures introduced classifications of waterfront development on the spectrum. For example, (Breen and Rigby 1996) discussed the waterfront development in six major themes, based on development modes: (1) commercial (providing public and social leisure via hotels, cafes and restaurants…etc.), (2) cultural and educational (via providing establishments such as museums, aquariums and modern amphitheaters that are easily accessible for social interactions), (3) historic (via emphasizing past structures and monuments and modernizing them to adapt and accommodate for modern needs), (4) recreational (by creating attractive landscapes facilitating parks, walkways and play areas), (5) residential (creating urban projects rather than resorts), (6) working waterfronts (by continuing the use of the industrial functions, that are usually forgotten when renewing the modern urban waterfront).

4 Case Study: Jeddah Waterfront

The city of Jeddah—nicknamed the Bride of the Sea—is the second largest city in the Kingdom of Saudi Arabia after the capital city of Riyadh, with an estimated population of 3.4 million (Jeddah Municipality 2021). Located on the eastern coast of the Red Sea, Jeddah is considered the largest port city on the Red Sea. It is also thought to be an important commercial, touristic, and economic hub of the Kingdom (Al-Fakahani 2005). The city acts as a key portal to the two holy cities of Makkah and Medinah, and a gateway port city welcoming the pilgrimage visitors during Haj and Umrah seasons. It is most known for its diverse culture, old traditional souks, unique Hijazi cuisine, and its attractive waterfront (Sahahiri et al. 2019).

4.1 The Development of Jeddah Corniche

The early development of the Corniche had some economic and social benefits such as attracting investors and contributing to the public realm. It also provided the city with spaces and facilities for the public use and community recreation. However, some environmentalists found that the early development of the Corniche was unsuccessful due to the overuse of artificial areas instead of sustainable recreational spaces, and the lack of design appropriation to the local climate and socio-cultural characteristics of the area (Bokhari 1978; Mandeli 2010). During September 2012, Jeddah Municipality announced the launch of renovation project for the Northern, Middle and Southern parts of Jeddah Corniche. The project is part of an urban regeneration program that aims to revitalize tourism in coastal cities. The latest regeneration plan aims to transform the appearance of the Corniche areas through providing new waterfront related leisure and entertainment facilities and services (Mandeli 2019).

4.2 The New Jeddah Waterfront

The new Jeddah Waterfront is part of the North Corniche. It's a national landmark and one of the most attractive tourist destinations in the Kingdom of Saudi Arabia. It is also considered to be the biggest regeneration project –thus far-that has been done on Jeddah Corniche. The new waterfront runs for a 4.5 km along the Red Sea, starting from Al-Nawras (Seagull) circle, reaching up north until Jubayr Ibn Al Harith Street (Fig. 2). The total area of the project is estimated to be around 730,000 m^2 with a capacity to accommodate around 120,000 users. The total cost of the project was estimated to be 800 million SR (Jeddah Municipality 2021).

4.3 Study Analysis

The process of developing an analytical evaluation framework was based on two main areas of knowledge, successful waterfront regeneration factors, urban morphology and perception theories, and the effect of these two aspects on the socio-cultural activities in the waterfront. In addition, the study will also conduct a survey inquiry, targeting the

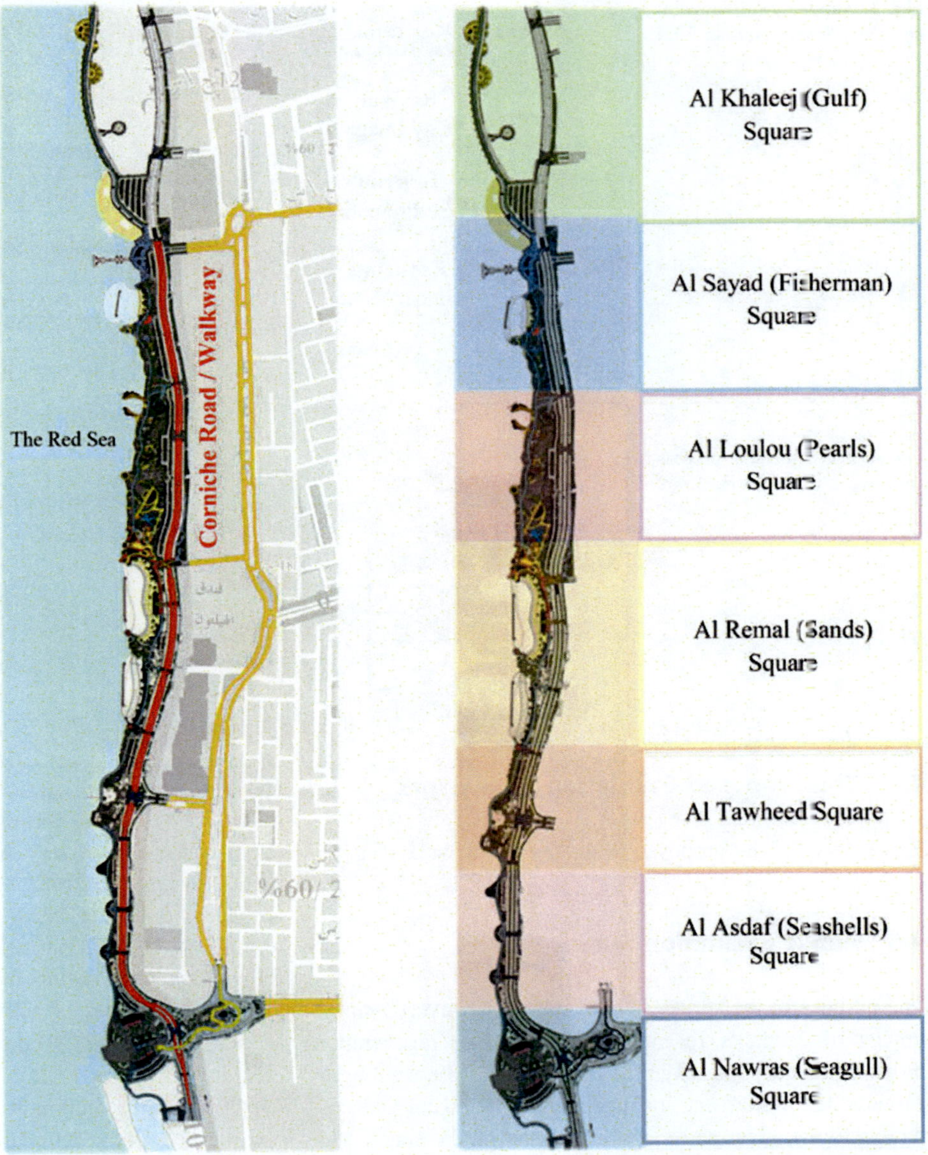

Fig. 2 Jeddah Waterfront map showing main access road (Corniche road) and the seven areas. *Source* Jeddah Municipality (2021), edited by Author

waterfront users in order to understand how they perceive and interact with the waterfront as a public space.

The diagram (Fig. 3) below summarizes and outlines the main concepts derived from the literature review on successful waterfronts. These concepts were used as a proposition for identifying the criteria of creating the initial evaluation framework and ergo, the analytical evaluation survey, which were narrowed down to selected success indicators and attributes.

In order to investigate Jeddah's waterfront, and the impact of the urban regeneration on the socio-cultural activities, the researcher is conducting an evaluation survey targeting the users of the waterfront. Following the literature review, the researcher identified the main success indicators and analyzed the criteria of success in order to create the initial framework. The initial framework is then used as a guideline in creating the evaluation survey. The findings and results of the survey are then studied in order to formulate a final proposal for an evaluation framework that can be used in the future to measure the success of urban waterfronts regeneration from a socio-cultural perspective. The process can be summarized in the following (Fig. 4).

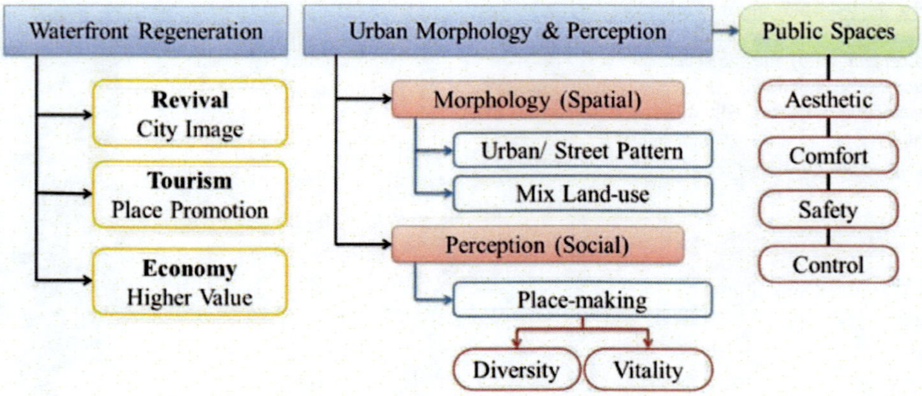

Fig. 3 Summary of the main concepts. *Source* Author, 2021

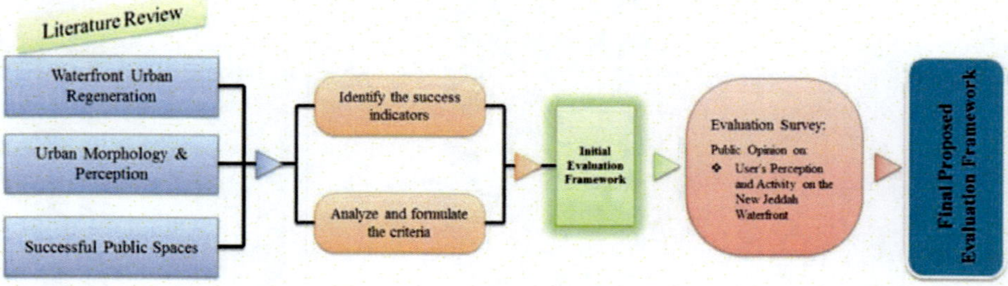

Fig. 4 The researcher process of devising the framework. *Source* Author, 2021

4.4 The Proposed Evaluation Framework

Break down of the proposed evaluation framework (Fig. 5).

4.5 Public Evaluation Survey

The purpose of the survey is to evaluate and investigate the impact of Jeddah's recent waterfront urban regeneration project. The questions examine how Jeddah waterfront users perceive the waterfront and understand the social and cultural activities they perform in the urban public space. The survey was conducted electronically, and its results helped in narrowing down the main elements in formulating the final framework.

Several factors were taken into consideration while designing the survey. Factors such as; the use of simple language to adhere to all types of users, providing a space to give written recommendations from the users' perspective, providing maps to help guide users, and lastly, maintaining anonymity of the users.

The survey was conducted during the months February and March of 2021. The total responses received were 145 responses. The survey was divided into five main sections. The questions variations included multiple choices, rating scale, and short answer questions. The first section of the survey is targeting the general information about the demographic of the users such as; gender, level of education, and age group. The second section is to find out the frequency of visitation. Questions included the preferred times to visit, duration of stay and the chosen methods of commuting to the waterfront. In the third section, the questions are specific to the latest regeneration happening at the waterfront. The section included questions such as; reasons for visiting or avoiding the waterfront, preferred areas to explore, they types of activities, variation in facilities and services. The fourth section is targeting the perception of comfort, safety and aesthetic scenery of the waterfront. Lastly, the fifth section was an addition dedicated to the period of world pandemic of Covid-19. The questions aimed to understand how the users viewed/interacted with the waterfront as a public spaces during the pandemic (Figs. 6, 7, 8, 9 and 10).

4.6 Survey Results Discussion

Revival

Out of the 145 users that answered the survey, 59 (40.7%) said they were encouraged to visit Jeddah waterfront more after the latest regeneration. On their perception of the image of the city of Jeddah, 111 (76.6%) said that the latest regeneration enhanced the image of Jeddah. With the majority of the responses being positive, it is possible to conclude that the new Jeddah waterfront achieved successful revival as an urban public space.

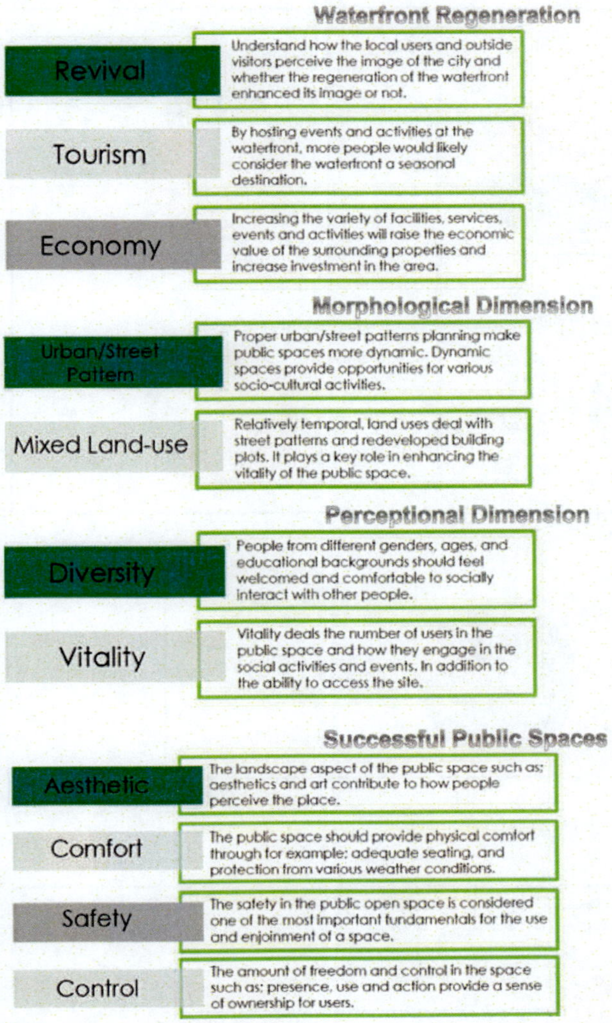

Fig. 5 Framework.*Source* Author, 2021

Tourism

According to the survey, out of the 145 respondents, 129 (89%) were local residents in Jeddah. Only 11% of the users were outside visitors. While respondents requested more events and social activities, the lack of such events might be due to the world pandemic of Covid-19 that is happening during the conduction of this study. Therefore, the aspect of tourism might not be an accurate indication during this year.

Economy

The variety of facilities provided at the waterfront seemed satisfactory to an extent to the users. However, there was a persistent request from the respondents to add more facilities and create more events. Similarly to tourism, the aspect of economy might not be an accurate indication during this year because of the world pandemic that suspended the possibility of hosting events that might lead to people gatherings.

Urban/Street Pattern

The urban pattern at Jeddah waterfront provided various themed areas to trigger excitement and change of the scenery as people walk along it. When the respondents were asked about the favorite areas to visit at the waterfront, the answers were relatively close. About 40 (27.6%) of the respondents said they frequent all the areas. The least visited area is Al-Khaleej (20%), a reason for this might be that the area is still under development and usually designated for events, which have been suspended at the moment due to the world pandemic. Overall, it can be concluded that the regeneration of Jeddah waterfront was successful in providing dynamic urban pattern.

Mixed Land-Use

Each area at the waterfront has a wide range of facilities and services. Despite the differentiation in the theme, the waterfront maintained a balanced coverage in the services and entertainment facilities in each zone. When asked about the satisfaction regarding the variety in the facilities provided, the range of answers was between 26.9 and 29% satisfactory. However, the suggestions provided by the respondents indicated that the users of Jeddah waterfront demand more facilities. It can be concluded that the mix land-use at the waterfront was successful to an extent, but needs to be improved through sustainable planning in the future.

Diversity

According the survey answers, female to male range of users is almost equal. Also, the age groups ranged from 18 to over 50. Although the majority of the respondents were bachelor degree holders, the waterfront managed to attract other users from different education backgrounds. Regarding the users' occupancy, the dominating majority were employed people. This indicates that Jeddah waterfront has successfully achieved diversity through attracting users from different age groups, genders, educational backgrounds and business spectrums.

Vitality

The survey shows that visitation frequency at the waterfront is relatively high. Most of the respondents indicated a weekly, monthly and occasional visitation. Also, about 60% of the responses showed that users are often accompanied by family members and friends for socializing at the waterfront, meaning group gatherings are very common. On accessibility and methods of reaching the site, the highest response was 44.8% found the site easily accessible, including for handicapped users and the elderly. The methods of commuting to the waterfronts were varied, but the use of personalized car was dominating at 88.3%. The reasoning of that might be due to the lack of public transportation system

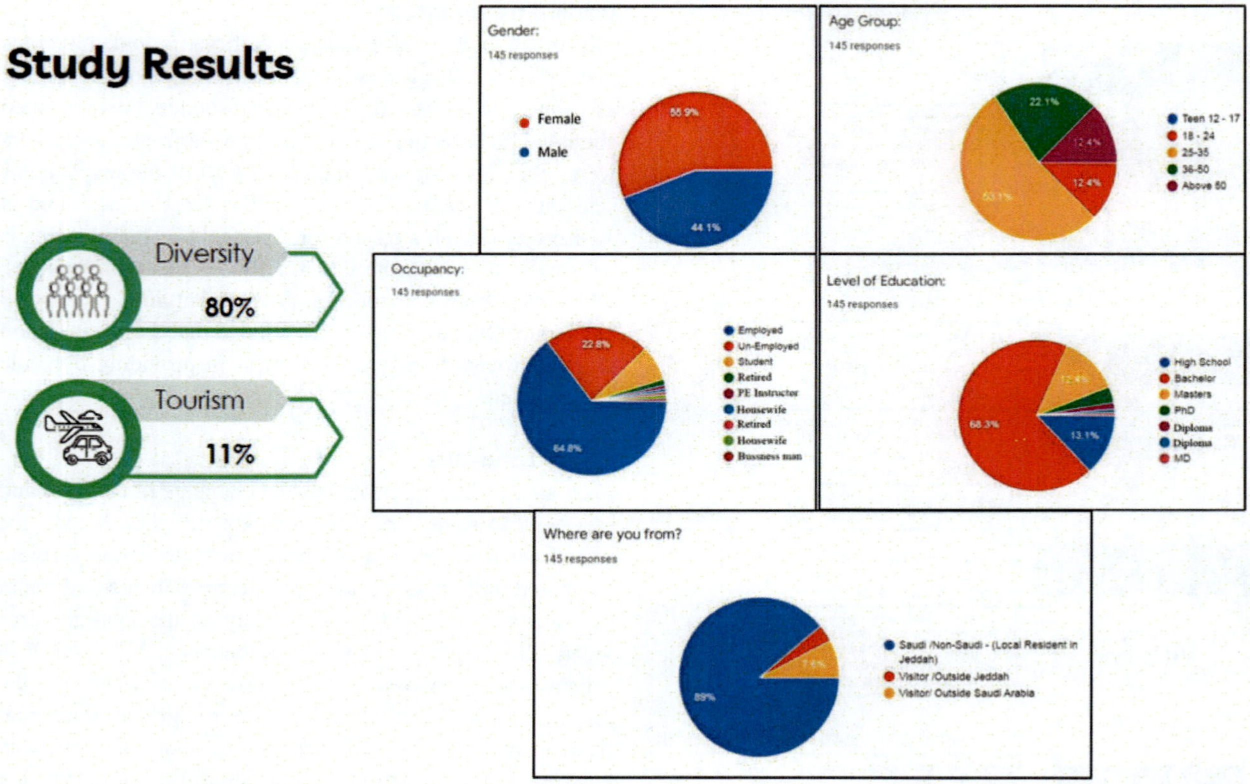

Fig. 6 Survey results. *Source* Author, 2021

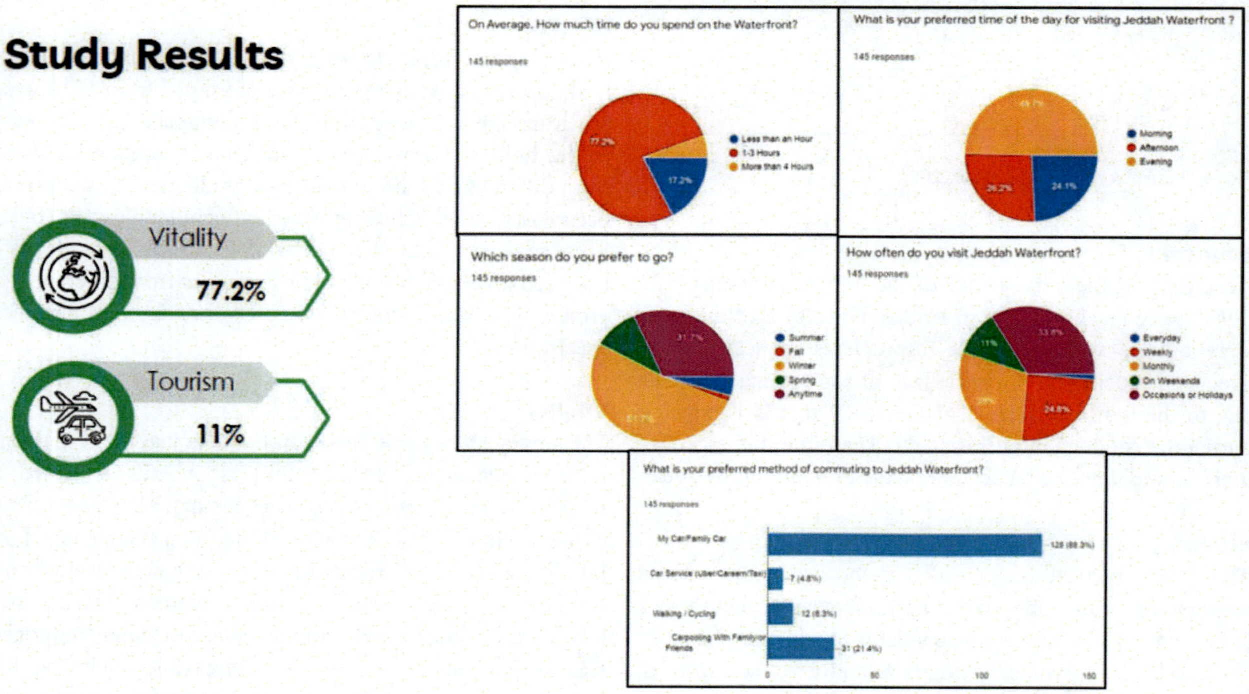

Fig. 7 Survey results. *Source* Author, 2021

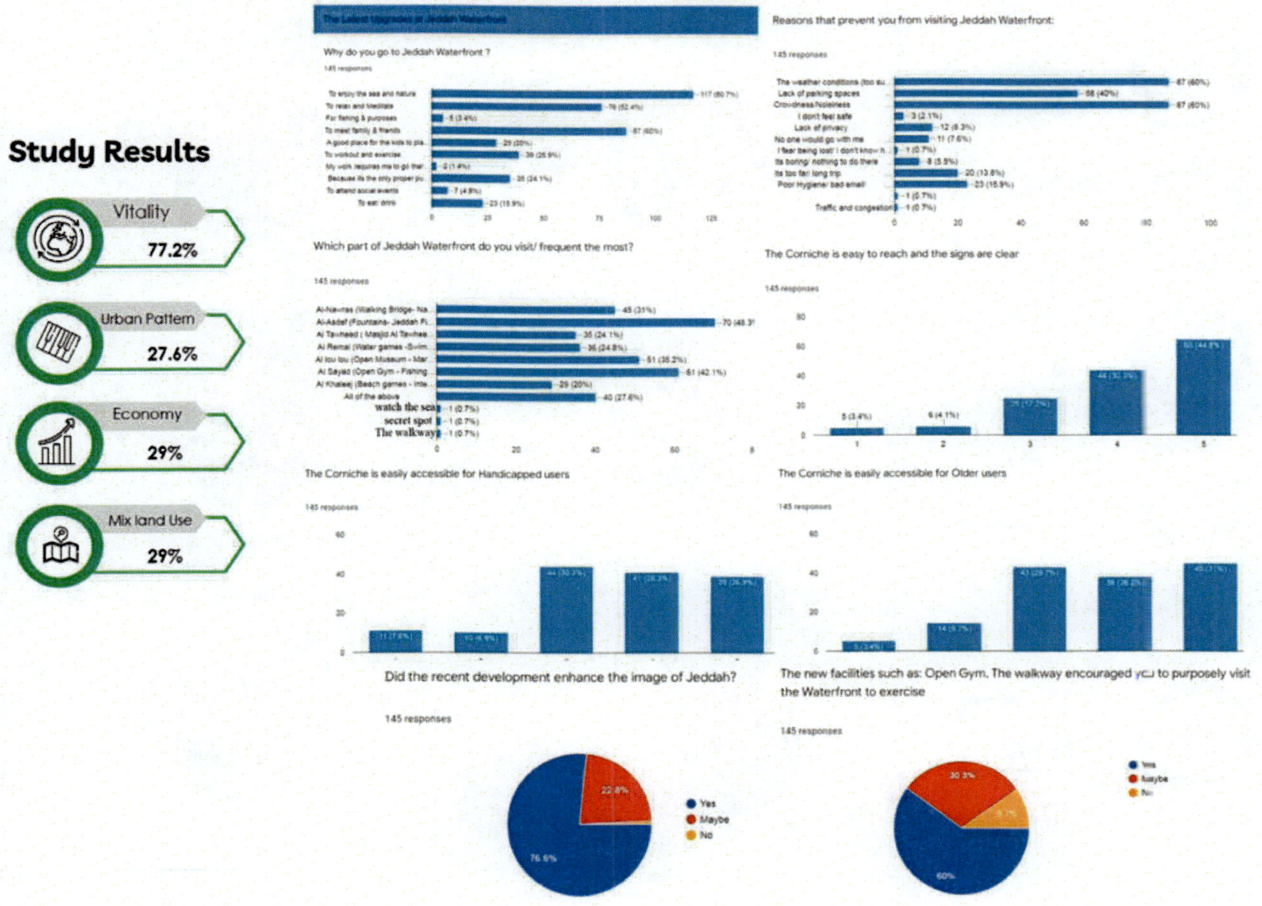

Fig. 8 Survey results. *Source* Author, 2021

in the city of Jeddah in general. Another reason might be that according to the survey, people like to go in groups, so it is easier for them use their own method of transportation.

With the harsh weather conditions being the dominating reason-60%- preventing users from visiting the waterfront, the respondents chose the winter season to be the best time to frequent the waterfront, followed by 31.7% chose anytime of the year. In regards to the amount of time spent at the waterfront, most of the users-77.2%-said they spent 1–3 h, during anytime of the day (morning, afternoon and evening). This indicates that the waterfront is vibrant for the majority of the day, with possible crowd-ness in the evening hours. Reason for evening crowding might be that the culture of Jeddah city more active during the evening hours. In conclusion, the regeneration of Jeddah waterfront has succeeded in achieving vitality.

Aesthetic, Comfort, Safety, Control

When asked about the aesthetic, comfort, safety and control of some of the areas at Jeddah waterfront, almost half of the responses (range of 51.7–45.5%) rated them as good. The rating was regarding the areas of seating, greenery, Marina and kids playgrounds. As for the recommendations provided by the respondents, some answers requested more lighting at night in the kids play areas. The specificity indicates that the lighting is fair in most of the waterfront, but they need a higher sense of safety at the children's play areas. Regarding comfort, some of the responses included requests for a more varied types of seating and an increase in the number of sitting areas. On the aesthetic side, the majority of the responses seem very satisfied with the natural beauty of the waterfront and the overall design of the waterfront. Some users responded with answers like "very calming" and "very charming atmosphere". The results also show a general sense of comfort and control is achieved. Most of the users indicated their ability to move freely on the waterfront and chose to participate in the activities they were drawn to, with the option of exploring new activities such as exercising. On that matter, the waterfront featured the new "Corniche Walkway", where 60% of the users said this new feature encouraged them to participate in health related activities. Overall, it can be concluded that the latest waterfront

Fig. 9 Survey results. *Source* Author, 2021

Fig. 10 Framework results. *Source* Author, 2021

Success Indicator	Evaluation Rate
Revival	76.6%
Tourism	11%
Economy	29%
Urban/ Street Pattern	27.6%
Mix Land-use	29%
Diversity	80%
Vitality	77.2%
Aesthetic	56%
Comfort	48%
Safety	50%
Control	62%

Final Evaluation Framework

regeneration succeeded in achieving aesthetic, safety, comfort and control to an extent, with a room for future improvement in that area.

World Pandemic of Covid-19

The study has been conducted during the world pandemic of covid-19, so the last section was an additional factor to evaluate waterfronts as public spaces during global outbreaks. While the majority of users said they did not frequent the waterfront more during the pandemic, other users felt that it was a safer alternative to be at an open public space rather than closed public spaces such as; shopping malls and cafes.

5 Conclusions

The main findings and contributions that can be concluded from the study are as follows:

- The reviewed literature on urban public spaces and their success factors identified the main aspects and concepts that can evaluate the quality of the urban public spaces such as; waterfronts.
- Within the context of Saudi Arabia, the study showed that there is no involvement of the local community in the decision making process when implementing a major regeneration for a selected urban area. This suggests that future regeneration projects can projectile higher chances of failure or not reaching the full potential of these public spaces.
- One of the main risks of waterfront regeneration is the loss of identity. The new Jeddah Waterfront site features managed to succeed in retaining the local traditional fishing seaport identity of Jeddah through naming the zones within the theme such as: Al Sayad, and Al loulou zones. In addition to Jeddah being an open air art museum, which was reflected in the use of art figures on site.
- The proposed evaluation framework was able to measure the perception and satisfaction of the users, and the results will provide future developers and stakeholders with an understanding on the variety of functions and facilities that can promote more diversity in the type of activities at the waterfront.
- The survey conducted suggests that most of the users in public spaces tend to act freely based on the proposed physical environment. One example was that their actions were mostly voluntary and none of the users were forced to perform a necessary activity, leading to the conclusion that the regeneration succeeded in providing social comfort, which was one of the main evaluation factors in the framework.
- The users of Jeddah waterfront reflected overall satisfaction on the services, facilities and amenities provided in the latest regeneration.
- Some of the services provided at the new waterfront seemed to be unknown by the users, using clear physical signage on site is necessary. For example, providing a clear sign for the site security cameras' locations will improve the sense of safety for the users.
- The initiation of social and seasonal events (such as; Jeddah Season 2019) will increase the vitality and revitalize the waterfront. It will also revive tourism on both the national and international levels.

Future Research

Waterfront regeneration is a long-term dynamic process. It is also considerably challenging to design a public space that can cater to human needs as well as represent the city image and identity. The process requires more supportive studies involving various disciplines to design successful and sustainable public waterfronts. Indeed, gaps still exist between the vision of designers and user expectations. Many other unknown aspects of the relationship between humans and waterfronts also exist, and such gaps require further research. The case of Jeddah Waterfront was only one part of Jeddah to be evaluated as a public space. Other parts of Jeddah Corniche such as the middle and Sharm Obhor Cornish can be further evaluated in the future. Moreover, the results and findings only reflect the situation and issues during a specific period of time. Further research can follow to explore more unique urban features and new findings related to human behavior and activity on the waterfront.

References

Al-Fakahani, H.: Jeddah: the bridge of the red sea: progress and development. The Arab Publishing House for Encyclopedias, Jeddah, KSA (2005)

Bokhari, A.: Jeddah: a study in urban formation. Doctoral dissertation. University of Pennsilvanyia (1978). https://repository.upenn.edu/dissertations/AAI7908710

Breen, A., Rigby, D.: The New Waterfront: A Worldwide Urban Success Story. Thames & Hudson (1996)

Bruttomesso, R.: Waterfronts: A New Frontier for Cities on Water. International Centre Cities on Water (1993)

Canter, D.: The Psychology of Place. Architectural Press, London (1977). https://fyduhyrijikikace.tula-music.com/the-psychology-of-place-book-12628ut.php

Carmona, M.: Public Places, Urban Spaces: The Dimensions of Urban Design, 2nd edn. Routledge (2010)

Carr, S., Francis, M., Rivlin, L.G., Stone, A.M.: Public Space. Cambridge University Press (1992)

Chang, T.C., Huang, S.: Reclaiming the City: Waterfront Development in Singapore. Urban Studies **48**(10), 2085–2100 (2011) https://doi.org/10.1177/0042098010382677

Gehl, J.: Life Between Buildings: Using Public Space. Island Press (1987)

Hoyle, B.: Urban waterfront revitalization in developing countries: the example of Zanzibar's Stone town. Geogr. J. **168**(2), 141–162 (2002). https://doi.org/10.1111/1475-4959.00044

Hussien, M.: Urban regeneration and the transformation of the urban waterfront: a case study of liverpool waterfront regeneration. Doctoral dissertation. University of Nottingham (2015)

Hutton, T.A.: Trajectories of the new economy: regeneration and dislocation in the inner city. Urban Stud. **46**(5–6), 987–1001 (2009). https://doi.org/10.1177/0042098009103852

Jeddah Municipality (2021). www.jeddah.gov.sa

Jones, A.: Issues in waterfront regeneration: more sobering thoughts-A UK perspective. Plan. Pract. Res. **13**(4), 433–442 (1998).

Lynch, K., Rodwin, L.: A theory of urban form. J. Am. Plann. Assoc. **24** (4), 201–214 (1958). https://doi.org/10.1080/01944365808978281

Mandeli, K.N.: Public space and the challenge of urban transformation in cities of emerging economies: Jeddah case study. Cities **95**, 102409 (2019). https://doi.org/10.1016/j.cities.2019.102409

Mandeli, K.N.: Promoting public space governance in Jeddah, Saudi Arabia. Cities **27**(6), 443–455 (2010). https://doi.org/10.1016/j.cities.2010.03.001

Marshall, R.: Waterfronts in post-industrial cities. Taylor and Francis (2004). https://www.perlego.com/book/1613956/waterfronts-in-postindustrial-cities-pdf

Montgomery, J.: Making a city: urbanity, vitality and urban design. J. Urban Des. **3**(1), 93–116 (1998). https://doi.org/10.1080/13574809808724418

Moudon, A.: Urban morphology as an emerging interdisciplinary field. Urban Morphol. **1**, 3–10 (1997)

Murayama, M., Parker, G.: Sustainable leisure and tourism space development in post-industrial cities: the case of Odaiba, Tokyo, Japan. In: Smith, M.K. (ed.) Tourism, Culture and Regeneration, pp. 69–84. CABI Publishing (2006). https://doi.org/10.1079/9781845931308.0069

Relph, E.: Place and Placelessness. Pion, London (1976)

Sahahiri, R., Arrowsmith, C., Alitany, A.A.: Mapping the historical places: a case study of promoting tourism in Jeddah, the Kingdom of Saudi Arabia. Cogent Arts Human. **6**(1), 1691315 (2019). https://doi.org/10.1080/23311983.2019.1691315

Shang, W.: Role of Waterfront in Shaping City Center Landscape: Perception of Tianjin Haihe Riverfront Landscape. The University of Hong Kong (2011). https://doi.org/10.5353/th_b4718657

Tibbalds, F.: Making People-Friendly Towns: Improving the Public Environment in Town, 1st edn. Taylor & Francis (2000)

Vallega, A.: Urban waterfront facing integrated coastal management. Ocean Coast. Manag. **44**(5–6), 379–410 (2001). https://doi.org/10.1016/S0964-5691(01)00056-4

Whitehand, J.W.R., Samuels, I., Conzen, M.P.: Conzen, M.R.G. 1960: Alnwick, Northumberland: a study in town-plan analysis. Institute of British Geographers Publication 27. London: George Philip. Progr. Hum. Geogr. **33**(6), 859–864 (2009). https://doi.org/10.1177/0309132509334948

Whyte, W.: The social life of small urban spaces. Urban Life **10**(4), 466–468 (1982). https://doi.org/10.1177/089124168201000411

Wood, R., Handley, J.: Urban waterfront regeneration in the Mersey Basin, North West England. J. Environ. Plann. Manage. **42**(4), 565–580 (1999). https://doi.org/10.1080/09640569911064

Wakefield, S.: Great expectations: Waterfront redevelopment and the Hamilton Harbour Waterfront Trail. Cities. **24**, 298–310 (2007). https://doi.org/10.1016/j.cities.2006.11.001.

Wrenn, D., Casazza, J., Smart, E., Urban Land Institute: Urban Waterfront Development. Urban Land Institute (1983)

Wyckoof, M.: Definition of Placemaking: Four Different Types. Better Cities & Towns (2014). www.miplace.org

Xie, P.F.: Waterfront Redevelopment and Urban Morphology. In: Industrial Heritage Tourism, pp. 1–254. Channel View Publications Ltd. (2015). https://doi.org/10.1016/j.tourman.2016.02.019

Theorizing the Construct of Place-Identity

Gayatri Nanda and Arvind Kumar Meel

Abstract

The intrinsic characteristics of built environments shape the way one identifies with any place. The dialectics of people and their physical settings over time negotiate and establish intricate bonds between the two; interweaved by belief systems, memories, experiences, and deep-rooted meanings. Peoples' association with place depends not only on its tangible aspects but are profoundly ingrained in its intangibles, temporalities, depiction in narratives, semiotics, diverse forms of cultural manifestations and one's varied experiences of such engagements. Accumulated meanings of place, passed on through generations, negotiated, partly eroded and transformed are the basis of a continuum of people–place ties forming 'Place-Identity' (PI). Scholars from across disciplines have defined and refined the theoretical construct of 'Place-Identity' (PI) and the processes of its formation. The objective of this paper is to theorize the construct of 'Place-Identity' through an in-depth literature review and deconstruct it into factors that can be observed and analyzed for documenting the same. A perception study conducted in various cities of the geographical subregion of Southern Asia has been used to evaluate these factors derived from literature. Extracts from this perception study are put forth in the paper to reinforce the theoretical narrative. In conclusion, this paper proposes a framework for systematically gathering the Place-Identity of a city.

Keywords

Place-identity • People–place association • Place-meaning • Place-attachment • Place-experience

G. Nanda (✉) · A. K. Meel
School of Planning and Architecture, Bhopal, India
e-mail: gayatri.nanda@spabhopal.ac.in

A. K. Meel
e-mail: arvind@spabhopal.ac.in

1 Introduction

Space is a physical entity which gets meanings and is given the status of being a 'place' when values are embedded by peoples' interaction with it. The dynamic processes of formation of people–place ties are based on both collective and personal interactions of people with place over time. The 'Soft City' as Jonathan Raban (1974) calls it plays a significant role in the construction of a city's identity. The subregion of Southern Asia, as geopolitically referred to consists of numerous ancient settlements, each having their own significant history as well as networked with other significant centers due to ancient trade connects. These historic places are complex entities having undergone the process of acculturation due to economic, socio-cultural and political exchanges. This paper enquires into the construct of Place-Identity by firstly undertaking a literature review to decode the construction of Place-Identity into measurable factors and secondly by evaluating peoples' perception of place for the case of the above-mentioned contiguous region along a part of the Silk Road by a perception survey to reinforce the understanding of the factors contributing to the formation of PI.

2 Materials and Methods

The explorations of the concept of 'Place' and 'Place-Identity' have not been limited to the urban studies field, but have been understood from various disciplinary perspectives. Therefore, the literature mapping conducted for this enquiry is a conscious attempt to outline the shifting paradigms in spatial thinking by reviewing works of key researchers from different disciplines associated with the shifts rather than providing an exhaustive list of research. Figure 1 shows the literature sources contributing to the evolution of these constructs across various disciplines. To support and evaluate the theoretical deductions, an online

S. Kostopoulou et al. (eds.), *Cities' Vocabularies and the Sustainable Development of the Silkroads*,
Advances in Science, Technology & Innovation, https://doi.org/10.1007/978-3-031-31027-0_4

perception survey was conducted with 84 participants from diverse occupational backgrounds (17 categories), gender and age groups (ranging from 23 to 86 years of age) residing in different cities (43 cities) located across the subregion of Southern Asia. The participants were given a short introduction of the objective of the survey and were given a short note as follows to articulate their perception.

> When one stays in a city for a long time, she/he develops a certain connect with the city. Many factors influence the way one thinks about the place in which she/he has spent so many years. The meanings one attaches to any city is very different for different people based on their experiences in the city. Please write a short essay/paragraph on how you have experienced your city.

Categorization of the contents of the essays under different factors derived from literature helped in not only understanding the ways in which PI was enhanced but also later in formulation of a structured questionnaire. To strengthen the theoretical review of PI formation, a few extracts from the surveys are quoted as part of this paper.

3 Literature Review on Diverse Disciplinary Perspectives on Spatial Thinking

There has been a paradigm shift in the discourses revolving around the definition of 'place' by thinkers from diverse disciplines in the recent decades. As can be seen in Fig. 1, initially in the seventeenth century CE defined by physicist and natural philosopher Sir Isaac Newton as an absolute entity which is mappable and empirical, the discourse shifted to 'place' being a 'relative' concept with 'time' playing an essential role by Leibniz in 1676 CE, E. Mach in nineteenth century and A. Einstein in 1905 CE (Madanipour 1996). Immanuel Kant in 1768 CE emphasized the role of human mind in shaping place experience and defined 'place' as a 'relational' construct (Smart 1988). In the Urban Studies field, the visual dimension of places which added legibility and memorability to places and urban forms were deliberated upon from the late nineteenth century till the last quarter of the twentieth century (Sitte 1889; Unwin 1909; Gibberd 1953; Lynch 1960; Cullen 1971; Krier 1979). Only in the mid twentieth century, 'place' started to be interpreted from the phenomenological, social, temporal, functional and psychological viewpoint also leading to paradigm change in the manner 'places' were discussed (Jacob 1961; Lynch 1962; Rapoport 1977; Appleyard 1979; Gehl 2011(first published 1971); Sack 1980; Tschumi 1990; Whyte 1980; Lynch 1981). From the Human Geography field, renowned researchers like Edward Relph and Yi Fu Tuan discussed about place and its experience. Relph (1976) writes about the nature of identity of a place from the perspective of insiders and outsiders and the concepts of 'sense of place' and 'placelessness'. Tuan (1974; 1980) discusses the ways in which people are attached to places through his concepts of 'topophilia' and 'rootedness'.

Environmental Psychologist Proshansky (1978) was first to define Place-Identity as a subset of self-identity. The values attributed to space as a consequence of peoples' interaction with the space and 'place' as an integral component of the identity process have been discussed by Stokols and Shumaker (1987), and Speller et al. (2002). Henri Lefebvre's (1991) approach to 'space' as a product of the spatial triad (consisting of 'spatial practices', 'representations of space' and 'Spaces of Representation') was an inference to rethinking of the philosophies of Marx (triad of affirmation-negation- negation of the negation in dialectic materialism), Hegel (concept of thesis- antithesis- synthesis) and Nietzsche (poetry of everyday life) (Schmid 2008). Lefebvre's theoretical insight which summed up the complex and contradictory nature of space enabled many researchers such as Harvey (1989) and Soja (1996) to take forward his proposition. Harvey (1989), Massey (1994), Soja (1996) and Thrift and May (2001) further deliberate on reasons for the construct of 'social space' put forward by Lefebvre They emphasize on the concept of time as inextricably associated to the above discourse. Manuel Castells (1996) deliberates on how 'space' is undermining 'place' in the information age referring to technological infrastructure as 'spaces of flows' which are annihilating culturally rooted localities of social activities or the 'spaces of places'. This line of thought on 'place' led to several discussions on the notion of 'non-places' conceptualized by Marc Auge (1995) from the anthropological domain. In recent times, an inquiry into ways to focus on cultural and place identities of cities has gained significance. This is a resultant of the globalized and homogenized cultural fabric of cities owing to time–space compression which have led to 'placelessness' (Harvey 1989). Thus, the shift in spatial thinking necessitates deliberation on the construct of Place-Identity at this point in time.

4 The Theoretical Construct of Place-Identity (PI)—An Analysis of Literature Review with Survey Results

The processes of formation of people–place associations have been categorized under diverse constructs by various researchers such as 'place meaning', 'sense of place', 'place-attachment', 'place memory', 'Place-Identity', 'place realization', 'place creation' and 'place intensification' to mention a few (Nanda and Khare 2018; Nanda 2019). By comparison of these concepts from across literature sources, it was deduced that these categories of people–place associations are co-created and have manifold overlaps.

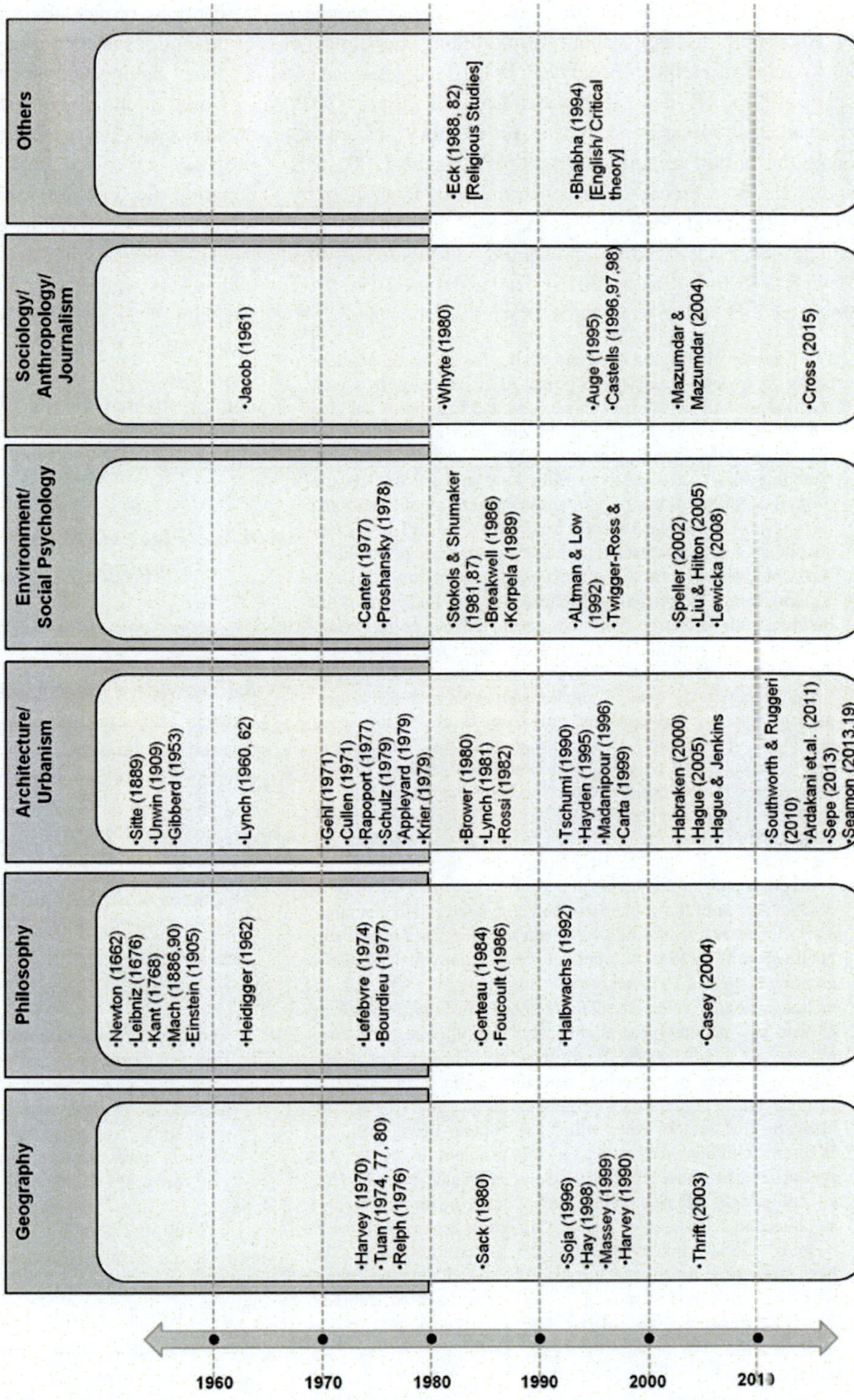

Note: The list is not exhaustive, but indicative of the evolution of spatial thinking across diverse disciplines.

Fig. 1 Evolution of diverse disciplinary perspectives on spatial thinking

Accumulated meanings of place, passed on through generations, negotiated, partly eroded and transformed are the basis of a continuum of people–place ties forming 'Place-Identity' (PI).

According to Proshansky (1978), Place-Identity is "a potpourri of memories, conceptions, interpretations, ideas and related feelings about specific physical setting, as well as types of settings. It is comprised of perceptions and comprehensions". The 'Identity Process Theory' proposed by Breakwell discusses about four guiding principles of construction of Identity (Breakwell 1983; Twigger-Ross and Uzzell 1996). Hayden (1995) and Liu and Hilton (2005) have talked about the concept of 'place memory', where they ideate the notion of 'urban reminders'. Hague (2005) discusses the three processes of construction of Place-Identity. 'Place attachment 'and 'place memory' inform the construction of 'Place-Identity' and should be analyzed together (Lewicka, 2005). Cross (2015) has elaborated on seven processes that add to a person's attachment to a place.

> I live in a neighborhood designed during the 1970s, in Modern Style, following Corbusier's 5 points of architecture. It is well planned with all kinds of urban services and facilities. I am also well connected with the rest of the city with a Metro Station nearby and other public transport. I live on the 8th floor with sweeping views of the Alborz Mountains and the city on the foothills. Our house has ribbon windows crossing all the rooms facing North. So even if I spend a lot of time at home I do not feel cooped up and am connected to the city with views of the traffic on the highway. Occasionally from my kitchen window I can hear a young neighbor practicing the Santoor. When I step out of the building, I am in a large well maintained garden. I go for a walk daily in the beautifully landscaped outdoor space which runs across two blocks. Seasonal flowers are planted twice a year, there are tall shady trees, fountains and pools with red fishes, garden benches, garbage bins, and even some outdoor gym equipment, table tennis tables, basket ball and mini-football courts. Cats are normally lazing around in the garden. Children have a place for cycling. Youth stroll or exercise in their favorite spot where they have created a graffiti wall. Often they are seen photographing themselves and their friends against the wall. An enterprising youth has loaned a corner from the housing cooperative to open a greenhouse and flower shop. It used to be a derelict area until the youth came and converted it into a popular spot with the residents. Old people can use the spaces for a walk, gentle exercise or to meet other people, without worrying about crossing a street. I have access to a large shopping area, within walking distance which has all that I need from daily shopping to clothes and personal goods, household equipment, furniture, repairs, restaurants, cafes, bookshops etc. The terraces every 100m are places of gathering, with live music in the evening (spontaneously organized by the neighborhood youth), a place for hawkers, and outdoor cafes which are popular in the evening. When my children were small, I had no problem finding them a school in the neighborhood. There are several clubs in the area for all age groups, offering gyms, game-courts, swimming pools, aerobics and several other physical fitness programs. The price range is wide and choices are available based on affordability. We have fairly good broadband speeds; I can watch Youtube online without disruption in streaming. (PI_003 from Ekbatan, Tehran)

The extract above is taken from the essay of PI_003 (Participant code) who has spent around forty years of her life in India and twenty years in Tehran. Having widely travelled, she compares and contrasts her built environment with other places experienced by her. As can be seen in the extract above, her evaluation of the place is based on the everyday experiences and smaller details which add up to the perception of the place. The process of accumulating perceptions of any urban setting and selective restructuring of the notions shaped around the place transforms space into a 'place' having unique characteristics.

Based on the literature review and perception surveys, the construct of 'Place-Identity' (PI) was classified into three processes of place cognition, place affection and place conation (as can be seen in Fig. 2). Attributes of built environment contributing to PI formation under each of the category were further deduced by eliminating overlaps and adding up similar descriptions to form a holistic understanding of the construct. Figure 3 shows the PI Framework consisting of fourteen distinct attributes contributing to the three dimensions of PI formation. This paper further details out the manner in which the attributes derived under the three dimensions contribute significantly to PI formation.

4.1 'Place Meaning' as the Cognitive Dimension Shaping PI

Cognition of spatial aspects is the process of accruing information and conceptualizing meanings of a place. Spatial knowledge is accumulated through numerous sources of written and oral histories, evolution of built fabric and diverse interactions of people with place. The perceived environment is sieved out of personal, cultural or other filters and is thus a filtered structure of the real-world facts (Rapoport 1977). The five attributes as seen in Fig. 3 contribute to the process of giving 'meaning' to spaces. The intricacies of each of the attributes is elaborated further with extracts from the perception survey and references to literature to elucidate the way in which they contribute to the cognitive dimension of PI formation.

> The scent of jasmine flowers, the smell of the hot *vadai* in the streets, the endless giggles of people relishing the street food, the enthralling chattering sounds of children in the nearby playground, the many hymns recited by the priests of temples, the ubiquitous clamor and fun, and many more alike, add to the different attires that the city wears every season, every month, all year round. (PI_002)

Living in the city of Madurai, India is a multi-sensory experience which is treasured by the residents of the city as evident from the extract taken from the essay written by PI_002 who is a student of Standard- VIII. Sensory experiences including vision, touch, auditory, olfactory and gustatory sense form the primary communication of a person with his surrounding environment. In the same physical setting each individual has a different sensory experience which can be owing to individual choices based on way of life, past experiences, frame of mind and other such parameters (Cross 2015). As a resultant, the meaning one ascribes to a place is also unique. When the sensory experience of a

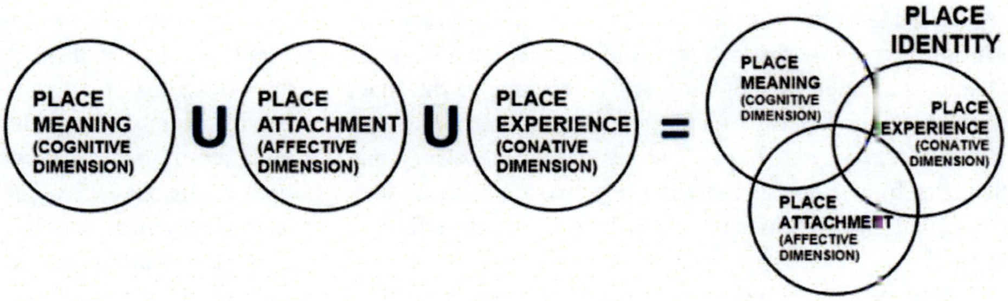

Fig. 2 The three processes constructing place-identity

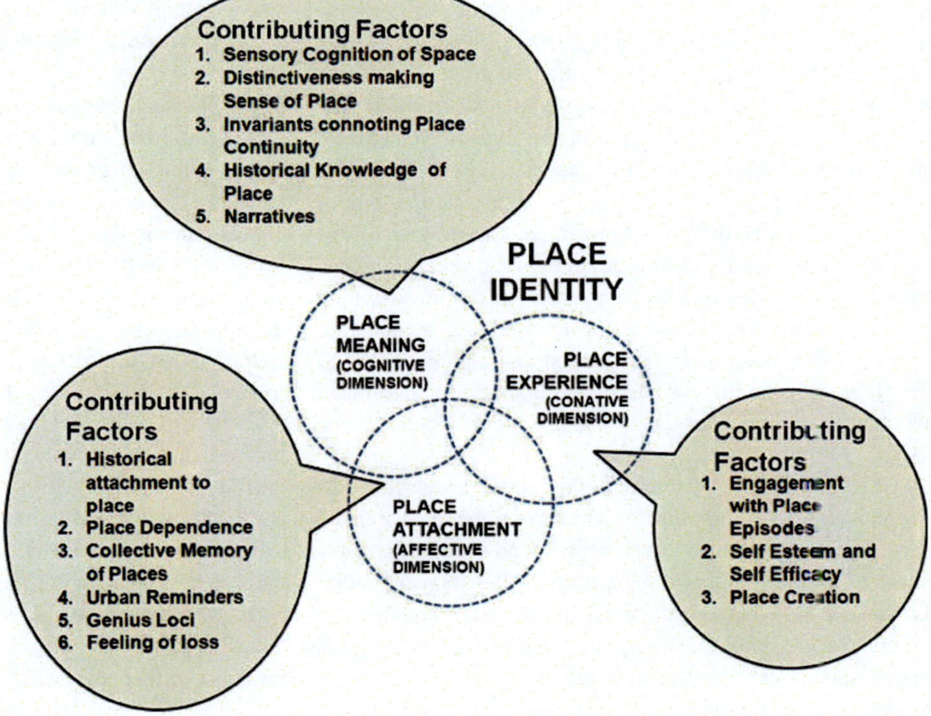

Fig. 3 The PI framework with 14 contributing factors

setting changes, the associated meanings of the place change as well; which either enhances or threatens the identity of the place. Although the visual dimension of places has been signified in global literature, the gustatory dimension has a considerable significance in the South Asian context as is observed from the perception study. The street food culture in most of our cities dominates the sensory perception as can be observed from the notes of PI_002. The distinct smells, myriad noises, contrasting images may all attribute a chaotic image to any city space, but for a person who engages with such a setting on an everyday basis, the places are loaded with diverse meanings, associations and values. Both literature review and perception study conclude on the notion that it is either the vibrant nature of the city providing diverse experiences or the 'restorative experiences' due to engagement with nature-scape or religious precincts which form Place-Identity as a result of sensory cognition.

'Sense of place' is the recognition of the distinctiveness of a place by separating the self from it (Tuan 1980). As an outcome of people's exposure to other places in today's time, inhabitants of a place are becoming aware and conscious of their home environments as they are now able to contrast and appraise their settings. Schulz (1979) discusses that the 'character of place' perceived due to the formal expression of the diverse elements of the place gives it a 'sense of place'. This character imparting the place with a unique identity can also comprise of its temporal variations due to factors like environment, flora, illumination or types of human transactions. The singularity of any place (*'locus solus'*) due to singular features in the urban continuum which impart distinctiveness is discussed by Rossi (1982). The 'cultural capital' of a place as an accumulation of elements symbolic of a place such as attire, ways of celebration, mannerisms, skill sets, and dialect to refer to some of the aspects adds to the uniqueness of a place (Bourdieu 1986). Thus, it is not only distinctiveness of a place, but also of the distinct characteristics of the people residing in the place which form the identity of a place (Twigger-Ross and Uzzell 1996). "I am a true Bhopali", writes one of the survey participants from Bhopal. Individuals who are proud of their city often use its name as adjective while referring to themselves.

Some specific elements in the built environment seem to have a continuous presence since eternity and impart meaning to physical setting. These elements are distinctive to the city, which each resident relates to or is associated with in some or other manner. These invariants weave generations of history because of their continued existence. As Sepe (2013) discusses, these invariants are conserved within the structure of a city and are central elements of place-identity. People have mentioned such elements like the port of Chittagong, River Ganga in Varanasi, the fort of Lahore, the marine drive in Mumbai, and the Howrah bridge in Kolkata to mention a few. It is crucial to recognize such invariants and strategies to adapt them to the transforming needs of the urban should to be devised. As a tool connecting generations, such 'invariants' can also be called as 'heritage specific to the place and thus form significant elements guiding a city's sustainable development (Sepe 2013).

Cities are in a continuous flux with layers of past accounts rooted in their history which continue to exist in the form of oral histories, archival records, and events to create a sense of pride and enhance people–place ties (Hayden 1995; Lewicka 2005). Historical knowledge of place enables residents to understand the finer points of their built setting and establishes a deep connect with the city.

Accounts of places such as myths, legends, folklores and poetries establishing people–place connects by forming notional territories impart a distinct identity to any place. Figure 4 shows an extract from the comic series which is extremely popular amongst youngsters in India since generations. In this case the legendary story of the queen of Jhansi is depicted which imbues children with a distinct perception of the city of Jhansi. Cross (2015) underscores that retelling of place narratives form a significant aspect in the process of the construction of history. Names of places used in narratives impart a sense of orientation and connect to a place. Often in the process of retelling of narratives, new connotations are added to adapt to the changing community ideologies. References in narratives to everyday places make people conscious of the historical aspects of the place and instill a sense of 'cultural insideness' (Altman and Low 1992). A survey participant from Port Blair talks about the way in which the light and sound show played every day in the Cellular Jail complex of the city re-enacting the episodes of Indian Freedom Struggle and role of Andaman Island helps in re-establishing connects of many elements of the present city to the past. As a compulsory part of school excursions, this show makes kids aware of the historical roots of the city. The value of 'Mythical Space' to which people are rooted and subconsciously consider part of their identity is discussed by Tuan (1977). Eck's (1982) book on Banaras elaborates on the place hymns or the māhātmyas which glorify the numerous elements of the physical fabric of sacred cities and aid in place cognition.

4.2 'Place Attachment' as the Affective Dimension Shaping PI

Many physical elements in the city fabric are storehouses of varied emotions and form potential elements to conserve and enhance the affective ties to place. Collective meanings attributed to any physical element as a result of shared affective ties to the built environment define the construct of 'Place-Attachment' (Altman and Low 1992). A number of comparable terminologies like 'Topophilia' by Tuan (1974), the intensity of 'insideness' and 'outsideness' of a person by Relph (1976), the concept of 'Rootedness' by Tuan (1980), 'place-belongingness' by Korpela (1989), etc. have been referred to in literature to define this construct. Figure 3 indicates the six attributes which have been deciphered from the literature review which shape place-attachment. These are further defined with extracts from the perception survey to validate the manner in which they influence an individual's place attachment. If the spatial elements which enhance place attachment are safe-guarded, conserved and made valuable to the changing society, they can help in rooting residents to their places.

Living in a city for a long time, having family connections in a place and belonging to a place culturally influences the ways in which people identify with the place. A sense of belonging to a place is often inculcated when family histories and experiences are embedded in a place and propagated in the form of stories. Numerous references such as "I used to visit the Lingaraj temple with my father and we used to play in its premises for a long time", "My grandmother used to tell us stories about the haunted house in the lake side", "I have a lot of memories attached to our old house in the old part of the city" showcase the ways in which these historical attachments are formed. Ancestral properties which gets transferred from one generation to the next form 'Genealogical place attachment' (Altman and Low 1992). Such associations not only signify places but also place names and symbolic representations which form a significant means for deciphering Place-Identity. They serve as 'social locator' of individuals in a particular space and time. Hay (1998) terms this affective dimension as 'ancestral sense of place' gathered through time as a person experience various

Fig. 4 A snapshot from 'Rani of Jhansi'. *Source* Singh (1974)

facets of life in a place to help form a 'cultural sense of place'. Cross (2015) terms this as 'historical process' by describing it as the method of accumulating and juxtaposing experiences pertaining to a place and thus constructing meanings which progressively gets enhanced with time.

Only after the fulfillment of the physiological needs which are the basic requisites of everyday living, there arises the aspiration to cater to other needs (Maslow 1943). Inhabitants of a place acquire a strong attachment to spatial attributes which assist in fulfilling such needs. The ways in which a person makes sense of a place of dependence is connected to the ways of assessment of the place quality in comparison with other places that contribute in satisfying similar needs (Stokols and Shumaker 1987). 'Economic place attachment' is the 'utilitarian relationship' which people have with places (Altman and Low 1992). Place ownership is a principal cause of such economic attachment as a person invests his assets for the place and undergoes the process of doing so by investing, negotiating with agencies, understanding the real estate scenario and finally decision making and upkeep of the property. Cross (2015) terms this as 'material dependence' which is a negotiation between the material aspects of a place and the social desires of an individual. PI_045 writes about his decision to move from Cumilla to Dhaka to avail the economic opportunities and fulfil his family's aspirations of international standard of education, health facilities and access to diverse opportunities. A strong feeling of ownership also develops with a place due to everyday interaction with the place upon which people are dependent. It often happens that when you ask for a coffee at the stall near your workplace, the vendor prepares it just the way you like it. In such a case the vendor and the shop becomes a small but significant part of the everyday routine of the person. Such place attachment generated by small but everyday interactions adds to the cognition of the place.

> I am so proud that I was a participant in the folk dance performance of Kullu called Nati in the year 2015 which was incorporated in the Guinness World records for the biggest group folk dance. This performance was specially organized for the occasion of Kullu Dussehra. Numerous *Devis* and *Devtas* from the entire region come to Kullu to meet and pay tribute to Raghunathji during this festival. Participating in this performance with many of my friends was not only a fun experience for me but it also filled me with pride for belonging to this small town which is visited by so many people. The entire programme was covered by the media and was shown in all television channels. I will forever cherish this experience. (PI_018)

Memories as preservators of significant incidents of the past are thoughts which can be knocked at and revived whenever desired (Ardakani and Oloonabadi 2011). After any incident, it is the place of its occurrence and its diverse features which strike a chord to recall the memory. Collective memories get constructed when people engage in shared experience of an event which further imparts a symbolic meaning to the physical place. Maurice Halbwachs (1992) conceptualized this collective mode of 'memory' and coined the name 'collective memory' to describe the shared contextualization of individual memories. The extract from the narrative of PI_018 is an example of such collective memory which enhances the way one identifies with the place. Hayden (1995) puts forth the concept of 'public memories' which get formed due to shared territories and time in the public realm. Casey (2004) further elaborates on the implication of 'public memories' formed in 'public places' which fosters 'public discussion' and hence bonding.

The cityscape is embedded with several elements such as place names, monuments, natural features, etc. which serve as mnemonic aids to one's memory. These connects in everyday spaces which provides links to past are regular reminders of the Place-Identity. The range of such markers could be a mere symbol depicted on a railing to a massive monument or an entire urban precinct. A visual engagement with such 'urban reminders' imparts richness to associations with the place. Donald Appleyard (1979) elaborates on the urban symbol system that adds to the imageability of a place. 'Imageability' here is not only associated to the aesthetic aspects of a place but is also a depiction of the values of the residents and the decision-makers. Eck (1981) calls these as 'visual texts'. Bourdieu (1986) describes such symbolic features which add meanings to the built environment as 'symbolic capital'. Rossi (1982) uses the term 'urban artifacts' and refers to places as historical texts. 'Place artifacts' serving as 'visual reminders' having 'symbolic meanings' is referred to by Mazumdar and Mazumdar (2004) in their narratives on religious cities. Lewicka (2008) describes 'Urban Reminders' as the bits and pieces leftover from past generations residing in the place which converse with the present populace by expressing meanings, historical information or simply prompting a curiosity to unravel the place history.

Genius loci refer to the spirit or soul of a place as unique, distinguishing and valued quality of the place (Schulz 1979). Communities and their engagement with the place co-create the genius of a place. Seamon (2013) refers to this aspect as 'Place Realization'. He writes about this distinctive character of a place which can be identified only by inhabitants who are familiar with, have frequent engagements with the place and consequently realize the place attributes. People often denote 'London-ness of London' or 'Indian-ness of India'. This realization of place enhances PI. Cross (2015) discusses this deep sense of belonging as 'spiritual attachment' to a place. In this type of attachment it is complex to identify the basis of this attachment as it forms of intricate relationships of the tangibles with the intangibles of a place.

Today, with the fast pacing urbanization transforming older fabric of places and hence place meanings, the residents of cities very often experience a disconnect to their own hometown. Relph (1976) calls such a scenario as 'placelessness' which detaches community from their traditional places. Only when such a feeling of loss is realized by people, they attempt to reconstruct places. When meanings attached to places accumulated over time get ruptured with only the recollection of the sense of place, a special form of place-attachment gets constructed. Low (1992) elaborates how this type of place-attachment recognized through realization of loss of place characteristics form a significant component of identity of a place.

4.3 'Place Experience' as the Conative Dimension Shaping PI

The meanings and attachments one constructs for a place, transforms into behaviour or conduct of inhabitants in a place. The reasons to do things in specific ways within a physical setting is a response to the accumulated place meanings and affective ties. People decide to partake in events, pick between forms of experiences, assess their place in contrast to other places, and sometimes even participate in the development process of the city. Hence, the conative dimension shaping PI has been categorized under the three aspects as seen in Fig. 3. Each of these attribute is further discussed in this paper.

The physical setting acts as a static backdrop while the activities within adds to the temporality of a place, hence making it episodic in nature. These episodes range from dramatically planned events such as grand festivals to a simple element of surprise in the everyday routine of a place. The degree of engagement one makes with an episode of the city is solely dependent on a person's choice. This aspiration or judgment to participate in an event in the city shows an affective bond with the place which is based on place cognition. Low (1992) discusses how people's aspirations to embark on pilgrimage or take part in cultural events establishes a special connect to a place due to the episode it organizes and its sacred connotations in majority of cases. Although the physical existence of such temporal events is short, its powerful impact attributes a lasting image of the event space, thus enhancing PI.

In the sphere of environmental psychology, self-esteem and self-efficacy have been specified as two indispensable attributes in the PI formation process. A person's positive self-evaluation or community assessment, defined as self-esteem (Twigger-Ross and Uzzell 1996). The superiority of the built environment which enhances a person's self-esteem also adds to the ways he experiences the place. To give an example, many residents of Madurai, India proudly speak of the fact that their city played an important role in the freedom struggle being the place where Gandhiji had initiated the khadi movement by throwing away machine-made fabric and wearing plain hand-woven attire. This past episode of the city inculcates a feeling of self-esteem in the citizens. Self-efficacy is the assurance one has of the complementary character of the built environment for an individual to accomplish his life goals (Twigger-Ross and Uzzell 1996). This realization is an exquisite emotion which contributes in reinforcing PI and changes the manner one engages with place.

'Insideness' and 'outsideness' and various in-between degrees of involvement in the way residents engage and experience places makes the place an integral part of their identity (Relph 1976). 'Empathetic insideness' is a degree of 'insideness' which means 'not merely looking at a place', but empathically contribute in the decision making process of the place which leads to 'place creation' (Relph 1976). A feeling of possessiveness for a place makes people express their territoriality and attempt to appropriate places (Brower 1980).

Twigger-Ross and Uzzell (1996) describe residents' realization of the place value due to which they desire to retain the essence of the built environment. They also elaborate the two types of continuity people aspire- one termed as 'Place-referent continuity' where people being sensitive to the built environment try to conserve and protect the place from transforming and the second termed as 'Place-congruent continuity' when people try to replicate and imbibe certain characteristics of an appreciated place to continue to feel associated to the place. Seamon (2013) describes as to how in these processes of place creation as a result of the conative dimension, residents have a considerable role in the decisions making and city building process.

5 Conclusion

The PI Framework which has been conceived as an operative framework to unearth PI of any built environment thus consist of fourteen distinct factors derived meticulously out of literature review and perception surveys which are typified under the three dimensions of place cognition, place affection and place conation as seen in Fig. 3. Application of this framework in several cities to draw parallels in diverse contexts has been taken up as part of further research pertaining to this inquiry. The PI framework was adopted to formulate a structured questionnaire in which each factor could be interpreted through a couple of questions. The manner in which the survey participants expressed their perception gave clues for articulating the questions for the structured questionnaire. This questionnaire formed the basis of collection of large scale data on the perception of the case cities taken up for further analysis. This paper thus fulfils the objective of elaborating the theorizing process of the construct of Place-Identity and introduces the PI Framework which was the basis of systematically documenting PI for multiple cities as part of further research.

References

Altman, I., Low, S.M.: Symbolic Ties that Bind-Place Attachment in the Plaza, Human Behaviour and Environment-Place Attachment, vol. 12. Plenum Press, New York (1992)

Appleyard, D.: The environment as a social symbol: within a theory of environmental action and perception. J. Am. Plann. Assoc. **45**(2), 143–153 (1979)

Ardakani, M.K., Oloonabadi, S.S.: Collective memory as an efficient agent in sustainable urban conservation. In: International Conference on Green Buildings and Sustainable Cities, pp. 985–988. Elsevier Ltd. (2011)

Auge, M.: Non-Places: Introduction to an Anthropology of Supermodernity. Verso, London (1995)

Bourdieu, P.: The forms of capital. In: Richardson, J.: Handbook of Theory and Research for the Sociology of Education, pp. 241–258. Greenwood, Westport, CT (1986)

Breakwell, G.: Threatened Identities. Wiley, Chichester (1983)

Brower, S.N.: Territory in urban settings. Environ. Cult., 179–207 (1980)

Casey, E.S.: Public memory in place and time. In: Phillips, K.R. (ed.) Framing Public Memory, pp. 17–44. University Alabama Press, Tuscaloosa (2004)

Castells, M.: The Rise of the Network Society, vol. I, 2nd edn. Wiley-Blackwell, Oxford (1996)

Cross, J.E.: Processes of place attachment—an interactional framework. Symb. Interact. **38**(4), 493–520 (2015)

Cullen, G.: The Concise Townscape. Architectural Press, London (1971)

Eck, D.L.: Darśan: Seeing the Divine Image in India. Anima Books, Chambersburg, PA (1981)

Eck, D.L.: Banaras: City of Light. Alfred A. Knopf Inc., New York (1982)

Gehl, J.: Life Between Buildings: Using Public Space. Island Press (2011(first published 1971)))

Gibberd, F.: Town Design. Architectural Press, London (1953)

Hague, C.: Planning and Place Identity. In: Hague, C., Jenkins, P. (eds.) Place Identity, Participation and Planning, p. 4. Routledge, Oxfordshire, New York (2005)

Halbwachs, M.: On Collective Memory. The University of Chicago Press, Chicago, London (1992)

Harvey, D.: The Condition of Postmodernity: An Enquiry into the Origins of Cultural Change. Blackwell Publishers, Oxford (1989)

Hay, R.: Sense of place in developmental context. J. Environ. Psychol. **18**, 5–29 (1998)

Hayden, D.: The Power of Place-Urban Landscapes as Public History. The MIT Press, Cambridge (1995)

Jacob, J.: The Death and Life of Great American Cities. Random House, New York (1961)

Korpela, K.M.: Place-identity as a product of environmental self-regulation. J. Environ. Psychol. **9**(3), 241–256 (1989)

Krier, R.: Urban Space. Cademy Editions, Great Britain (1979)

Lefebvre, H.: The Production of Space (trans: Smith, D.N.). Blackwell, Oxford (1991)

Lewicka, M.: Ways to make people active: the role of place attachment, cultural capital, and neighborhood ties. J. Environ. Psychol. **25**, 381–395 (2005)

Lewicka, M.: Place attachment, place identity, and place memory: restoring the forgotten city past. J. Environ. Psychol. **28**, 209–231 (2008)

Liu, J.H., Hilton, D.J.: How the past weighs on the present: social representations of history and their role in identity politics. Br. J. Soc. Psychol. **44**, 537–556 (2005)

Lynch, K.: Image of the City. MIT Press, Cambridge (1960)

Lynch, K.: What Time is this Place? MIT Press, Cambridge (1962)

Lynch, K.: A Theory of Good City Form. MIT Press, Cambridge (1981)

Madanipour, A.: Design of Urban Space: An Inquiry into a Socio-Spatial Process. Wiley, West Sussex, England (1996)

Maslow, A.H.: A theory of human motivation. Psychol. Rev. (1943)

Massey, D.: Space, Place, and Gender. University of Minnesota Press, Minneapolis (1994)

Mazumdar, S., Mazumdar, S.: Religion and place attachment—a study of sacred places. J. Environ. Psychol. **24**, 385–397 (2004)

Nanda, G.: Place-identity and its significance for a Heritage city: the case of Puri, India. J. Indian Anthropol. Soc. **54**, 74–87 (2019)

Nanda, G., Khare, A.: Spatial mapping of 'people–place ties' as an integral part of heritage management process. J. Herit. Manage. **3** (2), 192–208 (2018)

Proshansky, H.M.: The city and self-identity. Environ. Behav. **155** (1978)

Rapoport, A.: Human Aspects of Urban Form Towards a Man—Environment Approach to Urban Form and Design. Pergamon Press Ltd., Oxford (1977)

Relph, E.: Place and Placelessness. Pion Limited, London (1976)

Rossi, A.: The Architecture of the City. The MIT Press, Cambridge and London (1982)

Sack, R.D.: Conceptions of Space in Social Thought: A Geographic Perspective. Macmillan Education UK, London (1980)
Schmid, C.: Henri Lefebvre's theory of the production of space-towards a three-dimensional dialectic. In: Goonewardena, K., Kipfer, S., Milgrom, R., Schmid, C. (eds.) Space, Difference, Everyday Life—Reading Henri Lefebvre, pp. 27–45. Routledge, New York (2008)
Schulz, N.C.: Genius Loci-Towards a Phenomenology of Architecture. Rizzoli (1979)
Seamon, D.: Place attachment and phenomenology: the synergistic dynamism of place. In: Manzo, L.C., Wright, P.D. (eds.) Place Attachment: Advances in Theory, Methods, and Applications, pp. 11–22. Routledge, New York (2013)
Sepe, M.: Planning and Place in the City: Mapping Place Identity. Routledge, Oxon (2013)
Singh, M.: Rani of Jhansi. Mumbai: Amar Chitra Katha Private Limited (1974)
Sitte, C.: City Planning According to Artistic Principles. Phaidon Press, London (1889)
Smart, J.: Space, time and motion. In: Parkinson, G. (ed.) An Encyclopedia of Philosophy, pp. 256–274. Routledge, London (1988)
Soja, E.W.: Thirdspace: Journeys to Los Angeles and Other Real-and-Imagined Places. Wiley-Blackwell, Cambridge (1996)
Speller, G.M., Lyons, E., Twigger-Ross, C.: A community in transition: the relationship between spatial change and identity processes. Soc. Psychol. Rev. **4**(2), 39–58 (2002)
Stokols, D., Shumaker, S.A.: People in places: a transactional view of settings. In: Yadav, C.S. (ed.) Perspectives in Urban Geography-Perceptual and Cognitive Image of the City, pp. 91–153. Concept Publishing Company, New Delhi (1987)
Thrift, N., May, J.: Timespace: Geographies of Temporality (Critical Geographies). Routledge, London, New York (2001)
Tschumi, B.: Questions of Space. Architectural Association, London (1990)
Tuan, Y.-F.: Topophilia: A Study of Environmental Perception, Attitudes, and Values. Columbia University Press, New York (1974)
Tuan, Y.-F.: Space and Place—The Perspective of Experience. University of Minnesota Press, London (1977)
Tuan, Y.-F.: Rootedness versus sense of place. Landscape **24**, 3–8 (1980)
Twigger-Ross, C.L., Uzzell, D.L.: Place and Identity Processess. J. Environ. Psychol. 205–220 (1996)
Unwin, R.: Town Planning in Practice: An Introduction to Artistic City Planning. T. Fisher Unwin, London (1909)
Whyte, W.H.: The Social Life of Small Urban Spaces. Conservation Foundation, Washington, DC (1980)

Local Events as a Temporary Urbanism Tactic for Public Spaces Revival

Nadia Gohar and Tarek Saad Ragab

Abstract

One primary urban regeneration objective is to revive the city's declined and sluggishly animated public spaces. Nevertheless, due to a lack of funds, most of those projects target the city's principal public spaces, such as central parks, squares, downtown, and waterfronts. Whether selectively, consecutively, or collectively carried out, the usual magnitude of these projects drains the city's budget as they consume extended time and costly treatments, resulting in a few projects being implemented in primary spaces and fewer devoted to the city's secondary spaces. Tactical Urbanism is a viable way for the city's urban managers to implement their long list of regeneration projects and challenge the chronically limited budget the city chronically encounters. One of the common tools used is temporary events and activities. Through the arrangement of multiple-size and multi-nature events, the city brings life to the unappealing public spaces. This paper focuses on developing a strategic multi-tiered system of support framework to assist the city in reviving underutilized public spaces, considering the time, characteristics, and cost layers. Such a framework builds on the employment of local events as a viably flexible and low-cost "Temporary Urbanism" tactic that can be used in city spaces of different grades. The research employs a qualitative desktop analysis of the data collected from historical literature that includes classical theory and lessons learned from various international examples. In addition, a survey is carried out among the city's urban officials to validate the proposed framework structure. An analysis is then used to develop a framework that can be used as a tool assisting the city's managers in taking low-cost, quick, and efficient actions regarding the efficient activation of the city's spaces.

Keywords

Public spaces • Urban development • Temporary urbanism • Tactical urbanism • Local events • Public spaces animation

N. Gohar (✉) · T. S. Ragab
Department of Architecture, Effat University, Jeddah, Kingdom of Saudi Arabia
e-mail: nadiagohar@outlook.com

T. S. Ragab
Department of Architecture, Alexandria University, Alexandria, Egypt

1 Introduction

Maintaining a city of vibrant places has always been a target of urban designers, city planners, and city managers. Nevertheless, many public spaces in some cities surface as sluggish, inactive, and deserted for different reasons. In other cities, such as Jeddah, the multifactorial urban drivers branch out to form a loose spatial urban pattern full of vacant lands and urban voids. Such a situation pressures the city's management system, as these zones should be included in the city's agenda of development. The vacant beaches near Paris, the various squares and plazas in Central Eastern Europe (Balicka et al. 2021; Lehtovuori and Ruoppila 2012), and the inactiveness of neighborhood public spaces in Jeddah which raised complaints from residence (Mandeli 2019) are all examples representing a constant challenge to their cities' management systems.

Moreover, the city faces additional challenges in the form of the insufficiently allocated budget and the long time needed for the implementation of its redevelopment projects (Lehtovuori and Ruoppila 2012). As a result, most of the projects target the significant city's public spaces such as the main parks, the central squares, the principal streets, the downtown, the historical quarters, and the waterfronts. Due to the magnitude of these projects, which manifested in their high cost, a few projects are implemented, and a fewer are allocated to the secondary public spaces of different grades

S. Kostopoulou et al. (eds.), *Cities' Vocabularies and the Sustainable Development of the Silkroads*, Advances in Science, Technology & Innovation, https://doi.org/10.1007/978-3-031-31027-0_5

in the best case leaving many other parts of the city in need of development.

The contemporary urbanism theory discourse provides a development approach that brings forward the terms "*Temporary Urbanism*" and "*Tactical Urbanism*" asserting the need for multi-type temporary usages and events as a viable measure for vibrant spaces with low-cost stakeholders' sponsoring and focused short-term intervention (Haas and Olsson 2014). Temporary urbanism is introduced as the effective use and occupation of a place within time limitation and consideration (Lehtovuori and Ruoppila 2012). Such time-related intensity of use contributes to the activation and revival of poorly functioning spaces and can be used as an alternative solution for traditional, long-term redevelopment measures.

Reviewing historical literature reveals that the temporary urbanism tactic has been employed for an extended time and its validity has been tested in multiple scholarly studies (Martin and Hincks 2016; Lehtovuori and Ruoppila 2012). However, no clear strategy for contextual decision-making/taking and related managerial processes was noticed. Worthily noted within such a notion are the European cities' pioneering experiences that adopted temporary urbanism as a basic approach to urban regeneration. For example, Berlin played an essential role in utilizing the city's vacant areas in initiating cultural startups (Martin et al. 2019; Madanipour 2017). One important used tool depended on promoting cultural assets in public spaces through events, festivals, shows, and exhibitions (Richards and Palmer 2010). Also, there is the case of Bristol's Temple Quarter, in the UK, where temporary use stabilized the local land markets and roused real estate values (Martin et al. 2019). The temporary use of famous public spaces such as Trafalgar Square, Hyde Park, and Regent Street in accommodating events and occasional activities helped in reviving these spaces. Also, the same technique worked for less famous spaces including Gillett Square, Finsbury Park, and Berwick Street (Smith 2016). Another important example that shows the power of temporary uses is the NDSM wharf in Amsterdam North, which became a major cultural destination and initiated a wave of development, and in Magdeburg, Germany a simple lightweight open-air library and reading place, in place of a vacant land became a success. Also in Detroit, "Urban Farming" has evolved mainly as a community initiative in the city's public spaces, aiming to reduce poverty and support social inclusion (Lehtovuori and Ruoppila 2012). All these previous cities represented successful temporary urbanism tactics and projects, however, depended only on the vision and experience of their urban managers. Where they all missed a clear strategy or a managerial model that assists in the process of event selection based on socio-economic criteria, the space size, location, proportion, connectivity, accessibility, scalability, environmental suitability, and other influential factors.

This work focuses on developing a Multi-Tiered System of Support (MTSS) framework managing the processes and actions targeting the revival of inactive and underutilized public spaces. The framework builds on the employment of "local events" as a low-cost, short-term, collaborative, and flexible "Temporary Urbanism" tactic for decision-making and implementation considerations.

2 Methodology

Missing a clear and efficient mechanism of related decisions and implementation managerial actions such as selecting the event size, the hosting space physical specifications, and the space location, in addition to other attributes raises the main question of the research. How to devise a contextual decision-making framework aiming to induce, regulate, and facilitate the local events and their designated spaces selection? Answering this question necessitates carrying out multiple tasks. First, investigating the historical evolution of temporary events and activities in the outdoor public spaces. Within this task, the lens focuses on local events and their contribution to the revival of underutilized spaces. Second, identifying related spatial and management elements, and third, addressing the correlation of the two critical principles for revitalizing public spaces: the efficiency and appropriability of the public space and the local events' success drivers. The methodology includes a qualitative data collection through historical literature and urban managers' structured survey. Then, the researchers run desktop analysis to synthesize a framework that classifies and systemizes the space attributes, the event's nature, and the actions required.

3 Temporary Tactical Urbanism and Events

The concept of temporary urbanism goes hand-in-hand with the 'temporary use' of the space. The term stirs the very notion of what types of urbanism have evolved organically over many urban development cycles (Martin et al. 2019). Its origin is traced back to civilization's beginning, based on the flexibility and responsive behavior to the surrounding context. The scholarly attention rose in the early 2000s, through examining European cities and space production through the social act in Latin America and the Middle East, and South Asia (Martin et al. 2019; Madanipour 2017). Cities adopt short-term strategies to enhance the inception of Temporary Urbanism, which will result in the integration of temporary actions situated within the permanent plans (Smith 2016). As cited in Koyama (2017) Beekmans and de Boer, the authors of the book "Pop-Up City: City-making In a Fluid World" identify these short-term interventions as "momentum-oriented urban interventions" that allow for attraction in an area that is not within the urban planning project. As a result, Temporary Urbanism could be part of an ongoing permanent development or area until permanent planning plots are decided. Temporary-use significance can be presented in terms of social, economic, and placemaking (Khalil and Nasr 2021; Lehtovuori and Ruoppila 2012). The strategy aims to prevent misuse or foster activities that threaten the community, provide a better life quality for residents and visitors, and increase community participation. Depending on the popularity of temporary use and success, it can have an economic return, such as the support of local and new businesses. The clustering of shops or kiosks around a temporary use or event encourages more public spending, thus more financial returns for businesses and residents. Other benefits directed at placemaking, such as diversity of activities, reviving underutilized spaces, increasing the land value, and achieving a potential development goal, such as utilizing infrastructure and protecting the green area (Khalil and Nasr 2021; Kotval et al. 2010; Lehtovuori and Ruoppila 2012).

Which tactic of Temporary Urbanism is best employed to transform an inactive public space into an attractive, animating, and livable one? According to Martin et al. (2019), the rising international trend investigating temporary use can be divided into ordinary or extraordinary measures. The former indicates an ad-hoc development such as car parking, storage area, or construction compounds. Whereas the latter deals with the innovative/creative use of space such as building, reuse, or events (pop-up street markets, urban farming, musical event, street festivals, and street vending). Furthermore, according to Smith (2016), events in public spaces are a temporary phenomenon in the city and, therefore considered temporary urbanism tactic. It is not only an interim use to stop the gap, but it is oriented towards creative cultural activities and public participation. Martin and Hincks (2019), described events as spectacular temporary use to provide a more efficient, creative use of the space. Throughout history, public spaces were used for carnivals, fairs, circuses, and sports and markets.

Reviewing the chronological evolution of outdoor events in cities' public spaces (Fig. 1) shows that events were part of the city starting in the pre-industrial city. The concept starts as cities originally structured around event spaces, for example, agora in ancient Greek cities and the primate prayer plaza in the Islamic cities. Then medieval cities and industrial cities planned formal zones and spaces to host public events. Contemporary cities used to hold events on the city outskirts; however, now the city festival's branding is encouraged, and events are being staged and hosted in central public spaces. Throughout the theoretical discourse, events can be divided into several categories based on size, theme, and duration (Cudny 2016; Getz 2008; Smith 2012).

3.1 Local Events Typology

Throughout the theoretical discourse, events can be categorized based on a series of interconnecting factors: size, theme, and duration (Cudny 2016). They are also known to follow cultural celebrations including festivals, carnivals, and religious memorials. Other themes include art and entertainment, politics and state, business and trade, as well as educational and scientific themes (Cudny 2016; Getz 2008).

The duration of the event that will liven up the public space and intensify the lived experience, events in general have a limited duration, and mostly are recurrent. Their time limitation can be viewed as occasionally, yearly, seasonal, monthly, or weakly (Cudny 2014; Richards and Palmer 2010). Balicka et al. (2021), states that these events can be seen as temporary, seasonal, or permanent interventions. The temporary event duration can be limited and rare such as Two canal installations for the Bruges Triennial in Bruges, Belgium, or the "Between waters" art installation. The seasonal events for the summer can be dismantled such as Paris Plages, Paris, France, and Olive Beach, Moscow, Russia.

The size of the event is reflected in the location in which the event will be hosted. There are four main types of size-based classification: Mega, Hallmark, Major, and local events (Getz 2008) These events are equivalent to mega-events, hallmark, major, and minor described by Jago and Shaw (as cited in Cudny 2016; Lunga and Mihaela 2012). Local events target the residents and communities within one neighborhood or several ones, and each type differs in the number of participants in a city or a neighborhood. They are based on local social values and reflect their wants in terms of entertainment and social networking. Additionally, The flexibility of these localized movements and the fewer rigid requirements enhance the overall sense of belonging, social cohesion, and attachment to the community (Balicka et al. 2021; Getz 2008). The expected number of participants ranges from 50 to 5000 and typically occurs in neighborhood parks, pocket parks (parklets), streets, and connecting alleyways (Getz 2008).

Local events include but are not limited to the following (Getz 2008; Smith 2016):

- Small Festival/community celebration: Customarily staged based on a local community's unique aspect. They provide entertainment and activities for visitors and residents.
- Parade: Public march celebrates a special occasion. They are held in the city streets, accompanied by music and dance routines to encourage the residents' participation.

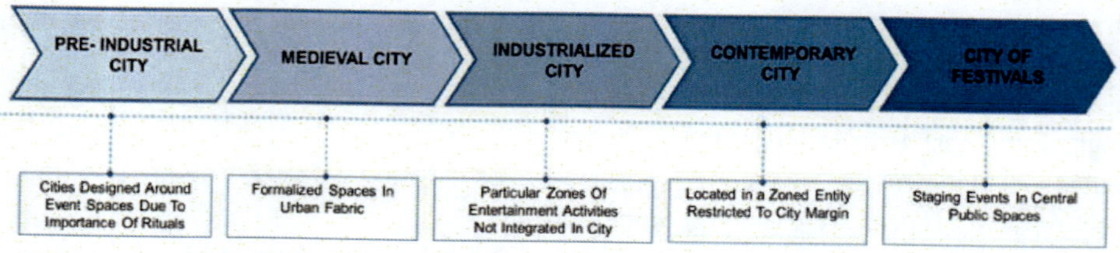

Fig. 1 Leisure activities spatial configuration timeline proposed diagram, adapted from text: Richards and Palmer (2010) and Smith (2016). *Source* Author

- Community market/Pop-up Street stalls or street Bazar: These are weekly or monthly markets that meet the community needs and take advantage of the seasonal demands.
- Sports Events: Introducing themed Event that Promotes a healthy lifestyle through encouraging the residents and visitors to participate in competitive sports
- Urban farming: A sustainable activity that encourages residents to explore urban farming ideas within their public spaces and form a social network for agricultural lovers'.

3.2 Successful Local Events and Revitalizing Public Spaces Key Criteria

Successful local events reflect a positive impact in public spaces, which include branding opportunities since it boosts the image and allows for higher exposure. This creates economic and tourism advantages. On a neighborhood scale, events create a nodal site and meeting point for residents. The growth of the audience will lead to Social interaction and improve the quality of life (Cudny 2016; Richards and Palmer 2010). For example, Jeddah's historical area has always been a major tourist attraction for visitors. Pilgrims made it almost part of the rituals to pay a lengthy visit to Al-Balad (Alawi et al. 2018). As part of the branding strategies is to label the cities, such as; "City of festivals" or "Eventful cities". These strategies freed cities from traditional spatial boundaries by having these festive events everywhere and, in contrast, transforming the space for a lasting memory. The concept of the urban festival was common in cities that were not affected by World War two, such as Edinburg, which was branded as the "City of Festivals" (Smith 2016). Furthermore, the revival and animation of public space, through staging events means the temporary animation of the public spaces via new uses, activities, and inviting new users. Moreover, the emphasis on flexibility of the space. Also, the enhanced Spatial experience through events highlights the cultural and social activities of the place; and allows for an experience-based value for the city. the public space will be enriched and leave a memory in the user's mind enhancing the image of the city on a macro scale and public space on the micro-scale (Johansson and Kociatkiewicz 2011), thus affecting the experience and attaching a memory to the location. For example, the city of Stockholm initiated the Stockholm Culture Festival, this festival took advantage of the history and culture within the city and the economic state to generate festivals within the city space (Johansson and Kociatkiewicz 2011).

The idea of the revitalization of the sluggish public spaces through temporary local events needs a strategy to create a balance between urban development and maintaining place identity, culture, and traditions. Utilizing the concept of temporary urbanism in form of local events requires formulating a set of criteria to create a thriving place that would help identify the space's quality and the needed measures required to revive it. The two sets of criteria studies in this paper are the "Place-Making" approach (Project for Public Spaces (PPS), 2018) and factors of successful events (Richards and Palmer 2010), these criteria will be the basis for building the framework.

Exploring the "Place-Making" approach in Fig. 2, which originated from the work of Jane Jacobs and William H. Whyte in 1960 (PPS 2018), focuses on the place's activities and connecting people through a collaborative process, which is a critical consideration in the developing a successful local event (Balicka et al. 2021; Cudny 2016; Smith 2012). The four main criteria of evaluating the Place-making diagram: sociability, uses and activities, access and linkage, comfort and image, are being used as a targeted factor for reviving the sluggishly animated inner neighborhood spaces, being achieved by the employment of successful outdoor local events (PPS 2018).

According to the model developers PPS (2018), to achieve successful places, cities need to enable Place-making at different scales; the city's identity and the image are reflected through its main destinations, places, and activities. These destinations can be in the form of a square, waterfront, or park that can help to draw more residents and visitors to the city. PPS specifies the importance of all cities to have ten destinations, which have ten places, and ten activities or events that connect people. It is important to consider the different types of public spaces, whether it is the leading destination for the city or the neighborhood. For

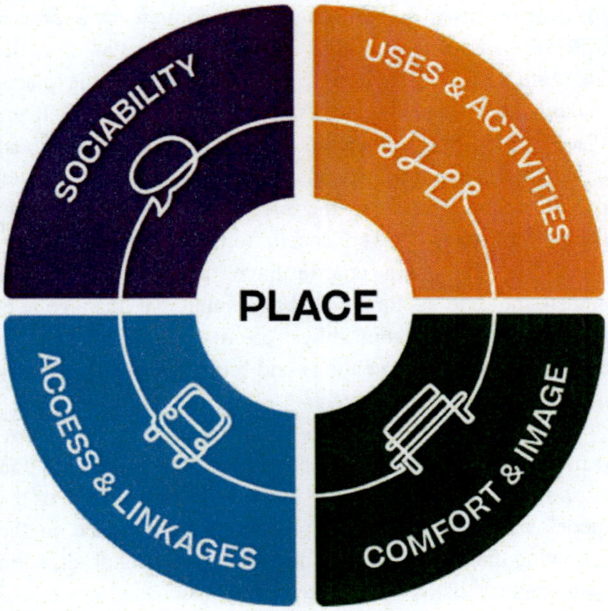

Fig. 2 What makes a great place. *Source* PPS (2018)

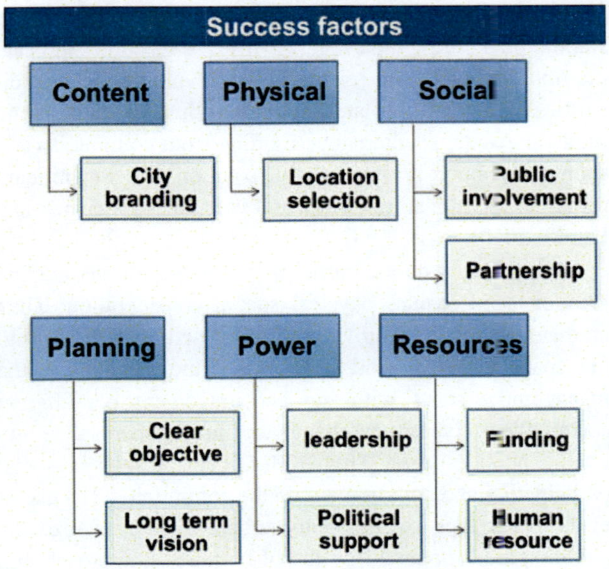

Fig. 3 The proposed combined Factors of successful Events, adapted from text: Richards and Palmer (2010) and Stevens and Shin (2012). *Source* Author

example, Congress Square Park in Portland, Maine, USA, the square was inactive in the early 2000s and the city was interested in selling it for profit however nonprofit organizations invited PPS and they devised a plan to introduce small events like movie nights and food trucks to attract locals (PPS 2015).

Using public space events as a temporary revival and animation tactic emphasizes the space's flexibility via new uses, activities, and inviting new users in public spaces. However, the events need to be successful to attract more people and enliven the spaces. Richards and Palmer (2010), stated the crucial factors to the success of events, these recurring factors are essential to event programs (Fig. 3). Starting with context, which reflects the city branding strategies, whether economic, social, cultural, or historic. The theme, location, public consent, political approval, and monetary fund reflect the context (Smith 2016). Space's culture and identity should be reflected in the event theme and space to ensure a high social interaction level. The context can be based on the current image and enhanced as a part of a new image. For example, Trafalgar square events were targeted to remove hostility due to the history of demonstrations (Smith 2012).

Next is the physical factor, which indicates events' physical location consideration to ensure efficiency and reduce negative impacts. Stevens and Shin, (2012) states the four essential spatial characteristics enclosure, axiality, centrality, and permeability. To create a lasting festive effect of freedom and adaptability. the social factor, which is commonly associated with the word "Event", along with other keywords like social gathering, occasions, competition, or adventure, all of which describe the lived experience and reflect the quality of life (Lunga and Mihaela 2012). Also, events are a platform for citizens so their success reflects and expresses culture, talents, and creativity. The most important outcome is social participation and experience; events are unique once missed; they will not be repeated with the same experience again and would be a lost opportunity (Richards and Palmer 2010; Getz 2008).

After that Planning, which is an essential factor to ease decision-making and contribute to city programs and future visions. Events should have a clear objective and a plan for the city's development process. Richards and Palmer (2010) stated the different scales of the event affect the long (15–25 years)—medium (5–15 years)—short (3–5 years) term plans. For example, mega-events require improved infrastructure (roads, supporting facilities, and buildings); therefore, such events' participation is part of long-term plans to accommodate events. In contrast, events on local districts' scale are part of short-term plans since they offer measurable short-term gains (Richards and Palmer 2010).

Another important factor is the successful use of government power is crucial for effective decision-making and coordination between stakeholders. Richards and Palmer (2010), stated a successful combination of the top-down approach, by which the municipal or higher authority dictates the formation of the basic guidelines. And the bottom-up involves the community and people's ideas and considers the social and cultural norms. And the involvement of the management team is another critical influencer since solid leadership, and a committed team will make a smooth plan during the event. Finally, the resources indicate the funding level from private investors and public municipalities' grants. Cities are cautious with allocating public funds to events, so alternative funding from community and sponsors' support is often a viable solution. Other significant resources are human resources and volunteers (Richards and Palmer 2010).

Events attract public participation to specific geographic locations and require physical spatial consideration. They animate the space by providing diverse activities that add to the space's identity and comfort. At the same time, the management starts with the encouragement of different stakeholders. Events can change any urban open space into livable public spaces (Smith 2016). This results in social participation and cohesion within the community. The use of events as a type of temporary urbanism to revitalize the public spaces supported by the place-making approach and crucial factors to the success of events will result in comprehensive criteria for the framework.

4 Proposing and Building the Framework

The research works on building a framework for implementation. Such a framework will aid decision-makers in ensuring the optimum selection of a particular public space that suits a specific event. The framework stands on two tiers: (a) The space selection decision-making/taking section, which provides criteria for city stakeholders to locate the appropriate locations for conducting events, and (b) The event management platform, which provides a set of recommendations for municipalities' officials to follow which relates event nature to space characteristics and help keep track of the individual incidents for control, monitor, and feedbacks.

Developing the framework involved linking three main areas of knowledge, temporary urbanism, events typology, and successful local events and revitalizing public spaces key criteria. The strategy is to establish a definite basis that covers a holistic border of applying temporary urbanism in form of events in public spaces. The framework approach follows a strategic sequence of key criteria stated in the literature review, a hybrid approach combining the place-making model and the successful events criteria. First, the Place-making diagram aspects include; sociability, uses and activities, access and linkage, comfort and image. Followed by the crucial factors to the success of events including; context, physical, power, social, resources, and planning. Therefore, five main criteria have been developed that embed the said factors including, municipal management and design consideration, physical criteria, events selection criteria, physical interventions and tools, and finally criteria for the management of events.

These criteria were reviewed and arranged in a recommended sequence followed by measurable indicators and attributes for an easy process, and a validation survey of the model was conducted. The process is summarized in the following diagram (Fig. 4). The proposed model criteria and indicators are listed below; their relationship, sequence, and measurable indicators are explained in the final framework (Fig. 11).

4.1 The Framework Structure

The initial basis of forming the framework (Fig. 5), comprises layers followed by attributes and indicators to achieve criteria that should be fulfilled to achieve a prosperous public space regardless of the variation that could occur from one place to another. The first criterion, municipal management serves the objective of a suggestive procedure for a governmental or municipal process that facilitates and encourages temporary urbanism. Followed by physical criteria, to understand the type of public space to be developed. This helps to identify the next criteria, which is the selection of the event.

In this research, the four domains of the placemaking model, including their aspects, will be adapted and explored under the placemaking checklist for temporary tools. This checklist represents temporary physical tools along with the social aspects that support creativity and cultural activities

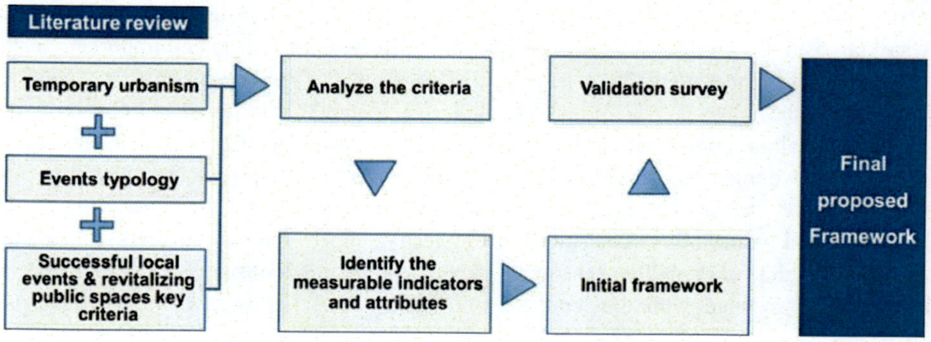

Fig. 4 The researcher's process of devising the framework. *Source* Author

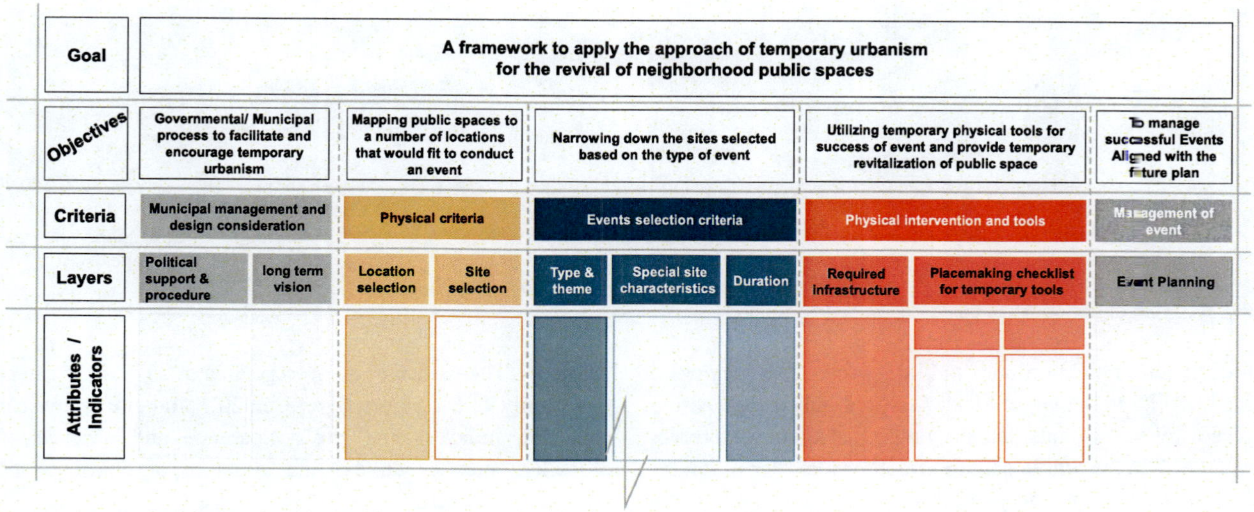

Fig. 5 The researcher's framework structure indicates the criteria, layers, and indicators sequence. *Source* Author

that aim of achieving a successful event. Consequently, this requires a final layer to manage a successful event by adding management criteria and indicators that are concerned with local events. The framework achieves each objective; this integration could lead to an effective approach to developing a customized event that can perform adequately in selected public spaces.

4.2 Analyzing the Criteria and Identifying the Indicators

Successful events require the primary identification of criteria that combines between literature's main findings including management, physical, social experience, and temporary tools. Therefore, the researchers adapted these criteria and definitions into the framework. Starting with the management, which is further divided into two main criteria, municipal management, and event management, or planning. The first criterion is "Municipal management and design consideration", which targets Governmental/Municipal processes to facilitate and encourage temporary urbanism. Consists of political support and a long-term plan. This former domain deals with the preparation of a quick approval system to encourage the repetition and start of the temporary events within several neighborhoods. That would entail a clear and understandable procedure, and a swift processing time for the application to obtain the necessary legal permission, producing a user's guide. And regulations for all categories of stakeholders (community members, private sector) and permits for the use such as Public Space Construction Permit, Public Space Occupancy Permit, Property Owner Approval, Parks and Recreation Permit, and Public Gathering Permit. As well as producing, a user's guide for each type/theme of event/site allowed for temporary urbanism and precaution, safety measures needed, and recommended official vendors for temporary place-making tools.

Whereas the long-term plan deals with plans that municipalities need to consider continuing the success of the public spaces and increasing the activity. This requires monitoring of spaces for maintaining the cycle of demand and supply. The reflection of any Permanent measures to be taken as a result of the temporary event, in some cases events allowed for the municipality to permanently enclose the streets to allow for easier access to the public space such as Putnam Triangle Plaza, New York (NYC Department of Transport 2010). Also integrating a bottom-up approach through engaging the public for knowing needs and in decision-making.

While the second is "Event management" which aims to list the impact and alignment of the event on city planning and directed to the event organizers. There should be a clear goal for an Event that Coincides with urban development plans for the neighborhood on a micro-scale and city on a macro scale. A plan that identifies the stakeholder, Funding, and costs (sponsor, community representatives, volunteers, local business), to minimize the risk of conflict between the stakeholders involved in a project. Following guidelines and procedures provided by the municipality and required permits as well as a clearly defined (social, economic, management) expected outcome defined by the organizer. This division creates a clear understanding of the management tasks for a swift procedure.

The second criterion is "Physical" which also can be referred to as location or context. The choice of location of public space within the city/the neighborhood is important to consider maximum accessibility and participation and adds

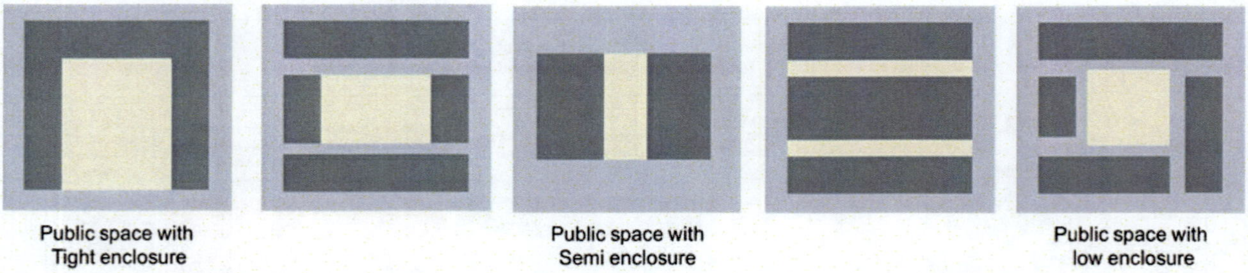

Fig. 6 Types of enclosures in public space. *Source* Author

life to the neighborhood public space. The degree of accessibility to the target location would indicate the ease of access by participants and the range that this event serves, whether it is the whole neighborhood or a cluster of houses. It is also a reference to pedestrian access and public transit access. It is vital to have a defined zone for events; either space is enclosed by a building (Fig. 6) or temporary measures such as; moveable barriers, barricades, or signs. This axial flow of spaces (Fig. 7) maximizes the exposure to people in the crowd alongside. It also encourages 'secondary activities'.

Events should utilize the central nodes in the neighborhood's public spaces to cover a wide range for the pedestrian to commute (Fig. 8a, b). These events could be determined following the "5 min walk" rule or "10 min walk", also referred to as "Pedestrian Shed." The 5 min walk is 400 m walking range (0.25 miles), while the 10 min walk covers a more extensive range of 800 m (0.50 miles) ("The 5-min walk" 2020). Permeability and connectivity through space and time describe the extension of the social atmosphere beyond the immediate time and space of the event. And the penetrable of the space within the neighborhood (Fig. 8c).

Another factor that is important to consider when selecting the site, is the area of space, which will help to decide the expected number of participants and the type of event. The number of occupants can be calculated based on the type of event. For example, the standing spectator 0.3 m^2/person-market or festivals 2.5 m^2/person (Crowd Safety 2020; Wolchover 2012). Besides, the distribution of functions affects the area required such as the market requires a prep, display, sales interaction, and social area. While farming area depends on the number of family members. The garden of an average area of 55–7 4 m^2 caters to a family of 5 or more (Eyerly 2020). And site's physical condition indicates the site's clearance and absence of obstacles, such as an existing structure or landscape or greenery availability.

The third criteria are the lived experience and the importance of the type of Event (Theme and aesthetic). This research focuses on Local events within neighborhoods that target the residents and communities. They reflect the locals' wants and social values in terms of entertainment and social networking. The local events can have many themes. However, the following themes are based most common in neighborhoods small festival/community celebrations: parades, community markets/pop-up street stalls or street bazaars, sports events, and urban farming (Getz 2008; Smith 2016).

Other elements to be taken into consideration are any special site characteristics that can be required for each theme or type of event. Of course, the duration and repetition of the event will liven up the public space and intensify the lived experience. Most importantly is the symbolic meaning behind the events, which tends to last in the visitor's memory. Weather Social, economic, political, or physical contexts all reflect on the neighborhood's identity, which must be represented in the event. Events can reflect the existing or new identity.

The fourth criteria are based on the four elements in the place-making diagram PPS diagram (uses and activities, comfort and image, access and linkage, and sociability). However, the researchers developed the elements in the form of a checklist. First, is the required infrastructure, these include the entire necessary infrastructure required to

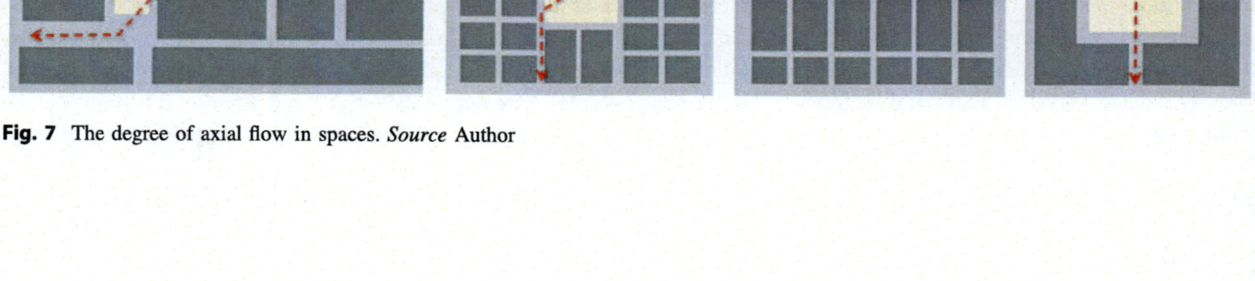

Fig. 7 The degree of axial flow in spaces. *Source* Author

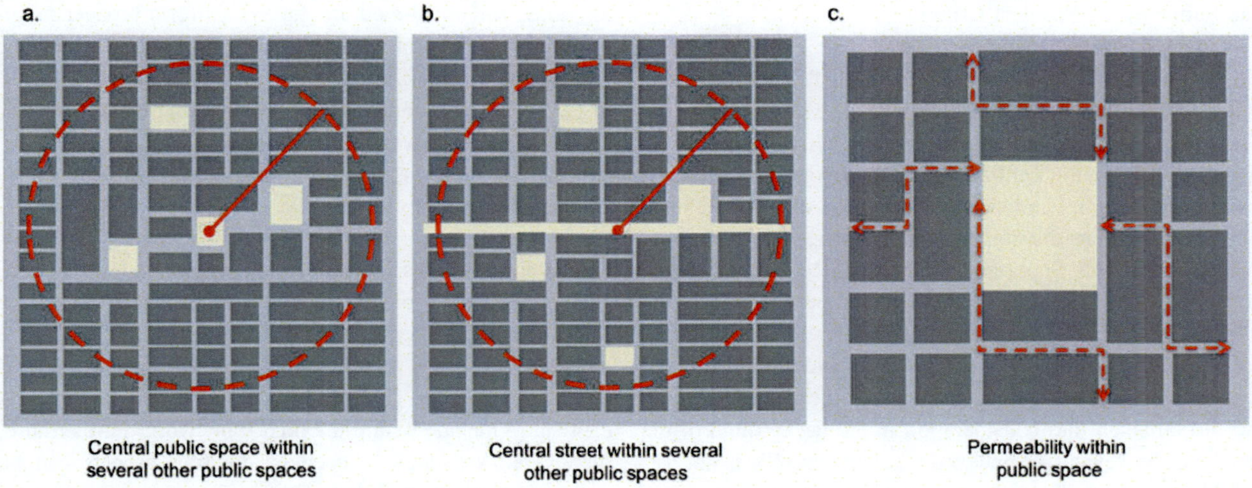

Fig. 8 Events utilizing centrality (**a**, **b**) and permeability (**c**). *Source* Author

establish the event, such as electricity (portable generator or provided onsite), and existing lighting (which will help decide the additional lighting location). The uses and activities, which List of various activities that the event provides, such as food kiosks, children's inflatable games, and any temporary structure such as theater, stage, or shading. Comfort and image, which requires listing the needed amenities to provide comfort for the event-goers, such as mobile seating areas/benches (such cover min of 10% of the expected site users), moveable lighting to provide safety. And movable trash containers for waste disposal, the event organizer should dispose of any waste according to the municipal procedure without overloading the existing waste collection system. Also, mobile shading devices (such as canopies, umbrellas, and tents), moveable flowering pots/artificial turf to enhance appearance, and portable amenities (food trucks/kiosks, toilet). The need to request official patrols to increase safety and avoid crimes. Then Access and linkage, require listing the needed amenities that would ease the site accessibility, such as onsite handicapped ramps, signage for public transit to the event location or wayfinding, or rerouting traffic. The allocating or painted parking spaces, speed calming tools and crossroads, and moveable barriers for the enclosure. Finally, Sociability, the social level can be indicated by the extended use of place during the event and the public volunteering and youth participant.

4.3 Validating the Framework Assumptions

The validation of the devised model was done through an online expert survey. This phase was essential to verify the structure of the model and that no factors were misplaced. The action followed was to weigh the importance of the aspects to assess the level of partners' contribution. It considers several factors, such as the designed questionnaire, selected participants, method of delivery, and the sample size. The survey combined the framework aspects (physical, management, governmental, and social) and aims to verify these aspects and weigh their importance. The survey was sent to experts in relevant fields, including municipality officials, urban planners, city planners, urban designers, architects, event planners, and event organizers.

The questionnaire is divided into four parts that reflect the framework criteria. The type of questions used were rating and open-ended questions. The rating questions use the numeric scale to calculate the importance weight for particular attributes/indicators of each criterion. The numeric scale from 1 to 5 where 1 is 'Least important' and 5 is 'very important'. This method is quick to analyze and process. The second type of question is open-ended questions, which are used for comments and suggestions to be considered.

4.3.1 The Survey Finding and Analysis

The questionnaire received 30 respondents from different professions, out of which architects and urban design were the majority. The answers to rating questions were analyzed using "GraphPad prism". Prism is a scientific 2D graphing and statistics software used for data organizing, presenting, and analyzing. Using Prism, the mean was calculated and presented in a bar chart representing the importance of the criteria. Following this technique for all the rating questions and taking into consideration the comments in the open-ended questions. Accordingly, the framework criteria would be updated to take into consideration the results of the questions.

The following part will demonstrate the questionnaire findings categorized according to four sections which represent the key criteria of the framework:

a. Public Space Physical Criteria

In the first Aspect (the location selection), Table 1 represents the result and the mean of each answer. The results show that the most important criteria are accessibility, followed by connectivity. In the centrality aspect, its rate is the lowest with a mean of 3.9 and standard deviation of 0.9, and since the answers were distributed equally on a scale of 3–4–5, which reduces its importance and focus. Regarding the site selection, the result shows that essential aspects would be the rate of visitors, place keeping while ownership and absences of physical obstacles show similar mean average. Additional remarks include the demographic of the district, which would be taken into consideration in the background of the analysis; however, it is believed that demographic is not part of the physical criteria of site or location selection.

b. Event Selection Criteria

The results in Table 2 of the Event selection criteria layers show that type and theme, and site characteristics are higher in importance than duration. Furthermore, the theme of the local Event proposed activities rated within medium to high importance, the highest in the community market since it would appeal to the most residents and culture while the least is parades. Additional events include art and cultural, religious, and small-medium gatherings for a broader range of activities. The unique site characteristics for each Event type can be summarized, as festivals require a higher degree of permeability, parade requires axiality, and market, sports, and farming require enclosure (Fig. 9).

c. Physical Intervention and Tools

The temporary tools utilized within temporary events were categorized based on the four successful place diagrams. Table 3 presents the mean of all four categories; they can be arranged to usage, comfort and image, sociability, and access and linkage. The result notes the importance of each element to include within the final framework. Furthermore, the overall sequence of each of the four aspects can be arranged based on their importance (Fig. 10).

d. Management

Management in terms of political support and procedure and planning both scored high results (Table 4). Especially the political support and procedure, three aspects scored higher than 4.5. Most comments are recommended to separate the management of events and city management into different categories since they both affect the Event's success and contribute to temporary urbanism.

Table 1 Physical criteria survey answers analysis

Physical criteria

Location selection

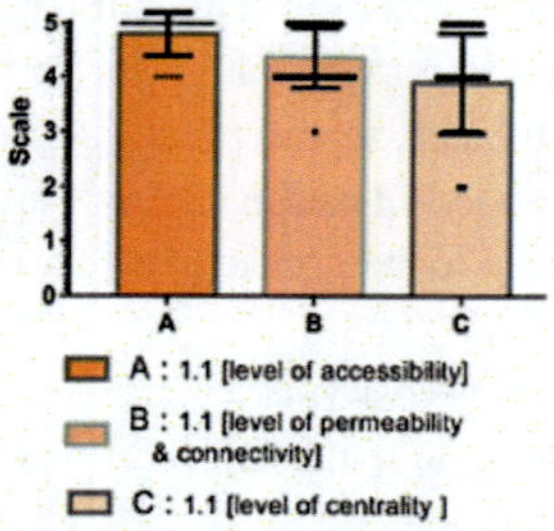

	A	B	C	
Number of values	30	30	30	
Minimum	4	3	2	
25% percentile	5	4	3	
Median	5	4	4	
75% percentile	5	5	5	
Maximum	5	5	5	
Mean	4.8	4.367	3.9	
Std. deviation	0.4068	0.5561	0.9229	

Site selection

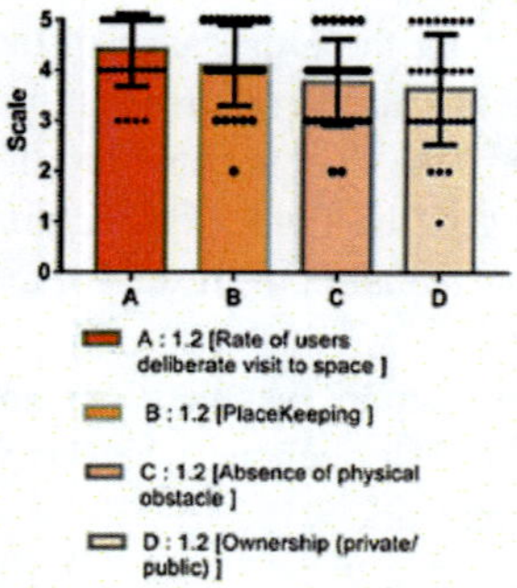

	A	B	C	D
Number of values	30	30	30	30
Minimum	3	2	2	1
25% percentile	4	4	3	3
Median	5	4	4	4
75% percentile	5	5	4	5
Maximum	5	5	5	5
Mean	4.4	4.1	3.767	3.633
Std. deviation	0.724	0.803	0.8584	1.098

Source Author

Table 2 Event selection criteria survey answers analysis

Event selection criteria

General criteria

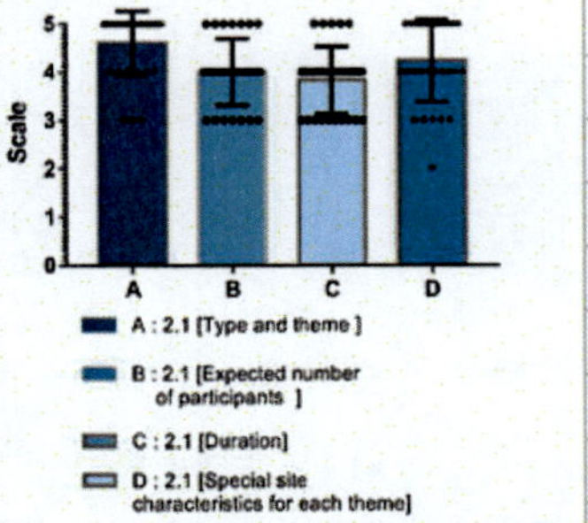

	A	B	C	D	
Number of values	30	30	30	30	
Minimum	3	3	3	2	
25% percentile	4	3.75	3	4	
Median	5	4	4	4	
75% percentile	5	4.25	4	5	
Maximum	5	5	5	5	
Mean	4.6	4	3.833	4.233	
Std. deviation	0.6747	0.6948	0.6989	0.8584	

Local event theme

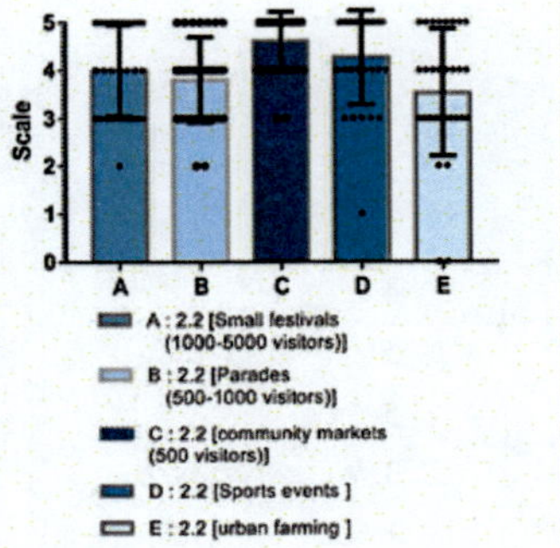

	A	B	C	D	E
Number of values	30	30	30	30	30
Minimum	2	2	3	1	0
25% percentile	3	3	4	4	3
Median	4	4	5	5	4
75% percentile	5	4.25	5	5	5
Maximum	5	5	5	5	5
Mean	4	3.8	4.6	4.267	3.533
Std. deviation	0.9469	0.8867	0.6215	0.9803	1.332

Special site characteristic

Degree of enclosure

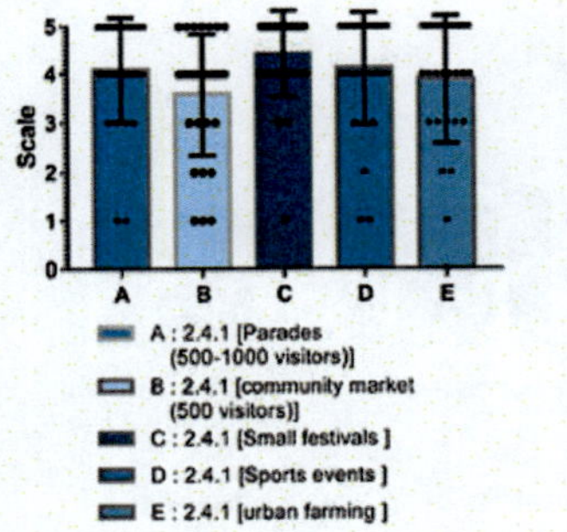

	A	B	C	D	E
Number of values	30	30	30	30	30
Minimum	1	1	1	1	0
25% percentile	4	3	4	4	3
Median	4	4	5	4.5	4
75% percentile	5	4.25	5	5	5
Maximum	5	5	5	5	5
Mean	4.1	3.6	4.433	4.133	3.9
Std. deviation	1.094	1.248	0.8976	1.167	1.322

Degree of axiality

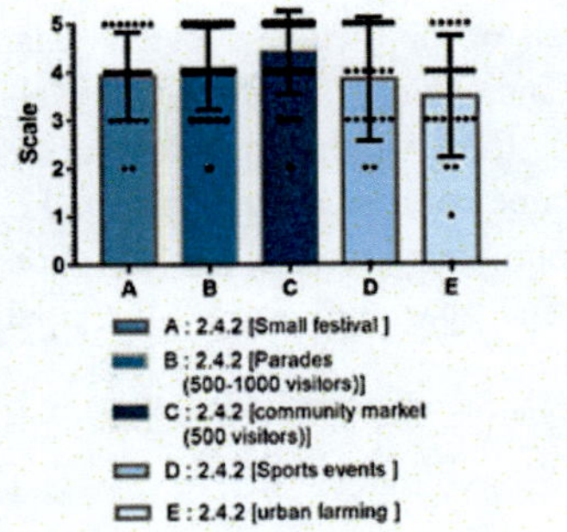

	A	B	C	D	E
Number of values	30	30	30	30	30
Minimum	2	2	2	0	0
25% percentile	3	3.5	4	3	3
Median	4	4	5	4	4
75% percentile	5	5	5	5	4
Maximum	5	5	5	5	5
Mean	3.92	4.08	4.4	3.84	3.48
Std. deviation	0.9092	0.8622	0.866	1.281	1.262

(continued)

Table 2 (continued)

Event selection criteria

Degree permeability

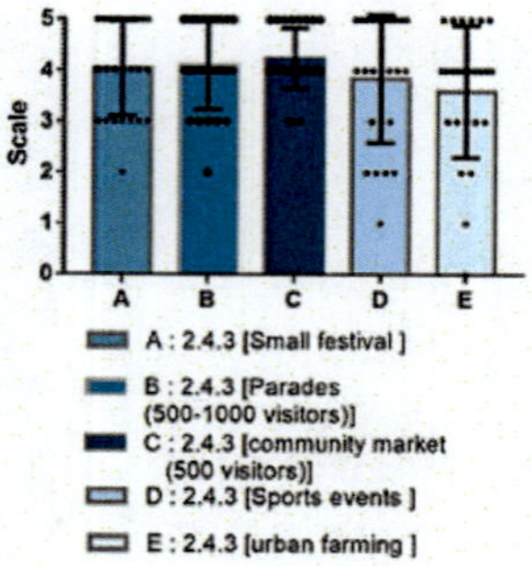

	A	B	C	D	E
Number of values	30	30	30	30	30
Minimum	2	2	3	1	0
25% percentile	3	3.5	4	3	3
Median	4	4	4	4	4
75% percentile	5	5	5	5	4.5
Maximum	5	5	5	5	5
Mean	4.04	4.12	4.24	3.84	3.6
Std. deviation	0.9345	0.8813	0.5972	1.248	1.291

Symbolic meaning

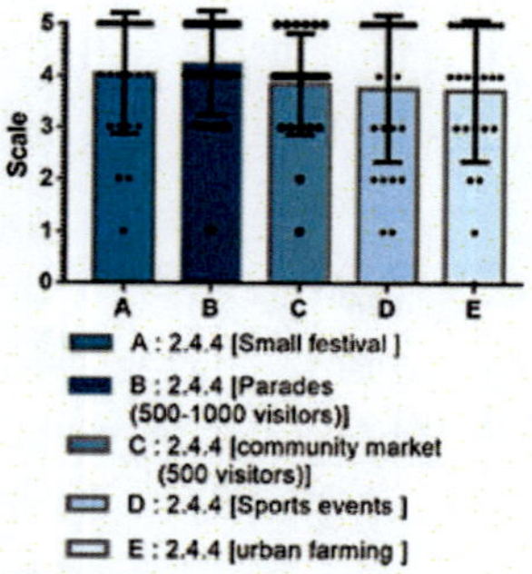

	A	B	C	D	E
Number of values	30	30	30	30	30
Minimum	1	1	1	1	0
25% percentile	3	4	3	2.5	3
Median	4	5	4	4	4
75% percentile	5	5	4.5	5	5
Maximum	5	5	5	5	5
Mean	4.04	4.24	3.84	3.76	3.72
Std. deviation	1.172	1.012	0.9866	1.422	1.37

Source Author

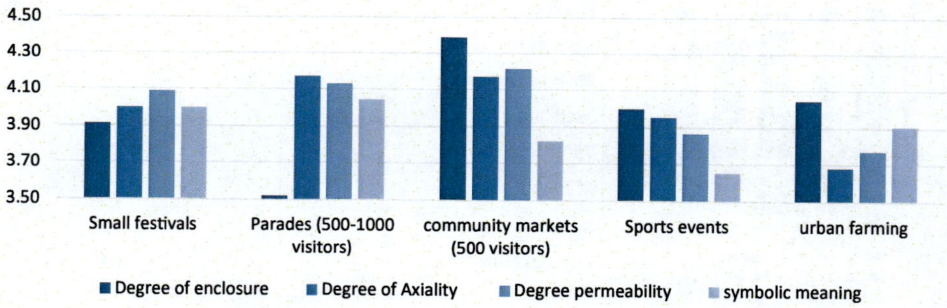

Fig. 9 Site characteristics of various events average values. *Source* Author

According to the previous analysis, the framework sequence was re-evaluated to consider all the comments. Including dividing the management into two aspects, the first is at the beginning that will initiate the idea on the city level. Followed by the site selection criteria, including physical and Event selection criteria. The proposed physical intervention and, finally, the management of event recommendation.

4.4 The Framework

Through the stages of analysis of the literature review data, identifying the criteria and indicators, followed by developing the initial framework then validation through the survey to reach the final framework presented in Fig. 11. The framework is divided into four sections; each section serves a purpose. The first part is for "Municipal

Table 3 Physical intervention and tools survey answers analysis

Physical intervention and tools

Uses and activities

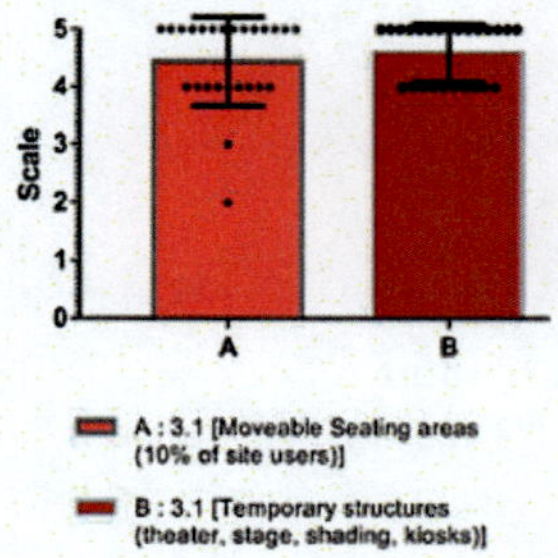

	A	B				
Number of values	30	30				
Minimum	2	4				
25% percentile	4	4				
Median	5	5				
75% percentile	5	5				
Maximum	5	5				
Mean	4.44	4.6				
Std. deviation	0.7681	0.5				

Sociability

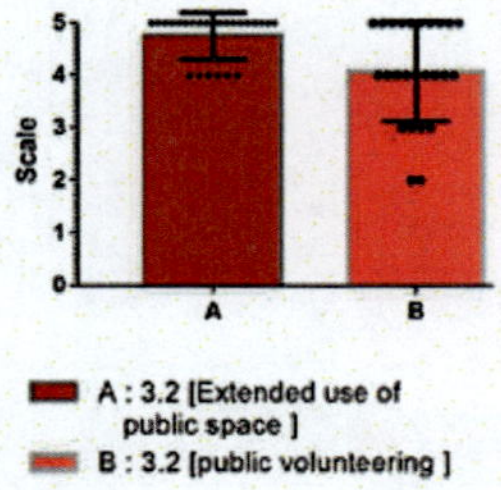

	A	B				
Number of values	30	30				
Minimum	4	2				
25% percentile	4.5	3.5				
Median	5	4				
75% percentile	5	5				
Maximum	5	5				
Mean	4.76	4.08				
Std. deviation	0.4359	0.9539				

Access and linkage

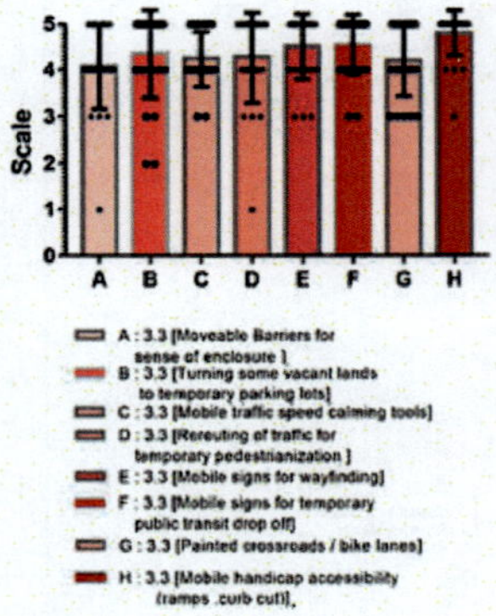

	A	B	C	D	E	F
Number of values	30	30	30	30	30	30
Minimum	1	2	3	1	3	3
25% percentile	4	4	4	4	4	4
Median	4	5	4	5	5	5
75% percentile	5	5	5	5	5	5
Maximum	5	5	5	5	5	5
Mean	4.08	4.36	4.24	4.28	4.52	4.55
Std. deviation	0.9092	0.9522	0.5972	0.9798	0.7141	0.6506

Safety and comfort

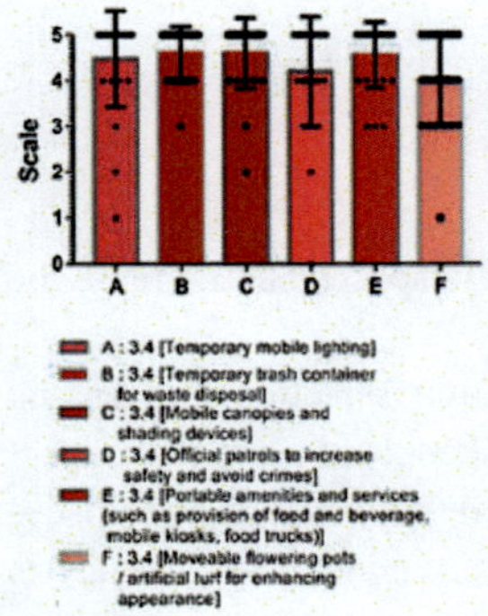

	A	B	C	D	E	F
Number of values	30	30	30	30	30	30
Minimum	1	3	2	0	3	1
25% percentile	4	4	4	4	4	3
Median	5	5	5	5	5	4
75% percentile	5	5	5	5	5	5
Maximum	5	5	5	5	5	5
10% percentile	2.6	4	3.6	2.6	3	3
90% percentile	5	5	5	5	5	5
Mean	4.48	4.6	4.6	4.2	4.56	4
Std. deviation	1.046	0.5774	0.7638	1.19	0.7118	1

Source Author

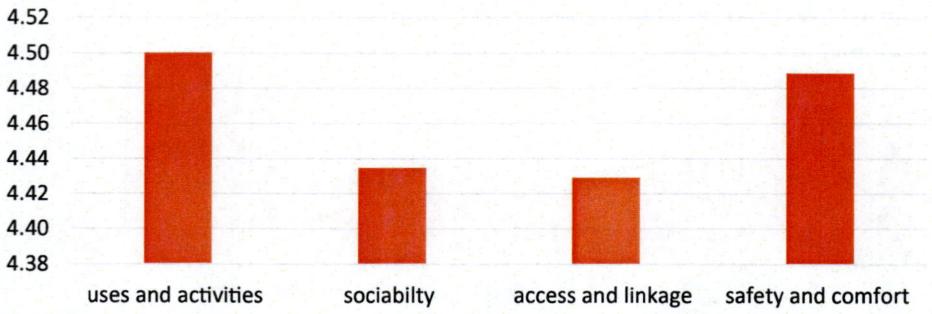

Fig. 10 Physical intervention and tools average value. *Source* Author

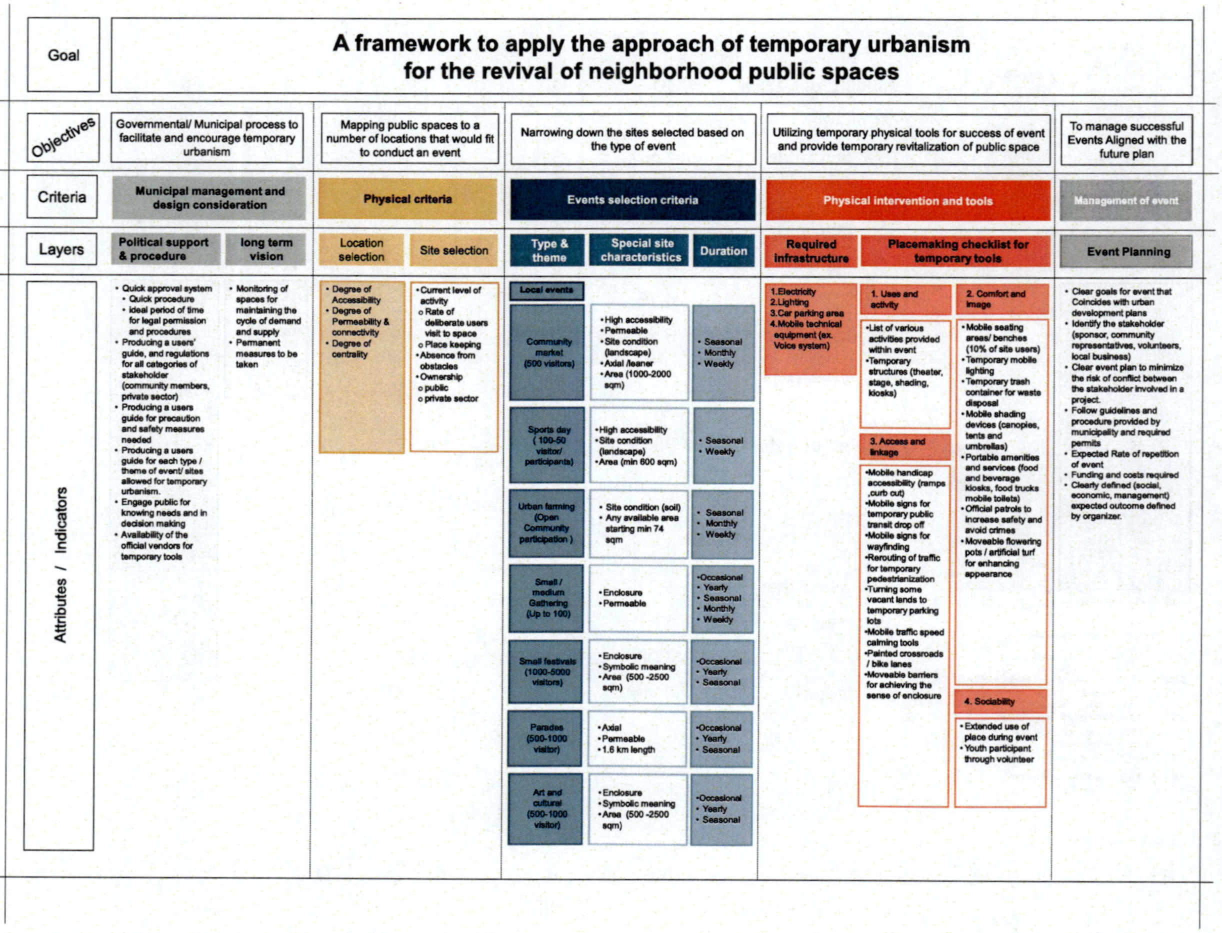

Fig. 11 The completed framework. *Source* Author

management and design consideration", which targets Governmental/Municipal processes to facilitate and encourage temporary urbanism. Consists of two layers. The first deals with the political support and procedure required from the municipalities. The second layer is the long-term plan, which deals with plans that municipalities need to consider to continue the success of the public spaces and increase the activity layer.

The second is a tool for location selection and spatial analysis of event spaces by all stakeholders, and this aspect consists of the macro-scale, dealing with identifying spaces and the accessibility, connectivity, and centrality analysis

Table 4 Management criteria survey answers analysis

Management

Political support and procedure

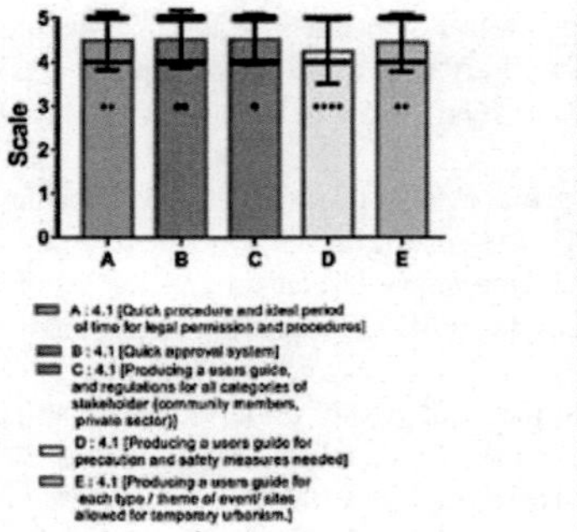

	A	B	C	D	E
Number of values	30	30	30	30	30
Minimum	3	3	3	3	3
25% percentile	4	4	4	4	4
Median	5	5	5	4	5
75% percentile	5	5	5	5	5
Maximum	5	5	5	5	5
Mean	4.48	4.52	4.52	4.24	4.44
Std. deviation	0.6532	0.6532	0.5859	0.7234	0.6506

Planning and long-term vision

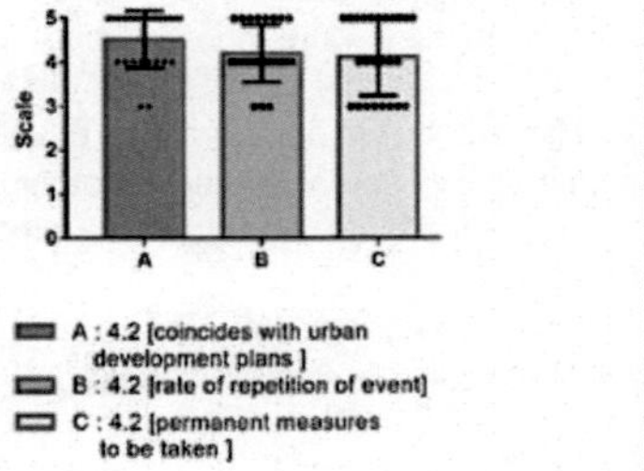

	A	B	C		
Number of values	30	30	30		
Minimum	3	3	3		
25% percentile	4	4	3		
Median	5	4	4		
75% percentile	5	5	5		
Maximum	5	5	5		
Mean	4.52	4.2	4.12		
Std. deviation	0.6532	0.6455	0.8813		

Source Author

within the district. And the micro-scale deals with site selection criteria, including site condition, shape and configuration, and the symbolic meaning. These help to indicate the sluggish public spaces and narrow down the selection to fit the event target to revive and activate it.

The third represents the tools that would be implemented by event organizers and designers for the success of public spaces. Represents the selection of the experience lived through the event. Starting with deciding the theme of the event and the number of users, and the special site characteristics for each Event. This would narrow down the selection from the first phase into sites that suit the proposed Event and duration.

The fourth part is for the management of successful events. Event management is the last criterion, which aims to list the impact and alignment of the event on city planning and directed to the event organizers. The framework can be used variously, firstly as a guideline to benefit the community and enhance the overall quality of life and public spaces. Secondly, to help the city stakeholders had better employ the temporary urbanism.

5 Conclusion

This paper presents Temporary Urbanism as a viable and cost-effective development approach to urban regeneration for unsuccessful public spaces. It discusses the utilization of public events as a major tool of attraction and place keeping in the city's poorly used public spaces. The paper's key findings outlined the development of a contextual decision-making/taking model of an executional managerial framework. The framework provides a systematic mechanism to assist the city's stakeholders, particularly urban officials, in considering the contextual multifactorial related attributes.

The framework development stages included the understanding of the local temporary urbanism approach and its historical literature, and the identification of the contextual spatial and managerial criteria and indicators that contribute to the success of local events and consequently. As well as the concerned public spaces, the identification of the contextual temporary events' additional attributes needed to revive public spaces, and finally, the development and validation of the sequence and the processes of the framework.

The spatial decision-making section resulted in criteria for city officials, event planners, and urban designers to be used for narrowing down the appropriate location for events within the city spaces and zones. While the management section concluded a set of recommendations for municipalities and officials to formulate as part of applications for events within the city, this would help keep track of the individual incidents and allows for control and monitoring.

References

Alawi, G., Jamjoom, H., Samir, H.: Enhancing the cultural tourism experience: the case of historical old Jeddah. In: Islamic Heritage Architecture and Art II, vol. 177. WIT Press, Italy (2018). https://doi.org/10.2495/IHA180041

Balicka, J., Storie, J., Kuhlmann, F., Wilczyńska, A., Bell, S.: Tactical urbanism, urban acupuncture and small-scale projects. In: Urban Blue Spaces, 1st edn., pp. 406–430. Routledge, London (2021). https://doi.org/10.4324/9780429056161-19

Crowd Safety: What considerations need to be made when defining occupant capacity? (2020). Retrieved 20 Mar 2020, from http://www.crowdsafety.org/2016/01/14/what-considerations-need-to-be-made-when-defining-occupant-capacity/

Cudny, W.: The phenomenon of festivals: their origins, evolution, and classifications. Anthropos **109**(2), 640–656 (2014). Retrieved from http://www.jstor.org/stable/43861801

Cudny, W.: Festivalisation of Urban Spaces: Factors, Processes and Effects, 1st edn. Springer. Switzerland (2016). https://doi.org/10.1007/978-3-319-31997-1

Eyerly: How Big Should My Vegetable Garden Be? (2020). Retrieved 7 Apr 2020, from https://farmandgardendiy.com/garden-size/

Getz, D.: Event tourism: definition, evolution, and research. Tour. Manage. **29**, 403–428 (2008). https://doi.org/10.1016/j.tourman.2007.07.017

Haas, T., Olsson, K.: Transmutation and reinvention of public spaces through ideals of urban planning and design. Space Cult. **17**(1), 59–68 (2014). https://doi.org/10.1177/1206331213493855

Johansson, M., Kociatkiewicz, J.: City festivals: creativity and control in staged urban experiences. Eur. Urban Reg. Stud. **18**, 392–405 (2011). https://doi.org/10.1177/0969776411407810

Khalil, M., Nasr, E.: Toward sustainability of temporary uses in public spaces: a case study muscat, sultanate of Oman. In: Lorens, P., Magidimisha, H., Balamiento, N. (eds.) 57th ISOCARP World Planning Congress, pp. 2268–2281. ISOCARP, Qatar (2021). https://doi.org/10.47472/ia9GQBH6

Kotval, Z., Machemer, P., Mullin, J.: Transformative temporary use. WIT Trans. Ecol. Environ. **142**, 233–241 (2010). https://doi.org/10.2495/SW100221

Koyama, M.: Can temporary urbanism be a permanent solution? Urban spaces in liquid times (2017). Retrieved 3 Oct 2021, from https://medium.com/@megukoyama/can-temporary-urbanism-be-a-permanent-solution-urban-spaces-in-liquid-times-643d423ae4fc

Lehtovuori, P., Ruoppila, S.: Temporary uses as means of experimental urban planning. Serb. Archit. J. **4**(1), 29–54 (2012). https://doi.org/10.5937/SAJ1201029L

Lunga, T., Mihaela, O.: Major special events: an interpretative literature review. Manag. Market. **7**(4), 759–776 (2012)

Madanipour, A.: Cities in Time: Temporary Urbanism and the Future of the City. Bloomsbury, London (2017). https://doi.org/10.5040/9781350014275

Mandeli, K.N.: Public space and the challenge of urban transformation in cities of emerging economies: Jeddah case study. Cities **95** (2019). https://doi.org/10.1016/j.cities.2019.102409

Martin, M., Hincks, S.: Temporary urbanism in England's core cities: looking beyond the iconic. In: Regional Studies Association Annual Conference. Austria (2016). Retrieved from https://vbn.aau.dk/en/publications/temporary-urbanism-in-englands-core-cities-looking-beyond-the-ico

Martin, M., Deas, I., Hincks, S.: The role of temporary use in urban regeneration: ordinary and extraordinary approaches in Bristol and Liverpool. Plan. Pract. Res. **34**(5), 537–557 (2019). https://doi.org/10.1080/02697459.2019.1679429

NYC Department of Transport: Putnam Triangle Plaza. NYC (2010). Retrieved from https://www1.nyc.gov/html/dot/downloads/pdf/2015-03-putnam-plaza-cb2.pdf

PPS: Congress Square Park Project for Public Spaces (2015). Retrieved 12 Apr 2022, from https://www.pps.org/projects/congress-square-park

PPS: PLACEMAKING what if we built our cities around places? (2018). Retrieved from https://www.pps.org/article/greatcitiesinitiative

Richards, G., Palmer, R.: Eventful Cities: Cultural Management and Urban Revitalisation. Elsevier Limited, Great Britain (2010). https://doi.org/10.4324/9780080940960

Smith, A.: Events and Urban Regeneration: the Strategic Use of Events to Revitalise Cities. Taylor & Francis (2012). Retrieved from https://books.google.com.sa/books?id=h1XFBQAAQBAJ

Smith, A.: Events in the City: Using Public Spaces as Event Venues. Routledge Advances in Event Research Series. Routledge, New York (2016)

Stevens, Q., Shin, H.: Urban festivals and local social space. Plann. Pract. Res. **29**(1), 1–20 (2012). https://doi.org/10.1080/02697459.2012.699923

The 5-minute walk (2020). Retrieved 24 Mar 2020, from https://morphocode.com/the-5-minute-walk/

Wolchover, N.: Why do we have personal space? (2012). Retrieved 20 Mar 2020, from https://www.livescience.com/20801-personal-space.html

Exploring Theories of Social Inclusivity and Creative Placemaking Analyzing Engaged Creative-Placemaking in Different Case Studies

Nourhan Mohamed Bassam Elhalawany, Hisham S. Gabr, and Ahmed Mostafa Abdelghaffar

Abstract

Arts and cultural initiatives in societies don't only promote social well-being, but they are indispensable elements of social well-being. Just as you can't strip out health or housing or transportation from social well-being, neither you can remove the arts. The built environment can contribute to a more equal, inclusive and cohesive society if the places where we live, the facilities we use and our community meeting places are designed to be accessible and inclusive. The Aim of this paper is to highlight engaged creative placemaking as a tool in creating places with everyone in mind. Covid-19 crisis which spotlighted the inequality and inaccessibility of a national culture to the idea of using public spaces. People cannot simply enjoy public spaces because they are either private or owned by the government. How can this change, the main research question of this paper is: "How can creative placemaking through community-led design exist to generate the resilient communities we dream of"? Methods used in the paper are based on multidisciplinary literature studies and detailed case studies of walls of connection project from 2017 till now and Asphalt art project 2021–2023 that shows how to improve civic-participation in our societies by analyzing the role of engaged creative placemaking as an enabling tool, creating a linkage between placemaking theory and practice. This linkage is due to studies that found how spatial planning and creative design be hand in hand through the design process, to stimulate the ideas of successful creative placemaking in the built environment to enhance social integration in open urban spaces. This paper offers a review that illustrates a re-orientation of engaged placemaking from being focused on physical (spatial) change of the environment (product oriented) created by designers (e.g. architects, spatial planners) towards placemaking as an enabling tool to be used by planners to facilitate the making of places by numerous people/role-players outside the planning profession.

Keywords

Community-led initiatives • Engaged-creative placemaking • Asphalt-art initiative • Walls of connection • Social inclusion • Belonging • Sociability

1 Introduction

"People have an intuitive desire to have art around" (Golden 2015). As cities grow due to the increased urbanization in major places around the world, raises the issue of inequality and segregation in the social life which leads to social exclusion to many citizens in the city. Cairo is suffering from this problem to a worrying level because the social differentiation is reaching high limits due to gated communities, separation that make different classes isolated which build a growing complexity of social life. Creation of civic infrastructure is necessary for healthy societies and collaborative problem-solving. Whyte (1989) emphasizes that Anthropology comes before Technology, which means that designs should be centered on humans and their needs and what they aspire for. This would suggest that people should lead any initiative within their community and should also be at the center of innovation process, as the engaged creative placemaking initiatives shown in this paper many of them is being led by the community to promote their ideas and aspires in their communities.

This paper highlights the issue of social exclusion and social segregation in cities and tries to highlight a solution in the community-led engaged creative placemkaing strategies as a solution, exploring how civic-organizations can support that interaction and build social integration within the local community in urban areas. As well as discussing the current

N. M. B. Elhalawany (✉) · H. S. Gabr · A. M. Abdelghaffar
Cairo University, Giza, Egypt
e-mail: nourhan.bassam@gmail.com

S. Kostopoulou et al. (eds.), *Cities' Vocabularies and the Sustainable Development of the Silkroads*,
Advances in Science, Technology & Innovation, https://doi.org/10.1007/978-3-031-31027-0_6

policies and strategies of the current urban development directing towards integrating the community as a vital part of decision making and how arts and culture are an integral part of creative placemaking projects.

This paper will introduce the concept of engaged creative placemaking, a term I have started to use in my research that is introducing arts-integrated urban projects produced by the community; this paper will focus on co-produced and performative art in the urban realm, art in the urban realm that is distinct from the public art that is usually done by artists and in the governmental plans but its "civic-art" which is a normal output of people in their communities as a method of self-reflection and expression. It will showcase some case studies of creative placemaking that was led by community and citizens were part of the design and execution process and analyzing how it affected the context it took place in.

2 Creative Placemaking then and Now

Placemaking is an interdisciplinary, multi-faceted concept that has been known and used since the 1970s when pioneers such as Jane Jacobs and William Whtye discussed the concepts of people–place relationship and the how the built environment affected the behavior of people has been widely researched since the mid-1970s and a lot of development was made by PPS by Fred Kent.

"Placemaking literature changed from pre 2000 from the discipline of spatial planning indicated that experts and professionals were empowered to make decisions regarding the making of place (Shibley 1998) to recent literature that suggests that spatial planners should be regarded as facilitators of the placemaking process with other stakeholders".

Creative placemaking is an umbrella term referring to the activation on a certain site, through physo-socio site interventions in a creative process as based on the toolkit of Valuing creative placemaking for public and private stakeholders (Cohen et al. 2018). Creative placemaking nowadays is an expression integrated in urban planning and urban development policies as a valuable part of enhancement on Micro or Macro scale of the city.

3 Arts and Culture as a Part of Creative Placemaking

Art and cultural activities have a high influence on society by the community being able to exoress itself and is a part of a community collective memory, Art represented in Painting, sculpture, music, literature and the other arts is a vital part of creative placemaking (Rembeza 2016).

Strengthening communities with arts and cultural interventions is the main aim of creative placemaking by enhancing a certain public space due to the community need by the community talents, and hence improve local assets and economic development.

Arts and creative expression play a substantial part in establishing sense of place, sense of belonging and economic growth for communities.

The main goals of integrating art, and cultural activities:

- Respectfully enhance urban space, relevant to local context and culture.
- Encourage neighbours to work together by involving them as active participants.
- Foster a sense of belonging and community identity.
- Build and enrich relationships between residents.

4 Creative Placemaking and Resilient Communities

Art helps in forming and strengthening community identity and build stronger relationships and social networks is a key asset in creative and resilient cities. Resilience of the community comes from adopting user-centric approaches in achieving social sustainability by increasing the quality of place based on Building resilient and sustainable cultural and creative sectors book by UNESCO (2022). Communities that strengthen cultural activities and art support and empower the citizens to develop cultural activities. Place is constructed and affected by multi intersectional factors as shown in Fig. 1.

Many disciplines are grounded to the particularity of space in creating self-sustaining resilient places, will lead to resilient communities (Turvey 2019). Creative Placemaking is an integral part to build social capital of places which guarantees economic sustainability as well as social belonging among the residents and that builds social sustainability (Cohen et al. 2018).

5 Inclusive Places and Sense of Belonging

As shown Fig. 2 to understand the difference between Inclusion, Belonging and Diversity is like being in a party so getting invited is diversity and being asked to dance is belonging and dancing like no one is watching is inclusion and being able to attend in the first place in Equity.

Having a sense of purpose provides direction and meaning to our work. It gives us a clear understanding of why we are doing what we do and what impact we want to make. However, simply having a sense of purpose is not enough. Without taking action, our purpose remains dormant

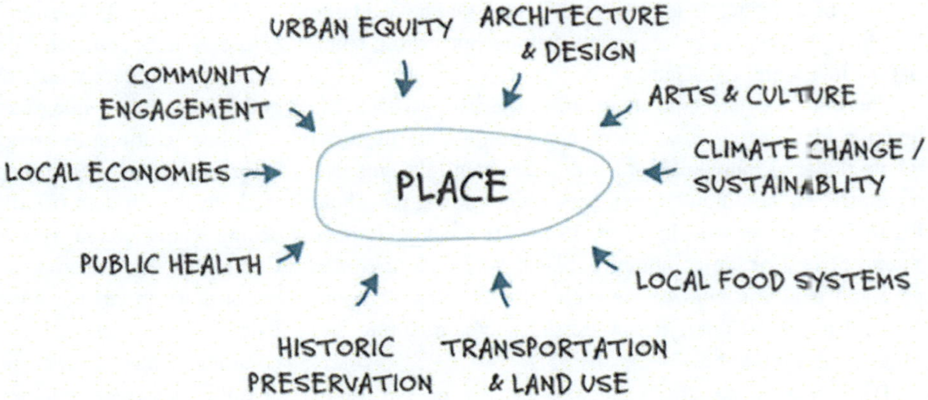

Fig. 1 Disciplines related to place (PPS project for public spaces 2018)

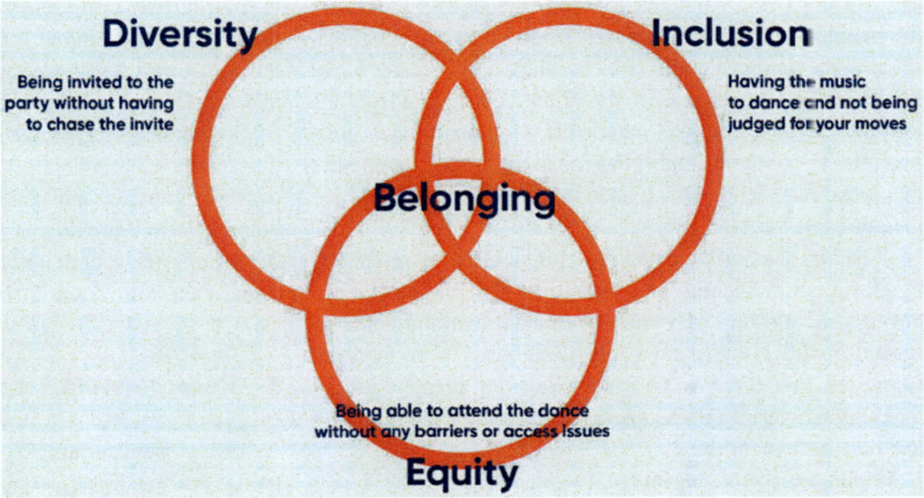

Fig. 2 Diversity, belonging, equity and inclusion (Waters, S. 2021) Here's how to build a sense of belonging in the workplace. BetterUp. https://www.betterup.com/blog/belonging

and unrealized. Sense of belonging is not only being related to people in a certain place, it's the integration of attention and acceptance people give to each other. As complex as our societies have become the Social challenges are being complex and should be solved by creative solutions and approaches which have lately proved success. The main advantage of placemaking is enhancing sense of belonging of the community by emphasizing citizenship and raising the perception of space to the individuals and the place ownership. As designing is not only for designers but goes beyond that boundary (Ellery and Ellery 2019).

This realization of the local community help achieve strong and resilient communities that through social interactions that can build equity and civic engagement. Currently a void is perceived between creative placemaking and the purpose of strengthening the community (Markusen and Gadwa 2010). This void lies because of the gap between civic organizations and community to speed up the design process and decision making and building a strong social capital (Dale and Newman 2010).

Places that are inclusive should be for everyone to use, they should be accessible and easy to use, with dignity and without undue effort or anxiety, they should promote our mental and physical health and well-being reflect the diversity of today's society while building on the history of local areas encourage mutual and harmonious relations between social groups and ensure that economic resources are evenly and fairly shared and divided by different sectors in the society such as the elderly and women and other vulnerable groups based on (UNESCO 2021). Reimagining our futures together: a new social contract for education.

6 Engaged Creative Placemaking in Post Covid-19 Cities

Over the past year, the pandemic has disrupted the social connections that used to be a vital part of everyday life, recovering from the pandemic in that sense gives many lessons to post-pandemic cities and communities and through past literature I have found the creative placemaking projects can positively influence the social interaction and can help in the recovery of our built environment (Ghanem et al. 2021).

"The basic principle is to allow activities that have a large benefit to the economy or human welfare but pose a small risk of infection" Bill Gates.

The big idea here is to maintain a flourishing social life under a challenging new set of conditions. This means a whole range of innovative practices, habits and designs that capitalize on community members own knowledge about life in their hometowns. In Covid-19 time it was vital to have shared places for the community, places where children can play and people socialize. Through difficult stress the pandemic has set on our cities it caused socio-economic and socio-spatial inequities.

The social crisis the pandemic causes in our cities exposes and highlights certain issues in the existing social shortcomings, the pandemic created unequal opportunities for people to integrate in spaces in order to protect themselves, as we have seen citizens started to create social spaces as the terraces gatherings, or the kite surfing from the top of Cairo roofs. The pandemic will still have great potential to continue amplifying social differences even after the lockdowns end, which lead to more inequalities and inequities in the long run.

Creative placemaking has the ability to develop spaces, by elevating the identity of a place by enhancing its essence through a collection of visual, cultural, social and environmental qualities that inspire the community to be engaged suggested that there is no ideal place for creative placemaking but everywhere people gather in can be the optimum location for enhancement.

Creative-placemaking doesn't work solely on the development of an infrastructure of a certain locale but it focuses on community uplift through listening to the community interests, needs and aspirations, and that's how it works to enhance place liveability and economic growth. Community engagement and creative-placemaking are human interactions being affected by social, political and economic factors and affecting these same sectors as well (Cohen et al. 2018).

Engaged creative-placemaking is therefore a collaborative process between different official sectors and the community which leverages the power of art and culture, making arts and people-centric spatial planning to shape the physical, social, cultural and economic identity of a community to stimulate economic development, social cohesion and inclusion for diverse societies.

7 Community-Led Design, Community Engagemnet in Creative Placemaking

Understanding the importance of inclusion of the local community in the process of design-led creative placemaking is vital to enable communities into the design process (Ramli and Ujang 2020).

Community led-design is a process where designers (professional) and the local community work aside with other stakeholders which help dissolves the power of local authorities and be integrated with the community.

When community engagement focuses on the local community (user-centric) approach in the design thinking, it leads to the process of community-led design. As this is a design characterized by a community focus, its approach is Community-led design is a practice whereby people take the opportunity to engage directly and creatively in the formulation of solutions for their own environment. In community-led design, professionals, stakeholders and communities come together to develop sustainable solutions to complex design and planning problems the role of design is transferred from civic organizations to communities, gradually, by building social capital and to develop and adjust the process as we evaluate the needs of our residents and the success and failures of the process (Stewart 2020).

Community engagement the second quality of creative placemaking is authentic and ongoing community engagement. Community engagement through the arts is important both because it provides a voice for residents in shaping the future of their community and because it contributes to social cohesion (Zitcer 2018).

8 Recommendations for Civic Organizations

As civic organizations help connect between designers and policy makers, Every civic organization needs to work on the community culture needs "Civic-footprint" and find the tools and approaches that work best for it. At some point when the local community and other stakeholders share the same vision the action starts to happen and projects start executions and problems being solved. Every community deserves revitalization projects in the built environment that work on the aspirations and dreams of the local community and transforming their reality (Lelieveldt et al. 2009).

Community Involvement Activities include volunteering, membership and participation in organizations, working with others to solve community problems.

It is essential to hear the community's thoughts as called in the literature "Civic-Footprint".

- Engaging volunteers and donors in the activities
- Bringing the community diverse sectors together
- Collaborating with organizations similar to (non-profit organisations and artists) within and beyond the community
- promoting community partnership
- Working on increasing education and awareness.

Civic organizations can develop a unique working model to understand the civic footprint of nonprofit organizations and the societal participation with methodological implications for future research that would seek to measure the extent to which nonprofits promote civic engagement, the challenge lies in establishing the civic organization as an external party to enhance the social dynamics of the community (Shier et al. 2014).

Gaining the trust of the local community from establishing connections is the first step to reduce the gap between civic organizations and the local community, then generating the need in the local community to work on the social capital by involvement activities and creating more interactions with the civic organizations and public sector.

As social impact is qualitative quality and usually over looked by the local community as well so civic organizations should work on effective engagements that lead to social interaction and build a need in people to change their mentality from passive users to being co-producers of spaces.

There two different of investments worldwide for creative placemaking:

Horizontal integration: Most of placemaking investments have been towards horizontal strategies to integrate creative placemaking into diverse similar organizations, systems, and sectors in arts and culture (Borrup 2015).

Vertical integration: means the connection between different sectors responsible of the decision making, but less emphasis has been placed on understanding how creative practices can be integrated vertically to encourage participatory citizenship and facilitate the cooperative development of public policy and give voices to those who are less engaged in the public policy or have access to power and resources. Kiley (2017) as shown in Fig. 3.

9 Policy to Achieve Engaged-Creative Placemaking

Engaged-creative placemaking project complications underline from the complexity of the different sectors involved in the placemaking project, from the civic community represented in civic-organizations to participatory of local governance (Dukanović and Zivkovic 2015).

The creative placemaking project to be created through the integration of many sectors such as artists, local government, designers, local community, architects and civic organizations (Zitcer 2018).

To set the right tone to such projects require the early engagement with collaborators. Artists should be involved in an artistic intervention for example to allocate the location of the installation and set the right expertise to build on the project with community.

Developing a holistic policy to execute the project is essential step before executing such project which includes the three components of space with the focus of the community in case of creative placemaking projects. Figure 4 shows that the space around us is composed of 3 main components.

In Egypt and many countries around the world where bureaucracies and complicated policies may prevent creative placemaking projects, people are trying a bottom-up plan to integrate such projects in the governmental policy. The integration between these different types of placemaking as shown in Fig. 5

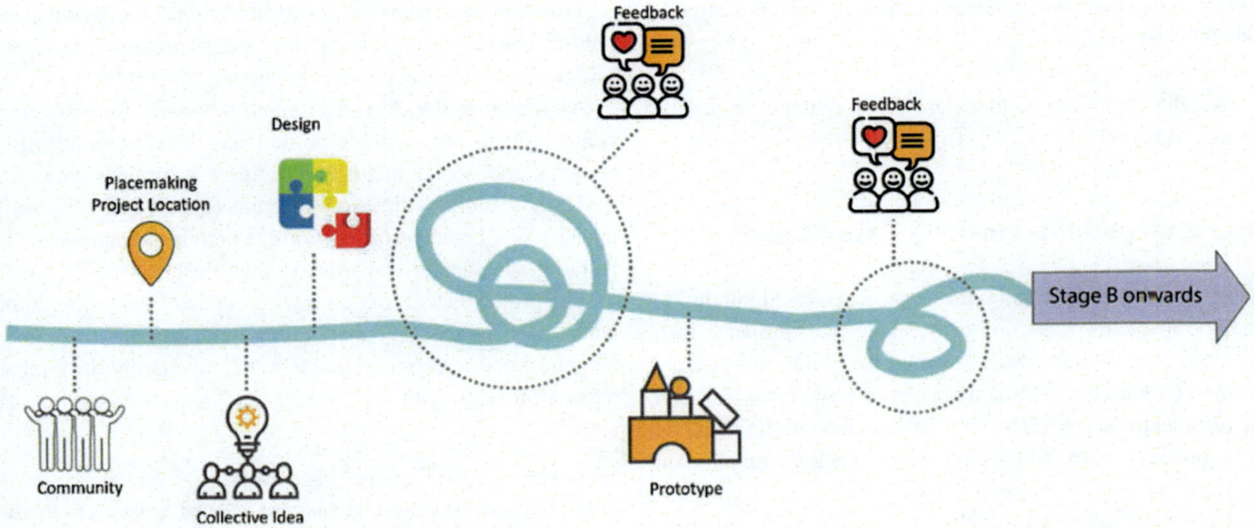

Fig. 3 Placemaking process integration (Tabrizi 2018)

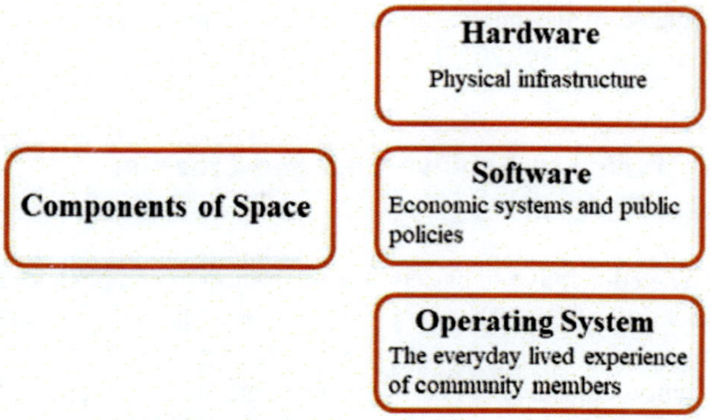

Fig. 4 Components of space ref: the author based on Lefebvre (1974)

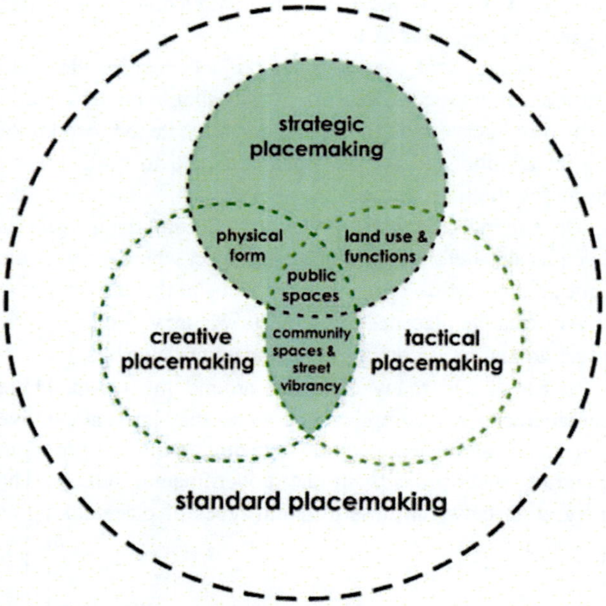

Fig. 5 Placemaking different sectors (Mallya, 2021). Urban ancdotes website URL https://www.the-urban-anecdotes.com/post/p-1-integrated-place-making

As shown in Fig. 6 there are three different sectors of placemaking are.

10 Engaged Placemaking Case Studies

This paper is showcasing these projects to highlight the methodology of these projects, The two projects were selected because both were executed by NGOs who listened to the community needs and started to raise funds and take governmental approvals where they lay behind the "Tactical placemaking" Idea which works on quicker, simpler and easier fixes.

Especially after Covid-19 art and cultural initiatives helped cities to rebuild and recover its industries after the pandemic by offering engaged community projects to take place, emphasizing the role of arts and culture in community engagement ad in building places for citizens to express themselves to coexist together, some of these initiatives that we can lean form in our societies is "Guerrilla-Style Art project Row houses Houston", "The Asphalt Art Initiative", "Walls of Connection Project", "Crossroads participatory urban art projects". In this paper I will look closely on two projects.

10.1 Walls of Connection Project

The idea of the Walls of Connection is developed by MasterPeace[1] Club leaders in 2016. It is a concept that covers the mission of Masterpeace to achieve social conversations and interactions through artistic interventions and cultural initiatives from concerts, street murals, and youth learning programs endorsed by more than 45 clubs around the world in different cities.

Walls of connection project contributes to Sustainable Development Goal (SDG) 11 that aims to make cities and human settlements inclusive, safe, resilient and sustainable communities, through community integration in the decision making of places. For this purpose, the main goal for studying walls of connection project is to stimulate successful creative placemaking in urban landscapes for social integration in open urban spaces through these walls from different parts of the world.

Walls of Connection project believes in the power of art to integrate communities in the decision making of the piece of art, the place of the wall, the way of execution the wall to enhance the community and raise the conversations and achieve the aspiration of the sectors included in the discussion and execution.

The plan was painting 100 walls around the world and now it's exceeding this number, transforming them into bridges that connect, not divide people.

As shown in Fig. 7, walls around the world for years are being built to separate people and idea that works against equality and social inclusion, its time as professionals to work in the other direction to endorse peaceful acts and equality of all citizens of the world, which is the main idea of painting specifically walls due to that reason.

The project is working on turning walls of separation into walls of connection, collaborating with people from the neighborhood creating a huge mural spanning the walls cities around the globe.

[1] NGO organization aims at creating peace based in the Netherlands and was established in 2011 with the aim to create value in the society and it has 52 clubs in 48 countries around the world.

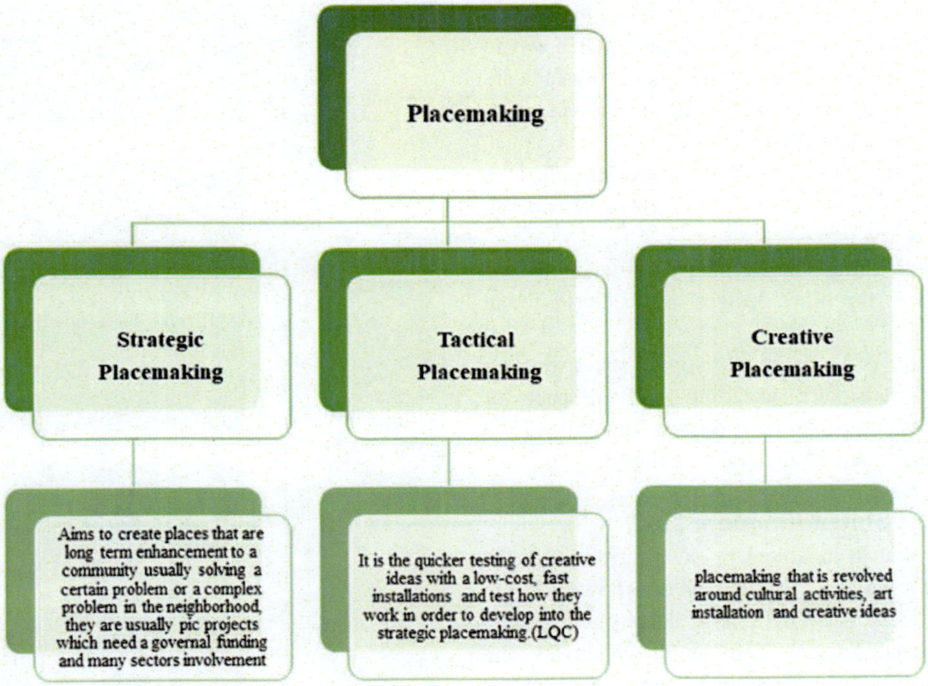

Fig. 6 Types of placemaking projects. Ref: made by author based on Ellery et al. (2020)

Fig. 7 Walls of separation from countries around the world (Masterpeace Organization 2021a)

The community come together, both ends of the wall to transform this wall from a symbol of separation into a way to connect, transforming colorless division into colorful connection, with the support of artists, volunteers, partners, businesses and local community.

The criteria are that the walls must have been built with the intent for separation, and with a relevant to its community in one way or the other.

These walls will feature shared elements in every design that is connected to the other designs across the different cities.

Siding the 'Walls of Connections' pop up street festivals and concerts will take place providing an open space for thousands of artists and performers to show their talent for a wide audience. The goals of the walls as Fig. 8.

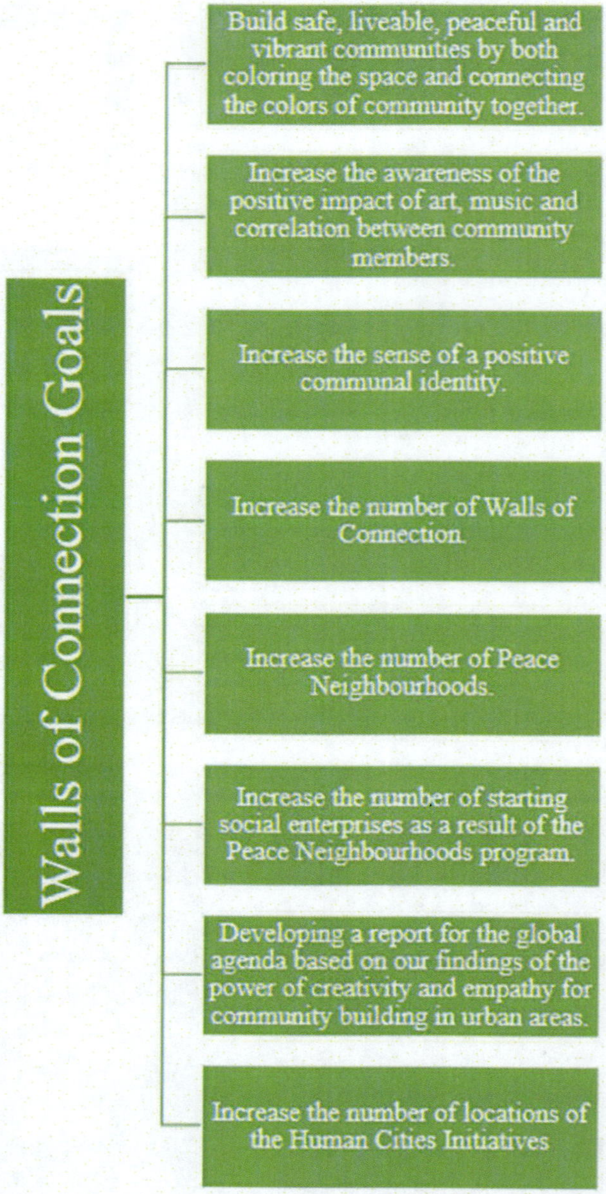

Fig. 8 Walls of connection goals (Masterpeace Organization, 2021b)

10.1.1 London, United Kingdom 2018

London is a city that in inhabited by people from different backgrounds that's why this wall has diverse flowers in the huge meadow to celebrate this diverse society and people from different backgrounds, The mural celebrates the diversity, the numerous sub-communities referring to London as their home, and gathering everyone to participate.

The process of coloring this specific wall due to the major historic value of the location attracted many tourists to visit and many artists and guests and volunteers were present while painting this wall, and turned the location to a place where people hangout and take pics and gather as a new temporary landmark of the city as in Fig. 9.

10.1.2 Utrecht, Netherlands 2017

This wall is located in a neighborhood in Utrecht that is called "Overvecht", what is different about this neighborhood that it consists of a diverse community diverse from refugees that call it their home to other nationalities that reached almost 105 nationality, which all different people were a part of coloring and planning this wall execution.

The word peace is written by different languages See Fig. 10 from the people living there and the hands signifies how peace can be carried by everyone and can spread worldwide. What's interesting and unique about walls of connection that the mural idea comes from the artists, citizens and the local community in where the wall exits, it is not a governmental plan that is handed to a certain artists but normal people are being a part of their built environment and be a part of this engaged placemaking projects.

Walls of connection continued its work even after Covid-19 pandemic by Pre-wall discussions collecting the voices of society through a Master Class Walls of Connection, which they lead a class as a pre-discussion with partners, students and invite different stakeholders.

There are main three steps to:

- First ask participants to paint their 'dream of connection' or so, which is making and sending their paintings for artists to start using it in their original painting.
- Then Artists will work with the input from participants, the different paintings submitted for them to come up with a virtual visualization as they create a 'masterpiece' out of it by using Google Street art endorse the project and support technologically the artists to co-create a virtual experience, They executed live session of street art painting where we can see/interact with artists painting and creating this art work.

Fig. 9 Walls of connection in London, UK (Masterpeace Organization 2021a)

Fig. 10 Walls of connection Utrecht, Netherlands (Masterpeace Organization 2021b)

- Lastly showcasing with an exhibition online, Using Google Street art virtually host the exhibition and asking participants to share and spread the artwork, since their painting/artwork is integrated

10.2 The Asphalt Art Initiative

As per Jan Gehl's states in his book in Life between Buildings in 1971, urban design can "integrate or segregate". Studies demonstrate that the presence of major roads may be a factor that changes people's walking behavior and social life. Traffic infrastructure can provoke physical and social segregation even within dense urban contexts, reducing the level of accessibility between neighborhoods. Breaking down traffic speed, improving pedestrian connectivity and increasing the number of crossing facilities are all measures that can integrate and encourage the idea of inclusiveness.

The Asphalt Art Initiative funds artistic interventions on the roadways, pedestrian spaces, and public infrastructure in different cities world starting with the United states, with the goal of improving street and pedestrian safety, revitalizing and activating underutilized public space, and promoting collaboration and civic engagement in local communities. Cities and citizens around the world are recognizing the

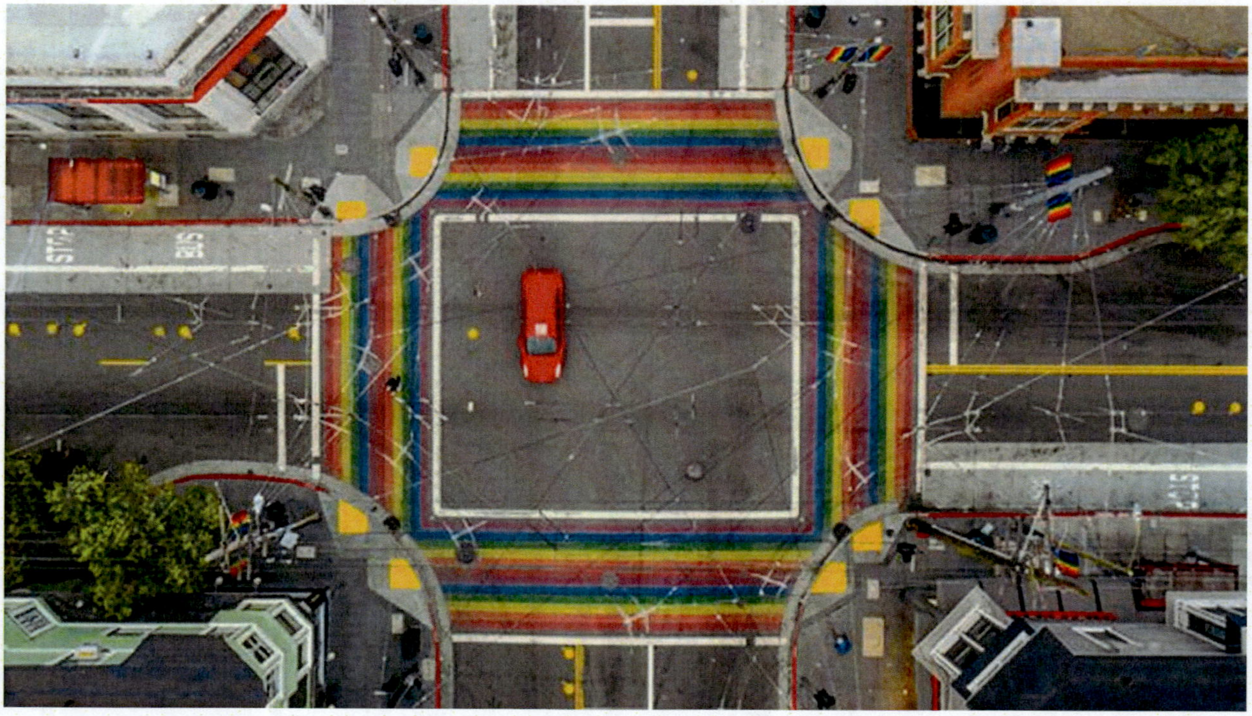

Fig. 11 A rainbow crossing in San Francisco, California, gives an idea of what the newly decorated crossings in London could look like

potential of art to reimagine roadways and vertical infrastructure, improving street safety, revitalizing public spaces, and bringing communities together. The increasing demand for these arts-driven transportation projects has inspired the creation of this project. Asphalt art project want to influence ideas and step-by-step tips for city agencies, civic organizations, and artists interested in executing the art initiative in different cities. Asphalt art projects can improve crosswalks and slow down car traffic with planters, bollards, and colorful markings that emphasize that the space is designated for pedestrians or plazas.

Michael R. Bloomberg, founder of Bloomberg Philanthropies in 2021 said *"Asphalt Art Projects can help cities rebuild from the pandemic by reinvigorating streets and making them safer, while also lifting spirits. As we've seen through our work in cities around the world, vibrant public artwork and smarter street design can inspire residents, build relationships between artists and the community, and help cities recover stronger than before"*.

The main "Asphalt Art" projects are located in (intersections and crosswalks), pedestrian spaces (plazas and sidewalks), and vertical infrastructure (utility boxes, electrical boxes and traffic barriers).

Taken together, these relatively low-cost, often short-term and scalable projects can create immediate positive impact and catalyze long-term improvements to the public realm, proving that the Lighter, quicker, cheaper (LQC) are not temporary projects but can build up for an outgoing project.

Projects that will be completed in 2022 and 2023 and the initiative is being spread from the US to the rest of the world as the (Open Plazas) "piazza aperte" project in Milan, Italy that was executed since 2019 to revitalize 50 different plazas around Milan in a collaborative societal acts endorsed by the municipality as well.

The Asphalt Art Initiative last year guaranteed 16 city to implement their artistic project and since then a combined 86,476 square feet of street space with art with the input of about 5000 residents and 170 artists. Moving the initiative to London as The City of London Corporation has a five year plan for its recovery after the pandemic. In the short term it is keen to encourage more visitors to enjoy some of the area's attractions such as Tower Bridge, St Paul's Cathedral and the Museum of London. The colorful streets plan is part of a major London wide Asphalt Art Initiative which will spread out from the City.

Cultural and visitor services manager Nick Bodger said: *"as an effective and relatively low-cost strategy to activate their streets, responding to the pandemic and improving the attractiveness of cities as places in which to dwell and spend time"*.

As shown in Fig. 11, and the coming figures an overview of some goals of this initiative such as (Fig. 12).

Fig. 12 Kansas city intersection courtesy of (Bloomberg philanthropies, 2021)

- Involvement of citizens in the decision making process
- Activating citizens in the (Artistic Intervention) change of the streets
- Improving Pedestrian and cyclist safety.
- Enhancing Pedestrian plazas and public.

10.2.1 Kansas City

As a shared project by both and the citizens improved an intersection in the city by reducing the vehicles overall speed 45% and shortening the length of pedestrian crossing distances by 52%. The reports of this enhancement came the pedestrians felt “very safe” due to this artistic colorful intervention, the percentage went from 23 to 63%. See Fig. 13 (Asphalt Art Report 2021).

10.2.2 Columbus, Indiana

Local leaders helped create a plaza with the local community to improve walkability, provide playing areas and places to gather a local goal to improve economic growth and social connections through placemaking and art See Fig. 13 (Asphalt Art Report 2021).

10.2.3 Glasgow, UK 2021

With the collaboration of local artists, a design team and the community residents this project was executed in Glasgow outside a metro station and the art coloration is inspired by the water fall in the area.

This colored project indicated some of the environmental issues the city faces in excessive rain times which lead to floods which leads to a rain garden that benefits from this amount of water and comes a s a low-cost, low-maintenance garden in front of the metro station with colors seated area which gave lively manner to the abandoned area and can be an area for youth to skate, play or gather, providing additional biodiversity and a greater sense of place to city center streetscapes.

Choosing sites to enhance the urban Asphalt in streets, crossroads, plazas and others in a certain area is the main

Fig. 13 Columbus, Indiana mike Higdon-downtown Reno partnership (Bloomberg Philanthropies, 2021)

Fig. 14 The **a** Vibrant colors **b** "The Asphalt Art Initiative" prior to COP26 in an abandoned concrete area close to the metro **c** Colored walkways (Source: https://www.reglasgow.com/vibrant-splash-of-colour-prevails-at-m8-concrete-jungle/)

goal of these artistic projects with the help and approval of the local community represented in the civic organizations with local artists and an example of engaged creative placemaking As shown in Fig. 14 some examples of enhancing the public transit in metro stations prior to COP26 (Bloomberg Philanthropies, 2021).

11 Conclusion

The studied cases highlighted that engaged creative placemaking generates benefits for community empowerment and building not only as a beautifying means for public spaces or streets but it is an indispensable part of the human well-being in the most complex societal fabrics. Adopting this approach in designing our neighborhoods, public spaces and streets can be beneficial for our societies for its citizens to be actively involved in shaping their cities and communities so their places can be safer, vibrant, because that environment was built with thought and understanding of the needs of local people.

Civic organizations have to work in favor of the local society and include them as a part of the change. The power of change lies in the citizens and in their role in decision making and involvement of the public projects. Every city desires an innovative different approach to transforming communities by creating and revitalizing spaces around the needs and aspirations of the community as there is no city similar to the other and there is no framework tailored for all cities. Engaged creative placemaking design projects shown in some shared examples hold many advantages rather than disadvantages as they promote innovative ideas, customized solutions, behavioral change, sense of belonging, funding opportunities, conflict prevention, positive environment and the relationship-building which builds the community's social capital. Giving attention to art within the communities as a way to re-shape the local area show positive results.

This paper opens further research on for the policy framework of engaged creative placemaking that has yet to be built in our society using the help of civic-organizations and community conversations in Cairo as there is "no one size fits all" policy so every place has their own set of challenges and learnings. Raising a question on how to execute an engaged creative placemaking project that can benefit our society and solve some of the current societal issues Cairo is facing. Community Involvement Activities will continue to spread from place to place, demonstrating that innovation is imperative and not optional for designing our cities and hence for designing our futures.

References

Asphalt Art Report: https://www.bloomberg.org/arts/supporting-public-art/asphalt-art-initiative (2021). Accessed 8 Nov 2021

Bloomberg Philanthropies: The Asphalt Art Guide. https://asphaltart.bloomberg.org/ (2021)

Borrup, T.: Creativity in urban placemaking: horizontal networks and social equity in three cultural districts. Dissertations and Theses (2015)

Cohen, M., Gajendran, T., Lloyd, J., Maund, K., Smith C., Bhim, S., Vaughan, J.: *Valuing Creative Placemaking: Development of a Toolkit for Public and Private Stakeholders*. NSW Government, Landcom, Sydney (2018)

Communities: Canadian Journal of Nonprofit and Social Economy Research (2014). https://doi.org/10.22230/cjnser.2014v5n1a158

Dale, A., Newman, L.: Social capital: a necessary and sufficient condition for sustainable community development. Commun. Dev. J. **45**(1), 28 (2010). https://doi.org/10.1093/cdj/bsn028

Đukanović, Z, Zivkovic, J.: Public art & public space programme: Learning, but doing!.**25**, 49–64 (2015)

Ellery, P.J., Ellery, J., Borkowsky, M.: Toward a theoretical understanding of placemaking. Int. J. Com. WB (2020). https://doi.org/10.1007/s42413-020-00078-3

Ellery, P.J., Ellery, J.: Strengthening community sense of place through placemaking. Urban Plan. **4**, 237 (2019). https://doi.org/10.17645/up.v4i2.2004

Ghanem, S., Ahmad, A.F., Aboualy, S.: COVID-19 bringing cairenes back to their streets. J. Urban Manag. **10**(4), 1–12 (2021). https://doi.org/10.1016/j.jum.2021.06.001

Kiley, K.A.: Creative policymaking: taking the lessons of creative placemaking to scale. Artivate **6**(2), 58–72 (2017). https://doi.org/10.1353/artv.2017.0000

Lefebvre: The Production of Space. University of British Columbia (1981)

Lelieveldt, H., Dekker, K., Volker, B., Torenvlied, R.: Civic organizations as political actors: mapping and predicting the involvement of civic organizations in neighborhood problem-solving and coproduction. Urban Affairs Rev. **45**, 3–24 (2009). https://doi.org/10.1177/1078087409332303

Markusen, A., Gadwa, A.: Creative placemaking. National Endowment of the Arts, Washington (2010)

Masterpeace Organization: Utrecht (2021a)

Masterpeace Organization: https://masterpeace.org (2021b). Accessed 16 Oct 2021b

Public Art Program Overview: https://racc.org/public-art/public-art-program-overview (2021). Accessed 8 Nov 2021

Ramli, N., Ujang, N.: An overview of creative placemaking as an enabler for a sustainable urban regeneration. Environ. Behav. Proc. J. **5**, 56 (2020). https://doi.org/10.21834/e-bpj.v5i13.2056

Rembeza, M.: Shaping places through art. The role of creative placemaking in Philadelphia. In: International Multidisciplinary Scientific Conference on Social Sciences and Arts/SGEM Vienna (2016)

Shibley, R.G.: The complete new urbanism and the partial practices of placemaking. Utopian Stud. **9**(1), 80 (1998). https://doi.org/10.1177/0739456X03255424

Shier, M., McDougle, L., Handy, F.: Nonprofits and the promotion of civic engagement: a conceptual framework for understanding the civicfootprint of nonprofits within local (2014)

Turvey, R.: Place-making and sustainable community development. Intell. Sci. Educ. Influences Sustain. Res. **49**, 253–272 (2019). https://doi.org/10.4018/978-1-5225-7302-9.ch011

UNESCO: Reimagining Our Futures Together: A New Social Contract for Education (2021)

UNESCO: Building Resilient and Sustainable Cultural and Creative Sectors. UNESCO, Paris (2022)

Whyte, W.H.: City: Rediscovering the Center. Doubleday and Company, University of Pennsylvania Press, Pennsylvania (1989)

Zitcer, A.: Making up creative placemaking. J. Plan. Educ. Res. **40**, 424 (2018). https://doi.org/10.1177/0739456X18773424

Using Cutout Materials While Reading Stories to Deepen and Embody Children's Identity

Mona Abd Elmoneam Ahmed Alwaraky

Abstract

Since the end of 2019, the world has been ravaged by a Coronavirus pandemic. The virus has altered many human behaviors, which must be dealt with and lived with during this pandemic. Art has evolved into a mirror for the documentation of this behavioral shift throughout this pivotal time in the history of the globe. This is because it is so closely tied to the vernacular of everyday life. The idea of art as a concept and how they achieve this concept of art sustainability so that its essence lies in the use of materials of all kinds to make various new shapes are consistent with the term "sustainability". This concept is based on the design of children's stories using various materials in order to expand the horizons of creation and artistic creativity through interaction. On the other hand, this contributes to the growth of the book arts industry in general, making it an appealing way to introduce children to the world of knowledge. This is not only a means of interaction and integration, and thus becomes a means of developing the child's awareness and acquiring multiple skills; it is also a means of distracting the child from the means of technology and its short-term damage, such as speech problems, autism, and long-term problems such as loss of concentration and hyperactivity, and don't expect the importance of this research to be limited to these only; it discusses more than that. It fosters a sense of identity and belonging in the child by featuring an article submitted through this book. This art was not created as a result of the Corona pandemic, but the events, developments, and ramifications of this virus have increased the need for it all over the world in order to protect our children from the dangers of technology.

Keywords

Identity • Sustainability • Cutout book • Coronavirus • Kids stories

1 Introduction

There are numerous records in human history that demonstrate how important art is to society. It is a special kind of honest translation of society's state, as a result of its easy and interesting nature, art is a necessity of life's necessities. So, it was expected, even natural, that after the coronavirus swept the world, art in its various fields would provide us with a record and different solutions to the consequences of this pandemic. The need to adhere to the social distancing system, as it is one of the means of preventing this dreaded virus, drove us to employ electronic solutions in all aspects of life. Unfortunately, in a short time, these electronic solutions affected our children negatively. As a result, the researcher in this study emphasizes how to design stories for children using various materials (cutout books) as a kind of attraction to the child and bringing him back into his world. On the other hand, we can instil the meaning of the concept of sustainability in the minds of our children through art.

1.1 Research Objective

- The researcher's goal with this study is to discover an association relationship between several axes (education, technical, and medical). This relationship can help our children avoid the drawbacks of excessive use of technology and give them a more balanced view of the world.
- The development of a child's sense of identity through artistic expression will allow him to be an active member of his community from the start.

M. A. E. A. Alwaraky (✉)
Faculty of Fine Arts, Helwan University, Helwan, Egypt
e-mail: alwaraky73@yahoo.com

S. Kostopoulou et al. (eds.), *Cities' Vocabularies and the Sustainable Development of the Silkroads*, Advances in Science, Technology & Innovation, https://doi.org/10.1007/978-3-031-31027-0_7

1.2 Search Problem

- The challenge of research is to find innovative, high-quality solutions to the problems arising in people's lives due to the effects of Coronavirus.
- By coordinating the axes of society with each other, art can serve as a means to clarify and consolidate the idea of identity during a child's development.

2 The Concept of Identity

Many researchers have attempted to explain the concept of identity. In the abstract, the concept of identity is primarily dependent on the self. The characteristics that the human self-possesses, whether individual or collective, are all unique. Psychology defines identity as an individual rather than a group, and individuality is the privacy that distinguishes one character from another.

There is a philosophical view of the problem of identity; we find that it is related to a duality inherent in philosophical thought since ancient times, a duality in the constant and the variable. There are fixed rules in the construction of identity, but it is the elements of this construction that are subject to change, and it is from them that the most significant problem arises. While this challenge can be viewed as one of identity preservation, with the tremendous technological advancement of the last few years, it has become much more difficult. This is especially true in the case of Coronavirus spread. One of the consequences of this virus has been a complete reliance on technological means to communicate and reduce virus spread. This has had a negative impact on our children, whether pathologically or intellectually. The question now arises: How can art play a role in countering these negative effects?

2.1 Sustainability

Many concepts and criteria have been explained and interpreted concerning the term "sustainability," from inclusivity to privacy. In its most basic form, sustainability implies continuity through what is available from the environment. It is considered an environmental term that confers the property of survival on various biosystems over time, and it is clear from this that the term "sustainability" is a broad term that can be applied to various aspects of life on Earth. Sustainability is also defined as the process of sustaining a lifestyle by adapting to the environment through the long-term exploitation of natural resources, thereby ensuring the continuity of life.

So, if we look at the ribs upon which the term "sustainability" is based, we can see that they are as follows:

- The field of sustainability is the social context in which sustainability is applied. Social, economic, and environmental factors can provide adequate support for the full scope of sustainability.
- Consumption means taking advantage of all the components of nature; this is an influential motivator of the term sustainability, and as the consumption index increases, the sustainability of life increases.
- Resources here mean natural and industrial sources that contribute to supporting the concept of sustainability. When resources match living organisms, this contributes significantly to maintaining continuity for as long as possible.
- Technology here means modern scientific influence on the nature of life leading to its development. The use of technology in different areas of life and the extent to which these modern methods are compatible with aspects of life helps to apply the real meaning of sustainability, which is continuity.

2.2 Sustainability in Art

The question is, how can art introduce the concept of sustainability to society, also make society taste it to achieve its goal of environmental protection? This question preoccupied the researcher greatly. The threats to the environment are numerous and diverse. Unfortunately, these dangers are now a part of almost everyone's daily life. The risks are not limited to environmental pollution but also excessive use of technology. One of the consequences of reducing Coronavirus spread has been this.

As a result, it became necessary to focus on the true meaning of sustainability. The idea is to do things in a way that doesn't waste our natural resources. We can use them in the future without harming the environment or our health.

The term "sustainable" has become widespread, and we can apply it to almost every aspect of life on Earth. So how can art be sustainable? Or in the sense that art can be considered long-lasting because of its essence, so this is based on the use of various materials in the creation of new forms, and this art style. The emergence of this style of art is not considered revolutionary or strange. This style stems from the advent of conceptual art in the late 1960s and early 1970s and the emergence of a new awareness of social and environmental issues. If the fundamental premise of all art, regardless of its physical structure, subject matter, or nature, is to compel us to meditate and interact. Then introducing sustainability concepts is a logical next step in the field of sustainable art, catalyzing analytical and critical thinking.

Community sustainability is a message everyone must work hard to spread. Therefore, incorporating the idea of sustainability into art benefits everyone. These will also provide us with more opportunities to investigate and develop practical methods for dealing with critical thinking. On the other hand, it focuses on community participation.

3 Folk Art and Identity

Over its long history, humanity has left behind an enormous heritage of stories of different kinds. Many people value them not for entertainment value only but also as a source of history, values, experience, and wisdom. Most importantly, it portrays a picture of human instinct that doesn't exist in any other historical era. This feature is one of the most important motivations that made the researcher benefit from the intellectual value of this art and inspired the vocabulary of her artwork presented in the form of a children's book through an exhibition will be held in the Faculty of Fine Arts' exhibition hall, Helwan University.

Folk art is one of the means of documenting identity. It is an innate art that stems from the conscience of its society, so this is a mark that distinguishes societies from each other.

Besides being a significant part of famous aphorisms or folklore, it is also creative material created by people to express their thoughts and consciousness and combines human experience with cultural heritage.

To illustrate the relationship between folk art and identity, we find that the intellectual value behind Folk Art is only an expression of its identity. In philosophy, identity is when people are differentiated from one another by their specific characteristics. However, in the culture of any nation, it is their culture that separates them from others.

Culture is of two types; formal is the creation of the elite and is the product of planning and conscious thinking, and it is practical and logical and needs official bodies to publish and circulate. Also, it does not distinguish society from others.

Populist culture can be defined as a lifestyle and the product of a group that expresses its own aesthetics and personality in its membership. It is spread among the public because it is popular and easy to understand. It's passed down through the generations with lip and mimicry and imitation, and its material symbols that express its identity are folk clothes, folk foods, and popular products. However, the immaterial symbols are customs, beliefs and norms, art translated this popular culture through symbols that formed an identity with a distinctive character that peoples are unique to each other.

The official and popular cultures influence the significance of any people's identity symbols. Popular culture symbols are always the foundation from which collective identity is derived, and popular culture is an active ingredient in making people's collective identities. It is the basis for maintaining this identity and ensuring its continuation and strengthening, especially in the face of modern technological challenges that have removed the boundaries between the worlds. It is easy to recognize the different cultures and customs of people when we are discussing cultural boundaries. Despite its importance and usefulness, it negatively affected the assimilation of our youth to these cultures and ideas. It introduced Western customs and cultures to us that did not exist, which led to the imposition of a Western identity on us. It has taken a toll on the children in their communities. The risk has grown with the spread of the new Coronavirus. As a result of its spread, the entire world has turned to electronic means of communication. Also, this is because it allows the world to continue without the need for social interaction. Furthermore, children and adolescents' addiction to social media and technological means in general poses pathological risks. Because these electronic devices have harmful psychological and mental effects on children, the prevalence of diseases such as autism, hyperactivity, and speech difficulties in children has increased in recent years.

3.1 Folk Tale Aesthetics and Child Psychology

Here the question arises why the researcher chose the direction of folk literature to present through it her experience and what is meant by the artistic and intellectual value and its relationship with the child. (The researcher's experience is creating a book for children whose theme is derived from folk literature, considered one of the most significant aspects of folk heritage).

Folk literature typically takes a verbal form, conveyed by tongues with anecdotes and singing, and its tales are only people's memories. Its heritage tells its stories, history, customs, and traditions, and it is a significant characteristic of this field (folk literature).

Psychologically, a child takes profound joy in stories that bring him into a fantasy world. It is common for heroes in these stories to embark on adventures or to seek magical tools, and at the end of the story, the good triumphs. However, in this case, the story is about his sense of belonging, his sense of community, and his sense of nationalism. Additionally, folk literature glorifies its heroes with some symbolism, inclusion, and exaggeration. This also makes it endearing to the soul, and this trait resonates with the child.

Contrary to The Legend, the folk tale does not carry the character of holiness, but rather is a heroism abounding with exaggerations and the paranormal, with its main heroes being humans or jinn. So the folk tale is considered a true rooting of the character because it has a psychological, intellectual, and dogmatic temperament. It highlights the habits and behavior acquired from the experience of fathers and mothers because of the actions of previous generations.

Legends, paranormal stories, and folk tales from everyday life are the three types of folk stories. The examination of folk stories reveals that it is an art not aimed at children, as it does not reflect their interests or meet their needs and the language of his narrative sometimes exceeds the child's linguistic abilities. In addition, to their abstract nature, children are drawn to fairy tales for their simplicity and the oddity of their events, which are full of superheroes and wonders. In this respect, writers have a role to play in selecting texts and paraphrasing them (treatments), so that they boost imagination and expand horizons in children based on their age and mental perception. The Arab heritage is rich in many stories that have been reworked to fit the standards of children's literature, such as "a thousand nights and nights," Sinbad stories, and others.

3.2 Folk Art Artist

Herein lies the role of folk art as a bridge through which folk stories are translated and simplified into drawings and symbols that express them. In general, folk art is distinguished by several characteristics that make it the genuine link between folk literature and the receiver (the child, the target group of research).

Most art forms are not governed by the visual logic of perspective; the size of the shapes varies according to what their semantics and connotations dictate to the artist. These pieces of art, which are mostly subjective projections of the subconscious, appear flat in their entirety, and their innate nature makes them resonate with a child. Colors used in folk art play a significant role in grabbing the attention of a child. This is because the use of expressive colors reveals a point of view. A culture of expression surrounds the artist's work, revealing exhilarating formulae. As for the artist, color is a symbol and an alternative to the expression of certain psychological states, and the folk artist deals in his works with two types of colors.

The first is neutral and the result from the use of natural materials (reflecting the quality of use in the artwork), such as wood, pottery, and leather. The second is prefabricated bright colors or the addition of various materials such as colored threads and metal parts. The materials play an active role in the completion of the artwork where, the folk artist composes in his works between more than one material with different surfaces to achieve more artistic richness to obtain diverse effects. The texture of the materials used here indicates the surface properties of them, which enriches the artistic work of the connection of visual and sensory perception together. All of these characteristics led the researcher to connect what this type of art can offer to the world of childhood in order to serve as a crossing for him (the child) from the world of technology with its negative effects and the conditions imposed by the spread of the coronavirus on the world to another world that works to spread the roots of sustainability within it in an easy and fun way while also documenting a rich part of its community's heritage.

4 Technique of Movable or Mechanical Books

This technique of the book design used to give a solitary character, which makes it a differentiated product, and this technique can be used in many different materials. This technique is usually used for children's books, and these books often contain themes that fit this technique.

Furthermore, this art has no borders, and its production varies from country to country. Its roots can be traced back to China, then to Japan, and finally to Europe and the United States. This art may sometimes interfere with the art of origami in folding and cutting paper, but the art of origami does not use any materials; it is only about folding paper.

The art of designing pop-up books is generally just one type of a larger collection of books known as moving or mechanical books; these types of books vary depending on the mechanics of their manufacture. It derives from the art of paper engineering. Historians discovered it in a text in 1306 AD. The Catalan Sufi author and poet Ramon Llull's astronomical theory was explained on a turntable he created. From this, it becomes clear that the original purpose of the creation of this art was not to help children, but for us to comprehend and understand the developing scientific theories. This trend continued until the eighteenth century when it was handmade by skilled craftsmen and then used machines in its manufacture, which saved a lot of time. Pop-up children's books have become so popular around the world that publishers are beginning to reprint traditional fairy tales to add new pop-up features, prompting many children to read.

Preparing a pop-up book requires a great deal of skill and planning, and it takes you through all of the stages of typical book work. When the story is finished and the illustrations are completed. The project is then handed over to the paper engineer. Pop-up books have many considerations and calculations, including the mechanics of the openings so they do not break over time, and the time it takes to pull the tabs, in addition to the fact that some include sound effects and other light so that the books have a visual and audio effect as

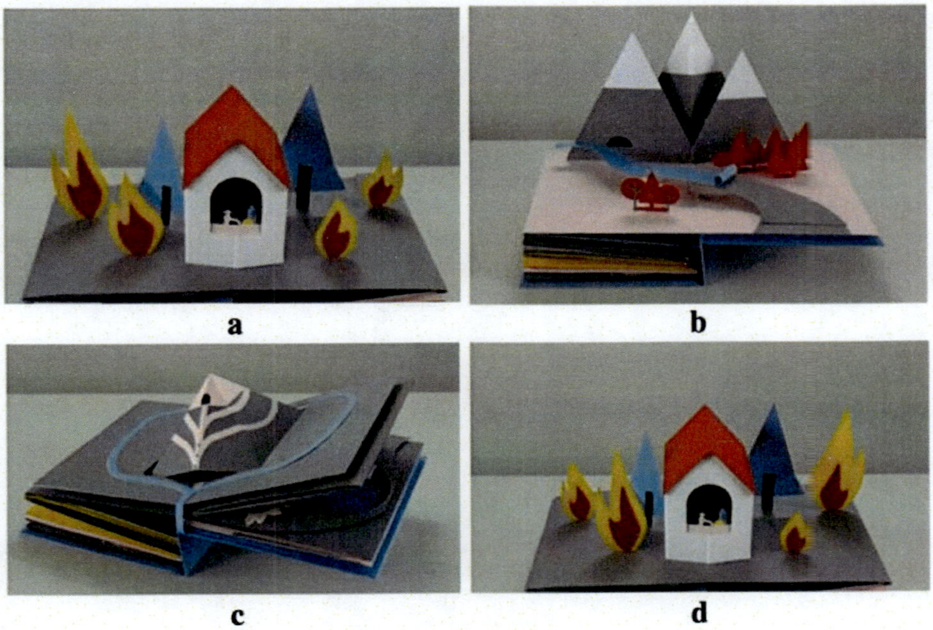

Fig. 1 A collection of photos showing a pop-up book titled life cycle of a drop of water designed by Chris Turner and produced year 2010

well. That's why this artwork requires a full staff of up to sixty people to assemble a single book.

Now, the researcher examines some examples of pop-up books to demonstrate some of the various directions of this art, as shown in Fig. 1. There are four different images in one book, and it is a kind of educational pop-up book that shows the significance of water for children. Additionally, we see the use of explicit colors that express nature, since the intellectual value behind this work is directed toward children, leaving no room for imagination.

Figure 2 the pop-up book titled Pop-up Garden is a type of entertainment geared toward children ages two to six years. The artist used bright colors to illustrate the world of the garden. While the piece is meant for children, he used a lot of explicitly colored elements as well as simple shapes without details. These shapes, after some time, become obvious to children.

Figure 3 it is the first pop-up book to emerge in Thailand, the country of origin, describing the famous sites of Bangkok and archaeological sites. The artist was keen to use the original colors of the tourist and historical places in the country to show the identity of the place, and we note the many details (as it is not aimed at children) as the mechanics of the shapes are more complex.

Figure 4 here the artist used layers of colored paper in skill to show the dimensions, which gives the work artistic value despite the simplicity, and the background colors appeared in harmony with each other, leaving room for the hero character to emerge, the artist then repeated the character's color range in the title, but this led to some confusion for the receiver.

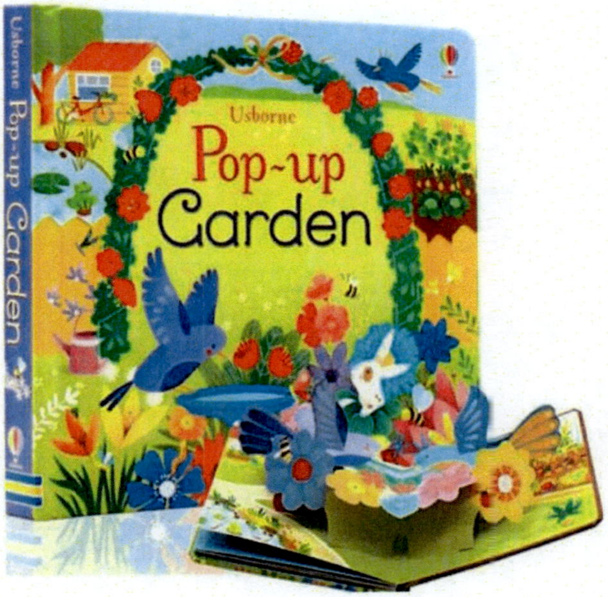

Fig. 2 A pop-up book titled, "pop-up garden," aimed to children from 2 to 6 years, country of origin England, size 18.8 × 18.5 × 1.9 cm

4.1 Researcher Experience

The theme shows two fundamental principles: sustainability and identity. They are bounded by a general framework, which is art. Art is a broad field with many branches and directions, so the researcher chose one of these branches, book design (the researcher's specialty), to illustrate the concept of sustainable identity in art. The researcher will

Fig. 3 A pop-up book of origin Bangkok, Thailand, the artist named Tawee Pong Limmakorn. It has Thai and English languages, and no date of release is indicated

hold an exhibition in which she will present her research idea in the field of book design aimed at children from the age of four to seven years, and the exhibition will be hold at the Faculty of Fine Arts in Cairo—Zamalek in 2022.

The exhibition includes several themes:

- Solidify the concept of identity in the minds of children by using an appealing artistic and cultural style.
- Applying the concept of sustainability in a simple practical way aimed at children in order to foster a sense of belonging and environmental preservation.

Fig. 4 A pop-up book with technology of paper layered art. The origin country is Korea. The artist named Sung Jin and it produced year 2017

- Provide a work of art that encourages the child to read and works to attract his attention from electronic devices that have increased their use in the recent period, especially since the spread of the coronavirus.

Accordingly, the researcher designed and created a children's story using the cutout book technique in a way that's similar to paper-layered art; this was to match the age group that was provided. The book contains several different stories to be a picture of the simple folk life of the Egyptian environment. The folk art contains many folk symbols with a special cultural fondness for the Egyptian environment. This art also includes a number of color groups with a special meaning of folk culture.

4.2 Practical Experimental Work

- Exhibition entitled (story of time).
- Number of works are twelve plates.

- Work experience in book arts, including a book cover and two inside pages.
- Book size 30 × 50 cm.
- Paper layered art is the technique used.

4.3 The Expressive Value of Morphological Vocabulary in the Researcher's Experience

The researcher drew her research vocabulary from folk art, and its symbols were reworked in simple configurations and presented to children as a pop-up book. In a comprehensive analytical view of the researcher's technical experience, therefore we find the following.

The researcher used the tokens of folk visual art in an innovative formulation. The symbol is the formative language that expresses the environment, feelings, emotions, thoughts, and beliefs of the identity of the place. The symbol here, in general, refers to the artistic unit used in this work to give it a uniquely local flavor that reflects the culture of the region.

One of the advantages of the symbol here is that it is a summation of several spatial data, in addition to the fact that it usually tends to abstract geometric shapes. (It is a feature that prompted the researcher to present her practical research in the form of folk art, albeit in a simplified form, to emphasize the child through a story).

These are one of the most important symbols that the researcher used here:

- Crescent and star: They are Islamic symbols used to express religious beliefs.
- Moon and Crescent: The moon has a special significance in the Pharaonic eras because it is a symbol of beauty. In addition to the relationship between human existence and the moon more generally. Both negative and good impacts result from the moon's external change in shape.
- The sun: It is a symbol of the center of life and energy, usually symbolized by the circle surrounded by Rays, a reference to the Pharaonic era. It is one of the most famous religious symbols of the time. Furthermore, the sun is a significant symbol in the Coptic and Islamic epochs, as it was dealt with by folk artists in different cases.
- The Triangle, the Circle and the Square: The researcher has dealt with the fragmented and overlapping forms of her artwork, and these forms return to the folk artist's culture as a symbol of sorcery and magic. In contrast, the square typically represents harmony, constancy, and holiness concerning the Kaaba.

The Triangle used the terms "mountains" and "plateaus" to describe natural features. If the Triangle's top points upward, it also represents optimism and hope; however, if it points downward, it represents water.

- Fonts: In folk artist culture, lines symbolize water, whether parallel, zigzag or spiral.
- Plants: The plant was used as a symbol in the folk art of livelihood and prosperity as it came to express friendship, love, and joy. The palm has a specific location where it signifies prosperity, fertility, and good fortune. Additionally, it was occasionally decorated with a fish to represent abundance and a lion to represent bravery and charity. It was used in conjunction with the bird as a representation of heaven, but in folk art, the tree represents motherhood and life, while the wheat spikes stand for hope, fertility, and livelihood.
- Animals: The researcher used several animals in his works. In general, the shape of the animal is a guide to a specific meaning and purpose. Therefore, the lion is a symbol of strength and courage, the camel is a symbol of patience and moderation, and endurance, and the horse is a symbol of bravery and courage, and fulfillment. The turtle is a sign of slowness and wisdom while the snake represents evil and enmity. The crocodile's shape is also utilised to ward against harm and provide protection from evil. Furthermore, the fish represents good fortune, mobility, and livelihood, the owl represents sadness and anguish, the crow represents gloom, death, and ruin, the dove represents harmony and goodness, and the rooster represents activity.
- The Palm: It is one of the most famous symbols of folk art and came to express rejection, fear, and envy, usually accompanied by the symbol of the palm of the eye, to protect against envy and push evil in her work.

The researcher used a color group specific to folk art that is characterized by primary innate colors. Red, yellow, orange, black, white, and blue are the dominant colors in her work, and the use of this color group is derived from the environment. This color group also has symbolic connotations and a psychological effect on the viewer. White symbolizes peace and serenity and is associated with religious belief. However, it is the pilgrimage as it represents goodness and generosity, while black is the opposite as it symbolizes sorrow, distress, and death.

Green comes to express faith, optimism, and prosperity, while Blue has multiple connotations to be in places a symbol of sadness, cunning, and pessimism and is sometimes used to repel the evil eye and nullify envy. The color red has more than one meaning, and its use is sometimes a sign of love, shame, modesty, and other times a symbol of death and blood. In an attempt, the researcher used these color symbols with the previous combination symbols to produce simple artistic compositions, as the child could taste them. Its composition has varied from a hierarchical composition, which is a symbol of stability, to a composition which represents assortment, which suits the general mood of the work.

In some works, rectangular compositions represent the glory and reverence of a work, and the contrast in those compositions highlights shocks and surprises. The researcher used all of those language elements, whether they were folk symbols, colors, or forms of compositions, to convey the idea expressed in Fig. 5.

From the above, after a review of the meanings of folk-art symbols and the connotations of their colors. The researcher used these indicators in new configurations suitable for the age period of the child to whom this book is presented. Although the researcher did not use the standard method of providing drawings for folk art, the paper layer art technique allowed the researcher to expose the child to the vocabulary of folk art in a practical, anthropomorphic manner. She took

Fig. 5 **a–m** Two pages of a child's book from the researcher's exhibition entitled "A story of time". **n** Full cover of a children's book from the researcher exhibition

Fig. 5 (continued)

layers of decorative compositions of folk art on top of each other and did not define them in black, as is customary as the nature of this art, but used black color in the formation of vocabulary and symbols of folk art. This has come with a white cut to allow space to put writing on it Fig. 5a.

As for the color set for the work, green tones were used through layers from dark to light, then using a degree of yellow to attract attention, and all this was done through layers of paper to give an intellectual value to the work that attracts the attention of the child. The researcher used the working vocabulary and redistributed it again to find different design solutions Fig. 5b.

The researcher added another formal vocabulary from the culture of folk art, and its meaning has previously been indicated, and this addition worked to emphasize the meaning. On the other hand, the folk decoration has been distributed in an updated style that serves the idea of adding new elements of folk symbols. We note here that the place dedicated to writing came in a light green tone to highlight the aesthetic of the composition. Also, the researcher used an orange tone in the decoration by a small percentage.

As for Fig. 5c, the researcher combined the plant and bird figures, and these symbols have its connotations in referring to development and prosperity.

Figure 5d the researcher presented other design solutions for the place of writing using another piece that works on the design balance, but there is another degree of hot colors used in two different degrees, and here we find the difference in the intensity of the green color on the receiver.

As for Fig. 5e, the design solution differed, where the writing area was light in color to occupy the central part of the work. However, on both sides of the work the green color came to emphasize the hot color (orange), which suggests nature symbols of greenery and flowers.

And, unlike folk art symbols, Fig. 5f is simply another model of Fig. 5e.

As for Fig. 5g, the researcher used the vertical distribution of all elements of popular symbols and decorations as well.

Figure 5h, comes to provide the researcher with another design solution for the vocabulary of folk art of the bird and the plant, but using green tones and degrees.

As for Fig. 5i, the researcher tried through the figurative symbols of folk art to formulate them in a different design formulation and participated in confirming the meaning of the symbols using a color combination that gives a sense of that to present a homogeneous work.

Figure 5j, the researcher took the contrast resulting from the use of yellow and blue and their degrees as a way to confirm the meaning around which folk stories mostly revolve. It is the relationship of the power represented by the knight riding on the back of a lion to beauty through the woman's destination. Through new design solutions to clarify the meaning easily, the age group to which the book is addressed can understand it.

Figure 5k, as for this painting, it is a design for two internal pages in which contrasting colors were used to express good and evil (or power), and the writing area came in yellow in the form of an oblique cut to attract attention.

Figure 5l, the researcher presented a symbol of the folk decoration through successive color tones, taking advantage of the paper layers technique to give a dimension that works to attract attention. It has been customary in folk art to use black to determine the decoration, but the researcher used this in the formulation of the symbols of the folk tale. As for this rectangle with a blue color, it took the foreground place to allow the opportunity to write the literary text on it, so that the researcher presented a model of the two opposite pages in the book, but with an updated vision.

The researcher formulated this design in another view in the form Fig. 5m, we note here that the difference of some tones led to giving a different intellectual and formative meaning.

Figure 5n is an updated design vision for the cover design, in which the researcher used the distinctive color range of plastic folk art accompanied by some symbols of this art.

5 The Results

- Reading has been and will continue to be a significant focus in human life, and its importance has been established since childhood and has its benefits in old age.
- The book cannot be dispensed with or replaced by any technological means.
- The overuse of technology affects all segments of society, whether physical or psychological.
- There are many treasures in our heritage that can be reworked for our children.
- Art is not only a way of having fun, it can also be a way of learning as well.

6 Recommendations

- The need to preserve our identity from the dangers of globalization.
- Create a work of art that encourages the child to read and draws attention from electronic devices, as these devices are being used more frequently lately because of the spread of Coronavirus.
- Coronavirus is a disease, but with all of our efforts, we can fight it and eliminate it.

- Encouraging children in schools through art education classes to make their art book, where they use materials that they collect from their environment to express themselves, their society, their language, and what is going on inside him.
- Expand this art activity to become a summer activity during the school holidays.
- Those interested in book arts can conduct workshops to encourage children to create artistic stories with different materials whose theme is inspired by heritage to give the child an opportunity to express how much he understands his identity.
- Urge to expand more skillfully in these courses and workshops to be hosted in commercial libraries as a kind of advertising for them under the supervision of educational bodies and with a promotion in various media.
- Urging the owners of various commercial projects to take advantage of the rest of their products as raw materials used by the child in making Cutout books, and from here, provide a real example of what sustainability means for the child.

References

Badalucco, L.: Kirigami. In: The Art of 3-Dimensional Paper Cutting. Sterling, Sholamadevi (2000)

Benet-Martinez, V., Hong, Y.Y.: The Oxford Handbook of Multicultural Identity. The Oxford Univeristy Press, Oxford (2014)

Jablonski, R.: The Paper Cut-Out Design Book. Stemmer House, Maryland (1976)

Lloyd, D., McDonough, S.: The self, the spirit and the social being, the formation of adolescent identity within community. In: Australian Association for Research in Education, Annual Confrence, Fremantle. (AARE 2001, LLO01455) (2001)

Mathews, G.: Global culture/individual identity. In: Searching for Home in the Cultural Supermarket. Routledge, New York (2000)

Moss, K., Faux, W.V.: The enactment of cultural identity in student conversations on intercultural topics. Howard J. Commun. **17**(1), 21–37 (2006). https://doi.org/10.1080/10646170500487905

Twigg, D., Garvis, S.: Exploring art in early childhood education. Int. J. Arts Soc. **5**(21), 1833–1866 (2010)

Winter, M.: Sustainable Living for Home, Neighborhood and Community. Westsong Publishing, Napa Valley (2007)

Zhang, D.: The Arts of Chinese Papercuts. Foreign Languages Press, Beijing (1989)

The "Duomo" Metropolitan Station in Naples: A Case of Urban Regeneration Through Public Art with a Focus on Spillover Effects on Real Estate Market

Irina Di Ruocco and Alessio D'Auria

Abstract

The paper explores the role of art as a tool for urban regeneration that can increase real estate market values in areas close to metro stations. In detail, the paper studies the effects of the so-called "art metro stations" of line 1 in the city of Naples, which since the 2000s has seen the development of new metro lines with "art stations" famous worldwide. The objective of the study is to estimate the "spillover effect" due to the construction of the Naples "Duomo" station, opened in August 2021, on the contiguous neighborhoods and to assess its impacts and externalities. Starting from the historical series of real estate values (€/m^2) in the neighborhoods of Naples for a period from 2010 the analyses observes an increase in economic activities near the station as of 2019. Finally, the paper has carried out an economic evaluation of the externalities deployed by "Duomo art station" and, generally speaking, a consideration on the role of art as an economic catalyst for urban regeneration.

Keywords

Real estate market • Naples art stations • Urban regeneration • Spillover effects • Econometric assessment

I. Di Ruocco (✉)
Uuniversità degli studi dell'Insubria, Varese, Italy
e-mail: idiruocco@uninsubria.it

A. D'Auria
University Suor Orsola Benincasa—Naples, Naples, Italy
e-mail: alessio.dauria@docenti.unisob.na.it

1 Introduction

Local development, carried out through different measures and policies on the territory, is an important key in the economic framework but also for infrastructures context, in which railway infrastructures can reduce territorial disparities, both on a micro and macro scale, increase cohesion as well as generate development (Gong and Li 2022).

Since about twenty years, urban regeneration of an area has not merely been only the result of infrastructural investments but has become a complex and multidimensional issue, that city brings out the need by public administrations and city managers to reach new most appropriate tools to answer the questions that the city expresses: enhanced possibility of using public spaces, increasing need to participation of inhabitants in the redevelopment processes, and, among all, a demand of greater liveability and better quality of life (Amin and Thrift 2002).

Within this progressive but rapid change, art has increasingly entered the redevelopment processes of cities. The modalities in which this connection takes place take various forms, from street-art to forms of integration of sculptural artistic expressions in squares, streets, stations and meeting places intended as forms of 'public art' (De Luca and Da Milano 2006; Sacco 2006). Artistic expression has become a pregnant element in urban regeneration, bringing out positive externalities, both on urban development and on the quality of space, making evident intrinsic relations between urban space, art and local development. Therefore, according to Cicciotti and Rizzi (2005), it is a matter of considering the development of a district or an area of a city in a more multi-sectoral way, integrating interdisciplinary aspects and attributing an improving function to the urban regeneration of infrastructures and stations through the use of art expressions.

The present work proposes a study method to evaluate the territorial economic development based on the initiatives and policies of local entities, encouraged by the development of

S. Kostopoulou et al. (eds.), *Cities' Vocabularies and the Sustainable Development of the Silkroads*, Advances in Science, Technology & Innovation, https://doi.org/10.1007/978-3-031-31027-0_8

the Metro Line 1 of Naples with a combination of urban regeneration measures, which can influence the territorial socio-economic structure. Existing studies in literature have already discussed the impact of railway and High Speed Rail on urban economic development (Fang 2021): in this paper we aim to discuss the real estate benefits of Metro system Line 1 from the perspective of economic development in metropolitan area (Table 1). Local development is also understood as the capacity of the local economy to create wealth for its residents (Bartik 2003). This is an attempt to reconstruct the genesis and characteristics of new large urban projects and their impacts, in terms of underlying political-economic dynamics and effects on liveability, by exploring the relationships between the economy, the housing market and transport.

As argued in a vast literature about urban development and investment in railway infrastructure (Chen and Haynes 2015; Healey 1998) the current phase of urban renewal involves important transformation processes at an economic, social and political level. In the case under consideration in this research, starting from the urban changes since the 1950s on the theories of local systems and the difference between exogenous and endogenous phenomena (Cicciotti and Rizzi 2005), an observation of endogenous development is proposed, as we want to evaluate the effects of real estate in terms of the development or increase of local businesses around a metro station. The area analysed is the urban context around the "Duomo" station of the Metro Line 1 managed by the 'A.N.M' know as Azienda Napoletana Mobilità Company and promoter of the so-called "Stations of Art". Therefore, neapolitan territory become a place of exchange in which to measure the effects of punctual regeneration actions that have an imprinting on the contiguous areas, in which the analysed territory present itself as complex of socio-economic relations that turn either into relational and social capital (Camagni 1996; Nanetti et al. 1994). The case study starts from a process started in the 1990s in which different actors act, forming a system of local governance (Camagni and Capello 2005).

The realization of a railway system has an impact on the economic development (Pagliara et al. 2017; Zhou et al. 2018), on real estate market and its relative residential and tertiary activities, increasing property values on housing, tertiary, commercial purposes (Du and Mulley 2007; Chen and Haynes 2015). The link between the real estate market and the metropolitan railway service has been investigated extensively in the literature. In particular in our case of study, the observation of growth in real estate market (Zheng et al. 2021; Bardaka et al. 2019; Diao et al. 2017) a result of public investment is found the construction of stations of art for the Metropolitan Line 1 of Naples. The paper aims to offer an additional point of view highlighting impacts due to urban regeneration, inside and outside the "Station of Art". The presence of the railway line Metro 1 on the real estate market implied several positive externalities on residential and commercial markets (Du and Mulley 2007). Moreover, the phenomenon analyzed for the Duomo station shows that metro construction can pilot real estate market phenomena as the example of real estate redevelopment represented by the "Nanà Palace" as a luxury palace, exactly located in the same area of the station and in particular above the square that hosts Duomo Station (Harjunen 2018).

The aim of the paper is to show that accessibility to the station plays a fundamental role in house pricing, and it is also interesting to understand how much real estate market is linked to the opening of the station by using both the estimate of the assets through hedonic prices and then using the "difference-in-difference" method, in order to understand how much the "housing price" in real estate is due to the aesthetic value of the station before and after its opening.

Considering that neapolitan project of metro line system is a process of urban metropolitan expansion to increase the transportation system and the economic strength of Naples, related to rail transportation investment choices that also generally tend to have more significant impacts in areas with a strong regional economy than in areas with a weak economy, where the metro line's development crosses strong urban and new start-up economic centres business districts

Table 1 Main aspects of case study, elaboration by Irina Di Ruocco

Main economic and transport factors in Naples	
Population (2019)	3085 Million (2021 update)
Dominant economic sectors	483,358 enterprises, of which 2/3 in the tertiary sector, 13,573 enterprises in the transport and warehousing, third Italian region for number of number of companies under 35 (12.6%)
Real estate supply	Good development in metropolitan area, well placed for tourism vocation (italian revenue agency)
Station Location	Historical city center
Transportation system	Bus + rail + car
Passenger	42,523,914 (2021 update)

(Cervero 2020). The paper draws on numerous contributions in the literature with the aim of developing an analysis of whether the role of art in stations, as a distinctive element of redevelopment (and consequently urban regeneration), can influence the real estate market, and from there propose reflections on the role of development and further investment, assessing the positive and negative externalities of policies to enhance the value of Local Public Transport infrastructure.

2 Materials and Methods

2.1 Real Estate Market in Intervention Area

Urban economics theories suggest that rail transit investments have the potential to stimulate urban and economic development (Wu et al. 2022; Sasaki et al. 1997). City accessibility is a relevant key of land rent in urban areas (O'Sullivan 2012). Also rail transit investments improve accessibility and increase land rents or phenomena of buying and selling. Many empirical studies have confirmed that price of residential and commercial properties increased around rail transit stations (Hess and Almeida 2007; Ko and Cao 2010). In anticipation of growing revenues around station, property owners have an economic incentives enhance the quality of their properties by remodelling and renovation, as realised for the case study of the luxury building Nanà Palace examinated in this chapter. In addition, increased land costs may cause investors to substitute capital for land in new development and redevelopment projects (O'Sullivan 2012). The existence of local spillovers and connection in real estate markets with railway system has been empirically proved in the literature (Zheng et al. 2021).

Property market is both an instrument and an effect of urban renewal, which is becoming a central issue for the city and raising it to a different stature, in which new patterns of consumption and the continuous search for innovation are emerging (Van den Berg and Pol 1998), and the development of stations helps to link the property market and stations, optimising accessibility. On-site investments result in the creation of space, infrastructure and new urban activities. Generic investments in the quality of space (Pol 2002) and railway investments in urban settings, have positively influenced land prices (Kaneko et al. 2019), improving the public space that becomes embedded in station areas (Bertolini 1996; Bertolini and Spit 1998).

According to the most recent real estate price listings (Borsino immobiliare Napoli 'BIN' 2020; Revenue Agency "Agenzia delle Entrate" 2020), the real estate market in Campania is represented by a concentration of 108,742 properties for sale and 34,913 for rent. The largest number of real estate advertisements is published in the metropolitan city of Naples (57%, of which 58,207 for sale and 24,377 for rent), which is also the most active in relative terms (27 advertisements per thousand inhabitants).

A further analysis has been carried out on the dynamics of prices in Campania and in the districts of Naples on real estate quotations by type (flat, penthouse, semi-detached house, two-roomed house, independent house, office, studio, etc.) from the Real Estate Campania (2021), analysing the historical series of the Revenue Agency (Agenzia delle Entrate section Campania) and the Naples real estate stock exchange BIN. The analysis of the real estate market was performed in two phases: the estimation of hedonic prices of real estate nearby to the station and in the nearby areas (Umberto Road, Garibaldi Square, Borsa Square, Duomo Street Northern side, Duomo Street Southern side) with an analysis of the historical series from 2008 to 2019 taking into account the phases of the Duomo station construction site and the renovation of Nanà Palace with the station opening in August 2021. During the analysis of the data from 2019 to 2020, loss values are observed following the Covid-19 pandemic event, where the spread of the virus caused effects on the real estate market. The data collected by the Statistics and Studies Office of the Central Directorate for Estimation Services and the Real Estate Market Observatory show that the volume of home purchases and sales in Campania amounted to 31,708 NTNs (number of normalised transactions) in 2020, with a decrease of 11.2% on an annual basis (Central Directorate for Estimation Services and the Real Estate Market Observatory, OMI publication on Regional Statistics Campania 2021). The analysis of the real estate market follows the fragmentation of the territory of the municipality of Naples into ten urban macro-areas understood as groupings of contiguous OMI zones with similar characteristics from a building, historical and social point of view. A detailed list of the macro-areas, with their names and the OMI zones that make them up, can be found on the real estate market Borsino Immobiliare website, and the 'Pendino' area that has already dealt with by Studi e Ricerche per il Mezzogiorno/Regional Studies for South Italy (SRM 2017) in which in the Report "Local Development And Urban Regeneration—Objectives And Values For A Sustainable Redevelopment Of The City Of Naples/Sviluppo Locale E Rigenerazione Urbana—Obiettivi E Valori Per Una Riqualicazione Sostenibile Della Città Di Napoli" the study deals with the theme of urban regeneration starting from a concrete case: the redevelopment and enhancement of the Mercato-Pendino district in the city of Naples (SRM 2017). The real estate market zone Pendino neighbourhood corresponds to the 'B8 and B9' area according to the zoning system of the Revenue Agency.

In the past the area has been the subject of studies for its redevelopment, as demonstrated by the participation in URBACT II of Municipality of Naples (Comune di Napoli

2011) and the redevelopment of Mercato Square, where the CTUR (Cruise Traffic and urban regeneration of city port heritage) project was approved by implementing the local action plan "The waterfront of the historic centre and port area from Municipio Square to Mercato Square: a sustainable development through the improvement of the cruise tourism impact".

2.2 The Metro System Line 1

Line 1 of the Naples metro is a light urban transport system that shares a character with high-speed trains in terms of cost, time and space savings. One result is that the urban transport system offers cities changes in terms of travel time, offering a reduction in costs and the connection of High speed rail (HSR) increases the choice of business and activities. The transport system, as will be analyzed in this paper, also has an important structuring effect on the decisions of companies/businesses to locate their business, becoming actors in the decision making process: for example a HSR in Paris showed that the market area was enlarged to include the regions around Paris in the HSR (Bonnafous 1987). The strategic location of stations is crucial for companies looking for a place for their activities and they prefer cities that are well connected to the main stations.

A similar phenomenon was also observed in terms of influences on the real estate market, where the sector anticipated the development of an HSR network, thus defining HSR as attractors of activity (Pol 2008) as described by Van den Berg and Pol (1998) in the case of Lille, where many private and public investments were attracted to the inner city after the new HSR station was completed (Van den Berg and Pol 1998). Connection to the HSR network can be a powerful incentive invest in the railway infrastructure (Pagliara and Di Ruocco 2018). The Duomo station can be expressed as belonging to a primary development zone (Pol 2008) with an accessibility radius between 5 and 10 min. According to Pol (2008), places around stations are redeveloped as attractive locations for commercial enterprises. Similar to high-speed railway stations, underground stations with a large catchment area are also intended to become attractive areas with offices, services and high-quality residential areas. Thanks to the transport service, light rail stations often become a core point of urban redevelopment, considering both residential and office functions (Papa et al. 2008) becoming emblematic cases to analyse the impacts of the transport system on the urban structure. Metropolitan areas served by a metropolitan transport service experienced a decentralisation of residents from the centre to suburban areas and an above-average increase in all types of property values near new light rail stations (Bertolini 2008).

The case of the Naples metro line 1 is worldwide known and exemplary because it is the location of the "art stations", for which €0.5 billion (Agenzia per la coesione territoriale 2020) has been allocated in the 2007–2013 "POR" Regional Operational Program and received unanimous international acclaim: its stations have been defined as the 12 most beautiful metro stations in Europe (Figs. 1 and 2).

Important and well-known all over the world architects such as Gae Aulenti, Massimiliano Fuksas, Domenique Perrault and Atelier Mendini have worked on them. The Naples Metro, with its technology, architecture and the art of its stations, has produced major transformations that have affected the urban and social fabric. Art stations combine the real estate market and the development of transport systems with a new aspect of urban regeneration as a cure for beauty, which is at the heart of the intervention in the construction of metro stations, giving the work both a functional and an attractive aesthetic role (Cascetta et al. 2014). For many stations, urban regeneration has taken place into as well as contiguous to the station area. The metro system transports around 45 million passengers a year. Art stations, as stated by ANM, have become attractive nodes, evaluating that 1.5 million of passengers are tourists, of which it has been

Fig. 1 Toledo station designed by Spanish architect Óscar Tusquets, source https://www.archilovers.com/

Fig. 2 Duomo station view as Fuksas' project, source www.fukas.com

Table 2 Variations in real estate, elaboration by Irina Di Ruocco on quotations by the Revenue Agency (2010–2021)

Year observation	Var % Yn/Yn − 1 (%)
2021	4
2020	− 3
2019	− 1
2018	− 1
2017	− 2
2016	− 8
2015	− 8
2014	29
2013	− 8
2012	− 5
2011	− 5
2010	

calculated that around 176,000 (126,000 as part of a tour and 50,000 as a sole purpose) visit the stations without using the service, a figure which is close to visitors of Capodimonte Museum or Royal Palace of Naples (ANM 2020).

The increase in passenger flows has also accompanied the decision to purchase a number of properties in the upper Pendino area (Duomo Street Northern side) where, in favour of the station opening in 2019 (postponed to 2021), many buyers have begun to follow the Nanà Palace's model. This is shown in Table 2 where it is evident that prices have increased in the Duomo Street North neighbourhood, close to the famous areas of the historic centre of Naples, which are already well established in tourism.

Moreover, ANM Company's aim is to transform the Stations of Art from transport nodes into a true underground museum, which requires articulated strategies for safeguarding and enhancing their value and calls for an active relationship with the "wider" public of its users. In particular, ANM with its Metro Art educational services, managed by a team of specialists in the sector, promotes and enhances this heritage with programmes of educational visits, itinerant lessons, creative workshops and personalised tours.

Line 1 is known for having received the Most Innovative Approach to Station Development award in London in 2009, winning out of over three hundred competitors, and for its art stations, which have won several awards including Europe's Most Beautiful Station for the Toledo stop. In 2016, it received the title of World tunnelling Congress "WTC". The ANM service has 412 km^2 of territory served with approximately 1,570,000 resident population (ANM 2018) with approximately 1,618,000 trips for urban and suburban surface transport services and approximately 72,000 trips for metropolitan transport services. Metro Line 1 with an extension of 18.8 km and 18 stops (ANM 2020), it is characterized that on weekdays in the year 2019, about 130 thousand passengers per day used the transport of Metro Line 1, while on holidays were about 45 thousand passengers and on Saturday days about 70 thousand. As of 2016, there were 150,000 daily passengers (weekdays) and 60,000 on holidays (2016) with an estimated N. passengers year 41,094,499 (ANM 2020) with 42,523,914 annual passengers (in 2017) against 38,056,560 (in 2016).

Passenger growth is seen as a factor in the development of stations and cities where there is a high-speed network connected to the European network, and this concept is reinforced by the evidence of the European HST network where many countries such as Germany, Spain, Italy, Sweden and the Netherlands are connected to this network, anticipating that stations where high-speed trains stop will have to cope with an increasing flow of passengers, some of whom are of high socio-economic status, who will bring purchasing power to the station building and its surroundings, thus strengthening both the urban and transport functions of the station (Bertolini 1996, 1998; Bertolini and Spit 1998; Serlie 1998). There is a third consideration that considerably strengthens the previously mentioned factors: the construction of light rail links and metro systems that improve public transport systems on a local and regional level (Raad voor Verkeer en Waterstaat 1996; Priemus and Konings 1999, 2000; Priemus 2004).

The Naples City Council has approved the regional program containing the indication of strategic objectives of primary importance for the socio-economic development of Campania with the expansion of the archaeological excavation and securing of the archaeological area "of the insula' of Nicola Amore Square site of the Duomo station, the enhancement of the archaeological finds of the Municipio station area. In an initial phase of this research, the Nanà Palace building was analysed using the hedonic pricing method (HPM), which is a common tool for the valuation projects, in which the HPM evaluates the implicit price of a non-market good or direct externalities on market prices (Tajani et al. 2017). The main intuition of HPM is that agents are willing to pay a premium price (monetary or not monetary) for a general building located close to attractive points (transit station, shopping center), accessibility feasility hubs (access to motorways or ztl zones) and the premium reflects the implicit price of proximity to their preference (Agarwal et al. 2015; Barkada et al. 2019).

The real estate market on Nanà Palace shows the confidence to buy the area in proportion to the construction phases and the opening of the station, that took place in August 2021. Figure 3 shows Square N. Amore and the renovated Nanà Palace, represented a famous heritage building in the historic centre of Naples that has been the subject of private buying and selling and overall renovation, attracting a particular type of luxury market.

2.3 Methodology

Early studies on the socio-economic and territorial impacts of public transport infrastructures (in particular, of the railway network) were conducted in the United States in the mid-sixties. The results of many scientific research and reports (Bernard and Terry 2002; Bowes and Ihlanfeldt 2001; Cervero and Duncan 2002) bring out that the

Fig. 3 Duomo station opening and Nanà Palace on the right, source www.Fuksas.com

reduction of the generalized cost of transport, and the increase in accessibility connected to the creation of new public transport exchange nodes, such as a new station (or a renewed and more attractive one), have often significantly influenced the increase in demand for spaces around these areas. In these cases, the increase in demand has reflected an increase in higher real estate values than other city areas. In particular, studies by Ryan (2005), and Chun-Chang et al. (2020) indicated an increase in housing prices near the metro and light rail stations.

The research methods used enhance the relationship between Metro System Transport and real estate development. First, the study utilizes a before-after analysis to determine whether metropolitan transport system project implementation impacted property houses and local tertiary activities, considering the different construction phases of Duomo station project. The property values of real market changed between 2010 and 2016 and from 2019 to 2021, including the covid-19 pandemic period. In this view, to compensate for post Covid-19 phase it has been uses a so called "Difference-in-Differences" (DID) approach, setting up a control and treatment group to evaluate before and after treatment. At this point it is necessary to make a clear observation downstream of the methodology used, i.e. the link between the counterfactual value and the study carried out.

Our causal identification is based on DID method, which is widely employed in policy evaluation to overcome the disadvantages of traditional econometric models (Görg and Strobl 2007; Zheng et al. 2021). It can address the endogeneity and produce casual evidence on HPR's spillovers on land markets. The DID method applied to the metropolitan system seeks to identify evidence that Line 1 may impact land markets in a certain way (positive or negative externalities), known as spillover effects. Externalities are referred to the residential land price base on the data set collected from BIN dataset.

This has implications in micro-zonal terms, where, the analysis was to identify the variation in terms of neighbourhoods, in order to understand which areas were advantaged or remained unaffected. In this analysis, a simplification of the DID was carried out, not using panel data and without identifying the independent variables; for this reason, latent or omitted variables inducing bias in the analysis were not considered. The main insight is to consider real estate market phenomena subject to heterogeneity and time-varying. Several DID methodologies consider the evaluation of the treatment with time fixed effects, in this case, we did not choose this type, despite the covid-19 factor, as the aim is to evaluate the role of art in the stations, indirectly analysing real estate market prices as a variable. The Duomo station and the art station programme are a clear public policy intervention, in which effects were seen following the opening of the other Line 1 stations. For the Duomo station, it is legitimate to ask what role art stations play where urban regeneration becomes a fundamental variable for the real estate market and the potential of the area? In anticipation of the DID method, the Covid-19 created a deadlock in which the study area was frozen by representing an unpredictable event that placed all the observation elements at the same point. The counterfactual approach to evaluating the effects of public policies aims to test the ability of a public policy to change the behaviour or conditions of a certain target population in the desired direction, i.e. to determine to what extent the intervention—rather than other factors—has contributed to the achievement of a certain result (Orr 1999). The result to be determined is the quantitative evaluation of the effect defined as the difference between among from what happened after the implementation of a policy (factual situation) and what would have happened if the policy had not been implemented (counterfactual situation) (Martini and Sisti 2009) starting from the deviation point T_D represented by the Covid-19 pandemic.

In order to estimate the spillover effect in spatial terms (DeFusco et al. 2018; Costello et al. 2011) the DID method has been applied to estimate the variation of real market price for the metro line. The DID technique, widely used in the literature to estimate the variation of a variable between two time scenarios, is a conventional approach that in the case study relaxes the main assumptions of the method and the spillover effect. Often, the DID is associated with the hedonic price method (Barkada et al. 2019; Dubé et al. 2014). Furthermore, the main argument is to identify how much the presence of art can actually influence the real estate market, rather than analysing the station as infrastructure, but as a concept of urban regeneration close to stations. The method is performed through the use of quantitative methods by identifying the "outcome-variables" (Martini and Sisti 2009). The use of the counterfactual approach requires that it is possible to identify the outcome-variables, against which the presence of an effect can be verified with the available analytical tools, a clearly identifiable intervention and circumscribed "treatment", i.e. it must be clearly identifiable, not only in its characteristics, but also in the consequences it is expected to produce. To the extent that the counterfactual situation can plausibly be reconstructed with other information, the effect can still be estimated as the difference between the observed post-intervention situation and the (plausible) reconstruction of the counterfactual situation. The game here is one of greater or lesser plausibility, not "scientific truth" (Agodini and Dynarski 2004). The experimental method comes into the scene, as a special case of the counterfactual approach because with this method the counterfactual situation is reconstructed by observing what happens to a control group made up of subjects very similar to those exposed to the intervention. This reconstruction is considered to be particularly plausible due to the absence of starting differences between the two groups: there in lies the essence of the "superiority" of the experimental method or or "fault" of the measure in the case of "unsuitable" effects (Campbell and Stanley 2015). The desirability or undesirability of the effects does not in itself have any consequences from a methodological point of view, it is a question of establishing a causal link in both cases. It can have consequences for the availability of data, as an undesirable effect is often also an undesirable effect (Martini and Sisti 2009).

It is a question of greater or lesser plausibility, not of "scientific truth" (Agodini and Dynarski 2004). The experimental method is used as a special case of the counterfactual approach, because with this method the counterfactual situation is reconstructed by observing what happens to a control group. This reconstruction is considered particularly plausible due to the absence of initial differences between the two groups: therein lies the essence of the 'superiority' of the experimental method (Campbell and Stanley 2015). The acceptance of the effect lies only in the reconstruction of the causal link between treatment and control group, it does not go back to methodological validity. In order to isolate the effects that are attributable to the public policy, one more term of reference enters the scene, which remains in the background: the counterfactual, consisting of the situation that we would have faced if the policy had not been activated (Palumbo 2001).

The comparison with the counterfactual, on the other hand, is about the extent to which the situation has changed as result of the intervention, and in the case of the pandemic being stationary. This question is actually divided into two distinct questions; the first concerns how the difference between the situation following the intervention and the (necessarily hypothetical) situation that would have occurred in the absence of the intervention can be measured. The second question concerns the degree to which the change can actually be attributed to the intervention, i.e. it can result from the confounding factors which are one of the main problems of impact assessment (Mohr 1998).

According to Wooldridge (2014), DID method should be used to estimate causal effects when discussing the impact of a certain event on the research subject. In the current paper, the DID method has been used as the empirical analysis method to understand the differences in the results caused by an event (the realization of a metro station). The data were divided into two groups: on one hand the treatment group, which consisted of the samples affected by the event, and the control group, which consisted of samples that were not affected by the event. The Covid-19 effect "assures" the plausibility of the analysis on the change in real estate shares.

With urban renewal activated by realization of metro station as the event, the data were divided into two groups by urban renewal delineation time.

The treatment effect was assessed by difference-in-difference estimation:

$$\widehat{\delta} = (\bar{y}_{11} - \bar{y}_{12}) - (\bar{y}_{21} - \bar{y}_{22}) \quad (1)$$

Changes in house prices are subject to the influence of many factors in a neighbourhood, including positive and negative externalities. Many studies have focused on neighbourhood environmental factors as a key influence on housing prices, discussing the changes caused by the externalities of neighbourhood housing (Boyle and Kiel 2001; Bourassa et al. 2005).

With this assumption, the methodology does not attempt to classically answer the impacts of rail transit on economic development after its opening, but to determine both the impacts on housing development before its opening and the difference between ex post and ex ante conditions. Using data on building permits in the city of Naples, this study investigates the effects of the line using hedonic price analysis and difference-in-difference models.

The evaluation on hedonic analysis of transport system influences on property values, that it has long been recognised that the provision of public infrastructure has a profound influence on the pattern of urban development and the spatial distribution of urban real estate values", has been carried out in order to reconstruct the real estate market of the area by evaluating the effect of the Duomo station and the spillover effect, phenomenon observed after the implementation of the HSR with consequent price increases in the adjunct areas (Zheng et al. 2021), as external effect in economic terms of flagship buildings on the urban environment (Oligschläger 2015) around Nanà Palace. The spillover effect, analysed widely as effect in real estate market (DeFusco et al. 2018), in economic terms indicates the effect that an intervention in one area produces cascading benefits in contiguous areas. Along the lines of the Anglo-Saxon approach, the spillover effect triggered by the Duomo station is connected with the urban regeneration of the surrounding area, in which the urban context affected by the intervention is benefited, as Musco (2009) suggests, by real estate transactions both in the short and long term and by various investments in the area making it attractive to a compound market. In this regard, the accessibility of the transport system and the increase of the attractiveness of the area are connected.

3 Case Study

According to Dattoli (2013), the development strategies that trigger the spillover process can be divided into two main types:

1. Creation and design of a strong magnet attractor that stimulates the process of future investments in the area;
2. Realization of a widespread mix of different functions such as to make the realization of subsequent, similar or different investments advantageous.

Both the strategic types of intervention must have in high consideration the connection infrastructure network that combines and binds different functions and several catalyst elements, so that they could be able to generate the spillover overflow effect.

Moreover, the urban core areas of interventions that can trigger a spillover effect, can be of iconic and symbolic type, or rather of functional type, based upon infrastructure-related activities.

The case under consideration falls into the first strategy, but its peculiarity lies in the conditions to be both an infrastructure-based intervention, and to have a strong iconic and symbolic value, due to its aesthetic and artistic characterization.

Subsequently, the Covid-19 effect on the change in real estate shares was determined using the difference-in-difference method and compared with the €/m^2 2020/2019 price change of Naples’ main city and the entire historic centre macro-area. According to the indications of the real estate stock exchange, the real estate quotations in Naples of the Pendino study area is the area defined as “West area of Duomo station, Umberto Street, Costantinopoli Street, Duomo Square” accordig to Borsino immobiliare.

The historical series of real estate prices (Table 2) is studied from 2010 to 2021 divided into two periods: 2010–2019 and 2019–2021. It was necessary to break the valuation trend in order to understand the Covid-19 pandemic sudden factor on economic and real estate dynamics as well as transport. The construction phases and the use of the transport system were suspended and resumed in 2020. The Pendino study area was analysed according to the areas identified by BIN (homogeneous with the Agenzia delle Entrate’s OMI zones), where the first observation was to analyse the variation of real estate prices in the area surrounding the station, noting an increase of €/m^2 in the areas in the proximity of Square Nicola Amore and matching with the buffer zone of the historic centre of Naples, while the area delimited by Street Duomo Southern side, a marginal area close to the port area not subject to regeneration phenomena, is decreasing. It is interesting to note that, thanks to the BINs, providing precise data with respect to the Agenzia delle Entrate, an increase in real estate prices can be seen until 2019 (presumed opening date pre Covid-19) in some points of Street Duomo Northern side of Square Nicola Amore, confirmed by surveys carried out in the area on the choices of buyers to purchase buildings (for residential and non-residential functions) around Nanà Palace. The value of real estate in a neighborhood will increase due to the presence of special buildings or environmental landscapes in the neighborhood, as Nanà Palace highlights (Chau et al. 2002).

The analysis of the historical (Fig. 5) series provided by BIN was also used to evaluate the intra-zonal variation of the area of intervention, where according to Fig. 4, it can be seen that the area near the Duomo station is growing compared to the Duomo Street Southern area (Fig. 5).

4 Results

The possible effects of the regeneration of the new metro stations have been assessed using high quality property market data with difference-in-differences (DID) estimation methods. Table 3 and analysis show that real estate markets begin to adapt to the policy on investment in infrastructure and historic buildings rapidly after the end of the 2021 construction period. First, buildings within 800–1000 m (i.e. Square Nicola Amore and Duomo Street North) from the new metro stations, where accessibility will increase the

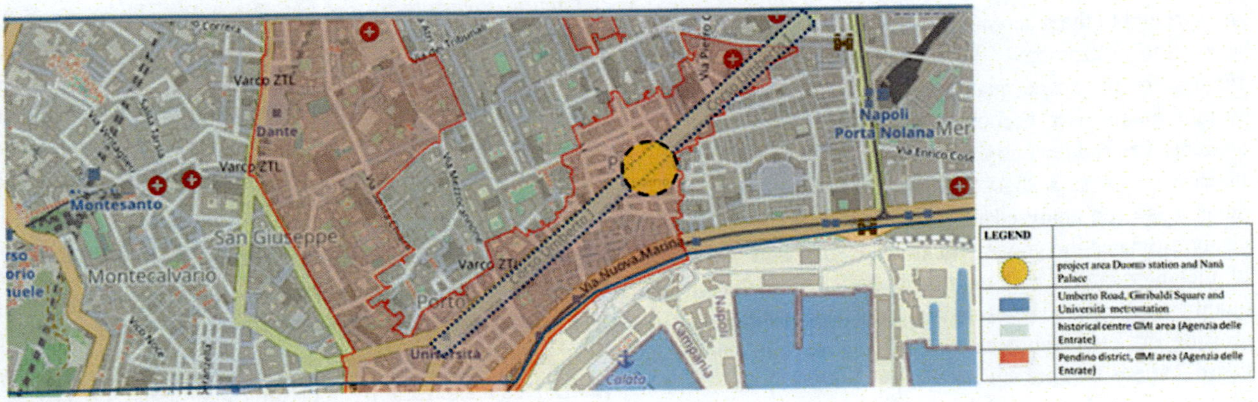

Fig. 4 Real estate valuation zoning, elaboration by Irina Di Ruocco on OMI macro-areas

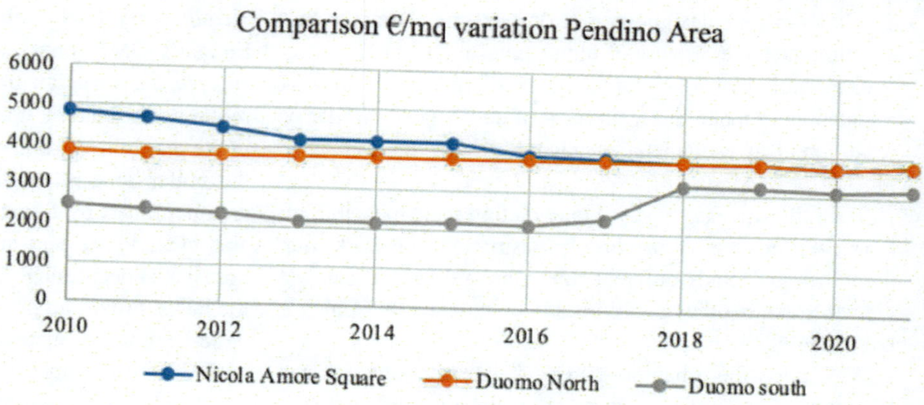

Fig. 5 Comparison €/mq variation Pendino Area, elaboration by Irina Di Ruocco

Table 3 Variations in real estate quotations by the BIN (2010–2021), elaboration by Irina Di Ruocco

Year observation	Var % Yn/Yn − 1				
	N. Amore square (%)	Umberto road (%)	Tribunali street (%)	Duomo street_north (%)	Duomo street_south (%)
2021	− 3	− 3	− 3	− 3	− 3
2020	− 1	− 1	− 1	− 1	− 1
2019	− 1	− 1	− 1	− 1	− 1
2018	− 1	0	− 7	0	38
2017	− 1	0	0	0	8
2016	− 7	− 6	0	0	− 1
2015	0	0	0	0	1
2014	0	− 1	0	0	0
2013	− 7	− 4	− 17	0	− 8
2012	− 4	− 5	0	0	− 3
2011	− 3	− 1	− 12	− 1	− 4
2010					

most, include a positive price premium expressed in €/m^2, that shows significant spillover effects.

At the end of the study, it is possible to evaluate how and in which way the real estate market reacts to a transport system enhancement and urban regeneration policies. The anticipation of the station opening, used as a "pre-announcement" event of the metro line acts as a background to the reaction of the housing prices to the future accessibility promoted by the Duomo station and the increase of the space quality thanks to urban regeneration actions, taking into account the big intervention of the art stations. The sudden entry of COVID-19 in the evaluation of housing prices as a "flat" indipendent variable, equal to all areas, has been analysed through the DID model, that precisely estimates the geographical extension (spillover effect) of the intervention and treatment, both in qualitative and quantitative terms. Therefore, the first deduction is that the housing market anticipates the forthcoming opening of the metro stations that are still in the construction phase, with a near vision of the end of the works. The area close to the Duomo station within 800 m as shown by the BIN data, which analyzing the €/m^2 values in a punctual way shows a high trend of the buildings near Square Nicola Amore and Street Duomo North, Corso Umberto, where the accessibility of the Duomo station still has a high value and allows the flow of tourists/locals for different reasons of movement, attracting investors in the area. Nanà Palace represents an example of heritage building attraction, which together with the Duomo station constitute an example of dual system of real estate market and artistic infrastructure. As described in the previous chapters, an instantaneous impact of the investment of the transport system is the improvement of the accessibility at urban scale: the proximity to the infrastructure is certainly an investment opportunity, especially if the area is prestigious (Historic Centre of Naples, Nicola Amore Square and Nanà Palace), interested from redevelopment phenomena of historic centre based on the project of UNESCO historic centre of Naples as reported in D'Auria (2009), characterized by tourism (represented by historic centre and port areas as showed in Fig. 4) and presence of

Fig. 6 Variation of prices pre and post treatment, elaboration of author from OMI data

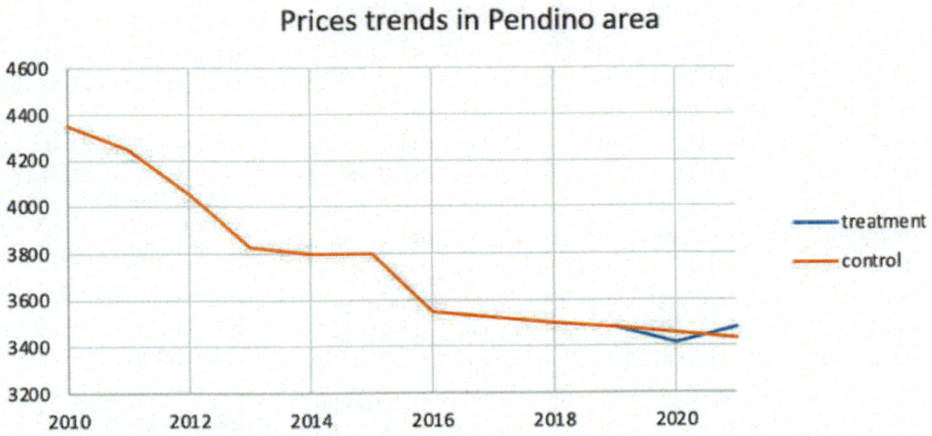

the biggest university in Southern Italy ("Federico II" campus on Corso Umberto).

The evaluation of the economic prices in the first part of the methodology allowed to observe a qualitative assessment of spillover effect, showing thah the areas around the station kept high their real estate value. The DID method made possible to measure the variation among three different areas: the intervention area, the historical centre and Naples main city in order to assess how much the COVID-19 impacted on the areas, representing both an exceptional event and an event of constant scale on the different real estate areas and their characteristics (Fig. 6).

According to the BIN outcomes (Table 4), the 2020/2019 percentage change is −1% compared to −2% for the same area according to the dataset provided by Agenzia delle Entrate. In addition, although the pandemic has affected the real estate market by causing several decreases in the study area, the results of DID method show that the historic centre of Naples would be negatively changed but with a less negative magnitude (−2.32% instead of −3.6%) in the same way for the Naples main city (2.29% instead of −4.6%). In the absence of Covid-19, the real estate market would have benefited from the price increase of the real estate market in the area. However, the increase in the €/sq.m. value of the real estate in the Pendino (Fig. 6) area induced a less negative change in the total area of the historic centre. Similarly, the impact of the Duomo station on the 2020/2019% change in the value of the city of Naples was compared and found to be less negative. The DID method was used for a comparison of the intervention area with the macro area, in order to assess the weight of variation also due to Covid-19 on the housing market, showing how much and in what way different elements respond to an event of global magnitude that impacted the whole real estate sector.

5 Conclusions

This study adopts, starting from the spatial hedonic method a difference-in-difference DID model (Boyle and Kiel 2001; Chau et al. 2002; Qiu and Tong 2021) to investigate the impact of particular measure of policy, station of art for metro line 1 in Naples, on land market. The real estate values represented the variable used to observe the price variation in different areas of the city in assessing whether, the urban regeneration (Conticelli 2011) around the 'Stations of art', may have influenced the price increase, following the redevelopment of metro line station (Duomo Station) and as a result the Nanà Palace is a spillover effect. The approach used was to support an evaluation system for urban econometrics (Qiu and Tong 2021; Wooldridge 2014) to identify the cause-and-effect relationship. The results, show a minimal price variation among the different areas.

With regard to the areas in the field of influence of the Duomo station, just some of them (north and west ones) are affected by the effect of the art station and the regeneration of the area (Conticelli 2011), and these are areas showing a stronger interest in re-investment in activities, as a result of the strong economic network, compared to the east area or the south one (close to the port area) where these phenomenon did not occur.

Table 4 DID evaluation results, elaboration by Irina Di Ruocco

		Var %	DID_BIN	DID_R.Ag
BIN DATA	BIN elaboration	− 1		
Revenue agency DATA	R.Ag. elaboration	− 2		
	Historic centre value	− 3.60	− 2.32%	1.29%
	Naples city value	− 4.60	− 3.32%	2.29%

Currently, the investment in the Duomo area is unlike that of any other station, particularly the redevelopment of Piazza Duomo and the Palazzo Nanà provides the remark that this is a singular case of a luxury building close to a light railway station in Campania. One implication is certainly the type of real estate market area analysed, where depending on the area the result is certainly different. The investment might be interesting in neighbourhoods where the price of real estate is already high (above 2500 €/m^2), which is different for neighbourhoods with a lower average price (either for purchase or for sale activities). Duomo Station is the only station of Line 1 with this characteristics. What attracted investors is an unique condition: proximity to access to the sea (shipping and maritime tourism), historical centre, proximity to universities, accommodation activities already present. The significance of a regenerated space and the design of Fuksas's structure transformed a railway investment condition into an economic development process with benefits for the whole local area. The increasing number of tourists and commuters boosts the transit in the area, which becomes more and more accessible for receptive activities. The market price originates from several factors: culture heritage, prestigious area, high mobility transit index, etc. The counterfactual projection of prices in the case of the absence of the Covid-19 scenario shows that the expected trend would have been slightly different from the initial situation, identifying an increase for stations already characterised by a positive growth trend. The Covid-19 factor has been taken into account for the two-year period (2019–2021), in which a note of caution has been applied in view of considering it not as an aggravating effect but as a external condition still in progress, as the study presented will be integrated with data up to 2022 in order to officially consider the effective ant total role of the pandemic.

Most importantly, the last table shows that urban regeneration can help reduce the gap induced by negative externalities. This result is interesting, and allows us to assess how other stations and neighbourhoods respond to shocks and how they recover. Our research point highlighted a spatial and economic correlation (omitted) in the average value of property market transactions (the values mentioned above). Contrary to what we would have expected, Covid-19 had less impact in neighbourhoods with regenerated stations than in other areas of Naples.

This result is not merely positive for infrastructure investments, a key factor for development, but regeneration measures are a non-monetary value added to the increase of local economic benefits. The adopted model shows a positive counterfactual effect here and is consistent with what has been analysed in the literature (Bardaka et al. 2019; Grimes and Young 2013; Zhang et al. 2020; Zheng et al. 2021; Zhou et al. 2018). For future studies, it will be relevant to benchmark the different stations and obtain a large area comparison in order to understand the effects of heterogeneity in the housing market.

Acknowledgements The research was conducted on the basis of the study carried out during the postgraduate course in Real Estate Market and Urban Regeneration at the Department of Architecture in Naples (2019) and on the basis of values provided by the Chamber of Commerce of Naples for BIN values.

References

Agarwal, S., Koo, S.M., Sing, T.F.: Impact of electronic road pricing on real estate prices in Singapore. J. Urban Econ. **90**, 50–59 (2015)
Agenzia delle Entrate.: Valori Immobiliari Campania. https://www.agenziaentrate.gov.it (2020)
Agenzia per la coesione territoriale.: Linea 1 della Metropolitana di Napoli - Le stazioni dell'Arte. https://www.agenziacoesione.gov.it/wp-content/uploads/2020/01/SCHEDA_POR_FESR_Campania_Stazioni_dellxArte.pdf (2020)
Agodini, R., Dynarski, M.: Are experiments the only option? A look at dropout prevention programs. Rev. Econ. Stat. **86**(1), 180–194 (2004). https://doi.org/10.1162/003465304323023741
Amin, A., Thrift, N.: Cities: Reimagining the Urban. Polity, Cambridge (2002)
Archilovers: Toledo metro station, Napoli|Oscar Tusquets Blanca. https://www.archilovers.com/ (n.d.)
Azienda napoletana mobilità A.N.M.: Carta della mobilità 2018. https://www.anm.it/images/stories/PDF/carta_mobilita_2016/carta_mobilit%E0_2018_a.pdf (2018)
Bardaka, E., Delgado, M.S., Florax, R.J.: A spatial multiple treatment/multiple outcome difference-in-differences model with an application to urban rail infrastructure and gentrification. Transp. Res. A Policy Pract. **121**, 325–345 (2019). https://doi.org/10.1016/j.tra.2019.01.028
Bartik, T.: Local Economic Development Policies. Upjohn Institute Working Paper No. 03-91 (2003). https://doi.org/10.2139/ssrn.400820
Bertolini, L.: Nodes and places: complexities of railway station redevelopment. Eur. Plan. Stud. **4**, 331–345 (1996). https://doi.org/10.1080/09654319608720349
Bertolini, L., Spit, T.: Cities on Rails: The Redevelopment of Railway Station Areas. Routledge, London (1998)
Bertolini, L.: Station Areas as Nodes and Places in Urban Networks: An Analytical Tool and Alternative Development Strategies (2008)
Bonnafous, A.: The regional impact of the TGV. Transportation **14**(2), 127–137 (1987). https://doi.org/10.1007/BF00837589
Borsino Immobiliare: Valori immobiliari di Napoli. https://borsinoimmobiliare.it/ (2020)
Bourassa, S.C., Hoesli, M., Sun, J.: The price of aesthetic externalities. J. Estate Literat. **13**(2), 165–218 (2005). https://doi.org/10.2139/ssrn.477782
Bowes, D.R., Ihlanfeldt, K.R.: Identifying the impacts of rail transit stations on residential property values. J. Urban Econ. **50**(1), 1–25 (2001). https://doi.org/10.1006/juec.2001.2214
Boyle, M.A., Kiel, K.A.: A survey of house price hedonic studies of the impact of environmental externalities. J. Real Estate Literat. **9**(2), 117–144 (2001)
Camagni, R., Capello, R.: Urban milieux: from theory to empirical findings. In: Boschma, R.A., Kloosterman, R.C. (eds.) Learning from Clusters. The GeoJournal Library, vol. 80, pp. 249–274. Springer, New York (2005)

Camagni, R.: Economia e pianificazione della città sostenibile. Il Mulino (1996)

Campbell, D.T., Stanley, J.C.: Experimental and Quasi-Experimental Designs for Research. Ravenio books, Cambridge (2015)

Cascetta, E., Cartenì, D.I.A., Henke, I.: Estetica delle stazioni e attrattività del trasporto ferroviario: evidenze empiriche e modelli matematici. Ingegneria Ferroviaria **69**(4), 307–324 (2014)

Cervero, R., Duncan, M.: Transit's value-added effects: Light and commuter rail services and commercial land values. Transp. Res. Rec. **1805**(1), 8–15 (2002). https://doi.org/10.3141/1805-02

Cervero, R.: Urban development on railway-served land: lessons and opportunities for the developing world. In UC Berkeley: Center for Future Urban Transport: A Volvo Center of Excellence. https://escholarship.org/uc/item/71v7m90b (2020)

Chau, K.W., Yiu, E.C.Y., Wong, S.K., Lai, L.W.C.: Hedonic price modelling of environmental attributes: a review of the literature and a Hong Kong case study. In Welfare Economics and Sustainable Development, vol. II. Encyclopaedia of Life Support Systems, UNESCO. http://www.eolss.net/Sample-Chapters/C13/E1-21-05-04.pdf (2002)

Chen, Z., Haynes, K.E.: Impact of high speed rail on housing values: An observation from the Beijing-Shanghai line. J. Transp. Geogr. **43**, 91–100 (2015). https://doi.org/10.1016/j.jtrangeo.2015.01.012

Chun-Chang, L., Chi-Ming, L., Hui-Chuan, H.: The impact of a mass rapid transit system on neighborhood housing prices: an application of difference-in-difference and spatial econometrics. Real Estate Manag. Valuat. **28**(1), 28–40 (2020). https://doi.org/10.1016/j.regsciurbeco.2017.08.006

Cicciotti, E., Rizzi, P.: Politiche per lo Sviluppo Territoriale. Teorie, Strumenti, Valutazione. Carocci Editore (2005)

Comune di Napoli: Incontro pubblico sul piano di azione locale URBACT: Focus piazza mercato. https://www.comune.napoli.it/flex/cm/pages/ServeBLOB.php/L/IT/IDPagina/16322/UT/systemPrint (2011)

Conticelli, E.: Assessing the potential of railway station redevelopment in urban regeneration policies: an Italian case study. Proc. Eng. **21**, 1096–1103 (2011). https://doi.org/10.1016/j.proeng.2011.11.2116

Costello, G., Fraser, P., Groenewold, N.: House prices, non-fundamental components and interstate spillovers: the Australian experience. J. Bank. Finan. **35**(3), 653–669 (2011). https://doi.org/10.1016/j.jbankfin.2010.07.035

CTUR project - Cruise Traffic and Urban Regeneration: Programme 2007–2013 URBACT II. Cruise Traffic and Urban Regeneration (physical and environmental regeneration of port-areas; cruise traffic and port-heritage as drivers to economic and social development; planning and management of cruise development). https://keep.eu (n.d.)

Dattoli, F.: Strategie urbane di rigenerazione: real estate e rigenerazione urbana attraverso le esperienze inglesi. Doctoral Dissertation, Università degli studi di Roma Tre. Università degli studi di Roma Tre, Roma Repository. http://hdl.handle.net/2307/4115 (2013)

D'Auria, A.: Urban cultural tourism: creative approaches for heritage-based sustainable development. Int. J. Sustain. Dev. **12** (2–4), 275–289 (2009). https://doi.org/10.1504/IJSD.2009.032782

De Luca, M., Da Milano, C.: Il patrimonio e le attività culturali nei processi di rigenerazione urbana. Economia Della Cultura Società Editrice Il Mulino **3**, 371–382 (2006)

DeFusco, A., Ding, W., Ferreira, F., Gyourko, J.: The role of price spillovers in the American housing boom. J. Urban Econ. **108**, 72–84 (2018). https://doi.org/10.1016/j.jue.2018.10.001

Diao, M., Leonard, D., Sing, T.F.: Spatial-difference-in-differences models for impact of new mass rapid transit line on private housing values. Reg. Sci. Urban Econ. **67**, 64–77 (2017). https://doi.org/10.1016/j.regsciurbeco.2017.08.006

Du, H., Mulley, C.: The short-term land value impacts of urban rail transit: quantitative evidence from Sunderland, UK. Land Use Policy **24**(1), 223–233 (2007). https://doi.org/10.1016/j.landusepol.2005.12.003

Dubé, J., Legros, D., Thériault, M., Des Rosiers, F.: A spatial difference-in-differences estimator to evaluate the effect of change in public mass transit systems on house prices. Transp. Res. B Methodol. **64**, 24–40 (2014). https://doi.org/10.1016/j.trb.2014.02.007

Fang, J.: Impacts of high-speed rail on urban smog pollution in China: a spatial difference-in-difference approach. Sci. Total Environ. **777**, 146153 (2021). https://doi.org/10.1016/j.scitotenv.2021.146153

Fuksas, M.: Duomo Station of Line 1 of the Naples Subway. www.fuksas.com (n.d.)

Gong, W., Li, V.J.: The territorial impact of high-speed rail on urban land development. Cities **125**, 103581 (2022). https://doi.org/10.1016/j.cities.2022.103581

Görg, H., Strobl, E.: The effect of R&D subsidies on private R&D. Economica **74**(294), 215–234 (2007). https://doi.org/10.1111/j.1468-0335.2006.00547.x

Grimes, A., Young, C.: Spatial effects of urban rail upgrades. J. Transp. Geogr. **30**, 1–6 (2013). https://doi.org/10.1016/j.jtrangeo.2013.02.003

Harjunen, O.: Essays in urban economics and housing market capitalization. Doctoral Dissertation, Aalto University. Aalto University Repository. http://urn.fi/URN:ISBN:978-952-60-8316-2 (2018)

Healey, P.: Regulating property development and the capacity of the development industry. J. Property Res. **15**(3), 211–227 (1998). https://doi.org/10.1080/095999198368374

Hess, D.B., Almeida, T.M.: Impact of proximity to light rail rapid transit on station-area property values in Buffalo, New York. Urban Stud. **44**(5–6), 1041–1068 (2007). https://doi.org/10.1080/00420980701256005

Kaneko, Y., Nakagawa, T., Phun, V.K., Kato, H.: Impacts of urban railway investment on regional economies: evidence from Tokyo using spatial difference-in-differences analysis. Transp. Res. Rec. **2673**(10), 129–140 (2019). https://doi.org/10.1177/0361198119845098

Ko, K., Cao, X.J.: Impacts of the Hiawatha light rail line on commercial and industrial property values in Minneapolis. J. Public Transp. **16** (1), 47–66 (2010). https://doi.org/10.5038/2375-0901.16.1.3

Martini, A., Sisti, M.: Valutare Il Successo Delle Politiche Pubbliche. Il Mulino (2009)

Metropolitana di Napoli: Progetto Metropolitana di Napoli. https://www.metropolitanadinapoli.it (n.d.)

Mohr, L.B.: Impact Analysis for Program Evaluation, 2nd edn. Sage Publications, New York (1998)

Musco, F.: Rigenerazione urbana e mercato: il caso di Madrid. In: Dialoghi Internazionali. Città del Mondo. Bruno Mondadori, London (2009)

Nanetti, R.Y., Leonardi, R., Putnam, R.D.: Making Democracy Work: Civic Traditions in Modern Italy. Princeton University Press, Princeton (1994)

Oligschläger, P.: Uitstralingseffecten van Iconen. Delft University Press, Delft (2015)

Orr, L.L.: Social Experiments: Evaluating Public Programs with Experimental Methods. Sage Publications, New York (1999)

O'Sullivan, A.: Urban Economics. McGraw-Hill, Hoboken (2012)

Pagliara, F., Di Ruocco, I.: How public participation could improve public decisions on rail investments? Reg. Sci. Policy Pract. **10**(4), 383–403 (2018). https://doi.org/10.1111/rsp3.12143

Pagliara, F., Mauriello, F., Garofalo, A.: Exploring the Interdependences Between High Speed Rail Systems and Tourism: Some Evidence from Italy. Transp. Res. A Policy Pract. **106**, 300–308 (2017). https://doi.org/10.1016/j.tra.2017.09.022

Palumbo, M.: Valutazione di processo e d'impatto: l'uso degli indicatori tra meccanismi ed effetti. In: Nicoletta, S. (ed) Valutazione 2001-Lo Sviluppo Della Valutazione in Italia, 1st edn. Franco Angeli, Milan (2001)

Papa, E., Pagliara, F., Bertolini, L.: Rail system development and urban transformations: towards a spatial decision support system. In: Bruinsma, F., Pels, E., Rietveld, P., van Wee, B. (eds.) Railway Development: Impacts on Urban Dynamics, pp. 337–357. Springer, Heidelberg (2008)

Pol, P.M.: A renaissance of stations, railways and cities. Economic effects, development strategies and organisational issues of European high-speed-train stations. Doctoral Dissertation, Erasmus Universiteit Rotterdam. Erasmus Universiteit Rotterdam Repository. https://lib.ugent.be/catalog/rug01:000744881 (2002)

Pol, P.M.: HST stations and urban dynamics: experiences from four European cities. In: Bruinsma, F., Pels, E., Rietveld, P., Priemus, H., van Wee, B. (eds) Railway Development. Physica-Verlag HD, London (2008). https://doi.org/10.1007/978-3-7908-1972-4_4

Priemus, H., Konings, J.W.: Stadsgewestelijk Openbaar Vervoer. Delft University Press, Delft (1999)

Priemus, H., Konings, R.: Public transport in urbanised regions: the missing pursuit of the economic of cities link in the vitality. Plan. Pract. Res. **15**(3), 233–245 (2000). https://doi.org/10.1080/713691899

Priemus, H.: Light rail: backbone of European urban regions. In: Beuthe, M., Himanen, V., Reggiani, A., Zamparini, L. (eds) Transport Developments and Innovations in an Evolving World. Advances in Spatial Science. Springer, New York (2004)

Qiu, F., Tong, Q.: A spatial difference-in-differences approach to evaluate the impact of light rail transit on property values. Econ. Model. **99**, 105496 (2021). https://doi.org/10.1016/j.econmod.2021.03.015

Ryan, S.: The value of access to highways and light rail transit: evidence for industrial and office firms. Urban Stud. **42**(4), 751–764 (2005)

Sacco, P.L.: Arte pubblica e sviluppo locale: utopia o realtà possibile? Econ. Cult. **3**, 285–294 (2006). https://doi.org/10.1446/23265

Sasaki, K., Ohashi, T., Ando, A.: High-speed rail transit impact on regional systems: does the Shinkansen contribute to dispersion? Ann. Reg. Sci. **31**, 77–98 (1997). https://doi.org/10.1007/s001680050040

Serlie, Z.: Stationslocaties in vergelijkend perspectief. PhD dissertation, University of Utrecht. University of Utrecht (1998)

SRM – Studi e ricerche per il Mezzogiorno: Sviluppo Locale E Rigenerazione Urbana Obiettivi E Valori Per Una Riqualificazione Sostenibile Della Città Di Napoli. https://www.sr-m.it/wp-content/uploads/woocommerce_uploads/2017/04/rigenerazione_web1.pdf (2017)

Tajani, F., Morano, P., Locurcio, M., Torre, C.M.: Data-driven techniques for mass appraisals. Applications to the residential market of the city of Bari Italy. *Int. J. Bus. Intell. Data Min.* *11*(2), 109–129 (2017). https://doi.org/10.1504/IJBIDM.2016.10001483.

Van den Berg, L., Pol, P.: The European High-Speed Train and Urban Development: Experiences in Fourteen European Urban Regions. Ashgate, Farnham (1998)

Wooldridge, J.M.: Introduction to Econometrics. Cengage learning, Boston (2014)

Wu, R., Li, Y., Wang, S.: Will the construction of high-speed rail accelerate urban land expansion? Evidences from Chinese cities. Land Use Policy **114**, 105920 (2022). https://doi.org/10.1016/j.landusepol.2021.105920

Zhang, X., Hu, Y., Lin, Y.: The influence of highway on local economy: evidence from China's Yangtze River Delta region. J. Transp. Geogr. **82**, 102600 (2020). https://doi.org/10.1016/j.jtrangeo.2019.102600

Zheng, X., Chen, X., Yuan, Z.: Exploring the spatial spillover effect of home purchase restrictions on residential land prices based on the difference-in-differences approach: evidence from 195 Chinese cities. Land Use Policy **102**, 105236 (2021). https://doi.org/10.1016/j.landusepol.2020.105236

Zhou, J., Yang, L., Li, L.: The implications of high-speed rail for Chinese cities: connectivity and accessibility. Transp. Res. A Policy Pract. **116**, 308–326 (2018). https://doi.org/10.1016/j.tra.2018.06.023

Construction of Islamic Feature Using Parametric Approach

Sherif Usama Nour El Din Zeid and Dina Mohamed Saadallah

Abstract

The identity of cities depend greatly on the vocabularies that it embodies through its architecture and arts, as they are the platforms of cultural, traditional and historical aspects. Lately, governments focused greatly on maintaining and fixing buildings in order to save their heritage. Investigating these approaches, the paper serves this discourse by using digital software to aid in providing a freedom of expression and to lay down the foundations to applying parametric and generative design to preserve the architectural elements which represent the vocabularies that give cities their identity and heritage, and to further apply innovative transformations to these elements while still respecting the historical, cultural and traditional aspects by which these elements were constructed. The paper forays into the study of historic ornamental details of Islamic architecture with a structure that exhibits the mediating relationship between the motifs' distinguishable geometric patterns and cosmological renditions through the Muqarnas. It signifies as one of the most original inventions of Islamic architecture, and with its construction and form being of complicated application, it inspired the paper's aim to investigate and introduce a new means of generating Muqarnas in the modern era. Parametric and generative design using Grasshopper, in the paper's context, help bring about a reformulated preservation of the historic architectural element Muqarnas. To achieve this, the paper studies the historical background of Muqarnas and its construction, to be of aid in the process of reformulating a parametric generative process for a new Muqarnas by using the hypothesis of the relationship between intricate geometrical and cosmological implications.

Keywords

Parametric • Algorithmic • Grasshopper • Islamic • Muqarnas

1 Introduction

Ornate repetition of geometric patterns in Islamic architecture's manifestations of facades and structures is characteristic of the style's authenticity, wielding both meaning and aesthetic value as kernel. Muqarnas, Arabic for stalactite vault, embodies the values and demonstrations argued. The Islamic ornamented vaulting replicates units arranged in tiers, each supporting another corbel atop. Supporting domes, the Muqarnas signified an architectural transition from a square base in early exemplifications of the structure. Later applications of the Muqarnas diversified the meanings carried out by the ornate structure. Innovations in the Muqarnas took on different forms ranging from cupolas to niches to arches, allowing different iterations of the architectural element in different Islamic countries since the eighth century (Fig. 1). Innovations declining since the fall of the Ottoman Empire allowed little room for the progression of Muqarnas in differing mediums and carried out meanings. Endowments to the architectural element is allowed more prospect in light of the twenty-first century's great technological advancements. The increasing advent of apparatus aiding in the forming of highly complex structures provides the paper's prospect and aim with tools to better envision and support a new Muqarnas in an increasingly digitized verse. Ensuring the preservation of the cosmological nature of the Islamic element and structure in context, the paper deems alluding to the original design processes necessary in reiterations of the architectural element.

S. U. N. E. D. Zeid (✉) · D. M. Saadallah
Faculty of Engineering, Alexandria University, Alexandria, Egypt
e-mail: sherifusa1988@yahoo.com

D. M. Saadallah
e-mail: dina.saadallah@alexu.edu.eg

S. Kostopoulou et al. (eds.), *Cities' Vocabularies and the Sustainable Development of the Silkroads*,
Advances in Science, Technology & Innovation, https://doi.org/10.1007/978-3-031-31027-0_9

Fig. 1 Different examples of applications of muqarnas (Korbendau 1997)

2 Islamic Architecture

Throughout its 1400-year life-span, Islamic architecture encompassed allusions and influences from varying architectural styles (Roman, Byzantine, Persian, Indian, and Chinese), serving the architectural style's affluence (Krautheimer 1984). The influences diversifying the Islamic architectural style relied on the widespread of Islam from North Africa to East Asia, extending to cover parts of Europe. The architectural style envelopes characteristics reliant on the favouring of geometry, intertwined with natural and cosmological implications, to convey spiritual and deeply symbolic style-manifestations. The manifestations attained physical, psychological, and climate comfort as the designers of the style based its vocabulary on the nature surrounding them, extracting proportions serving user comfort and attained place-spirituality (Skinner 2009). Understanding the fundamental and sacred truths underlying nature and its geometry marked the architects' and artists' interpretations and applications of Islamic architecture (Dabbour 2012).

2.1 Islamic Geometrical Patterns

Geometry, in its essence, is a science of properties and relationships of magnitudes; how lines, points, surfaces, or solids fit holistically as part of an object (Oxford Dictionary 1999). The Greek etymology of the word comes from a compilation of the words geo, meaning earth, and metry, meaning measurement. In Islamic art, moreover, geometry helped employ aesthetic and symbolic formations. The underlying design process of Islamic art and architecture grew dependent on the science of geometry and its sacred implications. Through these sacred implications, Islamic art and architecture can manifest the spirituality intended and desired to provoke users' mindfulness in the sanctity of the art produced, alluding to Creation. The geometric patterns, therefore, align with the prospects of Islamic architecture as the patterns and the mathematics involved draw introspections to the mathematics and science of the universe, man, and nature. Deployed as ornaments on walls, ceilings, doors, and domes, Islamic geometric patterns demand scale, accuracy, and architectural relativity. The aspects have been manifested for centuries in the art and architecture but seem to lack the guidelines and codes for the continuation of the aesthetic tradition and its progression.

2.2 Islamic Ornaments

Islamic ornaments manifest function and aesthetic value indispensably. The manifestation deems necessary in Islamic architecture and art; the vocabulary interpreted and meanings carried endow spirituality and comfort. Through repetition and decoration, Islamic architectural form weaves a context where form and function are inseparable, contrary to some held beliefs in Western architecture where form and function are separate categories that can be conversed separately and not necessarily manifested together. Sociological and philosophical dimensions are adhered to in Islamic ornaments; Islamic concepts and verses, furthermore, are incorporated into their architectural forms. The relationship weaved between the forms and the holistic concept conjoining them follows what we can call spiritual mathematics. The assigned relationship can be seen in Islamic architecture and art's spirituality bearing link to the variety of qualitative mathematics developed by Pythagoras. The essence of the universe, for a Pythagorean, is explicit in the divinity of numbers and their dynamic equations, rendering iterations of cosmological orders possible (Swinford 2001). Islamic ornaments and decorations are categorized into two main categories; two-dimensional and three-dimensional patterns. Two-dimensional patterns are used to ornate applications on flat or curved surfaces (e.g., walls and arches). Three-dimensional patterns adhere to the applications of the Muqarnas, an ornamental structure with aesthetic protruding surfaces.

3 Muqarnas

3.1 Origin and Geometry of Muqarnas

The Muqarnas is a unique architectural element exercised in Islamic vernacular. With its uses ranging from aesthetic to structural implementation, the Muqarnas derives its vocabulary and particularity its drawing from the contrasting and stark impacts of light and shadow. Its purpose serving a transition between a square base and a dome or aesthetic application to a wall or a ceiling, the Muqarnas audits varying techniques and styles in different Islamic regions (Fig. 2). A combination of convex and concave connotations of the motif and element of Islamic architectural vocabulary denotes, furtherly, the relativity and adaptability to varying architectural manifestations of the Muqarnas. Aligning the research's aim of contemporary manifestations of the unique archaic architectural element, works of architects such as Jørn Utzon (Fig. 3) endow insight into the possible applications and structural variations. The Muqarnas was first developed in North-Eastern Iran and Central-North Africa in the tenth century, marking the stretch of the Islamic regional influence. The element and its variations prominently influenced Islamic architecture by being part of almost all eleventh-century's iterations of the Islamic vernacular.

3.2 Structure and Construction Methods

The structure of the Muqarnas is that of wielded intricacy, bearing characteristics ensuring the relativity and authenticity of its variations. Defining its intricacy, the fifteenth century Timurid mathematician helped about setting a context unto which the Muqarnas is embedded; one that avid admirers of the structure and architects can acquaint themselves with. In his description of the structure, Al-Kashi (Creswell 1958) allows its basic characteristics distinguishable; describing the Muqarnas. "... *like a staircase with facets and a flat roof. Every facet intersects the adjacent one at either a right angle, or half a right angle, or their sum, or another combination of these two. The two facets can be thought of as standing on a plane parallel to the horizon. Above them is built either a flat surface, not parallel to the horizon, or two surfaces, either flat or curved, that constitute their roof. Both facets together with their roof are called one cell*" (Fig. 4). Characteristics of the Muqarnas are visible in its iterations: it is primarily three-dimensional and varying in the volume it provides, providing architects with the chance to employ it as structural support or as an ornamental element. The variation of its architectural interpretation (structural or aesthetic) denies definite mathematical boundaries; simple to the variation of which is that of a fractal, knowing no limitations but is ascribed to a simple algebraic equation. The Muqarnas, due to its varying volume and manifestation, allows the conversion of its three-dimensional units into a two-dimensional pattern plan (2DPP).

The kind of combinations and alterations possible in the manifestation of Muqarnas is determined in the mathematic variations and the angles of its facets: variation of the right angle. Demanding recognition as an interior corner support, the prerequisite geometry of the Muqarnas effectuates its aesthetic as proportionate and adhering to angular strength. The increase, therefore, of the structure's tiers renders greater surface area for support and thus an immediate response to the function employed by the Muqarnas (Grube et al. 1995). The structural benefit of the Muqarnas is iterated in varying construction methods: corbeled, superimposed and suspended. A corbeled Muqarnas is built using stone or wooden blocks; its cells are carved from the middle of the block face and outwards, before or after assembly (Fig. 5a).

Fig. 2 Different styles of muqarnas (Korbendau 1997)

A superimposed Muqarnas, moreover, demands a supporting surface be built first (a dome or a vault) and then the Muqarnas elements are built up against the concave surface (Fig. 5b). A suspended Muqarnas is more superficial in its dependence on the surface it's attached to eventually: the assembly of surface-decorated panels are first reconciled and then fixed to the surface through attachment ribs (Fig. 5c) (Hamekasi et al. 2011).

4 Parametric Design and Visual Programming to Generate a New Muqarnas

4.1 Investigating Digitally-Aided Muqarnas Application

Muqarnas is a tiered structure, consisting of overlaying layers; each layer marks a transition from the starting shape/structure to the final shape/structure. The differing size, form, or position of both final and starting structures encompass the tiered structure of the Muqarnas. Each layer of Muqarnas has layer lines, which are the final and starting shapes of each layer. The Muqarnas, moreover, constitutes a set of surfaces mediating consequent layer lines. Smaller patches constituting the mediating surfaces connect a point or line segment in a lower layer with a point or line segment in an upper layer. The mapping of points on the two layer lines is an important Muqarnas feature, resulting in the stark vertical appearance. Furtherly, the surface's profile allows a variation of the resulting design, endowing either straight or curved offsets at each end (Yaghan 2001).

Testing the variations of such an intricate architectural procedure and motif such as Muqarnas can, thus, be granted efficient medium in a contemporary digital-aided era of architectural manifestations. The vocabulary introduced can become testimonios to both historical significance in the character of the motif and novelty in its various interpretations given more broad and less time-consuming procedures. Parametric design, generative modelling, or computation design allows testing of such intricate procedures with ease and efficiency. Architects can be brought to acquaint

Fig. 3 Jørn Utzon's Jeddah Stadium, Saudi Arabia 1967. *Source* URL: https://www.flickr.com/photos/seier/3883097562

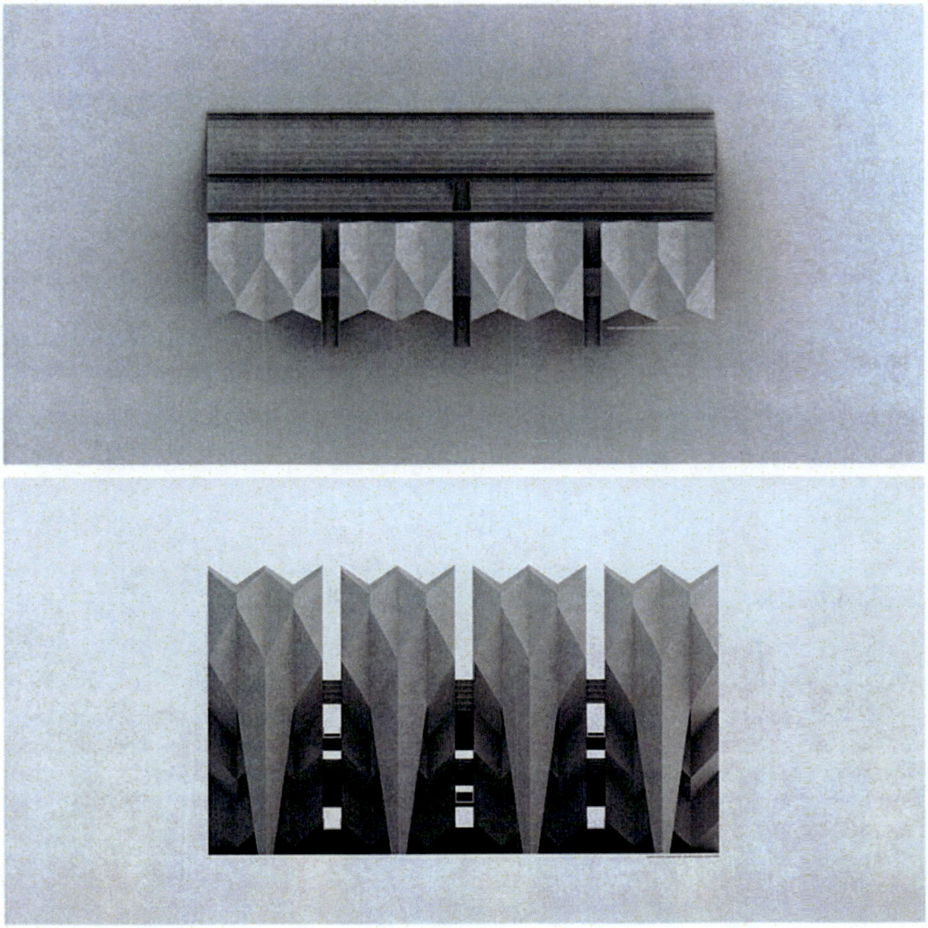

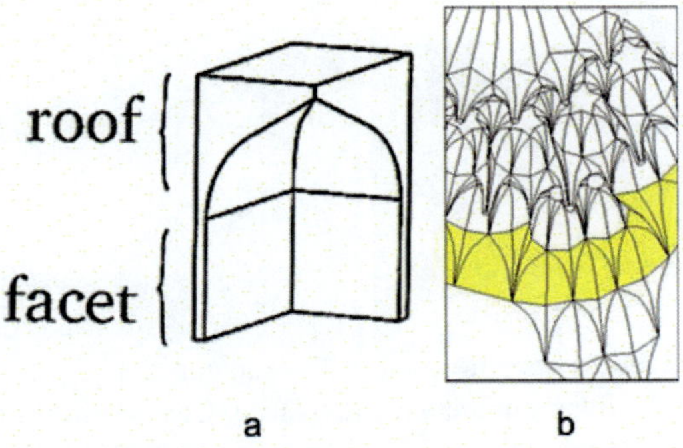

Fig. 4 **a** A muqarnas cell **b** Cells arranged in horizontal courses (Yaghan 2001)

themselves, furthermore, with the intricacy of the historic architectural element of Muqarnas with numerous iterations and reiterations of its aesthetic and structural implementations more efficiently.

Visual programming allows, moreover, user-manipulated logic elements graphically rather than textually. Textual programming languages such as C#, Visual Basic, Python and Rhinoscript require syntax code, whereas visual programming audits for the connection of functional blocks into a sequence of actions. The resultant syntax appropriates the blocks' data in an organized output to generate, finally, the desired result (Fig. 6).

4.2 Magic Squares

Generating a Muqarnas can stand implementation using the tool of visual programming, harnessing a base for the concept of design and its myriad applications. The method chosen for the paper's application is that of an original Islamic pattern concept called the magic squares. The Magic squares concept amuses the harmony of numbers and their interpretation of a cosmic order (Andrews 2004). William Andrews's work on Magic squares adopts a similarity to the traditional Islamic concept; the allocation of Divine origin to mathematics and the transcending of the quantitative aspect of numbers to achieve metaphysical demonstrations. Magic squares admonish a harmony, henceforth, transcendent of the terrestrial (Critchlow 1999).

Fig. 5 **a** Corbeled muqarnas **b** Superimposed muqarnas **c** Suspended muqarnas in the process of repair. Attachment ribs and empty spaces between the muqarnas surface and the walls are visible (Hamekasi et al. 2011)

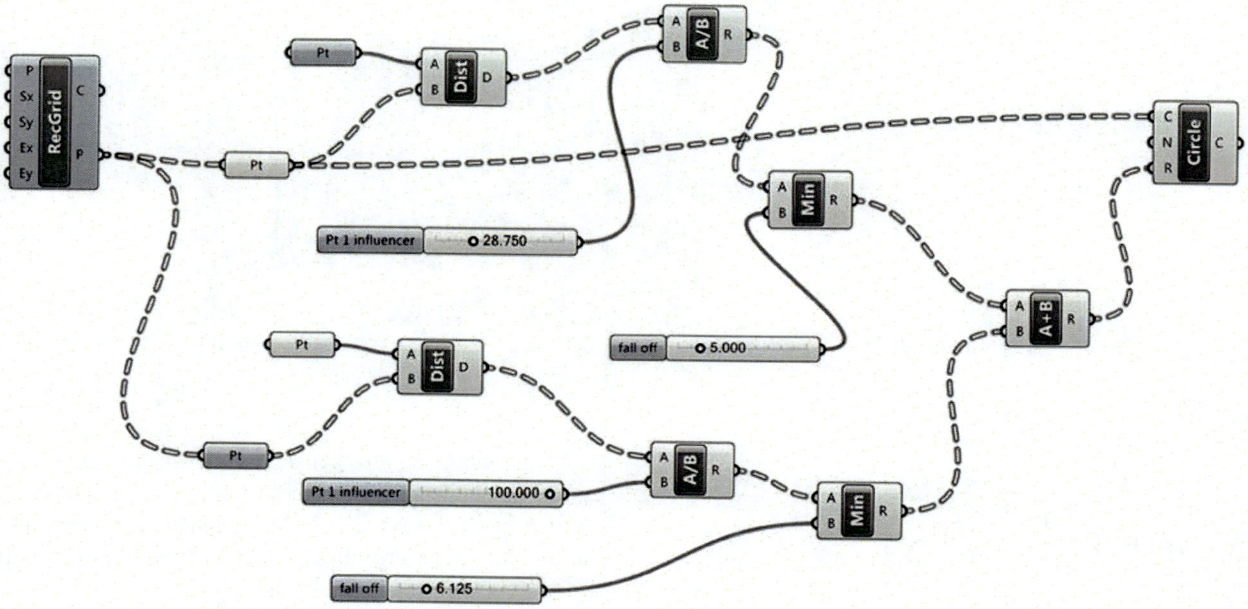

Fig. 6 A node diagram from the Grasshopper plug-in used in Rhino (Tedeschi et al. 2014)

Magic squares' iterations vary greatly, the 3 × 3 square of Saturn (Fig. 7), furthermore, is chosen for the paper's application of Muqarnas as it is one of the most basic iterations and will be easy to implement and comprehend. Worth noting, the paper's choice of a specific iteration doesn't limit its application to just the one iteration. The flexible nature of parametric design and applications permits the interchangeability of iterations, allowing instantaneous results. The 3 × 3 square of Saturn archetype, as illustrated, permits a symmetry between odd and even numbers (highlighted in tone). There are, henceforth, eight possibilities to the arrangement of numbers surrounding the central Fig. 5. Figure 8 demonstrates the varying arrangements: numbers from 1 to 9 are situated in three diagonal lines, resulting in five numbers encompassed in the square: 2, 4, 5, 6, and 8; while the other numbers are displaced. Curved arrows allow a manifestation of the differing arrangements that satisfy the rules of the 3 × 3 square of Saturn.

4.3 Generating Muqarnas Through Magic Squares

The parametric application of the aforementioned rearrangement would generate a geometric pattern. Furthermore, vertices marking the square's perimeter aid in the effectuation of parametric implementation (Fig. 9). An analogy of the magic squares is thus created in Grasshopper; this is

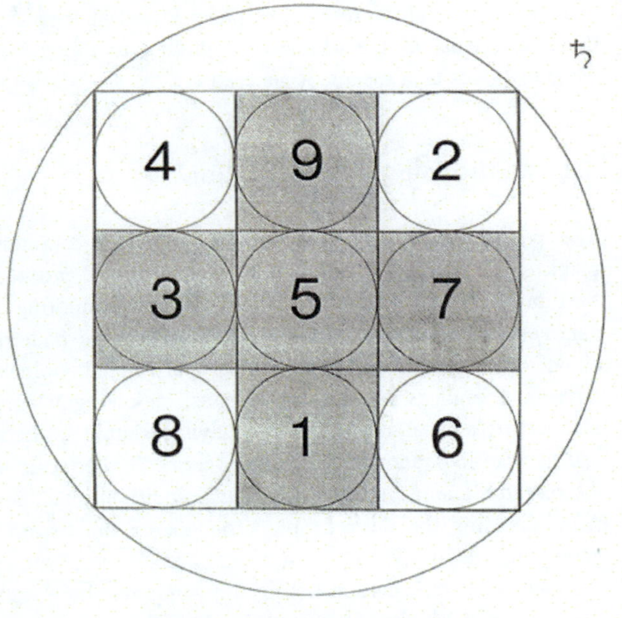

Fig. 7 3 × 3 square of Saturn (Critchlow 1999)

achieved by creating diagonal lines connecting only the odd numbered interchanging vertices (numbers are deployed on the perimeter vertices as a means of explanation only) (Fig. 10).

Allowing further clarification, a diagonal line connecting numbers 10 and 4 acts as an indication for the movement of number 3 to the grid of number 7; while the line connecting vertices numbered 5 and 11 demonstrates the movement of number 7 to replace number 3 in the grid. After arraying this pattern in a grid, a 2D geometric pattern is created (Fig. 11). The pattern manifested is henceforth extracted (Fig. 12a) to stand as a comparison with basic patterns of the geometrical art patterns of Islamic Architecture found in "Arabic Geometrical Pattern and Design" (Bourgoin 1973). Consequently, 2 geometric patterns where chosen for the resemblance in the relationship between the axis and the center point of the pattern (Fig. 12b), nominating the generated pattern to be considered an Islamic pattern.

Next, a 2D boundary is created using the geometric pattern generated (Fig. 13). A model is consequently created relying on two steps in its creation: an extrusion of the boundary with a height H and an extrusion of a cap through the midpoint of the longest edge with a height ½ H (Fig. 14).

Following the extrusion is an arraying of the base unit of Muqarnas to form, for example, a corniche or a part of a dome. A cuboid is created with its dimensions relative to the model and the thickness intended for the muqarnas base unit (Fig. 15). The cuboid is then subtracted from the extruded mass, thenceforth, generating a muqarnas base unit (Fig. 16).

To ensure the support system's vertical stability, a solid trim with an inclination angle is applied to the base unit of the Muqarnas (Fig. 17), before starting the array process. Arraying the resulting base unit in a linear path, grasshopper's node diagram definition allows the easy and efficient repetition and arraying of the base unit (Fig. 18a).

The repetition of the linear path is then applied vertically, with a displacement of ¼ the base's unit width, forming an upper tier (Fig. 19). Further repetition of the tiers with the same displacement relativity brings about a clearer vision of the resulting Muqarnas (Fig. 20).

5 Digitized Cultural and Architectural Asset: A Discussion

5.1 Resulting Vocabulary and Prospect of Muqarnas

Addressing the uniqueness and intricate application of Muqarnas, the research propagated a contemporary approach to the historic asset of Islamic architecture. Endeavoring upon such a quest required a harnessing and awareness of the authentic application of the Muqarnas through studying the origins of Islamic architecture's aesthetic. Imploring the cosmic implications and their intertwining with geometry

Fig. 8 Rearrangement of odd digits in magic square (Critchlow 1999)

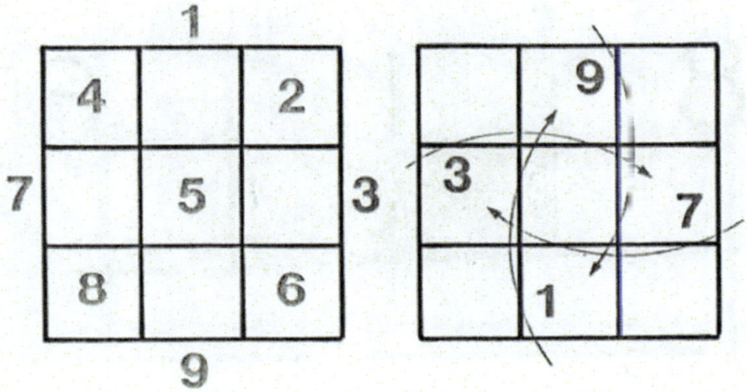

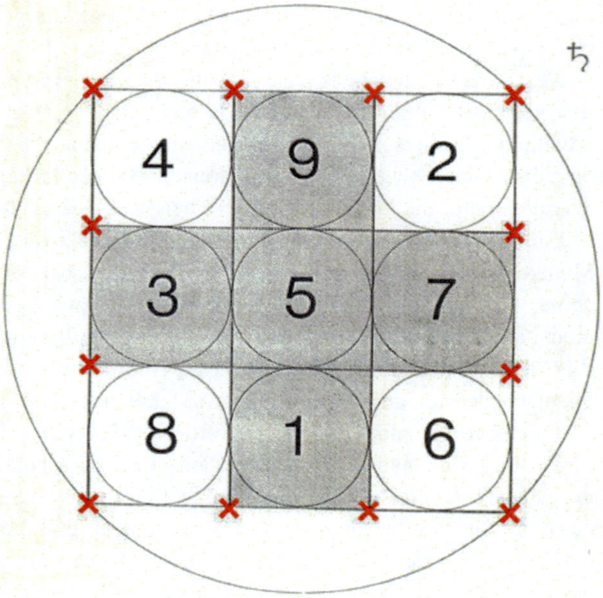

Fig. 9 Placing vertices on the outer perimeter grid of the magic square (Critchlow 1999)

served as kernel to the paper's methods. Muqarnas persevered and has been handled, often, through trial and error, making innovations to the vocabulary posed by the integral historic element unlikely. Reiterating and exploring the Muqarnas generation process, the paper aimed at analyzing the constituents of the Muqarnas, alluding to the use of Magic squares as a means of generation.

5.2 Methodology of Generation

The paper's methodology employing the 3 × 3 Saturn magic square pattern is but an example and offered prospect for future applications, endeavoring through the imploring of parametric apparatus. A myriad of 2D geometric patterns could be explored following the paper's methodology, arriving at endless creative innovations; endowing Islamic architecture's vocabulary greatly. Studying Islamic aesthetic and geometric patterns is an essential asset, furthermore, to the research's application, placing authenticity at its center. After deciding the desired geometric pattern and investigating its relativity, the pattern is transformed into a 3-dimensional unit demonstrating the Muqarnas base unit. Arraying is, thenceforth, allocated in either a linear or radial path. Linear arrays consist of tiers with relative displacement, constituting linear niches and alcoves. Radial arrays, moreover, constitutes the inside of constructed ceilings or domes. The paper addressed linear array as exemplar as its application allows for simple structural allocation and form, serving as an introduction to innovation in Muqarnas generation.

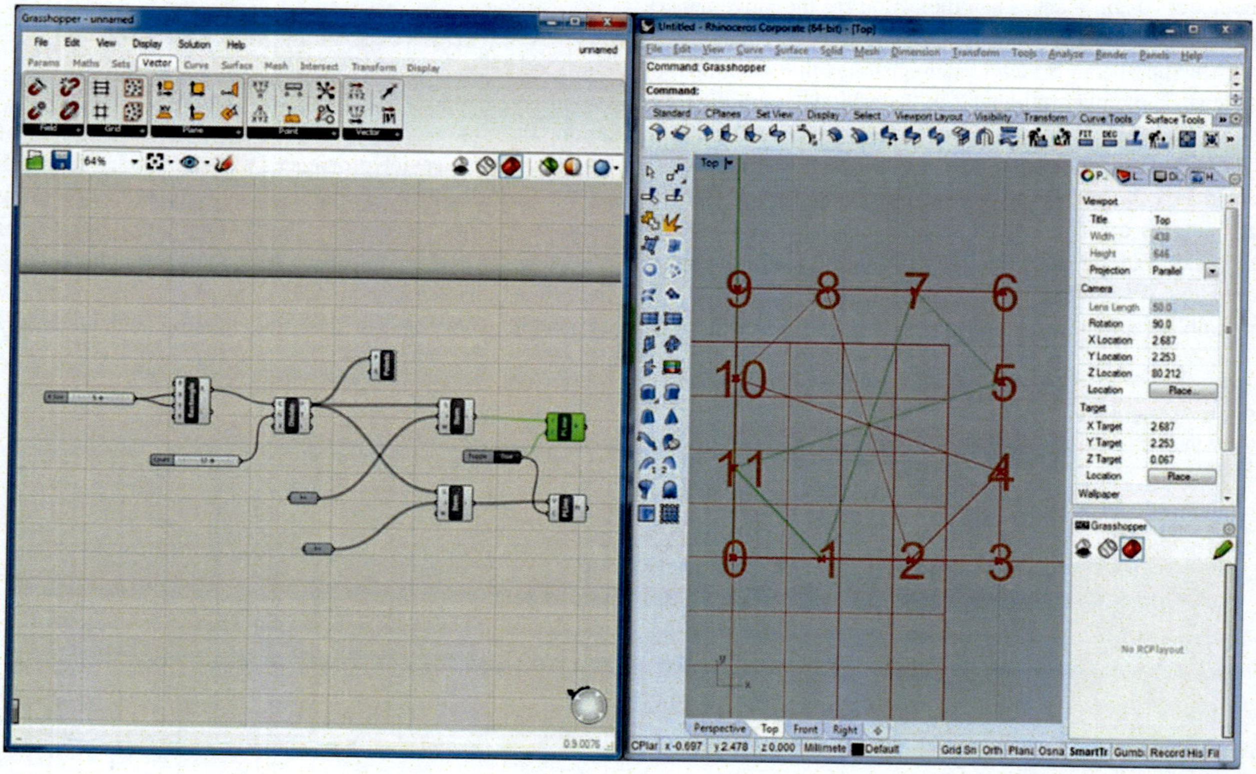

Fig. 10 Generating pattern as an analogy of the rearrangement method of the magic square. *Source* Generated by researcher from grasshopper

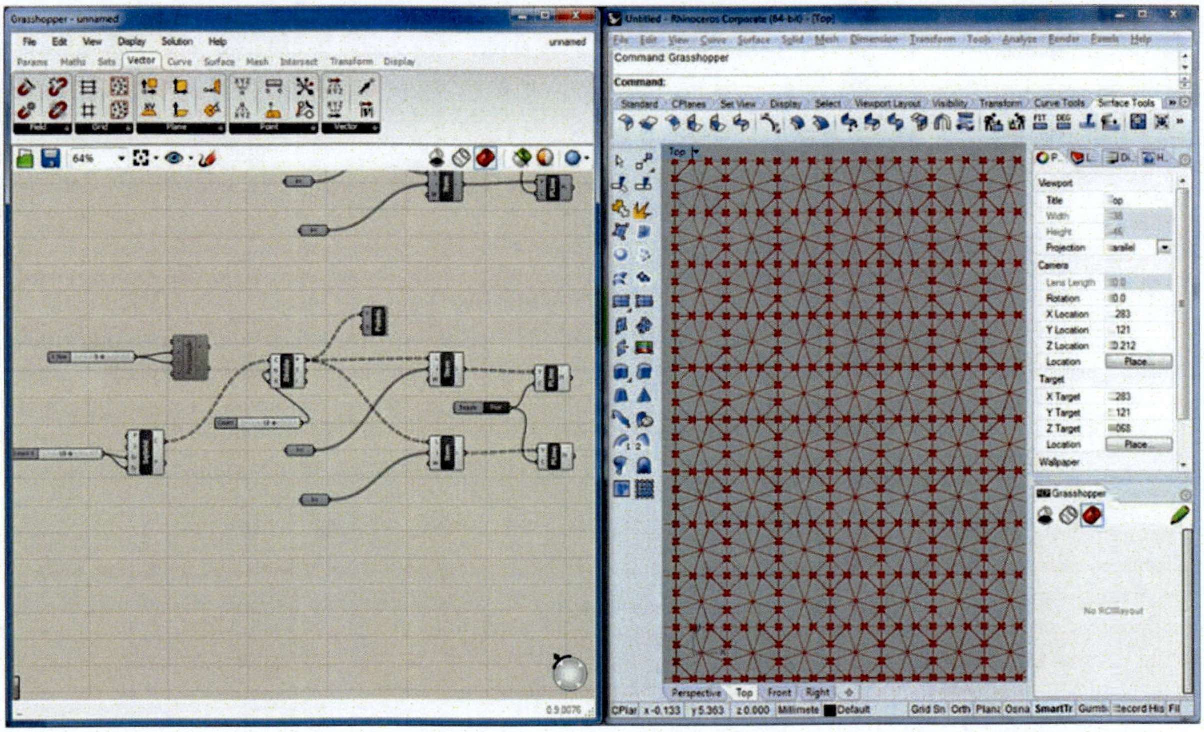

Fig. 11 2D geometric pattern achieved from the repetition process. *Source* Generated by researcher from grasshopper

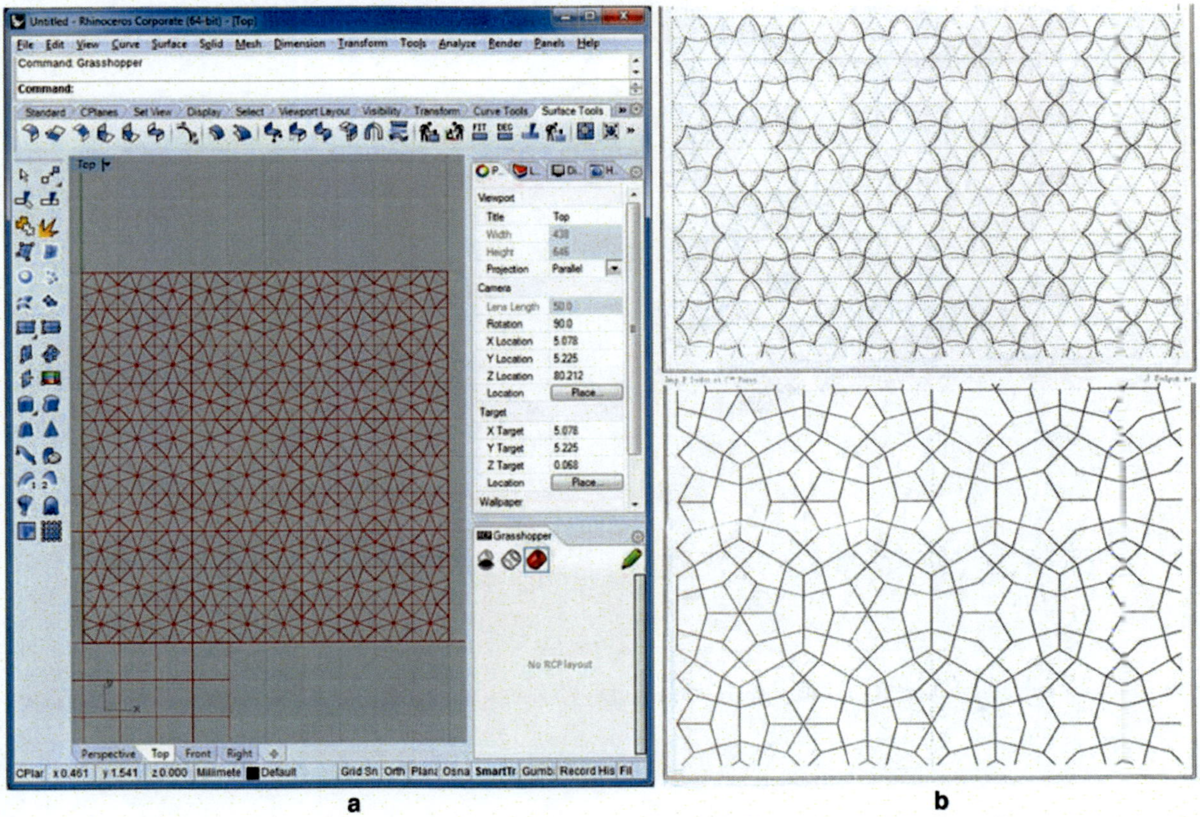

Fig. 12 **a** 2D geometric pattern achieved from the repetition process. *Source* Generated by researcher from grasshopper **b** Islamic geometric patterns resembling the generated grasshopper in basic concept (Bourgoin 1973)

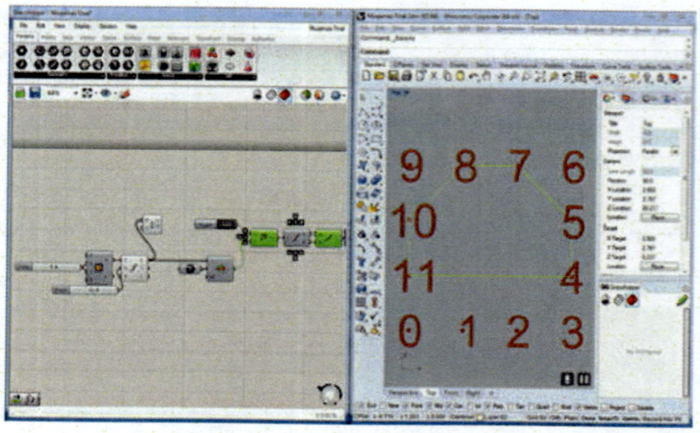

Fig. 13 2D boundary created from the geometric pattern. *Source* Generated by researcher from grasshopper

6 Conclusion

As discussed above, the research's objective was to formulate a parametric generative process for a new muqarnas. The methodology conceived in this paper could be used to generate several new designs of muqarnas forms based on Islamic architecture concepts relating to the cosmological order of the universe and its relation to the beautiful objective nature of geometry. In the paper's attempts to reiterate a muqarnas generation process are deliberations to preserve an architectural element and vocabulary. Allowing the historic architectural element contemporary augmentation, parametric endeavors abbreviate weary original processes of trial and error to efficiently amount to contemporary design processes' aesthetic.

Worth noting, the research alludes to the original design processes of muqarnas, allocating importance to preserved identity of the motif and structural element. Furthermore, the research's methodology employs parametric design and hypothetical geometric patterns in linear or polar arraying directions, producing a clearly defined muqarnas. Amending the research's findings, future studies should address the examination of the structure identified and construction methods of ceiling transformation of architectural spaces, performing structural analysis on the generated form to test its structural limitations. Further application in varying digital softwares might also endow the research's scope and appropriation of the architectural element, motif, and resulting milieu.

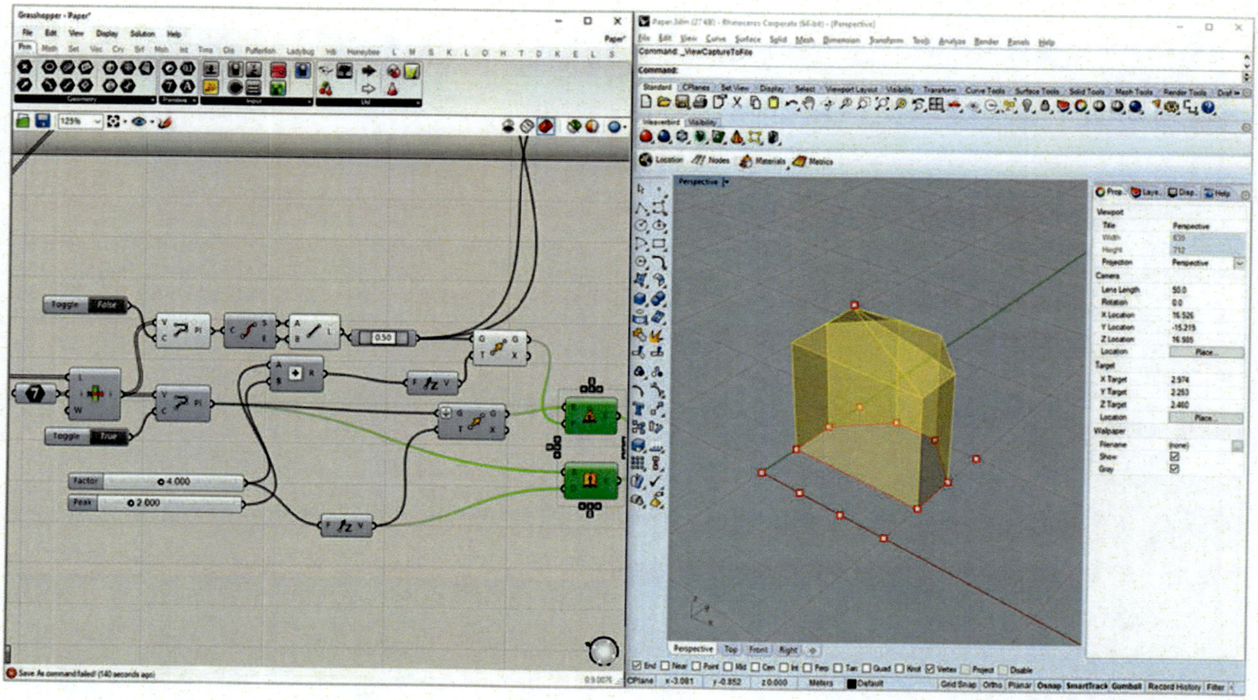

Fig. 14 3D model extruded from the 2D boundary. *Source* Generated by researcher from grasshopper

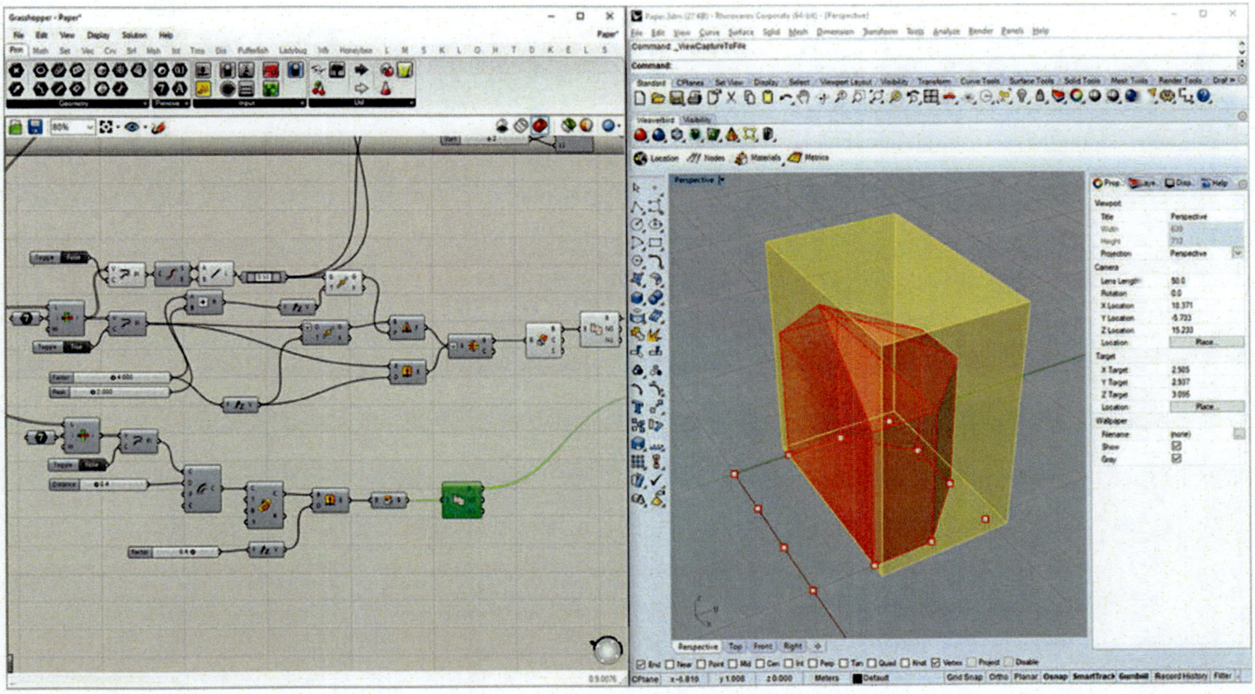

Fig. 15 Cuboid to subtract the model from. *Source* Generated by researcher from grasshopper

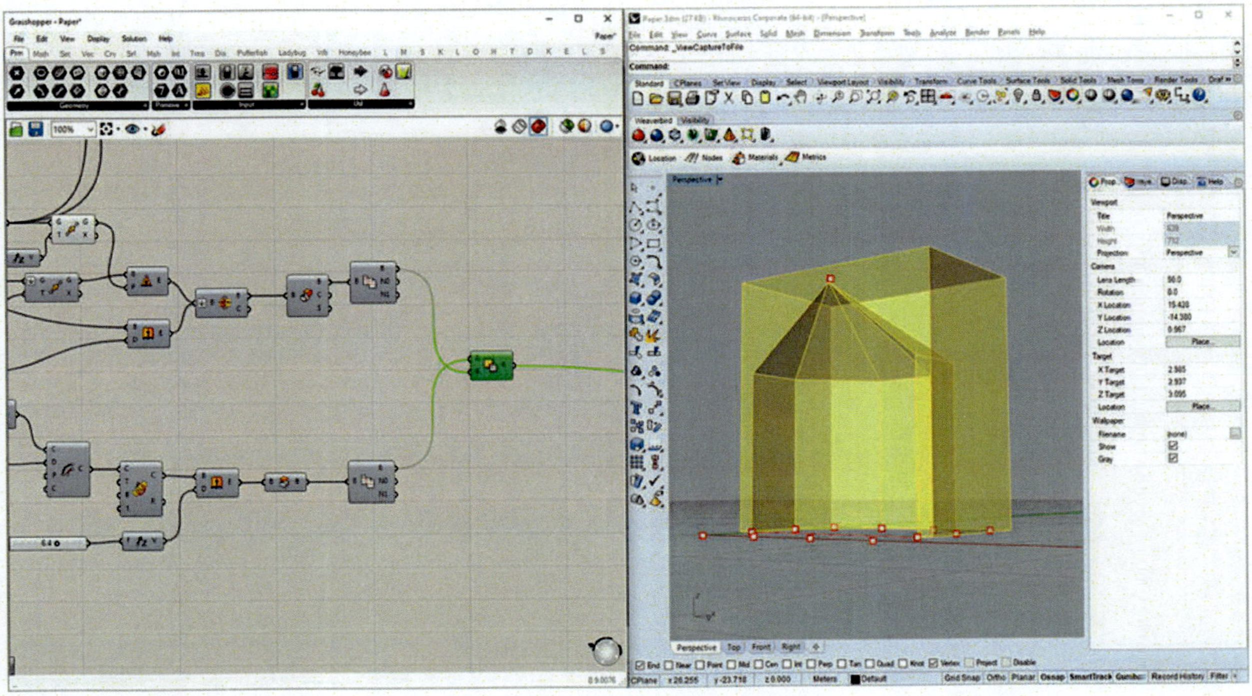

Fig. 16 3D Muqarnas base unit generated after subtraction. *Source* Generated by researcher from grasshopper

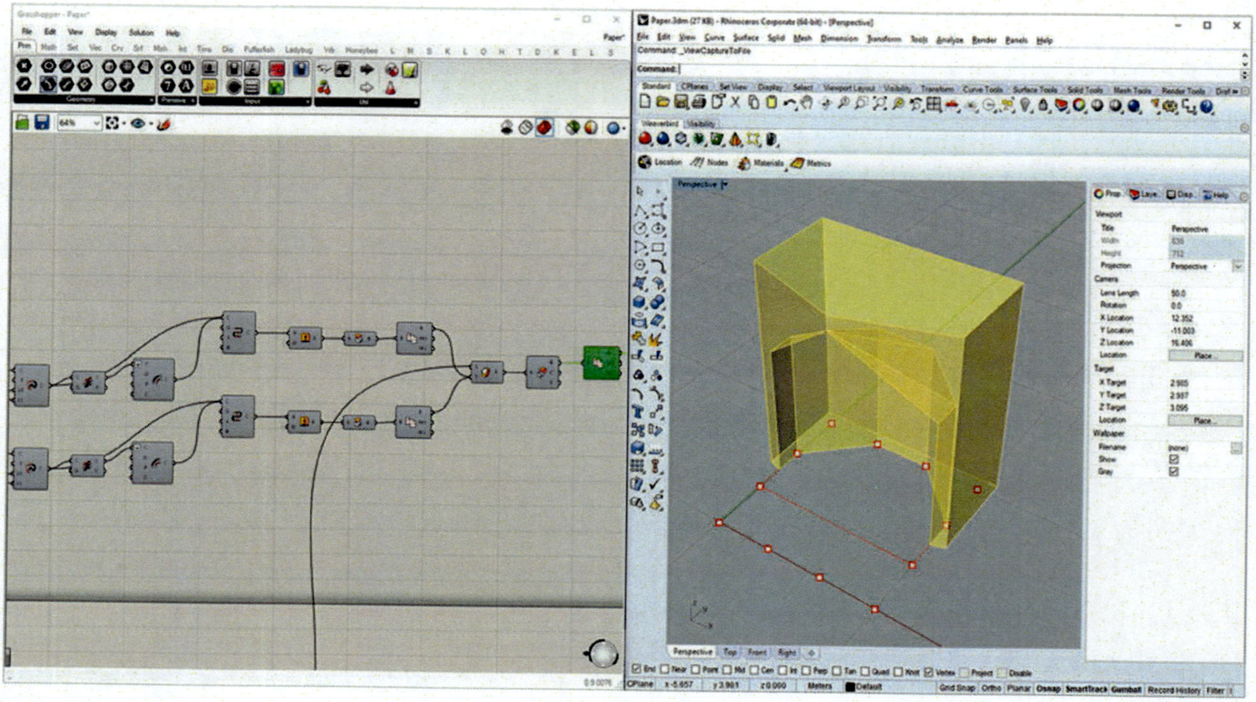

Fig. 17 Solid trim to muqarnas base unit. *Source* Generated by researcher from grasshopper

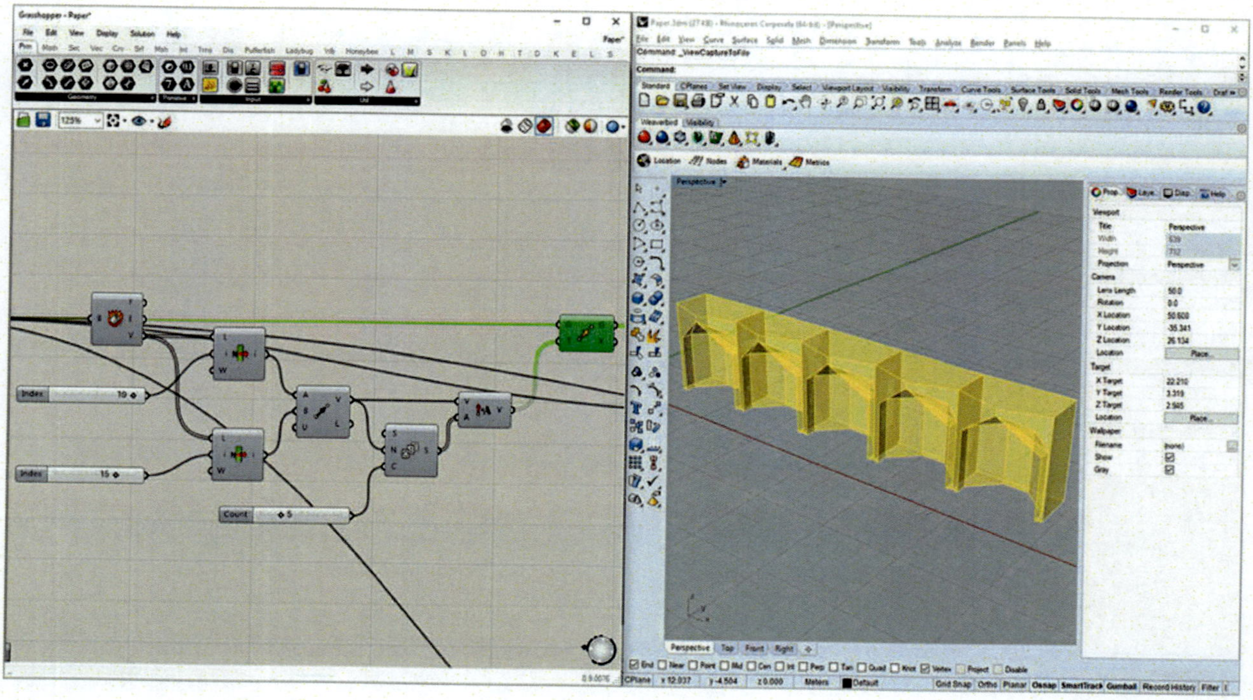

Fig. 18 Arraying first row of muqarnas. *Source* Generated by researcher from grasshopper

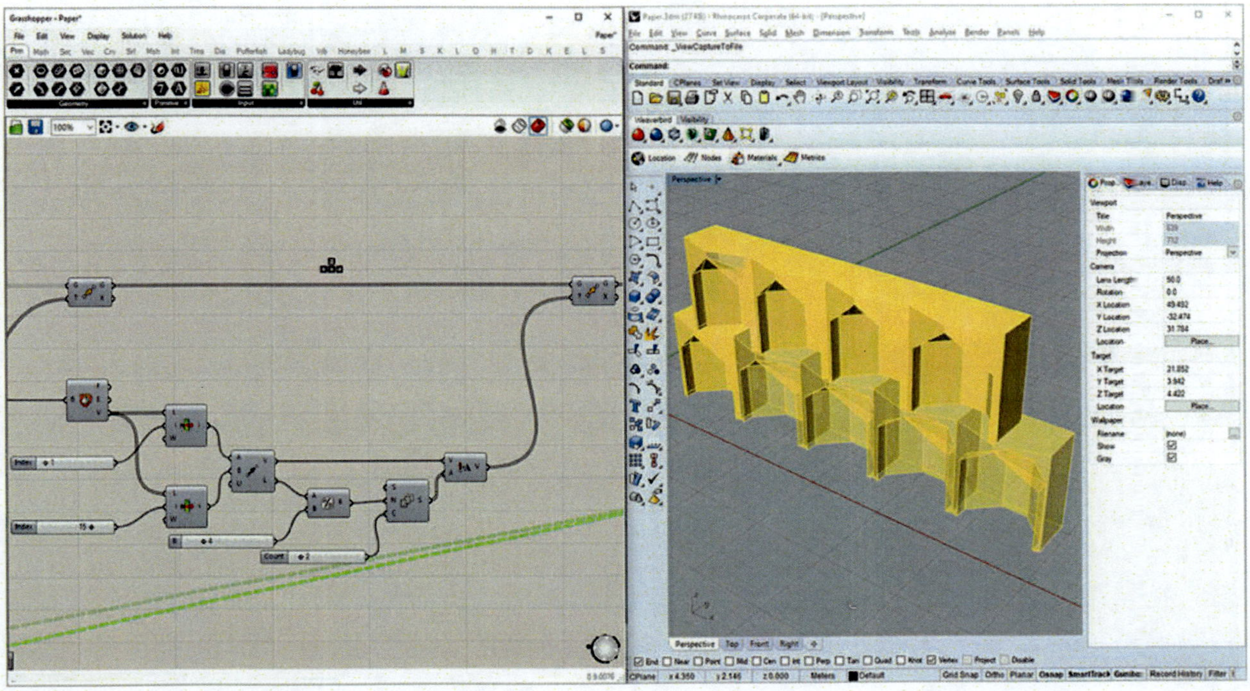

Fig. 19 Arraying top row and displacing it. *Source* Generated by researcher from grasshopper

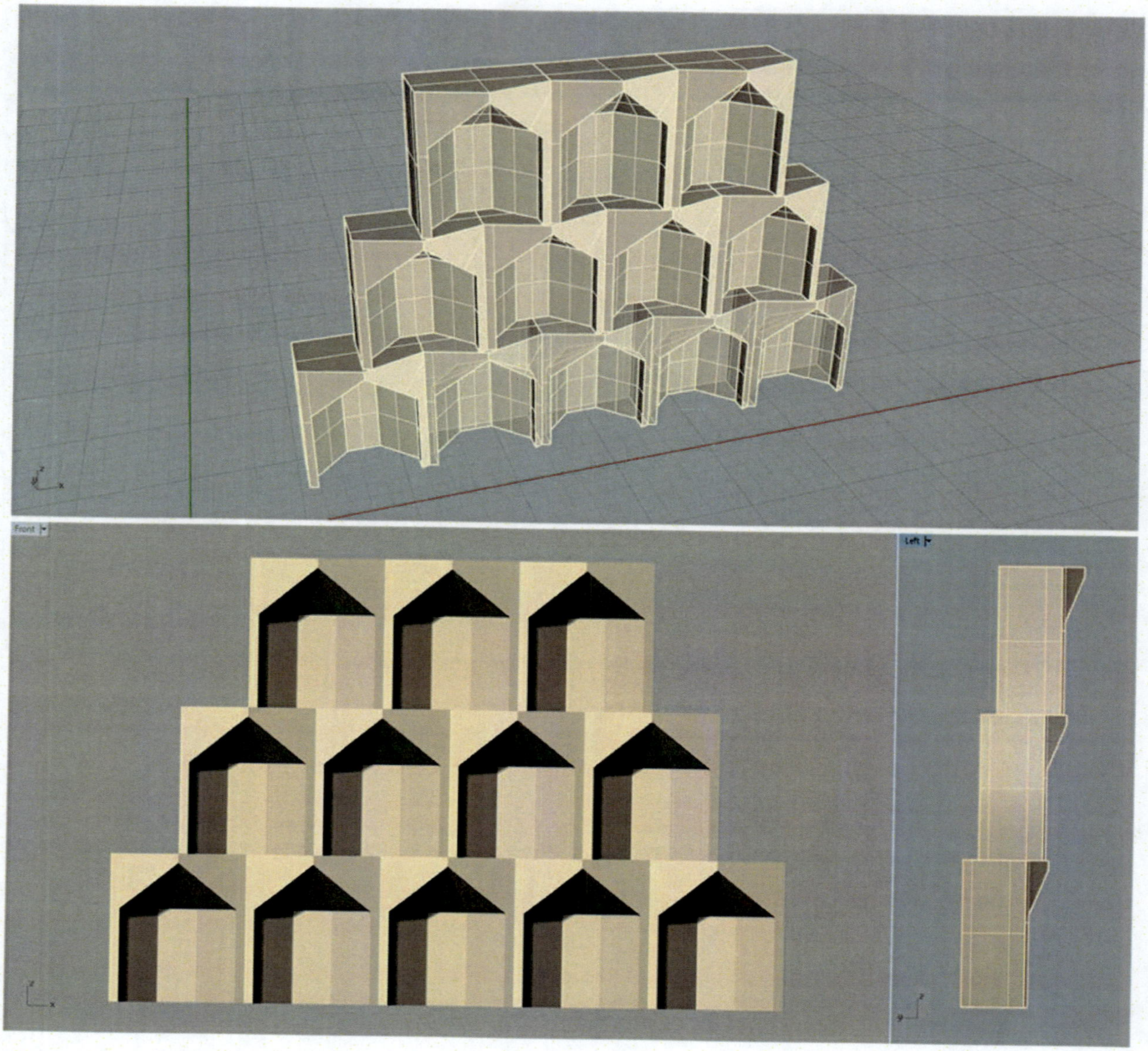

Fig. 20 Generation of Muqarnas. *Source* Generated by researcher from grasshopper

References

Andrews, W.S.: Magic Squares and Cubes. Cosimo Classics, New York (2004)

Bourgoin, J.: Arabic Geometrical Pattern and Design. Dover Publications, New York (1973)

Creswell, K.A.C.: A Short Account of Early Muslim Architecture. Penguin Books, London (1958)

Critchlow, K.: Islamic Patterns: An Analytical and Cosmological Approach. Inner Traditions, Rochester (1999)

Dabbour, L.M.: Geometric proportions: the underlying structure of design process for Islamic geometric patterns. Front. Archit. Res. **1** (4), 380–391 (2012). https://doi.org/10.1016/j.foar.2012.08.005

Dictionary, O.: GEOMETRY|Meaning and Definition for UK English. Oxford University Press, Oxford (1999)

Grube, E.J., Dickie, J., Grabar, O., Sims, E., Lewcock, R.B., Jones, D., Pethenrbridge, G.T., Michell, G.: Architecture of the Islamic World: Its History and Social Meaning: With a Complete Survey of Key Monuments and 758 Illustrations, 112 in colour. Thames and Hudson, London (1995)

Hamekasi, N., Samavati, F., Nasri, A.: Interactive modeling of muqarnas. In: Proceedings of the CAe 199: International Symposium on Computational Aesthetics in Graphics, Visualization, and Imaging (2011). https://doi.org/10.1145/2030441.2030469

Korbendau, Y.: L'architecture Sacrée de l'Islam. Art Creation Realisation, Rueil-Malmaison (1997)

Krautheimer, R.: Early Christian and Byzantine Architecture (The Yale University Press Pelican History of Art). Yale University Press, New Haven (1984)

Skinner, S.: Sacred Geometry: Deciphering the Code. Sterling, Sholamadevi (2009)

Swinford, D.: The Muqarnas: A Key Component of Islamic Architecture. https://www.encyclopedia.com/science/encyclopedias-almanacs-transcripts-and-maps/muqarnas-key-component-islamic-architecture. Encyclopedia.com (2001)

Tedeschi, A., Wirz, F., Andreani, S.: AAD, Algorithms-Aided Design: Parametric Strategies Using Grasshopper. Le Penseur, Paris (2014)

Yaghan, M.A.J.: The Islamic Architectural Element Muqarnas: Definition, Geometrical Analysis, and a Computer Generation System. Phoibos, Vienna (2001)

Cultural Offerings of the Silk Road

Between China and the Mediterranean: Detecting Intercultural Communication Before the Silk Road

Asterios-Evangelos Kechagias

Abstract

From the late second century BCE onwards, when Chinese trade with the West becomes more evident, historical research acquires an important interpretive tool to explain intercultural contacts and influences between Asia and the Mediterranean. The problematic term "Silk Road," which first appeared in the nineteenth century CE, has been used to emphasize these intercultural contacts between East and West. However, things are more difficult for researchers focused on intercultural contacts and influences before the time of the Silk Road, i.e. during the era of the first great civilizations of antiquity. Even in that early period, the exchange of ideas, peoples, and goods is apparent. But their exact communication networks are difficult to detect accurately, for we lack an interpretive tool similar to the "Silk Road(s)." More concretely, ample evidence supports the presence of ongoing communication and trade between the Sumerian and the Indus Valley civilizations or between the Indus Valley Civilization and civilizations of the Iranian Plateau and Central Asia. This study seeks to achieve three key aims: (1) to show that there was extensive mobility of ideas, people and goods between East and West, long before the second century BCE, already present from the 3rd millennium BCE onwards, (2) to show that many geographical locations pivotal during the Silk Road era were already playing an equally important role as early as the 3rd and 2nd millennium BCE and (3) to correct the arbitrariness and the problematic issues inherited from our traditional understanding of the Silk Road. The following presentation will connect all this information, assembling a comprehensive puzzle of ancient intercultural communications. Based on studies such as those of Toby Wilkinson, Philippe Beaujard, and Charles Clifford Lamberg-Karlovsky, this study will try to construct an updated version of our understanding of the Silk Road.

Keywords

Intercultural contacts • Ancient routes • Harappan civilization • Lapis lazuli • Silk road

1 Introduction

The term "Silk Road" famously first appeared in 1877, when the German geologist Ferdinand von Richthofen used the expression "Seidenstrasse/Seidenstrassen" to describe the commercial route(s) passing through the Tarim Basin in Northwest China to Transoxiana, in lower Central Asia (Richthofen 1877, p. 97ff.). As one of the main products transported on this road was silk, a product to this day considered a trademark of China, this road was called the "Silk Road." Conventionally, the Road is said to have emerged in the second century BCE, during the expansion of the Han Dynasty (Liu 2010, p. 4ff.).

However, as is almost always the case, the reality is more complicated. From quite early on, serious critical voices questioned the historical reality of the Silk Road (Whitfield 2007; Rezakhani 2010). Was there ever a Silk Road? At this point, we face a historical and a philosophical problem at once. What exactly does it mean to ask "Was there ever a Silk Road?" If, following a materialistic approach, we consider "matter" to be a necessary component of existence, the existence of such a road depends on a specific, constructed geographical path beginning in China and ending in the Mediterranean, and proving its existence depends on discovering material evidence of that path. In that case, the

A.-E. Kechagias (✉)
Faculty of Theology, Aristotle University of Thessaloniki, Thessaloniki, Greece
e-mail: asterioskechagias@yahoo.com

A.-E. Kechagias
School of Ancient Language and Text Studies, North-West University, Potchefstroom, South Africa

S. Kostopoulou et al. (eds.), *Cities' Vocabularies and the Sustainable Development of the Silkroads*, Advances in Science, Technology & Innovation, https://doi.org/10.1007/978-3-031-31027-0_10

answer is definitely "no." Such a path has never existed. But, if we consider the phenomenon of existence more broadly, as the existence of a coherent idea or a symbol, the answer becomes trickier. We know that, indeed, there was a wider intercultural network of communication between the East and the West,[1] which could symbolically be described as a road or a route. In this light, the question of the existence of the Silk Road seems more philosophical than historical.

The best way to define the Silk Road more accurately is to investigate the core features of this concept. It has two such characteristics: geography and the main product. Concerning geography, the Silk Road was understood as a bridge or road between the East and the West, specifically connecting China with Central Asia, Upper Mesopotamia, Anatolia, and the Mediterranean. Concerning the main product, as the name indicates, it was the *Silk* Road, a road where silk played a central role. However, both of these main features are historically problematic.

In terms of geography, we encounter two problems. First, as Rezakhani rightly points out, the geography of the regions usually included in the Silk Road contains several mountains, mountain ranges, and deserts, and as a result, it precludes an uninterrupted, continuing road (Rezakhani 2010, p. 420). These regions are characterized by geographical patchiness. Second, looking back at the history of the areas that played an active role in the ancient East–West trade, one notices the existence of regions that are unjustifiably excluded from the Silk Road, as it is commonly understood (Rezakhani 2010, pp. 424–425). Where is India? Where is the Persian Gulf? Where is the Arabian Peninsula? Where is the Levant? Where is Africa? We know that all these areas were pivotal for ancient trade and communication (Boivin and Fuller 2009; Fuller and Boivin 2009; Laursen and Steinkeller 2017; Mutin and Lamberg-Karlovsky 2020). Their unjustified absence has led some researchers to suggest that it would be more accurate to use the plural number and to talk about many Silk Roads, instead of a single Silk Road. So, terms such as the "maritime silk road" have emerged (Bellina 2014; Guan 2016).

The second basic characteristic of the Silk Road, the primacy of silk, is no less problematic. From several recent studies, we now know of multiple traces of silk outside China long before the second century BCE. As will be discussed below, these discoveries call into question the established perception of silk as an exclusively Chinese product (Good 2002, 2015; Good et al. 2009). Moreover, we possess several Chinese lists of products traded between China and western countries in which silk is a single item, and sometimes not even the most important one, among many other goods (Schafer 1963, p. 28ff.).

In short, both characteristic features of our common image of the Silk Road evince arbitrariness and asymmetry. First, we invent a road that connects the East with the West, emphasizing some geographical points while arbitrarily excluding others which also play a crucial role. Next, we arbitrarily name this road after a product that was certainly neither the only one nor perhaps the main one that was traded. Finally, quite as arbitrarily, we date this road to the second century BCE, whereas the East–West relations and the existence and circulation of silk began long before. We are faced with two options. Either we abandon the "Silk Road" as a term referring to trade between the East and the West, or, to allay the arbitrariness, we modify its content and definition. For reasons that will be explained later, the second option is preferable. To carry it out, we ought to first date this Road more accurately. That is, we must more carefully discern when the most extensive contact between the East and West began and when silk first appeared as a staple.

For reasons of convenience, to track contacts between the East and West, we may split the traditional Silk Road into smaller geographical portions and examine each separately. Since the classical conception of the Silk Road refers (approximately) to the connection between China and the Mediterranean, the following quadripartite division can be made:

1. From China to Central Asia.
2. From Central Asia to South Asia.
3. From South Asia to Mesopotamia.
4. From Mesopotamia to the Mediterranean.

This division is made based on one main criterion, viz. adequate evidence of intercultural contacts. More concretely, only those geographic locations from which we have sufficient evidence of intercultural communication were included. Connections of peoples or geographical areas that remain enigmatic or need special research to be demonstrated were not included in the above division. For instance, we have ample evidence of the relations between ancient China and Central Asia, justifying the inclusion of the first section. On the other hand, we possess little evidence of direct relations between China and India, at least in very early times, and so, such a section will not be discussed. This means that this study will not beg the question in favor of our traditional idea of the Silk Road. Rather, it will be driven by the available data. It will also try to include those geographical territories which are usually arbitrarily omitted, thus correcting the previously arbitrary selection of geographical points.

[1] The terms "East" and "West" are commonly used to denote the cultures of Asia and the cultures of Europe, respectively. That is not how I use them here. By "East," I mean primarily China and its environs, while I use "West" to denote anything west of China, starting in Central Asia and reaching as far as the Mediterranean.

In addition, to provide less random dating for intercultural contacts between East and West, which is one of the goals of the present study, dating will also be based on the evidence available. So, if we find evidence that India had been in contact with Mesopotamia and more western territories since the 3rd millennium BCE, this means that we should start our research from that point in time at which the first extensive examples of intercultural contacts emerge. However, it is important to note that if we are to provide a more accurate updated picture of the Silk Road, the date chosen should be as widely applicable as possible. That is, it should be a common feature of the histories and activities of the peoples who will be discussed. There would have been little use in presenting incidental connections between peoples communicating with each other at different times. It would not make sense to "construct" a Silk Road of peoples who did not communicate with each other at the same time.

Finally, once the research on ancient intercultural contacts is established and we arrive at a date that is as reliable as possible, we will consider whether these contacts are related to silk, justifying talk of a Silk Road, or whether the new version of this "road" does not justify that arbitrary product choice. More simply, the following pages will suggest an updated version of the Silk Road, correcting arbitrariness and asymmetries.

2 From China to Central Asia

Starting at the furthest east and heading west along the Silk Road, we first stop to explore ancient China's relations with Central Asia. As Shuicheng Li observes, "Northwestern China, a region between Central Asia and the Central Plains of China, is of crucial importance to the exploration of cultural interaction between East and West" (Li 2002, pp. 172–173). In his recent study, *The Worlds of the Indian Ocean*, Philippe Beaujard cites a series of archaeological findings which attest to China's relations with its west and north as early as the 4th millennium BCE (Beaujard 2019, p. 198ff.). It is well-known that from 4000 until 3000 BCE, the Yangshao culture, the main culture of the Shaanxi region in northwestern China, gradually expanded westwards (Li 2002, p. 173). The findings from Shaanxi include ceramic figurines of bearded men with high hats and long noses, undoubtedly representing westerners (Beaujard 2019, p. 198). Throughout the 3rd millennium BCE, this westward movement was continued by the Majiayao culture in the Upper Yellow River region (Beaujard 2019, p. 202), while almost simultaneously, in eastern China, we encounter traces of a specific variety of domesticated wheat, which probably arrived from regions of modern Pakistan (Crawford 2006, p. 79; Stevens 2016, p. 1543). About the same period, sheep and taurine cattle were introduced to the region from western Asia (Beaujard 2019, p. 202; Stevens 2016, p. 1544). A similar and equally important discovery included a series of mace-heads that have been excavated across China, dating primarily from the 4th but also the late 3rd and early 2nd millennia BCE, during the Qijia culture (Li 2002, p. 181). Since mace-heads had previously spread largely throughout western Asia and Iran, they undoubtedly reached China through Central Asia (Beaujard 2019, p. 202).

Relations between China and Central Asia developed further from the late 3rd millennium BCE onwards, coinciding with the flourishing of the Bactria-Margiana Archaeological Complex (BMAC) in modern Turkmenistan. During that period, several mixed populations became established along the Gansu corridor, the most important corridor for intercultural contact between China and Central Asia (Beaujard 2019, p. 204). Koryakova and Kohl noticed that the bronze and copper found in western China from the early 2nd millennium BCE present many similarities with the bronze and copper from Central Asian regions, like the ones of the Andronovo culture (Koryakova and Kohl 2000, p. 641). Additionally, the discoveries at Lin-Ya cemetery, where the skulls of several people with Caucasoid characteristics have been unearthed, confirm that western Gansu was a critical region for the development of the initial cultural interactions between the East and the West (Li 2002, p. 179). During the time of the Siba Culture (1950–1550 BCE), new bronze forms emerged. Specifically, the use of arsenical copper became dominant (Li 2002, p. 180), a material that was already widely and extensively used in the Near East and Central Asia during the 3rd millennium BCE (Li 2002, p. 180).

So far, only archaeological finds of objects that have reached China from the West have been reported. But goods also arrived in the West from China. The most typical example is millet, which was first domesticated in China. Chinese millet reached Central Asia, especially present-day eastern Kazakhstan, in the late 3rd millennium BCE (Spengler 2015, p. 216; Stevens 2016, p. 1545). Soon after, before the mid-2nd millennium BCE, broomcorn millet from Central Asia spread to modern Afghanistan, southern Pakistan, and India, before finally reaching territories as far as Iraq, Turkey, Syria, Yemen, and Sudan via the Persian Gulf (Stevens 2016, p. 1546). In addition to millet, many other products which are speculated to have been domesticated in the wider region of China spread westward, into Central Asia, India, and Pakistan, at the beginning of the 2nd millennium BCE. These include apricots, peaches, hemp, and rice (Stevens 2016, p. 1546ff.).

From this historical evidence we may draw the conclusion that China's relations with Central Asia started very early, as early as the 3rd millennium BCE. However, these relations became more extensive towards the end of the 3rd millennium and the beginning of the 2nd millennium BCE.

The example of domesticated millet is characteristic of this East–West interaction: we see it starting in China and then passing through Central Asia to parts of South Asia, Anatolia, Africa, and the Mediterranean. Of note, too, are Lü Qin-Qin's recent findings of natron glass beads from the Mediterranean, excavated in territories of western China. However, these beads are dated to the early 1st millennium BCE (Lü et al. 2021, pp. 6–7).

3 From Central Asia to South Asia

The picture of the relations between Central and South Asia, to which we now turn, is substantially clearer. Numerous archaeological findings attest to the relations between Central Asia and regions of present-day Afghanistan, Iran, and India from the middle to late 3rd millennium BCE. This is the era of some of the most important known ancient civilizations. These include the Oxus Civilization or BMAC in today's southern Turkmenistan and northern Afghanistan, the Elamites in western Iran, the Indus Valley Civilization in western India and present-day Pakistan, and the Sumerians in southeastern Iraq.

Lamberg-Karlovsky, a leading scholar on the Oxus Civilization, presents a series of BMAC findings, like statuary, seals, and other objects, from several locations across Iran (Tepe Hissar, Tepe Yahya, Shahdad, Jiroft, Susa) and India (Mohenjo Daro, Harappa), in addition to Indian seals and inscriptions that have been unearthed in Turkmenistan and northern Afghanistan (Lamberg-Karlovsky 2013, p. 24, 41). Henri-Paul Francfort describes Elamite inscriptions unearthed in Central Asia (Francfort 2005, p. 284). Shereen Ratnagar provides a detailed list of products and raw materials, including lapis lazuli, turquoise, and ivory, among others, which were transported from Central Asia to South Asia and vice versa (Ratnagar 2020, p. 592ff.). As Lamberg-Karlovsky remarks, "there can be little doubt that the BMAC influenced the indigenous cultures of the Iranian Plateau and the Indus Civilization, while in complimentary fashion numerous artifacts of the Indus and Iranian plateau are known from Gonur Depe" (Lamberg-Karlovsky 2013, p. 24). Findings representative of the BMAC have been unearthed as far as the Persian Gulf, most notably in Bahrain, among the remains of the ancient Tilmun[2] civilization. Conversely, Bahrain's stylistic artifacts, such as stamp seals and chlorite containers, have been discovered in Central Asia and Iran (Lombard 2020, pp. 609–615). A more recent study by Lamberg-Karlovsky and Benjamin Mutin informs us of several BMAC remains recorded in northern and southeastern Iran, especially at Shahdad, and in the Indo-Iranian borderlands (Mutin and Lamberg-Karlovsky 2020, p. 552, 554). However, the most significant note about our present study is the expert consensus that the vast majority of these sites date from the end of the 3rd to the beginning of the 2nd millennium BCE (Hiebert 1994, pp. 163–164; Francfort 2005, pp. 260–262; Mutin and Lamberg-Karlovsky 2020, p. 553).

Despite the fact that South Asia's contacts with Mesopotamia, which will be discussed below, are well documented, Central Asia's contacts with Mesopotamia present several problems. One of them is that, although Mesopotamian material has been found in the BMAC (including a seal with an Akkadian inscription, which is regarded as the most eastern testimony of cuneiform we possess), BMAC finds appear absent from Mesopotamian sites (Lamberg-Karlovsky 2013, p. 25; Guichard 2020, p. 66). Guichard addresses the problem as follows:

> The great distance between Margiana and Mesopotamia is not in itself an obstacle for movements of goods and men. The greatest difficulties were neither technical nor stemming from topographic impediments such as mountains or deserts. Camels were fundamental for the transport of goods through Margiana and Mesopotamia and they were already domesticated by the end of the 3rd millennium... Real barriers were largely human and political (Guichard 2020, p. 66).

Another problem with Mesopotamia and Central Asia is due to the controversy over how the Mesopotamians called the BMAC. Various solutions have been proposed, like Shimaski, Marhashi, or Tukrish, among others, but no consensus has been reached (Potts 2008; Francfort and Tremblay 2010; Fuchs 2014).

On the subject of camels, it is noteworthy that, although we cannot be sure of the exact chronology of their domestication, camels seem to have spread extensively by the end of the 3rd millennium BCE, reaching as far as Oman and Bahrain (Wilkinson 2014, p. 50). Steinkeller identifies a textual reference that attests that the Third Dynasty of Ur (UR III) possessed a herd of camels, which were probably Bactrian camels (Steinkeller 2009, p. 416). Similarly, though horses had been imported to the Near East during the 4th millennium BCE, it is only in the late 3rd and early 2nd millennium that we encounter traces of horses in Central Asia, Iran, and India (Wilkinson 2014, pp. 49–50).

This very extensive communication between the cultures of Central and South Asia at the end of the 3rd millennium BCE led Gregory Possehl to coin the term "Middle Asian Interaction Sphere (MAIS)" in order to capture the vast extent of these intercultural contacts (Possehl 2002,

[2] The exact transcription of the ancient Sumerian name of the Bahraini civilization is not certain. Most scholars prefer "Dilmun." I prefer "Tilmun", because, in later Greek sources, such as in Arrian, we find Bahrain called "Tylos," with "t" (see Arrian, *The Anabasis of Alexander*, 7.20.5–7.20.6). Although this is certainly not proof of the ancient pronunciation, I see no reason to deviate from these sources, which offer a sounding equivalent of the name.

p. 215ff.; Possehl 2007, pp. 40–42). The example of the spread of millet, to which I referred earlier, is by itself proof of the existence of this wide network of communication between Central Asia and its neighbors to the south and west. Commenting on Possehl's term, Lamberg-Karlovsky notes: "the constituents of the term MAIS (Central Asia, Eurasian Steppes, Indus Valley, Persian Gulf, Iranian Plateau) had extensive relations, political and economic with Mesopotamia, which in turn, had continuous interactions throughout the Bronze Age with the eastern Mediterranean" (Lamberg-Karlovsky 2013, p. 48). In the same vein, Prof. Massimo Vidale emphasizes that "the increasing interaction of the Oxus with the Indus worlds, in the last centuries of the 3rd millennium, was part of a more generalized intensification of cultural contacts and interregional links. Some scholars have recognized in this process the opening of a prehistoric Silk Road, controlled at its western end by Oxus merchants and political leaders" (Vidale 2017, p. 56).

4 From South Asia to Mesopotamia

As with the relations between Central and South Asia, we face no particular problems establishing the relations between South Asia (mainly India and Iran) and Mesopotamia. Their existence has been repeatedly confirmed. They began in the middle of the 3rd millennium BCE and continued until the first centuries of the 2nd millennium, which witnessed the almost simultaneous collapse of several great civilizations of South Asia and Mesopotamia, like the Sumerians and the Indus Valley Civilization.

Our understanding of Mesopotamia's relations with the wider East changed radically in 1925 when the British archaeologist Ernest MacKay published his article "Sumerian Connexions with Ancient India." In this article, MacKay described Indian-style seals found during excavations in the city of Kish in Mesopotamia, dating to before 2000 BCE (MacKay 1925, p. 697). The issue escalated when similar seals were later found elsewhere in Mesopotamia, in Lagash, Eshnunna, and Umma, and also in Susa in southwestern Iran (Gadd 1932, pp. 4–5). When, in 1932, the British archaeologist Cyril John Gadd published his article "Seals of Ancient Indian Style Found at Ur," in which he identified the existence of Indian seals as far south in Mesopotamia as Ur, archaeologists began to consider the possibility that the two large civilizations communicated not by land but by sea. Two decades later, in 1954, Leo Oppenheim brought to light textual reports proving the constant communication of the city of Ur in southern Mesopotamia with merchants and merchant ships from the Persian Gulf, primarily from Bahrain (Oppenheim 1954, p. 7). A decade after the study of Oppenheim, the Indian archaeologist Shikaripura Ranganatha Rao proved, in his article "A Persian Gulf Seal from Lothal," the existence of trade between the Harappan civilization and the Persian Gulf (Rao 1963, pp. 96–99). Taking the next step by proving the existence of trade between the Sumerians and the Indus Valley Civilization (probably called "Meluḫḫa" by the Sumerians) was not very difficult: several Indus-style findings have been discovered in Mesopotamia (Marshall 1931, pp. 385–389; During-Caspers 1973, p. 6ff.; During-Caspers 1992, p. 6ff.; Possehl 1996, p. 137; Laursen and Steinkeller 2017).

The same conclusions can be drawn by investigating the relations between Iran and Mesopotamia. There is ample textual and material evidence of Mesopotamian contact with the Elamites and other Iranian groups (Potts 1999, p. 380ff.). The best proof of these relations is the diffusion of lapis lazuli. Lapis lazuli is undoubtedly the most famous kind of precious stone used during the 3rd and 2nd millennia BCE (Wilkinson 2014, p. 125). Luckily, there seems to be only one main source from which the Indians, the Iranians, and the Mesopotamians imported lapis lazuli: the deposits of Sar-i Sang in Badakhshan, modern northeastern Afghanistan (Herrmann 1968, p. 21; Majidzadeh 1982, p. 59; Francfort 1984, p. 302; Wilkinson 2014, p. 125). Lapis lazuli was distributed throughout an area stretching from Afghanistan to Iran, India, Mesopotamia, Egypt, and the Aegean. Since the only source of this mineral was in Afghanistan, lapis lazuli is a very clear and important sign of intercultural exchange and commerce (Herrmann 1968, p. 53; Tosi and Vidale 1990, p. 98; Wilkinson 2014, p. 126). The Mesopotamians imported the stone from the Indus Valley or directly from Iran, especially from its northwestern part (Majidzadeh 1982, p. 59). A few kilometers from the location of the lapis lazuli mines, at Shortughai, there stood a Harappan settlement, dated to the end of the 3rd millennium BCE. The settlement's artifacts undoubtedly originated in the Indus Valley (Dupree 1981, p. 108; Francfort 1984, p. 303). Shortughai was, in the words of Dupree, "a manufacturing and commercial center, one of the many links which connected the luxury trade route which thrived in the 3rd–2nd millenniums BC, stretching from the eastern end of the Mediterranean to Central Asia (and China?)" (Dupree 1981, p. 108).

Modern research has proposed at least five different possible ways of transporting lapis lazuli and other goods. Wilkinson summarizes these as follows[3]:

1. A mostly maritime route: across the passes of the Afghan hills to the Indus and then via coastal or maritime transport up the Persian/Arabian Gulf;

[3] Majidzadeh proposes three ways. For these, see Majidzadeh 1982, 59–61.

2. The southern route: through Baluchistan and/or Seistan along inland or coastal southern Iran;
3. The former "Khorasan road": through Bactria and/or Khorasan, probably south of the Elburz across central Iran into north–west Mesopotamia;
4. The Caspian route: again through Bactria, either to the south of the Karakum desert, around the Caspian and then southwards, or else westwards into the Black Sea or beyond;
5. The steppe route: an alternative, very long and rarely-considered connection from Bactria north along the Oxus to Transaralia and Transcaspia across the southern Russian steppes to the Caucasus or the Black Sea. (Wilkinson 2014, p. 130)

Remarkably, two millennia later, we find that Darius the Great declared that the Persian Empire imported its lapis lazuli (kâsaka hya kapautaka) and carnelians (sikabruš) from Sogdiana, that is, from about the same areas of Afghanistan and Central Asia.[4]

5 From Mesopotamia to the Mediterranean

Having connected China with Central Asia, Central Asia with Iran and India, and the last two with Mesopotamia, connecting Mesopotamia with the wider Mediterranean does not seem so difficult. First, there is sufficient evidence from the early 2nd millennium of Assyrian trade outposts in Anatolia, like in Kanesh in today's Turkey. As Michel observes: "At the end of the 20th and beginning of the nineteenth centuries BCE, Assyrian merchants from Assur developed a long-distance trade in Central Anatolia and settled there progressively in some 40 towns… The center of their operations was located at Kanesh (Kültepe), where, according to written records dated to the first half of the nineteenth century, several hundreds of them lived more or less permanently" (Michel et al. 2014, p. 69). From there to the shores of the Mediterranean in Asia Minor, the route was neither long nor difficult.

Moreover, textual and archaeological evidence supports the existence of a road that, according to written sources like the biblical Numbers 20:17, passed from Mesopotamia to Egypt crossing modern-day Jordan. This road was called the "King's Highway." Although we cannot be sure of the historicity of this road, or of when and how it was "constructed," Fawzi Abudanah, in his recent study on the ancient roads of Jordan, has noted ample relevant evidence of ancient roads which crossed Jordan reaching more northern areas (Abudanah et al. 2015, p. 158, 183–184). Among ancient roads, we should not omit a reference to the so-called "Royal Road" mentioned by Herodotus and other late Achaemenian sources (see Herodotus, *Histories*, 5:52). According to Herodotus, this road started at Sardis in western Asia Minor and reached Susa in southwestern Iran. Herodotus even informs us of various stations (σταθμοί) along the route. Wilkinson must be right in his thought that the Persians' construction of this road must have been based on a pre-existing road that connected at least Iran with Anatolia (Wilkinson 2014, p. 78).

At first sight, more direct contact between ancient Mesopotamia and Egypt as far back as the 3rd millennium BCE seems difficult to discern. Indeed, Leemans is skeptical of such early contacts between these two remote cultures (Leemans 1960, p. 36). Nevertheless, Luc Watrin presents fascinating evidence for a connection between Mesopotamia and Egypt as early as the 4th millennium BCE. Watrin divides these contacts into two phases, corresponding to the Late Chalcolithic period 4 (3550–3300 BCE) and the early Late Chalcolithic period 5 (3300–3150 BCE). During the first phase, the intercultural contacts are reflected by the introduction of precious stones like lapis lazuli and obsidian in Egypt, but also by the discoveries of ceramic containers typical of Uruk in Mesopotamia (Watrin 2004–2005, pp. 81–82). The second phase involves primarily the introduction of Mesopotamian motifs on weapons possessed by pharaohs of Upper Egypt (Watrin 2004–2005, p. 82). Payne (1968, p. 58) emphasizes that there is "a striking resemblance between the use of lapis lazuli in Iraq and that in Egypt", and Leemans (1960, p. 36) seems very confident that Egyptian lapis lazuli came from Mesopotamia. The finding of lapis lazuli from Badakhshan in Egyptian territories is in itself an indication of the (at least indirect) contacts of the Egyptian culture with those of Central Asia and Mesopotamia. However, as Watrin (2004–2005, p. 82) notes, these contacts were "unidirectional," from Mesopotamia to Egypt. The absence of Egyptian material in the Mesopotamian archaeological record means that, at least in this early period, communication between these two peoples was carried out through intermediaries like Levantine merchants (Watrin 2004–2005, p. 82). Contacts between Mesopotamia and Egypt apparently did not end in the 4th millennium BCE, but the chronological point that interests us, the end of the 3rd millennium BCE, unfortunately, coincides with the enigmatic and dim First Intermediate Period, about which one can find more information in the work of Thomas Schneider (Schneider 2017).

[4] Darius the Great, Susa DSf, 37–38: "*kâsaka hya kapautaka utâ sikabruš hya idâ karta hauv hacâ Sugudâ.*" Translation: "The stone which is blue and the carnelian which was wrought here, this (was brought) from Sogdiana." The translation is mine, following transliteration and notes of Roland Kent (Kent, 1953, pp. 143–144). The "stone which is blue" is undoubtedly lapis lazuli. We encounter the same logic in Egypt, where the word for the lapis lazuli (xsbD) denotes simultaneously the blue color.

While we cannot speak with certainty about Egypt's contacts with Mesopotamia in the late 3rd millennium BCE, we have greater certainty about East Africa's relations with the Arabian Peninsula and India from the same period. In a previous section, I referred to the spread of millet from Central Asia and India via the Persian Gulf to Yemen and Africa during the early 2nd millennium BCE. At precisely the same time, sorghum, pearl millet, and hyacinth bean are observed to have spread in the opposite direction, from Northeast Africa to parts of India (Boivin and Fuller 2009; Fuller and Boivin 2009; Stevens 2016, p. 1546). As Stevens stresses, the fact of this diffusion is based "on secure finds" (Stevens 2016, p. 1546). This observation is important, because, so far, the study has been concerned with the connection of northern regions with each other. Now, we see that the Arabian Peninsula and Africa can be added to the pattern of intercultural contacts during the early 2nd millennium BCE.

The involvement of the Persian Gulf in this patchwork of intercultural relations does not stop there but extends to Syria and Anatolia. Briggs Buchanan observes that, after the collapse of the Indus Valley Civilization in the early 2nd millennium BCE, representations of horned bullhead known as *bucranium*, which was very common in earlier periods throughout the Persian Gulf, disappeared almost entirely from the iconography. However, the *bucranium* reappears in Syria and Cappadocia in the nineteenth century BCE and remains there throughout the 2nd millennium. As he argues, "it seems possible that around 2000 B.C. the Persian Gulf merchants had a relationship, other than one involving the trade, with some ethnic elements in Syria (merchants or colonists)" (Buchanan 1965, p. 207). Poul Kjærum compared early 2nd millennium BCE seals found in Bahrain with seals from the wider region of Lebanon and Syria. In the same vein, he concluded that the Bahrain-Tilmun seals bear a "strong ethnic element of western origin," and that the depictions of the motifs prove that they must have served the same ritual goal in both geographical locations. He concludes: "This influence is so strong and of such a nature that it cannot solely be due to relations like trade and sporadic personal contacts of various kinds" (Kjærum 2010, p. 269, 272, 275). Kjærum's article offers many examples of seals strikingly similar to those of the Mediterranean or of Syria and Cappadocia. For example, seals depicting Aquarius together with astral symbols have been unearthed in Bahrain and Failaka (Kjærum 2010, pp. 269–275).

This subchapter will close with an even more provocative scenario: the possibility of contact between the Indus Valley Civilization and the Mediterranean Cretan culture. Although it is well known that Aegean cultures like the Minoan had contacts with parts of the Near East such as Mari (Guichard 1999, p. 165ff), its possible relations with India during the Bronze Age have escaped the attention of most researchers. This possibility, however, was hypothesized by scholars like During-Caspers, a prominent Dutch archaeologist and scholar of Sumerian and Indus Valley archaeology. It was prompted by some bizarre sealing amulets and especially some prism seals, unearthed by excavations at Harappa, Mohenjo-Daro, and Chanhu-Daro (During-Caspers 1985, p. 435). According to During-Caspers (1985, p. 443), the decorations of these prisms are "neither Indian nor Near-Eastern in origin". After careful study, she concludes that these prisms are strongly reminiscent of the square prisms of Early Minoan Crete (During-Caspers 1985, p. 451). During-Caspers' observations were not taken much into account by her contemporary colleagues because until the beginning of the twenty-first century the dominant view of most researchers was that Indo-Mediterranean relations began in the sixth century BCE and onwards. Nevertheless, in the last twenty years a number of new studies have emerged that offer much more evidence in favour of an Indo-Mediterranean connection as early as the Bronze Age. Geoffrey Ludvik, Mark Kenoyer, Magda Pieniążek and William Aylward recently brought to the fore specimens of Indus style carnelian beads undoubtedly originated in India which were found in various parts of the Aegean and the eastern Mediterranean (Ludvik et al. 2015; Ludvik 2018, pp. 550–567; Ludvik et al. 2020). In the same spirit, Marie Nicole Pareja and her colleagues argued that the famous iconography of monkeys in Room 6 of the Building Complex Beta at Akrotiri in the Greek island of Thera, dated before the seventeenth century BCE, depicts monkeys originating from India and even from places where the Indus Valley Civilization was active (Pareja et al. 2020). Prof. Arnott (2022) from the Oxford University in his very new book on Bronze Age contacts between India and the Aegean, has collected much material which shows that objects and commodities undoubtedly originated in India, somehow reached the Mediterranean. Even though the exact way by which they came westerly is not known, the findings indicate that there was a kind of contact. Prof. Arnott (2022, pp. 83–84) concludes that the Indo-Aegean contacts during the Bronze Age seemed to have been mostly indirect. The research on the most direct contacts of the Indus Civilization with the Mediterranean during the Bronze Age is still quite nascent but at the same time highly promising.

6 Early Silk Beyond China

In the introduction, I referred to two basic characteristics of the Silk Road: its geography—the areas that participated in this intercultural exchange network—and its main, eponymous product, i.e. silk. So far, the more demanding geographical part has been addressed. From the previous analysis, we can conclude that to retain the term 'Silk Road" and correct the arbitrariness that has accompanied it so far, we might redate the Road's earliest development to the turn of the 3rd millennium BCE. This chronology seems to be common among all the contacts studied. The updated version of the Silk Road concept that has so far been presented includes peoples and geographical locations that were unjustifiably excluded from the classic concept, including India, the Persian Gulf, the Arabian Peninsula, and Africa. However, a Silk Road cannot be so called if silk plays no special role in it. Therefore, we should investigate whether silk did play an important role at such an early time. This investigation requires searching for silk outside of China during the 3rd millennium BCE.

To study the presence of silk beyond China before the Han Dynasty, one must turn to Irene Good. Good has published several in-depth studies on silk findings outside China (Good 2002, 2015; Good et al. 2009). As she points out, the study of ancient silk presents an inherent difficulty: silk leaves no strong traces in the archaeological record, and, therefore, is difficult for an archaeologist to detect (Good 2015, p. 959). Nevertheless, silk samples have been found in present-day Germany, the Mediterranean, Egypt, Turkey, and India, among others, but most of them are of much later date, primarily from the middle 2nd to late 1st millennium BCE. Moreover, the presence of silk in a given area does not suffice to imply trade with China (Good 2002, p. 7ff.). As Good rightly observes, in most cases, the silk that was found was produced from a unique variety of silkworms different from those prevalent in China (Good 2002, p. 7ff.). Chinese silk is made from *Bombyx mori* silkworms, whereas silk outside of China came from silkworms of the genus *Pachypasa*. Good warns strongly against believing that all silk that is found comes from China (Good 2015, pp. 966–967).

According to Good, one way to distinguish Chinese from non-Chinese silk is by examining whether the fibers come from degummed silk or raw silk. Chinese silk is mostly degummed, for the production process consists of boiling the cocoons before the insect emerges. When the insect does emerge from the cocoon, it must eat a part of the cocoon to open a way out. As a result, a part of the cocoon is damaged and needs spinning (Good 2015, p. 960). This method of degumming is a characteristic of Chinese silk production. In India, however, it was taboo, because of the belief in saṃsāra, the cyclicality of life, which demands the protection of every living thing, even of a silkworm which should not be boiled alive (Good 2015, p. 960). Nevertheless, Good warns, again, that it is not always possible to check with certainty whether a silk fiber comes from cocoons that have been degummed or not (Good 2015, pp. 966–967).

In any case, Good's most important finding, together with Kenoyer and Meadow, is the discovery of silk in several locations in the Indus Valley, dating to the middle of the 3rd millennium BCE, c. 2450–2000 BCE (Good et al. 2009, p. 1). So far, the earliest evidence of silk in China comes from a Neolithic site in Qianshanyang, which is dated to the middle of the 3rd millennium BCE (Good 2002, p. 10). As a result, the new findings from the Indus Valley show that silk threads were being produced in India during the period of prosperity of the Indus Valley Civilization and very close to the date of the earliest Chinese silk (Good et al. 2009, p. 2). The researchers emphasize that "this is the earliest evidence in the world for any silk outside China" (Good et al. 2009, p. 1).

From this finding, one might infer the existence of pretty early relations between China and India, and surmise that Indian silk came from China. The reality, however, seems even more interesting: the specific fibers found in India were *not* degummed, as one would expect of silk originating in China (Good et al. 2009, p. 2). This observation seems to contradict the suggestion of very early contacts between China and India but simultaneously provides new motivations for exploring India's early contacts with its western neighbors. Stated differently, since we have different types of silk in China and India within a very short time span, and since silk findings outside China before the Han period are, like Indian silk, not degummed, we cannot exclude the possibility that India and its separate silk production were the source of the western spread of silk (Good et al. 2009, pp. 7–8).

Apparently, silk poses no major problem for the updated version of the Silk Road that is proposed here. There may be few silk samples from the end of the 3rd millennium BCE, but the dating of the Indus Valley finds from the middle to the end of the 3rd millennium BCE proves the early presence of silk outside China. In other words, the Silk Road of the late 3rd millennium not only associates the cultures already connected by the classical concept of the Silk Road, in addition to several more cultures that the latter concept excludes, but it can also be linked with the silk product without great difficulty. The only problem one could detect is that silk was by no means so central to the communication of the peoples presented. Throughout this analysis, one can identify far more important products which moved from people to people and which, in fact, help us to more easily identify the dynamics of intercultural contacts in those ancient times. Commodities such as lapis lazuli or millet would certainly serve as justified alternatives for this updated Road: as a Lapis Lazuli Road or Millet Road. Nevertheless, I think that we may retain the traditional name without any problem: whatever single object or commodity we choose, it will underrepresent the complete picture of the specific intercultural relationships. As we unproblematically talk about "World Wars," knowing that many countries did not participate in them, so, for the sake of convenience, we may talk about the Silk Road, realizing that many other products played an equally important role in the trade between China and the West.

7 Conclusions

As can be seen from the data presented, most of the problems with the traditional concept of the Silk Road can be solved by simply moving its start date at least two millennia into the past, to the late 3rd to early 2nd millennium BCE. Even though far earlier intercultural contacts may be noticed, the date I suggest is a common starting point for extensive and almost simultaneous intercultural contacts stretching from China to the Mediterranean. This way, we address the arbitrary chronology of the classical Silk Road concept. We know of particularly extensive relations between China and the West during the end of the 3rd millennium BCE. Around the same time, we find silk nearby, but outside of China. Following the example of Alfred Crosby, who named the early contact between the Old and New World (the Americas) "the Columbian Exchange," Andrew Sherratt names early China's contact with the West "the Trans-Eurasian Exchange," emphasizing the breadth of this early intercultural communication (Sherratt 2006, p. 31). In the same vein, Vandkilde describes this same period as a time of ancient globalization or bronzization, as it mostly coincides with the so-called Bronze Age (Vandkilde 2016).

Moreover, in this updated version of the Silk Road concept, significant areas for ancient trade such as India, the Persian Gulf, the Arabian Peninsula, and North Africa are included, thus correcting the arbitrary selection of the civilizations which were included in the classical concept. One could easily add to the puzzle the Levant civilizations (Israel, Ugarit etc.), which were important bridges between the Mediterranean and Mesopotamia. I have not referred much to these civilizations in this research because their possible connection with India or China remains unclear and more research is needed. One could perhaps cautiously mention the vanilla finds at Megiddo. Of course, in the present study, I referred only to some of the evidence of ancient intercultural relations. More could be added, such as the "movements" of obsidian, copper, carnelian, and other materials. But such an extensive study would have been beyond the scope of this project. More importantly, this study has the additional contribution that it replaces the former abstract concept of the Silk Road with a considerably more studied approach based on a plethora of archaeological evidence, which gives greater coherence to the puzzle of the interactions of ancient civilizations. It goes without saying that the discovery of archaeological finds and organic or inorganic commodities in many different places should in no way be taken as the result of direct contact between the various cultures under discussion. Some of these finds may have resulted from indirect contact. However, in both direct and indirect contact the protagonists are always people moving from one place to another, and therefore the conclusion that contact was possible in some quite remote areas of the then known world from very early times, does not change.

An additional advantage of the position presented in this study is that it enables other scholars to further integrate many other geographical territories that, until now, have not been part of the classical Silk Road—regions such as Korea, Japan, Southeast Asia, or even Mongolia. According to Mark Hudson and his colleagues, the millet from northern China spread to Korea and regions of eastern Russia between 3500 and 2700 BCE (Hudson et al. 2021, pp. 121–122), whereas the Korean and Japanese cultures of the 3rd millennium BCE seem to have been influenced by the globalization spreading to their westernmost regions (Hudson et al. 2021, pp. 121–123). Likewise, as Bellwood describes, rice and millet spread to Taiwan around the early 3rd millennium BCE, contemporaneously with the arrival in the Philippines of the Proto-Austronesians, who had inhabited Taiwan and would arrive in Indonesia in the 2nd millennium BCE (Bellwood 2005, p. 134ff.). It is of great interest that a few decades ago, the archaeologists Buccellati and Kelly-Buccellati (1983), during their excavations at Tell-Ashara in present-day Syria, found traces of cloves, dating back to around the eighteenth century BCE. The interesting thing with this particular story is that clove is a native plant of the Molucca Islands of Indonesia. We know that the Indians used it as a spice, and although it is a very big and probably mistaken leap to assume that there were such early direct contacts between Indonesia and Mesopotamia, one could assume that these cloves reached Mesopotamia via India. Could we have had such early contacts between India and Indonesia as early as the beginning of the 2nd millennium BCE? It is not impossible that later studies, based on archaeological finds that are not currently available to us, will succeed in presenting evidence which is capable of moving the date of expanded intercultural contacts well beyond the proposed choice, presenting us with an even more fascinating picture of the relations between ancient civilizations.

The only unavoidable arbitrariness is the suggestion, engendered by the use of the term *Silk* Road, that silk emerged as a staple in such early times. This was certainly not the case, and, in both the classical and the updated version of the concept, we should definitely avoid the impression that silk was the only or the most basic product that participated in this extensive trade. However, we can keep the name conventionally without causing that much damage. The existence of silk outside China during the 3rd millennium BCE is an indisputable fact. It is clear that the name of the Silk Road, even in this updated version, remains symbolic: there has never been a single road that connected these cultures. But this only makes the symbolism more fertile and productive. Even though, in the materialistic sense of the word, the Silk Road "has never existed," this non-existent entity has nevertheless become one of the few symbols of unity and cooperation between many different cultures at a time when much more tangible tools have failed to bring about such a productive outcome. Although research on intercultural communication has a long way to go, and more data will certainly come to light, it is already clear that the ancient civilizations moved far more than they have been credited with.

References

Abudanah, F., Tarawneh, M., Wener, S.E.: The legend of the 'King's Highway': the archaeological evidence. Zeitschrift Für Orient-Archäologie **8**, 156–187 (2015)

Arnott, R.: Crossing Continents: Between India and the Aegean from Prehistory to Alexander the Great. Oxbow Books, Oxford (2022)

Beaujard, P.: The Worlds of the Indian Ocean, vol. 1. Cambridge University Press, Cambridge (2019)

Bellina, B.: Southeast Asia and the early Maritime Silk Road. In: Guy, J. (ed.) Lost Kingdoms: Hindu-Buddhist Sculpture of Early Southeast Asia, pp. 22–24. The Metropolitan Museum of Art, New York (2014)

Bellwood, P.: First Farmers: The Origin of Agricultural Societies. Blackwell, Bromsgrove (2005)

Boivin, N., Fuller, D.Q.: Shell middens, ships and seeds: exploring coastal subsistence, maritime trade and the dispersal of domesticates in and around the Ancient Arabian Peninsula. J. World Prehist. **22**, 113–180 (2009)

Buccellati, G., Kelly-Buccellati, M.: Terqa: the first eight seasons. Les Annales Archéologiques Arabes Syriennes **33**(2), 47–67 (1983)

Buchanan, B.: A dated "Persian Gulf" seal and its implications. In: Güterbock, H.G., Jacobsen, T. (eds.) Studies in Honor of Benno Landsberger on his Seventy-Fifth Birthday, pp. 204–209. The University of Chicago Press, Chicago (1965)

Crawford, G.W.: East Asian plant domestication. In: Stark, S.T. (ed.) Archaeology of Asia, pp. 77–95. Blackwell, Amsterdam (2006)

Dupree, L.: Notes on Shortugai: an harappan site in northern Afghanistan. In: Dani, A.H. (ed.) Indus Civilization, New Perspectives, pp. 103–115. Centre for the Study of the Civilization of Central Asia, Samarkand (1981)

During-Caspers, E.C.L.: Harappan trade in the Arabian Gulf in the third millennium B.C. Proc. Semin. Arab. Stud. **3**, 3–20 (1973)

During-Caspers, E.C.L.: A possible Harappan contact with the Aegean World. South Asian Archaeol. **23**, 435–452 (1985)

During-Caspers, E.C.L.: Intercultural/Mercantile contacts between the Arabian Gulf and South Asia at the close of the third millennium B. C. Proc. Semin. Arab. Stud. **22**, 3–28 (1992)

Francfort, H.P.: The Harappan settlement of Shortughai. In: Lal, B.B., Gupta, S.P. (eds.) Frontiers of the Indus Civilization: Sir Mortimer Wheeler Commemoration Volume, pp. 301–310. Books & Books, Navalpattu (1984)

Francfort, H.P.: La civilisation de l'Oxus et les Indo-Iraniens et Indo-Aryens en Asie centrale. In: Fussman, G. (ed.) Āryas, aryens et iraniens en Asie central, pp. 253–328. Collège de France, Paris (2005)

Francfort, H.P., Tremblay, X.: Marhasi et la civilisation de l'Oxus. Iran. Antiq. **45**, 51–224 (2010)

Fuchs, A.: Tukriš. In: Meissner, B. (ed.) Reallexikon der Assyriologie und Vorderasiatischen Archäologie, vol. 14, pp. 174–175. De Gruyter, Berlin (2014)

Fuller, D.Q., Boivin, N.: Crops, cattle and commensals across the Indian Ocean: current and potential archaeological evidence. Études Océan Indien **42–43**, 13–46 (2009)

Gadd, C.J.: Seals of ancient Indian style found at Ur. In: Gadd, C. J. (ed.) The Proceedings of the British Academy, vol. 18, pp. 3–22. Humphrey Milford Amen House, New York (1932)

Good, I.L.: On the question of silk in pre-Han Eurasia. Antiquity **69** (266), 959–968 (2015)

Good, I.L., Kenoyer, J.M., Meadow, R.H.: New evidence for early silk in the Indus civilization. Archaeometry **50**, 1–10 (2009)

Good, I.L.: The archaeology of early silk. In: Silk Roads, Other Roads: Proceedings of the 8th Biennial Symposium of the Textile Society of America September 26–28, 2002, pp. 7–15. Northampton. https://digitalcommons.unl.edu/tsaconf/388/ (2002)

Guan, K.C.: The maritime silk road: history of an idea. NSC Working Paper 23, 1–30 (2016)

Guichard, M.: The oxus civilization and mesopotamia: a philologist's point of view. In: Lyonnet, B., Dubova, N.A. (eds.) The World of the Oxus Civilization, pp. 66–81. Routledge, New York (2020)

Guichard, M.: Les mentions de la Crète à Mari. In: Caubet, A. (ed.) L'acrobate au Taureau. Les découvertes de Tell el-Dab'a et l'archéologie de la Méditerranée orientale, pp. 165–177. La Ducumentation francaise, Paris (1999)

Herrmann, G.: Lapis Lazuli: the early phases of its trade. Iraq 30(1), 21-57 (1968)

Hiebert, F.T.: Origins of the Bronze Age oasis civilization in Central Asia. Peabody Museum of Archaeology and Ethnology (1994)

Hudson, M.J., Bausch, I.R., Robbeets, M., Li, T., White, J.A., Gilaizeau, L.: Bronze age globalization and Eurasian impacts on later Jōmon social change. J. World Prehist. **34**, 121–158 (2021)

Kjærum, P.: The Dilmun seals as evidence of long distance relations in the early second millennium. In: Al Khalifa, H.A.S., Rice, M. (eds.) Bahrain Through the Ages: The Archaeology, pp. 269–277. Routledge, New York (2010)

Koryakova, L., Kohl, P.L.: Complex societies of Central Eurasia from the 3rd to the 1st millennia BC: regional specifics in the light of global models. Curr. Anthropol. **41**(4), 638–642 (2000)

Lamberg-Karlovsky, C.C.: The Oxus Civilization. Cuadernos De Prehistoria y Arqueología De La Universidad Autónoma De Madrid **39**, 21–63 (2013)

Laursen, S., Steinkeller, P.: Babylonia, the Gulf Region, and the Indus: archaeological and textual evidence for contact in the third and early second millennia B.C. Eisenbrauns (2017)

Leemans, W.F.: The trade relations of Babylonia and the question of relations with Egypt in the Old Babylonian Period. J. Econ. Soc. Hist. Orient **3**(1), 21–37 (1960)

Li, S.: The interaction between Northwest China and Central Asia during the second millennium BC: an archaeological perspective. In: Boyle, K., Renfrew, C., Levine, M. (eds.) Ancient Interactions: East and West in Eurasia, pp. 171–182. McDonald Institute for Archaeological Research, Cambridge (2002)

Liu, X.: The Silk Road in World History. Oxford University Press, Oxford (2010)

Lombard, P.: The Oxus Civilization/BMAC and its interaction with the Arabian Gulf: a review of the evidence. In: Lyonnet, B., Dubova, N. A. (eds.) The World of the Oxus Civilization, pp. 607–634. Routledge, New York (2020)

Lü, Q.Q., Henderson, J., Wang, Y., Wang, B.: Natron glass beads reveal proto-Silk Road between the Mediterranean and China in the 1st millennium BCE. Sci. Rep. **11**, 1–13 (2021)

Ludvik, G., Kenoyer, J., Pieniążek, M., Aylward, W.: New perspectives on stone bead technology at Bronze Age Troy. Anatol. Stud. **65**, 1–18 (2015)

Ludvik, G., Kenoyer, J., Pieniążek, M.: New evidence for interregional interaction in the 3rd millennium BCE Aegean: indus-style carnelian beads at Aegina-Kolonna, Greece. Paléorient **46**, 161–178 (2020)

Ludvik, G.: Hard stone beads and socio-political interaction in the Intermediate Bronze Age Southern Levant, ca. 2500–2000 BCE. Unpublished PhD thesis, University of Wisconsin-Madison (2018)

MacKay, E.: Sumerian connexions with ancient India. J. r. Asiat. Soc. Great Br. Ireland **4**, 697–701 (1925)

Majidzadeh, Y.: Lapis lazuli and the Great Khorasan Road. Paléorient **8** (1), 59–69 (1982)

Marshall, J.: Mohenjo-Daro and the Indus Civilization, vol. 2. Arthur Probsthain, London (1931)

Michel, C.: Considerations on the Assyrian settlement at Kanesh. In: Atici, L., et al. (eds.) Current Research at Kültepe-Kanesh: An Interdisciplinary and Integrative Approach to Trade Networks, Internationalism and Identity, pp. 69–84. Lockwood Press, Nottingham (2014)

Mutin, B., Lamberg-Karlovsky, C.C.: The relationship between the Oxus Civilization and the Indo-Iranian borderlands. In: Lyonnet, B., Dubova, N.A. (eds.) The World of the Oxus Civilization, pp. 551–589. Routledge, New York (2020)

Oppenheim, A.L.: The seafaring merchants of Ur. J. Am. Orient. Soc. **74**(1), 6–17 (1954)

Pareja, M.N., McKinney, T., Mayhew, J.A., Nash, S.D., Heaton, R.: A new identification of the monkeys depicted in a Bronze Age wall painting from Akrotiri, Thera. Primates **61**, 159–168 (2020)

Payne, J.C.: Lapis lazuli in early Egypt. Iraq **30**(1), 58–61 (1968)

Possehl, G.L.: Meluhha. In: Reade, J. (ed.) The Indian Ocean in Antiquity, pp. 133–208. Routledge, New York (1996)

Possehl, G.L.: The Indus Civilization: A Contemporary Perspective. AltaMira Press, Walnut Creek (2002)

Possehl, G.L.: The middle Asian interaction sphere: trade and contact in the 3rd millennium BC. Exp. Mag. **49**(1), 40–42 (2007)

Potts, T.F.: Patterns of trade in third-millennium BC Mesopotamia and Iran. World Archaeol. **24**(3), 379–402 (1993)

Potts, D.T.: Puzur-Inshushinak and the Oxus Civilization (BMAC): reflections on Shimaski and the geo-political landscape of Iran and Central Asia in the Ur III period. Zeitschrift Für Assyriologie **98**, 165–194 (2008)

Rao, R.S.: A 'Persian Gulf' seal from Lothal. Antiquity **37**(146), 96–99 (1963)

Ratnagar, S.: Interaction between the worlds of South Asia and Central Asia. In: Lyonnet, B., Dubova, N.A. (eds.) The World of the Oxus Civilization, pp. 590–606. Routledge, New York (2020)

Rezakhani, K.: The road that never was: the silk road and the trans-Eurasian exchange. Compar. Stud. South Asia Afr. Middle East **30**(3), 420–433 (2010)

Richthofen, F.: Über die zentralasiatischen Seidenstrassen bis zum 2. Jh. n. Chr. In: Boguslawski, G. (ed.) Verhandlungen der Gesellschaft für Erdkunde zu Berlin, vol 4, pp. 96–122. Verlag von Dietrich Reimer, Berlin (1877)

Schafer, E.H.: The Golden Peaches of Samarkand: A Study of T'ang Exotics. University of California Press, Berkeley (1963)

Schneider, T.: What is the past but a once material existence now silenced: the first intermediate period from an epistemological perspective. In: Felix Höflmayer (ed.) The Late Third Millennium in the Ancient Near East, pp. 311-322. The Oriental Institute of the University of Chicago, Chicago (2017)

Sherratt, A.: The trans-Eurasian exchange: the prehistory of Chinese relations with the West. In: Mair, V.H. (ed.) Contact and Exchange in the Ancient World, pp. 30–61. University of Hawaii Press, Hawaii (2006)

Spengler, R.N.: Agriculture in the Central Asia Bronze age. J. World Prehist. **28**, 215–253 (2015)

Steinkeller, P.: Camels in Ur III Babylonia? In: J.D. Schloen (ed.) Exploring the Longue Durée: Essays in Honor of Lawrence E. Stager, pp. 415–419. Eisenbrauns, Ann Arbor (2009)

Stevens, C.J.: Between China and South Asia: a middle Asian corridor of crop dispersal and agricultural innovation in the Bronze Age. Holocene **26**(10), 1541–1555 (2016)

Tosi, M., Vidale, M.: 4th millennium BC lapis lazuli working at Mehrgahr. Pak. Paléorient **16**(2), 89–99 (1990)

Vandkilde, K.: Bronzization: the bronze age as pre-modern globalization. Praehistorische Zeitschrift **91**, 103–123 (2016)

Vidale, M.: Treasures from the Oxus: The Art and Civilization of Central Asia. I.B. Tauris, London (2017)

Watrin, L.: From intellectual acquisitions to political change: Egypt-Mesopotamia interaction in the fourth millennium BC. De Kêmi à Birît Nâri: Revue Internationale de l'Orient Ancien, de l'Égypte à la Mésopotamie 2, 8–94 (2004–2005)

Whitfield, S.: Was there a Silk Road? Asian Med. **3**, 201–213 (2007)

Wilkinson, T.C.: Tying the Threads of Eurasia: Trans-Regional Routes and Material Flows in Transcaucasia, Eastern Anatolia and Western Central Asia, c. 3000–1500 BC. Sidestone Press, Leiden (2014)

Central Asian Rock Art on the Silk Road

Viktor A. Novozhenov

Abstract

For thousands of years, petroglyph's sites in the Central Asia have been holy and sacred places, reflecting spiritual and religious ideas, any narrative structures as part of a communicative network and markers of migrations. Historical and epic events are reflected on the rocks in the form of remained rock art sites and preserved for centuries in a relatively holistic and practically intact state, which allows the recreation of information on the particular events in the prehistory and history of Silk Road. Methods of this study is connected with reviewing, collecting, classification and dating evidence and its possible interpretation from points of communication and globalization theories. As a result, the development of new trade routes, numerous migrations of population and emerging identities of local communities, were most clearly reflected in the graphic series/traditions, structure of sites and especially petroglyphs themselves. Rock art sites have become an important element of mobile culture and the entire communication infrastructure that has been forming since the Eneolithic period on the routes of migrations, which mark and reconstruct the geography of path on Eurasian Silk Roads.

Keywords

Arts history • Rituals • Communications • Petroglyphes • Migrations • Pictorial traditions

V. A. Novozhenov (✉)
UNESCO Center for the Rapprochement of Cultures, Almaty, Kazakhstan
e-mail: vnovozhenov@gmail.com

1 Introduction

The rock art of Central Asia is a stable tradition of visual activity historically determined by the natural landscape, ways of cultural adaptation and other factors that influenced and partly retained their impact on the lifestyle, traditional forms of economic activity, and social and religious practices of ancient and modern communities of the region. Rock art monuments are an integral part of the historical and cultural landscape of most modern Central Asian countries (Brooks and Wakankar 1976; Bhatnagar et al. 2001; Chen 2008; Chenchenkova 2004; Jettmar 1975; Jacobson-Tepfer 2020; Kovaleva 2011; Komissarov et al. 2021; Kubarev et al. 2005; Sher 1980; Sukhbaatar 2015; Rogozhinski (ed.) 2004; Clotte (ed.) 2011; Jacobson-Tepfer and Novozhenov (eds.) 2020; Rogozhinski and Novozhenov 2018; Novozhenov and Rogozhinski 2019; Novozhenov and Sydykov 2019; Novozhenov 2020; Volkov 2002; Chugunov et al. 2020; Cheremisin et al. 2019; Rozwadowski 2020).

This phenomenon is a part of Eurasian Rock Art, presented from Europe and Scandinavia till China and India, should be check from points of date, iconography and styles synchronization, structure of sites, macro and micro topography, rituals, mythology and trans continental migrations.

2 Methods

From the standpoint of communication theory, the tribes that settled in the steppes of Eurasia have developed a verbal-pictorial internal communication channel that is different from writing, but no less informative—a verbal-pictorial internal communication channel, presented in the form of petroglyphs (petroglyphs communication channel itself), drawings and other pictorial monuments (statuary, ornamental, others) that are found everywhere on continent and fix on rocks or stones specific events and facts from the

S. Kostopoulou et al. (eds.), *Cities' Vocabularies and the Sustainable Development of the Silkroads*, Advances in Science, Technology & Innovation, https://doi.org/10.1007/978-3-031-31027-0_11

life of the ancient inhabitants of Eurasia and in some cases—more clearly and informatively, in comparison with written sources.

Comparative and analytic methods allow to reconstruct the structure of rock art sites, rituals, any narrative structures of surfaces, codes and art traditions, iconography, mythology in time and space. In this case the rock art sites are the complexes of petroglyphs and other archaeological objects, concentrated within a single landscape; the boundaries of these sites are identical to the territory of habitat and vital activities of ancient communities interlinked by the landscape.

Together with other archaeological monuments (settlements, burials, roads, crossings, etc.), rock art indirectly reflects the lifestyle of ancient communities, their attitude to the environment, the functional significance of its individual components depending on the level of sociocultural and technical development in a community and exact date of its existence.

3 Results

Today, there are more than 500 identified and documented rock art sites on the territory of Central Asia (see Fig. 1): Uzbekistan—more than 150 (Khuzhanazarov 2018), Kyrgyzstan—around 50 (Sulaimanova et al. 2016), Tajikistan—around 100 (Ranov 2016), Kazakhstan—more than 200 (Lymer 2015; Maksimova et al. 1985; Margulan 2003; Maryiashev 2011; Maryiashev and Goriachev, 2001; Merz 2002; Merz and Antonov 2019; Novozhenov 2002; 2012; 2014; 2018; Samashev 2006; 2012; Samashev et al. 2011; 2013; 2014; Sala and Deom 2005; Shvets 2012; Rogozhinski 2011).

The development of the petroglyphic tradition demonstrates specific economic and cultural adaptation forms: wide regional contacts, co-existence of various cultures and ethnic groups, and predominance of mobile forms of animal husbandry over settled agriculture. In return, the main economic and cultural activities of the local population were naturally oriented towards the permanent extensive development of the landscape (Clarkson et al. 2017; Khazanov 2018; Masanov 2011; Masson 2006; Yang et al. 2020; Wu 2013; Wu 2015; Qi and Wang 2008; Semenov 2015).

1—Baikonur; 2—Zyngertau; 3—Tasutkel; 4—Terekty-Aulie; 5—Tesiktas; 6—Kesteletas; 7—Semizbugy; 8—Shamatai, Kilybai, Baizhan; 9—Iymurundy; 10—Saiak; 11—Badran; 12—Kokpekty; 13—Nikitinka; 14—Akbaur; 15—Sagyr; 16—Menovnoe; 17—Kanai, Sentas, Tamurashi; 18—Zevakino; 19—Kalbak-Tash; 20—Dzhuramal; 21—Berel; 22—Moinak; 23—Kurchum; 24—Tulkune; 25—Dolanaly, Narbota; 26—Bolgar-Tabaty; 27—Okey; 28—Kipeli; 29—Dzhansugurov; 30—Bayan Zhurek, Kapal; 31—Yeshkiolmes; 32—Terekty; 33—Usek; 34—Taigak; 35—Karaespe; 36—Karashoky; 37—Tamgalytas; 38—Uzun Bulak; 39—Aksai; 40—Satty; 41—Assy; 42—Turgen; 43—Uzun Kargaly; 44—Maibulak; 45—Cholpon Ata; 46—Akterek; 47—Sergeyevka; 48—Asyk; 49—Serektas; 50—Tamgaly; 51—Karakyr; 52—Kulzhabasy; 53—Chokpar; 54—Oizhailau; 55—Ungurli; 56—Khantau; 57—Zhambyl; 58—Mirki; 59—Kurgantash; 60—Urmaral (Zhaltyraktash); 61—Aksy Zhabagly; 62—Karasai; 63—Maimak (Ters); 64—Gabaevka; 65—Eliksai; 66—Boraldai, Borisai; 67—Bayalder; 68—Arpauzen; 69—Koibagar; 70—Run, Ksan, Suiundyksai; 71—Karasoir; 72—Bala Sauyskandyk; 73—Tamgaly Lake; 74—Koksai; 75—Karakiyasai, Khodzhikent; 76—Varzik; 77—Saimaly Tash; 78—Uchterek; 79—Unek (Kochkor); 80—Akshyirak; 81—Zhaisan 1–4.

The history captured in stone by our ancient ancestors in this form of art continues to live, while the ethnic groups, political borders and social foundations disappear and collapse. In this capacity, the rock art of Central Asia provides an outstanding example of the eternal cultural values of many ancient and modern communities of the region developed over 5000 years between the main centers of the human civilization of the Old World. The content and artistic form of this tradition of fine art reflects the mutual influence and cultural synthesis of settled agricultural and mobile pastoralist cultures (Fang 2013; Fang et al. 2010; Frankopan 2015; Gao and Wang 2016; Hansen 2012; Rtveladze 1999).

The number of rock art sites in Central Asia and the number of images themselves still can't be accurately calculated; today we can talk about one thousand locations and more than one million individual images in total still not good documented. Concentrations of a large number of petroglyphs of different periods vary from 1000–5000 to more than 10,000 images in each site.

The geographical range of rock art monuments of Central Asia covers the foothill, mountain and lowland, and steppe areas, including the Sayans and Altai, Tarbagatai and Dzhungarsky/Zhetysuisky Alatau and the Tien Shan Mountains, the spurs of Alai and Gissar, Pamir and Kopetdag Mountains, as well as the Ustyurt Plateau and the Mygodzhary, Karatau and Chu-Ili Mountains, and the Kazakh steppe. The eastern and north-eastern borders of this area include the territories of modern countries of Central Asia and Kazakhstan, the Xinjiang Uygur Autonomous Region (XUAR) of the People's Republic of China, Mongolia, southern Siberia in Russia, in the Altai Mountains, and the southern and south-western borders cover the Western Himalayas, Karakorum and Hindukush Mountains, Tibet and the Indus River Valley.

The main types of landscapes common for the rock art sites are the desert-steppe lowlands, small hills (up to 200 m above sea level), low mountains (200—up to 1500 m), medium mountains (1500–2500 m) and high mountains

Fig. 1 Map of main Rock Art Sites in Central Asia. Created by the author

(more than 2500 m). The main types of substrates (rocks, texture and surface type) are closed rock surfaces (cave, grotto and canopy), open surfaces of sedimentary rocks (shale and carbonate rocks), open surfaces of intrusive rocks (granites, diorites etc.) and open surfaces blocks/boulders of sedimentary rocks.

The dating of rock art remains the primary task in the archaeology of the rock art of Central Asia. Over the past 50 years, specialists have developed basic methods for the classification and age determination of petroglyphs and on the basis of a series of subregional chronological schemes proposed by researchers, an idea has been formed of the periodization of the rock art of Central Asia as a whole.

The age of the region's most ancient sporadic experiments in the creation of drawings on stone is assumed to be within the limits of the Early Holocene (Mesolithic). The existence of a stable tradition of rock art depicts the whole late Holocene, with an apogee of creative activity in the Bronze Age and the early Iron Age (the late 3rd—1st millennia BCE) to the Middle Ages (the 1st millennium AD) inclusive. The late manifestations of this phenomenon identify the pictorial traditions of the modern indigenous peoples of Central Asia—the Kazakhs, Kyrgyz, Tajiks, Turkmen, and Uzbeks of the 16th—early twentieth centuries, as well as the West Mongolian (Oirat, or Dzungarians) the 17th——eighteenth centuries. The pictorial activity on the rocks fades away in the industrial era (the late 19th—twentieth centuries), but among the population that preserves the forms of traditional economy and life, it continues to exist to this day in most of its area.

The problems of chronology, periodization and cultural attribution of the isolated series of petroglyphs are the most complex and debatable in the archaeology of the rock art of Central Asia. Speaking about visual traditions, researchers, as a rule, characterize the main identified plot-stylistic groups of images, substantiate their dating and periodization, and determine the distribution areas. At the same time, researchers recognize that the selection of graphic (cultural) traditions of rock art requires an assessment of such qualitative characteristics of plot-stylistic groups of petroglyphs as the integrity of the graphic/pictorial series, the attitude of their creators to the works of predecessors, the order of grouping petroglyphs in the landscape, and others

(Rogozhinski (ed.) 2004; Clotte (ed.) 2011; Jacobson-Tepfer and Novozhenov (Eds.) 2020; Rogozhinski and Novozhenov 2018; Novozhenov 2020; Yatsenko et al. 2019).

Most confidently identified are certain stylistic groups of petroglyphs of the Bronze Age, Early Iron, the Medieval period, and also the New Age (the 17th—early twentieth century). An analysis of their geographical distribution allows to determine the ranges of local types that are specific to individual territories within the region, as well as regional, widely represented within the whole or a significant part of the region. In the latter case, there is a discrepancy between some areas and the borders of post-Soviet Central Asia (the style of Deer stones, Saka-Scythian's animal style, the ancient Turkic tradition in southern Siberia, Altai, Mongolia and XUAR in China) or overlapping (links) with local variations of the Central Asian types identified outside the region (the Animal style in the Indus Valley). Mapping the ranges of identified pictorial traditions, which will be possible on the basis of updated data, will make it possible to more accurately determine the spatial and temporal boundaries of the phenomenon of rock art in Central Asia.

The earliest examples are the murals in several grottos found in the south of the region. The Shakhty Grotto is located in the Eastern Pamirs at an altitude of about 4200 m above sea level and is one of the world's highest mountain cave art monuments from the Stone Age.

Findings of the stone industry here show that the grotto could have been a short-term shelter for primitive Mesolithic hunters. On the other hand, the possibility of human habitation in the shallow rock canopies of the Zaraut-sai Valley in the Kugitang Mountains in southern Uzbekistan is excluded, and the drawings painted here on the walls and arches of stone niches are open to the outside. The paintings are monochrome, made of red-brown mineral paint. The repertoire is limited to images of wild animals and anthropomorphic figures in hunting scenes, as well as complex schematic figures or objects.

Monuments dating from the Neolithic are more numerous and more widespread in the region. They are presented as paintings on the closed surfaces of grottoes and canopies, as well as engravings on open-air rocks. As in the previous epoch, in the high mountainous regions there are known refuge-grottoes with cultural deposits from different periods of habitat and with wall paintings.

The repertoire of paintings still includes images of various wild animals, reptiles—single figures or in compositions, in hunting scenes, as well as simple geometric figures and their combinations; the repertoire of anthropomorphic images is supplemented with engravings in the form of masks with surreal features. The repertoire of animal images is associated with their natural habitat: horse, kulan and deer images are characteristic of steppe and mountain-steppe landscapes; bull images are ubiquitous.

The rock art of Eneolithic and the early Bronze Age (approximately the 4th—3rd millennia BCE) remains poorly studied, although a significant number of drawings could be dated to this time, when the proto-urban and urban culture flourished in the settled agricultural areas of Central Asia. This period represents the high point of rock art activity. Pictures in the so-called *bi-triangle style,* some images of Bactrian camels, carts drawn by bulls, camels, a number of typical objects related to known artifacts appeared at this time.

The period of the Bronze Age (the 2nd—the early 1st millennia BCE) became the time of the widest distribution of rock art and the time of its flourishing. Monuments are located in different landscapes, but are predominantly found in the zone of low mountains and middle mountains, actively mastered by the pastoral-agricultural tribes of the Bronze Age; in highland areas, the number of monuments is also increases. Never before or after this period the repertoire of rock art was so diverse and rich in content.

Along with traditional animalistic and hunting scenes, an important place in the theme of rock art is occupied by scenes of cattle-breeding, military and ritual human activity with a corresponding variety of realities. The object world (transport, weapons, household items), represented in the petroglyphs of the Bronze Age, is well recognizable and finds many archaeological parallels, which makes possible the cultural and chronological identification of individual pictorial traditions.

Their ranges are generally consistent with the distribution map of the archaeological cultures of the Bronze Age of Central Asia, and sometimes vividly reflect the directions of subregional relations and external impulses. In the rock art of this epoch, along with realistic characters, fantastic anthropomorphic and zoomorphic images, solar symbols and astral signs are created, indicating the existence of a developed mythology, worship and ritual practices of the pastoral and agricultural tribes of the region.

At the same time, the extremely widespread practice of creating petroglyphs, among which there are many mediocre works in the artistic sense, testifies to the availability of this type of activity for different individuals and the multifunctional significance of the cultural tradition itself. Finally formed in the Bronze Age, the tradition of rock art work continued developing in subsequent historical periods as an important element of the culture of successive ancient tribes and peoples of Central Asia (see Fig. 2).

The absolute majority of the rock art monuments of the early Iron Age (the eighth century BCE—the fifth century AD) is an integral part of the petroglyphs of the previous historical period; the murals of this time have not been discovered, but petroglyphs carved on closed rocky surfaces are known. The locations of petroglyphs of the early Iron Age are common in all landscape zones where rock

Fig. 2 Kazakh Steppe. Tamgaly. Altar's surface. Sun headed Gods, animals and ritual dancing. Late Bronze Age. Photo by the author

paintings are found, but their range is wider than in the Bronze Age, due to the greater territorial development of middle mountains and high mountains by ancient nomadic cattle breeders. It is here that the rock art of various groups of nomads of this long historical epoch appears in its original form.

The most significant difference in the content of rock art is a return to animalism and hunting topics, which reflects profound changes in the economy, material culture, social structure and ideology of significant groups of the region's population. Many pictorial traditions of this era are convincingly identified due to the relatively well-studied archaeological cultures of the region, the abundance of pictorial analogies to petroglyphs in dated artistic objects from archaeological complexes, as well as the presence of narrative sources.

Visual traditions with transregional distribution (style of *Deer stones, Saka animal style*), as well as local variations with specific cultural-chronological order (Saka, Pazyryk, Syanbi style, etc.) can be distinguished. The image of tribal marks—*tamga* (tribal/clan's signs)—arises as a new phenomenon in the rock art. In this epoch, the main features of the rock art of the Central Asian nomads as a cultural tradition are formed, which retained their importance in subsequent historical times (Savinov 1998; Rogozhinski and Yatsenko 2015: 109–125).

The nomadic culture phenomenon is strikingly apparent in the repertoire of rock art, where early images (the early Iron Age) are dominated by images of wild fauna and hunting, and military-political symbols and plots of pastoral themes also spread to the later (the Middle Ages and modern times). This reflects the genesis of the nomad culture and its subsequent transformations, when, following the period of the development of nomadic cattle breeding and reconquering the ecological niche from the wild, there was an assertion throughout the steppes of the *Horse nomadic formation* with its specific forms of material production, social relations and ideology (Deopik 1979).

Another important feature of the rock art of nomads is the activity in the transformation of works of art from previous eras (renewal, redoing and overlapping of earlier drawings), caused both by the desire to establish their own aesthetics and cultural identity (own methods of communication) and by the real conditions of frequent habitat changes and the permanent development of the landscape nomadic pastoralists. It was a period of changing the previous system of communication to a new one.

The medieval rock art of Central Asia (the 6th—sixteenth centuries AD) is marked by significant changes associated with the spread of statehood (especially the Turkic-Mongolian), writing, and world religions, including Buddhism, Christianity, Manicheeism and Islam. The area of monuments in this era is even wider than in the previous time. In some areas (Fergana, Nuratau and Kopetdag), the practice of creating paintings on closed surfaces is sporadically been revived, but the tradition of creating engravings continues to dominate in the main territory.

Two complementary areas coexist in rock art: *secular*, connected by sources, on the one hand, with the traditional content of nomad art, and on the other, with heroic epic, military aesthetics and tribal (ethnic) symbolism; and *religious-cults*, due to missionary, pilgrimage activities, and the adaptation of traditional holy places to new denominations. A common innovation for both areas is the organic combination of figurative activity with the creation of written (epigraphic) texts of diverse contents.

The rock art of the Middle Ages of the pre-Mongol period (the 6th—twelfth centuries) is filled with nomadic pastoral and battle themes. The heroic image of a warrior-rider (often with a banner) and images of tribal signs (tamga) of the Turkic clans form the main elements of propaganda in the content of the rock art of this era. The repertoire of petroglyphs of certain territories of the Kazakhstan and Central Asia reflected the participation of nomads in the operation of the caravan trade along the Silk Road routes. As in previous periods, the relationship of the rock art of Central Asia with the art of sedentary farmers is clearly manifested and deserves special study.

Overall, the rock art of Central Asia demonstrates a variety of natural substrates and landscapes, a great similarity of methods of using the natural features of macro- and micro-relief, visual techniques and for many historical periods—the unity of themes, range and stylistics of rock art. Along with other historical (written and archaeological) sources, petroglyphs provide important material for understanding many processes of cultural development, the interaction of communities of Central Asia and neighboring regions in Antiquity and in the Middle Ages, and can be considered as evidence of progressive development of a single Central Asian rock art tradition.

In the history of the Central Asian rock art several main graphic/art traditions stood out. They were directly related to key stages (eras) in the history of the local population: the Yamnaya-Afanasyevo fine art tradition in the 3rd millennium BCE (including statues and megalithic traditions); the Andronovo in the 2nd millennium BCE (within which independent artistic styles and traditions developed, the early Andronovo (Seima-Turbino) style and the Andronovo proper (late stage), a bright ornamental tradition; the Karasuk or Begazy-Dandybai—the (15th) twelfth—seventh centuries BCE (Proto-Turk period); the Saka in the seventh—second centuries BCE (the classical Scythian-Saka animal style was formed and represented in their crafts and artifacts), as well as the unique Turkic (fifth—eighth centuries AD) and any local ethnic art traditions (14th——nineteenth centuries) (Novozhenov and Bedelbaeva 2018; Rogozhinski and Novozhenov 2018; Novozhenov 2014; 2020).

4 Discussion

An important feature of communication's points is their fixed geographic location. There is a significant similarity of some images in the petroglyphs of the Bronze Age throughout the vast territory of Central Asia. Perhaps this is explained, on the one hand, by the common ideological basis of the peoples who left these monuments, and on the other, by the movement in space and time of the carriers of this communicative/pictorial tradition themselves.

Rock art sites will be considered as markers of migrations and relocations of the ancient inhabitants of Central Asia due to their geographical position fixed on the motionless rocks. The territorial distribution of similar signs, images, styles, same pictorial series, artistic traditions and, as a result, similar mythologies, concepts, beliefs, and rituals may indicate the geographical movements of their carriers. In turn, the determination of the chronological position of the distinguished graphic series, traditions, styles, and images which have been identified, and their comparison with the archaeological finds, allows us to date eventual migrations and population movements.

Development of new trade routes, numerous migrations of population and emerging identities of local communities were most clearly reflected in the graphic sites and especially petroglyphs (expressed in graphic traditions), tamga signs and epigraphies. Due to their unique and strictly fixed position on the rocks, they were demonstrating the presence of certain ethnic groups in certain regions and determining specific areas of resettlement of individual clans or groups of blood relatives. Rock art sites have become an important element of mobile culture and the entire communication infrastructure that was being developed (see Fig. 3).

They are located in the most areas of the steppe—on the rocky outliers along the tops and slopes of hills, and on the steep banks of small rivers which dried up in the summer. These points are visible from the far distance. As a rule, numerous local legends are associated with them, and they are considered as the holy places—*Auliye*. These points were used by the cattle breeders as important orientation signs in the vast steppe and played an important role in a complex system of ancient communications (the network of roads, river crossings, warehouses, repair centers, caravanserais and other logistic and infrastructure elements) that was used both by the local communities and by various population groups, and entire nations and countries located in the steppe at considerable distances from each other. It is no

Fig. 3 Scene of Clan migration, wagons, hunting and domesticated animals. Baikonur River valley. Kazakh Steppe (Betpakdala Desert). Early Bronze Age (Beg. of 2-nd mill. BCE). Sketch by the author

coincidence that even today the cattle breeders are building stone masonry *obo* on the tops of hills not only for the cult but also for the orientation purposes. This is how they mark the most convenient ways to move herds to the pastures rich in grass and water.

Located on the mountain passes, in the gorges and on the most convenient pastures and movement routes, petroglyph sites were probably played the same role in Antiquity. This feature and the impossibility of taking rocks away transformed pictorial sites into a unique and holistic source for studies of the most ancient communications and interactions of the population in the steppes and on the foothills. Moreover, these signs on the rocks and primitive astronomical knowledge were forming a certain cult-informational complex of the ancient cattle breeders, and served as the open-air temples where the Sky represented a dome and the World represented the surrounding wall. Mobility and inclusion in the environment have formed the sacred landscapes and a country marked by sacred places and the routes between them.

In the absence of writing among the early steppe communities, there was a limited number of channels for transmitting information between their members. Communication possibilities turned out to be limited by verbal and visual traditions, which reflected the knowledge accumulated by a community in the only form available at that time—mythopoetry.

Stable channels of communication revealed during different historical periods of Central Asia require other types of sources such as archaeological artifacts and rarely preserved written sources. Together they formed stable and repeating structures, the explanation of which leads to the comprehensive model of interaction between the Central Asian communities in Antiquity and the Middle Ages. It eventually resulted in the emergence of such a historical phenomenon as the Silk Road, which was so successful for many centuries as a model of communication reflected in various types of historical sources.

5 Conclusions

In general, the rock art sites are one of the most numerous and widespread types of the cultural heritage sites of the peoples of Central Asia and many territories of the region; their number prevails over the famous sites of ancient settlements and burials.

Obviously, all of them are markers of migration and located along the routes of ancient communications in the most significant strategic points of the surrounding landscapes. At the same time, fine art artifacts situated at a high altitude demonstrate the ways, in which the population of lowland areas developed the most convenient mountain passes to reach other fertile valleys by using the surrounding natural landscapes and consuming resources. It is not surprising that such difficult routes in the mountain areas were completely naturalized and mythologized: a mental map plotted a path through both specific geographical spaces and through the mythical worlds of the local universe.

In most of Central Asia, the archaeological landscapes with rock art represent the most typical and informative way to characterize the cultural heritage of several historical periods from the Bronze Age to the Early Iron Age, and even to the Middle Ages. Rock art sites are integral parts of the historical landscape in most of the countries in the region. Most of Central Asian rock art sites currently remain in their original cultural (archaeological) landscapes. However, today many sites or their individual sections with petroglyphs are significant components of sacred landscapes (Holy places) and are integrated into the practice of their worship.

The vast majority of the steppe rock sanctuaries or temples in the open-air functioned for a number of periods, as evidenced by the proximity to the same site of engravings dated to different chronological periods. This allows conclude that the functions of the rock art sites as a type of communicative center remained unchanged for millennia.

Currently, in the countries of Central Asia, in a number of rock art sites, the practice of irregular creation of new images by the local population has been recorded, whose conditions of traditional forms of living and farming remain important: this is a modern rural population practicing mobile livestock farming. At the same time, examples of creating rock art for religious purposes or in connection with the performance of religious rites were not registered anywhere.

A special group of rock art sites is integrated into the modern religious and cult practice of the indigenous population of the region. As a rule, rock art is not of independent cult significance, but is a significant component in the system of revered cultural and natural objects (Holy places) united by religious tradition. Some of the known holy places with revered objects of rock art are related to the cult of Caliph Ali and his horse Dul-dul.

There is also isolated evidence of the ritual veneration of sites of this type by adherents of Buddhism (Lamaism); for example, Tamgalytas in Kapshagai (Ili River) and the Kelinshektas stone suburgan near Tekeli in Kazakhstan, as well as stones with Tibetan texts and a Buddha image in the gorges of Tamga and Issyg-Ata in Kyrgyzstan or in Karakorum in Pakistan.

A separate group is rock art sites, the entire territory of which, or some parts of the landscape, are traditionally revered as a Holy place, but without of any religious connection, ritual or interpretation with the petroglyphs (for example, the central part of the Tamgaly canyon or the granite hill of Terekty-Aulie). In some cases, veneration of such places is associated with cultural objects or archaeological sites of a different kind: the graves of khojas, legendary or historical figures, with the ruins of religious buildings (mosques), etc.

Development of new trade routes, numerous migrations of population and emerging identities of local communities were most clearly reflected in the graphic sites/traditions and especially petroglyphs (expressed in graphic traditions), tamga signs and epigraphies. Due to their unique and strictly fixed position on the rocks, they were demonstrating the presence of certain ethnic groups in certain regions and determining specific areas of resettlement of individual clans or groups of blood relatives. Rock art sites became an important element of mobile culture and the entire communication infrastructure that was being developed in the steppe from the Eneolithic period till the Silk Road period, which became the classic model of population communication in Central Asia.

References

Bhathnagar, S., Chodhary, A., Pancholi, R., Tillner, E., Wakankar, V., Wanke, L.: Indishe Felsbilder. Jahrbuch der Ge.Fe.Bi, 2nd ed. Damböck. (Original work published 1977/1978) (2001)

Brooks, R., Wakankar, V.: Stone Age Painting in India. Yale University Press (1976)

Chen, Z.: Zhongguo yanhua faxian shi. A history of rock art's discoveries in China. Shanghai renmin chubanshe Press (2008)

Chenchenkova, O.P.: Stone sculpture of forest-steppe Asia of the era of paleometal III - I millennium BCE. Thesis (2004)

Cheremisin, D.V., Komisarov, S.A., Solovyov, A.I.: Images of chariots of Central Asia as a marker of mobility and migration. In: Molodin, V.I. (ed.) Mobility and migration: concepts, methods, results, pp. 215–228. Siberian Branch of Russian Academy of Science Press (2019)

Chugunov, K.V., Rawson, J., Grebnev, G.A.: Companions and victims of Shan China. About localization of the steppe component in the Shang culture. Stratum plus. **2**, 409–438 (2020)

Clarkson, P.B., Santoro, C.M., Levy, T.E., Núñez, L., Nielsen, A., Rosen, S., Förster, F., Capriles, J.M., Khazanov, A.M., Frachetti, M., Valenzuela, D., Standen, V.G., Cases, B., Pimente, G., Lecoq, P., Medinacelli, X., Briones, Tripcevich, N., Riemer, H., O'Ryan, E., Loayza, X., Lynch, T.F., Woldekiros, H.: A worldwide network for comparative studies on caravans: past, present and future. Revista de Antropología Chilena. **49**(3), 297–307. (2017). https://doi.org/10.4067/S0717-7356201700030000

Clotte, J. (ed.).: Rock Art in Central Asia. A Thematic Study. ICOMOS (2011)

Deopik, D.V.: Horseman culture in the upper Yangtzy and the eastern version of the Animal style. In: Culture and art of the peoples of Central Asia in antiquity and the Middle Ages. Nauka (1979)

Fang, M.: Red China: The Christmas Tree Route. Publishing House of the Huang Shan Book Club (2013)

Fang, S.Y., Wang, X.Y., Li, D.C.: Silk Road: rise and fall. Chinese construction Press (2010)

Frankopan, P.: The Silk Roads: New History of the World. Ox-Press (2015)

Gao, Y.F., Wang, L.J.: Silk Road Understanding Scheme. Publishing House of the Book Company John Hua (2016)

Jacobson-Tepfer, E.: The anatomy of deep time: rock art and landscape in the Altai Mountains of Mongolia (elements in environmental humanities). Cambridge University Press (2020). https://doi.org/10.1017/9781108855518

Jacobson-Tepfer, E., Novozhenov, V.A. (eds.): Rock Art Chronicles of Golden Steppe: from Karatau to Altai, vol. 2. UNESCO Center for the Rapprochement of Cultures Press (2020)

Jettmar, K.: Die Religionen des Hindukusch. Kohlhammer (1975)

Hansen, V.: The silk road: a new history. Oxford University Press (2012)

Khazanov, A.M.: Steppe nomads in Eurasian trade. In: Baipakov, K.M., Novozhenov, V.A. (eds.) Central Asia on the Silk Road: a dialogue of cultures and faiths from antiquity to modernity, pp. 12–28. UNESCO Center for the Rapprochement of Cultures Press (2018)

Khuzhanazarov, M.: Sarmishsoy koyatosh razmlari, idmiy tadkikotlar (Sarmishsay: rock art, scientific research). Nurfayz Nashriyotti Press (2018)

Komisarov, S.A., Solovyov, A.I., Cheremisin, D.V., Kudinova, M.A.: Study of rock art sites in China archaeology. Vestnik RFFI. Humanitarian sciences, 109–120. (2021).https://doi.org/10.22204/2587-8956-2021-103-01-109-120

Kovaleva, O.V.: Rock art of the Late Bronze Age of the Minusinsk Basin. Institute of Archaeology and Ethnography of Siberian Branch of Russian Academy of Science Press (2011)

Kubarev, V.D., Tsewendorzh, D., Jacobson, E.: Petroglyphs of Tsagaan-Salaa and Baga-Oygur (Mongolian Altai). Publishing House of Institute Archaeology and Ethnography of Siberian Branch of Russian Academy of Science (2005)

Lymer, K.: The petroglyphs of Terekty Aulie in Central Kazakhstan. Expression. **8**, 96–101 (2015)

Maksimova, A.G., Ermolaeva, A.S., Maryashev, A.N.: Petroglyphs of Tamgaly. Oner (1985)

Margulan, A.K.: Petroglyphs of Saryarka. Engravings depicting a wolf totem. Stone sculptures of Ulytau. Nauka, vol. 3–4 (2003)

Maryiashev, A.N.: Rock Art of Kazakhstan: results of a 20-year study and problems. In: Baitanaev, B.A., Beisenov A.Z. (eds.) Archaeology of Kazakhstan in the Era of independence: results and prospects. Materials of the international scientific conference dedicated to the 20th anniversary of Independence of the Republic of Kazakhstan and the 20th anniversary of the Institute of Archaeology named after A. K. Margulan vol. 1, pp. 36–40 Institute of Archaeology named after A. K. Margulan Publishing House

Maryiashev, A.N., Goriachev, A.A.: Petroglyphs of the Semirechye. Fund XXIst Century Press (2001)

Masanov, N.E.: Nomadic civilization of the Kazakhs: the foundations of the life of a nomadic society, Print-S (2011)

Masson, V.M.: Cultural genesis of ancient Central Asia. Institute of History of Material Culture of Russian Academy of Science Press (2006)

Merz, V. [Viktor]. K.: Rock paintings of the region Kereku-Bayan. Pavlodar State University Press (2002)

Merz, I. [Ilia]. V., Antonov, M. A.: Grotto Enbek with rock paintings in the vicinity of Taldykorgan. Otan Tarikhy. **2**(86), 200–214 (2019)

Novozhenov, V.A.: Petroglyphs of Saryarka. Margulan Press, Institute of Archaeology named after A. K (2002)

Novozhenov, V.A.: *Communication Miracle and the Earliest Wheeled Transport of Eurasia.* Taus (2012)

Novozhenov, V.A.: The Great Steppe: a human in the system of ancient communications. In: Epimakhov, A.V. (ed.) Sacrament of the Ethnic History of the Earliest Nomads of the Steppe Eurasia, pp. 18–267. Krym's Island Press (2014)

Novozhenov, V.A.: Communications and wheeled transport on the Great Silk Road. In: Baipakov, K.M., Novozhenov, V.A. (eds.) Central Asia on the Great Silk Road: a dialogue of cultures and faiths from antiquity to the present, pp. 241–263. UNESCO Center for the Rapprochement of Cultures Press (2018)

Novozhenov, V.A.: The Rock Art Chronicles of Golden Steppe: model of communication in Antiquity and Middle Ages, vol. 1. UNESCO Center for the Rapprochement of Cultures Press (2020)

Novozhenov, V.A., Bedelbaeva, M.V.: Figurative Monuments and Rock Art traditions of the Kazakh Steppe. Art of the Orient. 7, 119–130. In: Malinowski, J., Łakomska, B. (eds.) Proceedings of the International Conference in Kracow. Polish Institute of World Art Studies (2018)

Novozhenov, V.A., Sydykov, A.Z.: Bronze Age Trans Eurasian Communications. Otan Tarikhy. **3**(87), 183–204 (2019)

Novozhenov, V.A., Rogozhinsky, A.Y.: New chariot plots in the petroglyphs of the Koksu river valley (Yeshkiolmes). In: Goriachev, A (ed.) History and Archaeology of Semirechye: collection of articles and publications vol. 6, pp. 170–196. Institute of Archaeology named after A. Margulan Press (2019)

Ranov, V.A.: Running on the rocks: Rock Art of the Pamirs. Donish (2016)

Rogozhinski, A.Y. (ed.): Rock art monuments of Central Asia: public participation, management, conservation, documentation. Iskander (2004)

Rogozhinski, A.Y.: Petroglyphs of the archaeological landscape of Tamgaly. Signet Print (2011)

Rogozhinski, A.Y., Yatsenko, S.A.: The Ancient Tamga-signs of South-East Kazakhstan and their owners: the Route from East to West in the 2nd century BCE—2nd century CE. Silk Road. **13**, 109–125 (2015)

Rogozhinski, A.Y., Novozhenov, V.A.: Cultural landscapes with petroglyphs of Central Asia: questions and answers. International Institute of Central Asian Study (IICAS) Press (2013)

Rozwadowski, A.: Rock Art as a Source of History in Central Asia. In: Encyclopédie des historiographies: Afriques, Amériques, Asies, vol. 1, pp. 1520–1535. Presses de l'Inalco (2020)

Rtveladze, E.: The Great Silk Road: Antiquity and the Middle Ages. Uzbekistan Mile Encyclopedia (1999)

Qi, X., Wang, B.: The Ancient Culture in Xinjiang along the Silk Road. XUARICEA (2008)

Sala, R., Deom, J-M.: Rock Art of Southern Kazakhstan. Publishing House of Kazakh Institute of Culture (2005)

Samashev, Z.: Petroglyphs of Kazakhstan. Oner (2006)

Samashev, Z.: Rock Art of Zhetysu. Margulan Press, Bayanjurek. Branch of the Institute of Archaeology named after A. K (2012)

Samashev, Z., Zhang, S.H., Bokovenko, N., Murgabaev, S.: Rock art of Kazakhstan. Fund for the History of Northeast Asia. Branch of the Institute of Archaeology named after A. K. Margulan Press (2011).

Samashev, Z., Baytleu, D., Kurmankulov, Z.H.: Petroglyphs of Terekty Aulie. Branch of the Institute of Archaeology named after A. K. Margulan Press (2013)

Samashev, Z., Murgabaev, S., Eleuov, M.: Petroglyphs of Sauyskandyk. Margulan Press, Branch of the Institute of Archaeology named after A. K (2014)

Savinov, D.G.: Karasuk traditions and Arzhan-Mayemir style. St. Petersburg University Press, In Ancient cultures of Central Asia and St. Petersburg (1998)

Semenov, V.A.: The Art of Barbarian tribes. St. Petersburg Printing house NP-Print (2015)

Sher, Y.A.: Petroglyphs of Central Asia. Science (1980)

Shvets, I.N.: Studien zur Felsbildkunst Kasachstans. Materialien zur Archaeologie Kasachstans. Bd.1. Philipp von Zabern (2012)

Sukhbaatar, D.: The Petroglyphes of Mongolia. Ulanbaatar Publishing House (2015)

Sulaimanova, A.T., Zholdoshov, C.M., Duyshanalieva, A.T.: Archaeological landscape of Saimaluu - Tash. Bishkek Press (2016)

Volkov, V.V.: Deer stones of Mongolia. (2nd ed.). Scientific world. (Original work published 1991) (2002)

Wu, H. [Hsiao-yun].: Chariots in Early China: Origins, cultural interaction and identity. BAR Int. Series (S2457). Archaeopress (2013)

Wu, H. [Hong]. W.: New Silk Road and Central Asia. Beijing Publishing House of Social Sciences (2015)

Yang, J., Shao, H., Pan, L.: The metal road of the Eastern Eurasian Steppe. Springer, Singapore. (2020). https://doi.org/10.1007/978-981-32-9155-3_1

Yatsenko, S.A., Smagulov, E.A., Rogozhinski, A.Y., Tabaldyev, K.S., Baratov, S.R., Ilyasov, J.Y., Babayarov, G.B. In: Suleymanov, R.K (ed.): Tamgas of pre-Islamic Central Asia. International Institute of Central Asian Study (IICAS) Press (2019)

Mythological Itineraries Along the Western Silk Road: Finding Myths in Visits to Eastern Macedonia and Thrace Today

Anna Vacalopoulou, Anna Mastrogianni, Charilaos Michalopoulos, Despoina Tsiafaki, Natasa Michailidou, Ioannis Mourthos, Paraskevi Botini, and Gregory Stainhaouer

Abstract

This paper presents work within the framework of the R&D project "Mythological Routes in Eastern Macedonia and Thrace", which showcases the cultural wealth of the region of Eastern Macedonia and Thrace as the basis for the development of the tourism industry. The starting point of the project is the systematic recording, mapping, and promotion of ancient myths that are, in some way, related to this region. The area is of key commercial value, as it forms the north-eastern part of Greece, bordering two countries: Bulgaria and Turkey. In addition, it has always been an important part of the Western Silk Road map, with major centres such as Soufli, and Alexandroupolis. Apart from its geopolitical and economic significance, the area has substantial historical and cultural value, as it has been inhabited continuously since the Prehistoric era. It also boasts a variety of diverse landscapes and ecosystems as well as a rare long-lasting mosaic of different cultures. This project marks the first dedicated collection of myths and mythological references concentrating exclusively on this geographical region. At another level, the project allows the creation of personalized itineraries in the area for the support of the full tourist experience through a website and a mobile application (https://mythotopia.eu/). Starting from the promotion of the mythological identity of the region, its past and present are connected to enhance its tourist and cultural elements. Visitors of the region will enjoy the possibility of designing their own itineraries based on a variety of personal criteria, be them mythological, archaeological, cultural, natural, gastronomic, recreational, etc. In addition, the project has developed a specially designed educational application for the pedagogical utilization of the research outcomes in secondary and higher education.

Keywords

Western silk road map • Eastern Macedonia and Thrace • Technologically enhanced tourism • Cultural tourism • Classical mythology

A. Vacalopoulou (✉) · D. Tsiafaki · N. Michailidou · I. Mourthos · P. Botini · G. Stainhaouer
Institute for Language and Speech Processing, Athena R.C., Xanthi, Greece
e-mail: avacalop@athenarc.gr

D. Tsiafaki
e-mail: tsiafaki@athenarc.gr

N. Michailidou
e-mail: amixaili@athenarc.gr

I. Mourthos
e-mail: jmourthos@athenarc.gr

P. Botini
e-mail: pbotini@athenarc.gr

G. Stainhaouer
e-mail: stein@athenarc.gr

A. Mastrogianni · C. Michalopoulos
Department of Greek Philology, School of Classics and Humanities, Democritus University of Thrace, Komotini, Greece
e-mail: amastrog@helit.duth.gr

C. Michalopoulos
e-mail: chmichal@helit.duth.gr

1 Introduction

The project presented in this paper, entitled "Mythological routes in Eastern Macedonia and Thrace", has been implemented by a consortium of two Greek organizations, namely the Institute for Language and Speech Processing of Athena Research Center based in Athens and Xanthi and the Department of Greek Philology of the Democritus University of Thrace. This partnership reflects the interdisciplinary nature of the project, which combines extensive research that has been carried out by the two main organizations of excellence in Eastern Macedonia and Thrace; an area which also features major Silk Road centres. Partners saw this

S. Kostopoulou et al. (eds.), *Cities' Vocabularies and the Sustainable Development of the Silkroads*, Advances in Science, Technology & Innovation, https://doi.org/10.1007/978-3-031-31027-0_12

project as a chance to explore interdisciplinary research to an innovative end. The groundbreaking element of the project is the fact that it marks the first systematic scientific research in the country that combines the areas of mythology and tourism with a focus on this geographical area. Its outcomes, comprising a website and a mobile app, as well as an online educational app, provide a useful tool to support the contemporary tourist experience with the added value of being utilized in education.

A major goal of the project was to build on interdisciplinary research in order to highlight the cultural wealth of Eastern Macedonia and Thrace, seeking to contribute to the tourist development of the area. This aim is reached by systematically recording, mapping, and highlighting the myths related to the area, as well as attempting to connect them with local cultural and tourist treasures. The final outcome is a multi-faceted view of the region, which goes beyond this starting point. Therefore, apart from mythology, other aspects such as history, folklore, architecture, and gastronomy are linked to the contemporary environmental and social landscape of the district, attempting a connection between past and present. The main vehicle of communicating this outcome to the public is a bilingual (Greek and English) website and a mobile app. In order to provide end users with ways of putting all these diverse types of data in use to serve their own needs and preferences, they are offered the opportunity to create personalized routes, with parameters of their choice, thus enjoying a complete tourist experience. The second goal of this project was to develop a specially designed educational app to utilize this content in the context of secondary and higher education. To this end, several tools are offered online, including educational scenarios, lesson plans, exercises, and games, for use both by students and by educators.

Combining sightseeing and educational services, both of which are supported by a robust background of knowledge and research in several disciplines, the end product of this multi-faceted project asprires to be a trip in history and myth, imagination and reality, addressing a rather broad audience. Its tourist part is a useful tool to any Greek- or English-speaking visitor of the region, literally anyone who has an interest in the area and is looking for an alternative, personalized experience, combined with a wealth of information beyond the superficial tourist level. Furthermore, through the educational application, the project addresses the educational community—both students of multiple levels and ages as well as educators.

1.1 Research Background

In terms of the research pertaining to this project, the main challenge was to find ways to combine scientific research results with the technological know-how from different disciplines so as to provide each of them with added value while enhancing tourism in the area.

The main disciplines that contribute to create the outcomes through research results and technological expertise are mythology and philology, archaeology and relevant artefacts, data science (with the collection and interconnection of diverse data), metadata schemata, educational technology, as well as technologically enhanced tourist applications. This paper will summarize and explain different aspects of these areas and the way in which they are combined to bring project outcomes to life.

2 The Case of Eastern Macedonia and Thrace

Located beyond the northern periphery of the Greek world, ancient Thrace never ceased to spark the interest and fuel the imagination of ancient writers (both Greek and Roman) about its geography, its climate, its people and their way of life. The historical borders of this vast and mostly unexplored land were never settled or accurately deciphered, but were subject throughout the ages to political upheavals and military action (for more on the boundaries of ancient Thrace see Bouzek and Graninger 2015, 12–15). Broadly speaking, ancient Thrace covered a territory in the southeastern Balkans bordered by the Black Sea and the Sea of Marmara to the East, the Axios (Vardar) region to the West, the Balkan Mountains and the river Danube to the North, and the Aegean Sea to the South. In modern terms, this extensive area comprises southeastern and northern Bulgaria, southern Romania, northeastern Greece, the European part of Turkey, and some eastern regions of North Macedonia.[1]

2.1 The Area in Classical Literature

In ancient literature, Thrace appears as early as in Homer's *Iliad* where we read about the renowned Thracian riders and the snow-covered mountains of the region (*ἐφ' ἱπποπόλων Θρῃκῶν ὄρεα νιφόεντα*, Hom. *Iliad* 14.227) and in Greek lyric poetry (e.g., Archilochus fr. 28; frr. 51, 40–3, and Anacreon. fr. 2.1.10). Homer (again) and pseudo-Euripides make special reference to the "whiter than snow" steeds of Rhesus, a mythical king of Thrace (Hom. *Iliad* 10.437; Eur. *Rhesus* 304), while Virgil, one of Rome's greatest poets, comments on the famous horses of Thrace (*Aeneid* 5.565,

[1] The part of ancient Thrace belonging to Modern Greece (Western Thrace) also includes Modern Eastern Macedonia. Today, both areas form the administrative 'Region of Eastern Macedonia and Thrace'. In this paper by using the term "ancient Thrace" we refer both to Western Thrace and Eastern Macedonia.

9.49). The loss of Book 7 of Strabo's *Geography*, which was dedicated to Thrace, is very unfortunate, since it deprives us from a wealth of information. In any case, classical literature is dotted with references not only to the geography and the natural features of Thrace (mountains: *Haemus*, *Rhodope*, *Pangaeon*; rivers: *Strymon*, *Nestos*, *Hebrus*, *Angites*; lakes: *Bistonis*, *Cerinitis*, *Stentoris*), but also to what had become almost stereotypical of Thrace, namely, its extreme weather conditions. Thrace was considered one of world's almost inhumanely cold and frozen regions, perhaps with a certain degree of poetic exaggeration (and inaccuracy). The image of the river Hebrus getting frozen is often evoked in Latin poetry as a manifestation of harsh icy weather (Virgil, *Aeneid* 12.331; Horace, *Epistles* 1.3.3).

It seems as if the extreme conditions of the place had a huge impact on the behavior of its inhabitants as well, judging from the surviving literary sources that almost unanimously portray ancient Thracians as ferocious, warlike, and barbarous. Herodotus (5.3), who had himself visited large parts of Thrace, claims that the Thracians were the second most numerous nation of his world after the Indians. The majority of Greek and Roman historiographers offer stories of unprecedented brutality with Thracian leaders as protagonists (e.g., Herodotus 8.116; Thucydides 7.27; Xenophon *Anabasis* 7.4.6; Tacitus *Annals* 2.64ff.). The amount of truth in such descriptions is debatable, especially in light of the most recent archaeological finds, which offer glimpses of a much more advanced and sophisticated culture than the one conventionally thought of. Perhaps this perceived primitiveness may be related to Thracians living simple lives in open villages. Although their exact origin is unclear, the Thracians must have been of Indo-European descent and their society was tribal (Pliny the Elder in his *Natural History* 4.18 mentions approximately 26 different tribe names). Their first contact with the Greeks must have taken place well before the seventh and sixth centuries BCE, when the Greeks planted various colonies on the shores of the Black Sea and the southern Aegean coastline, probably already in the Late Bronze Age (Graninger, 2015; Tsiafaki, 2018, 2020).

In terms of religion, we know that one of the deities most worshipped by the Thracians was Dionysus (also called Sabazius) who had an important cult center and oracle on mount Pangaeon. It is surely not haphazard that that wine was one of most famous products of Thrace throughout antiquity. From Homer we hear that Odysseus got Cyclops Polyphemus drunk with wine from Maroneia, a Thracian city on the Aegean coastline (*Odyssey* 9.196–207). This wine with its exquisite colour and aroma maintained its reputation even during Roman times, according to Pliny the Elder (*Natural History* 14.6). In addition, the area often appears in literary sources also for its rich mineral deposits. Thrace was renowned for its gold and silver, but copper, iron, and salt should also be mentioned. Mount Pangaeon hosted the most productive gold mines of Thrace (Herodotus 7.112; Appian, *Civil Wars* 4.106), which made Philip II the Macedon occupy the nearby settlement of Krenides and refound it as a new city, named after him (Philippi).

2.2 Mythology as a Key Element

The presence of Thrace in mythology is equally strong (Tsiafaki 1998). Situated at the edges of the known world, this unexplored territory of wonders and violence, lingering between reality and fantasy, quite understandably earned for itself a special place in the imagination and the mythological unconscious of both Greeks and Romans. The very origins of Thrace were traced back to myth. According to Andron of Halicarnassus (a 4th c. BCE Greek historian), Thrace together with Europe were the daughters of Ocean by Parthenope (Jacoby 1995, 162 = *FGrHist* 10 F7). We also hear that Thrace was the residence of Thrax, one of the sons of god Ares and patron of the region. Following Euripides (*Alcestis* 498) a golden shield of Ares, who is called Thracian, was kept in his temple near the lake Bistonis. The Thracians, led by Akamas and Peioros, appear in Homer's *Iliad* (2.844, 5.462) as allies of the Trojans. In the *Iliad* we also read about the Thracian king Rhesus, who came to help the Trojans, but was killed by Diomedes and Odysseus, and his famous horses were stolen during a night raid on his camp (*Iliad* 10.469–502).

Beyond doubt the most famous mythical figure of Thrace is Orpheus, the legendary musician and poet with the unique ability to charm all living beings (even stones) with his music. The Thracian land with its mountains (Pangaeon, Rhodope) and rivers (Strymon, Hebrus) offers the background for Orpheus' tragic story: the accidental loss of his wife, his desperate (and almost successful) attempt to bring her back from the underworld, and his cruel death at the hands of the Bacchants (cf. e.g., Virgil, *Georgics* 4.453-523; Ovid, *Metamorphoses* 10.1-85, 11.1-66). Diomedes is another important mythical figure of Thrace. He was the son of Ares and Cyrene, king of the Thracian tribe of the Bistones, renowned for his man-eating horses. One of Hercules' Twelve Labours was to bring to Eurystheus at Mycenae the savage horses of Diomedes. Hercules stole the horses to complete the task and killed Diomedes.

Greek mythology is replete with Thracian heroes: Lycurgus, Phineus, Eumolpus, Polymnestor, Tereus, and Oeagrus to name but a few populate the strange region of Thrace and with their bizarre and (most often) disturbingly violent stories transform Thrace from a faraway region into an unusual, dark and mysterious mythical place.

2.3 The Area in the Silk Road

Placed at the crossroad between Europe and Asia, the Balkan hinterland and the Aegean Sea, Eastern Macedonia together with Thrace were traversed by a network of roads connecting the West and the East. The fact that Constantinople (modern Istanbul), the capital of the Eastern Roman Empire, was built on the Eastern Thracian shore, made this road network one of the most travelled and surely one of the most important (economically, politically, culturally) bridges of communication between Europe and Asia.

Thrace possessed two main highroads, one at the south and one at the north (for details, see Smith 1854 s.v. *Thracia*; Bouzek and Graninger 2015, 17–18). The southern one, which was also named "the King's road" (Herodotus 7.115; Livy 39.27), because it was used by Xerxes during his first invasion of Greece, started from Byzantium and following a coastal route (crossing rivers Hebrus, Nestus, the city of Philippi, river Strymon, the city of Amphipolis) through inland Macedonia ended to Dyrrhachium on the Adriatic sea. The same route was later followed more or less by the famous Via Egnatia, which was constructed by the Romans in the 2nd c. BCE. The northern highroad started also from Byzantium and following a course from the upper valley of Hebrus river, reached Adrianopolis (modern Edirne) and Philippoupolis (modern Plovdiv) and from there continued to Sophia, Niš and ended in Belgrade (Serbia).

This system of ancient routes became part of a wider network of trade routes, the so-called "Silk Road", which for centuries facilitated not only the trade of goods (e.g., silk, paper, gunpowder, pepper, glass, gems) but more importantly the exchange of people and cultures, customs, ideas, technology, science, and art. Since the time of Alexander the Great (356–323 BCE) and more emphatically during the time of the Han Dynasty in China (c. 202 BCE-220 CE), the routes of Eastern Macedonia and Thrace played a vital role in the economic, cultural, political, and religious interaction between West and East as part of the Western Branch of the Silk Road. Archaeological sites (Amphipolis, Philippi, Abdera), commercial stations along Via Egnatia (Anastasioupolis, Komotini), ancient and modern ports (Amphipolis, Neapolis (modern Kavala), Abdera, Maroneia, Alexandroupolis), historical centres of the tobacco industry (Kavala, Xanthi), silk production centres (Soufli), museums (archaeological, folklore, and thematic of tobacco and silk production), architectural heritage (churches, mosques, warehouses, caravanserais) form a dense network of historical links of the region to the Silk Road and offer a unique opportunity for a critical re-appraisal of its assets (natural, cultural) within the framework of cultural tourism (for a detailed discussion of the potential of the region for economic growth and social development through the promotion of cultural tourism, see Kostopoulou et al. 2016, 1–30; World Tourism Organization 2018, 10–33; 41–43). From the above, it becomes clear that the routes (historical, mythological) of Eastern Macedonia and Thrace, as a varied and colourful mosaic of cultural, linguistic, and religious diversity, constitute an integral part of the legacy of the first transnational contact between people of West and East, while offering at the same time today the opportunity for historical and cultural research, interregional cooperation, travel facilitation, and tourism innovation.

2.4 State of the Art in Relevant Research

Classical literature (both Greek and Roman) feeds on references to the myths of Thrace and/or mythical figures associated with Thrace, e.g., the unfortunate love story of Orpheus, the Thracian bard, and his descent to Hades or the story of Rhesos, the Thracian king, who helped the Trojans, or the man-eating horses of the Thracian Diomedes, which were tamed by Hercules belong to a reserve of well-known mythological narratives. Indeed, Thrace as a land of myth has its own special place in classical scholarship. These references, however, generally concern individual figures. Moreover, even if we are not dealing with dedicated studies on specific heroes, most comments on mythical Thrace are usually scattered in secondary literature, like commentaries or individual studies of literary texts, mythological textbooks and dictionaries. As a result, myth is treated within the interpretative context of particular literary texts. Hence, the mythology of Thrace per se is yet to be studied. It is indicative that even in the most recent and comprehensive Companion to Ancient Thrace (Graninger 2015) mythical Thrace lacks its own, distinct presence, since it does not receive a dedicated chapter. It thus becomes clear that a monograph, a study or a project dedicated to Thracian myths is still a desideratum. In this respect, our project through the collection, systematization and collective presentation of the available literary evidence about the myths of Thrace, combined with the relevant archaeological data, constitutes a major and innovative step towards a more comprehensive approach of the mythological identity of the region. Finally, the pedagogical utilization of the research outcomes adds a modern twist to conventional educational application by enhancing learning through play and IT.

3 Archaeology and Artefacts

3.1 Studing Myths via the Greco-Roman Visual Arts

Whether ancient Greeks believed the stories of these myths has been an intrigue question for researchers of

Greco-Roman past (Veyne 1983; Burgess 1989; Wiebe 1994; Naas 2008; Buxton 2009, 207–209; Johnston 2018, 18–22, 299). However, it is widely accepted that myths were considered to be a distant but acceptable past, often used to legitimize choices of the present (Boardman 2007, 15–32, 74–75). In fact, antique landscape was, among other things, a mythological landscape (Boardman 2007, 103–115). That being said, it is no surprise that a major part of Greek art revolves around mythical past. Marble and clay, the two most durable materials that managed their way through more than 2000 years to modern era, very often present gods, heroes and their deeds. They weren't the only ones, however. Mythical scenes were also depicted on mosaics, metal or paintings (the bibliography on the topic is vast; a few examples: Ackermann et al. 1981–2009; Shapiro 1994; Kondoleon 1995; Dunbabin 1999, 298–300; Junker 2012, esp. 96–119; Sourvinou-Inwood 2014; Newby 2016).

Being one of the most common sceneries for many myths, as well as the geographical and cultural area of a "significant Other" (for versions of the Thracians' "Otherness": Hartog 1988, 86–88, 190–191, 245, 334 no 71; Hall 1989, 101–110, 133–138, 151–154; Cartledge 2002, 45–46, 68–70, 153–155; Xydopoulos 2007a, esp. 171–178; Xydopoulos 2009; Gruen 2011, 203, 321–323; for less "Otherness" in Herodotus' narrative: Xydopoulos 2004; Xydopoulos 2007b; Tsiafakis 2000), ancient Thrace was not ruled out from the canon. Iconographical studies are a huge field that interests many archaeologists (the titles are practically uncountable, especially in Attic pottery; a few examples: Tsiafaki 1988; Kefalidou, 1996, 2009; Tsiafaki, 2002, 2016; Franceschini, 2021). Studies are related complementarily to other archaeognostic sciences, too, with philology being a primary one. Hence, the amount of archaeological discourse produced on iconography is directly proportional to the massive body of data available that comes to light by numerous excavations. Thus, the gathering and processing of representations for the project's myths took place among a variety of artefacts decorated with related mythical scenes. The existence of various publications and corpora, some of them digital, facilitated this procedure.

First of all, there is the *Lexicon Iconographicum Mythologiae Classicae*, a multivolume encyclopedia abbreviated as LIMC. The printed version (1981–2009), as well as its most recent, but not so extended, digital one are the guides for every researcher interested in classical mythology and its presence in visual arts, an "indispensable research instrument" (Hansen 2004, 244). The digital version (Digital LIMC, weblimc.org) comprises three components: Digital LIMC, LIMC France and LIMC Greece. Another valuable resource is the *Beazley Archive* (Classical Art Research Centre—Beazley Archive, ox.ac.uk), a repository implemented by the *Classical Art Research Centre* (CARC), a research unit of the *Oxford University Faculty of Classics*, where one can have access to and search via various terms the *Beazley Archive Pottery Database* (BAPD, Pottery ox.ac.uk), as well as an archive of engraved gems (Gems, ox.ac.uk). Other (digital) resources available are the *Corpus Nummorum Online* (Corpus Nummorum, corpus-nummorum.eu), useful since many times gods and heroes linked with specific *poleis* (ancient cities) were presented on the obverse of coins as well as many websites of museums, foreign and domestic, with their rich collections available in online repositories. Whereas mythographic texts or selected ancient artefacts regarding Thracian mythology can be found separately in other research projects[2] and publications (e.g., Hawes 2017; Thomopoulos et al. 2022), the combination of both with a full collection of artefacts focused on the myths of Thrace is unique (at least to our knowledge). The dual purpose to enhance relevant joint academic research but also to reinforce and visualize a trip in the region (in the periphery of ancient and modern Greece) with ancient stories and heroes (turning the intangible myths to tangible artefacts and suggested routes), brings added value.

3.2 Artefacts and the Thracian Myths

As far as the iconography per se is concerned, Thracian myths were quite a popular subject in visual arts, especially in Attic decorated pottery. However, not every Thracian story or even the various episodes of a myth, as documented by the project's philologists, enjoyed the same popularity (compare, for instance, the various cases of Attic pottery in Tsiafaki 1998). According to our first general findings, among the thirty six (36) myths that have been selected for inclusion in the project, seven (7) of them, are not so far connected with any ancient represantation. More specifically, they are the myths regarding Acamas, Eyphemus, Cisseus, Oeagrus, Pangaeus, Sithon and Phyllis. As one would expect, the least widespread myths with only regional connotations or the less prominent heroes belong to this group.

On the other hand, scenes with world-known Orpheus are, expectedly, on the top of this list. Being a famous musician, poet, a divine singer, prophet, and mythical founder of orphism, a system of mystery cult rituals and

[2] Like in the digital initiatives of the Mythlab (https://www.manto-myth.org/) or within ToposText (https://topostext.org/), which combine literature and topography, or in the COSMOS project (https://cosmosapp.gr/) that will visualise myths, artworks and the connections between them.

doctrines (Tsiafaki 1998, 41–44), it is no surprise that he can be found in more or less narrative scenes and in a variety of art forms throughout the Greco-Roman antiquity. Orpheus is not depicted on Attic vases before the fifth century BCE, despite the fact he was considered older than Homer (Tsiafaki 1998, p. 45 ff). We find Orpheus singing among the Thracians[3] as a victim of the outraged Thracian women,[4] or even his severed head giving oracles on the island of Lesbos.[5] On the other hand, we see him as one of the Argonauts on a marble metope of the Sicyonian treasury at Delphi dated to the second quarter of the sixth century BCE.[6] The most well-known incident of Orphic mythology is, of course, Orpheus' descent to the underworld and his attempt to retrieve Eurydice. This deed is the subject of a marble roman relief, considered to copy a fifth century BCE original made by Alcamenes.[7] In one of the few preserved wall-paintings from early first century AD Orpheus is depicted among other deities and heroes,[8] whereas Orpheus wearing a Phrygian cap and playing his lyre is a very common subject in Roman mosaics[9] and appears even in coins.[10] The scene with Orpheus playing the lyre surrounded by animals charmed by his music was popular among the early Christians as well, since the mythical musician became an allegory for Christ[11] (for Christian correlations of Orpheus in art and ecclesiastical discourse, see Huskinson 1974, 69–73, 82–85; Jourdan 2011, 2014a, 2014b).

One of the most famous myths concerning Thrace is, of course, the aforementioned Mares of Diomedes. This myth, although not very popular, is also depicted in various materials and periods. As one of Hercules' labours the capture of the Mares is found at one of the metopes of the temple of Zeus at Olympia, dated to the early Classical period.[12] The exact same scene, Hercules capturing a horse, also decorates the interior of an Attic black-figure kylix (drinking cup) made by Psiax around 520 BCE.[13] The subject is still represented on Attic decorated pottery until the end of the fourth century BCE, as shown by a red-figure oinochoe from the national archaeological museum of Taras.[14] The Mares of Diomedes occur also on sarchophagi[15] and mosaics[16] of the the Roman period.

These are, of course, only indicative examples of the many Thracian myths appearing in Greco-Roman art. A list with just the most representative artefacts would have been far from the scope of this paper. However, we feel that we shouldn't fail to mention one more artefact: the famous Lycurgus cup, a rare artefact from late Roman times, made of glass and silver, now in the British museum, depicting the death of King Lycurgus who was punished because he didn't accept the worship of Dionysus.[17] The Lycurgus cup is of course an extravaganza. It is, however, indicative of the range of materials used to depict various episodes of the Thracian mythology.

---

[3] E.g., Museum für Kunst und Gewerbe, Hamburg, 1968.79 (red-figure column krater ca 450 BCE; BAPD 275004; Digital LIMC 9489); Antikensammlung, Staatliche Museen, Berlin, V.I. 3172 (red-figure column krater, ca 440 BCE; BAPD 216168; Digital LIMC 9490); Metropolitan Museum of Art, New York, 24.97.30 (red-figure bell krater, ca 440 BCE; BAPD 14496; Digital LIMC 9513).

[4] E.g., Museum of Fine Arts, Boston, 13.202 (red-figure lekythos, 475–425 BCE; BAPD 214046; Digital LIMC 9545); Louvre G 416 (red-figure stamnos, 470–450 BCE; BAPD 205400; Digital LIMC 9528); Louvre, G 436 (red-figure amphora 450–440 BCE; BAPD 214178; Digital LIMC 9549); Harvard Art Museums 1960.343 (red-figure bell-krater, 440–430 BCE; BAPD 213593; Digital LIMC 9557).

[5] E.g., Antikenmuseum und Sammlung Ludwig, Basel BS 1411 (red-figure stamnos, ca 500–450 BCE; BAPD 275231; Digital LIMC 9523) and BS 481 (red-figure hydria, 475–425 BCE; BAPD 3735; Digital LIMC 9608); Otago Museum, Dunedin, New Zealand, E48.266 (red-figure hydria, 450–400 BCE; BAPD 215555; Digital LIMC 9610). The subject also appears on the reverse of a set of Etruscan mirrors dated to the 300 BC (De Puma and Guthrie 2001).

[6] Archaeological Museum of Delphi, 1322, 1323 + 1323a + 1210, 1345, 1547; Digital LIMC 6564.

[7] Naples, National Archaeological Museum, 6727 (copy of the Augustan era); Digital LIMC 17746.

[8] E.g., Pompei, II 2, 2–5 (House of Loreius Tiburtinus); Digital LIMC 117.

[9] E.g., Antikensammlung der Staatlichen Museen zu Berlin—Preußischer Kulturbesitz, Ident.Nr. Mos. 72 ("Orpheus Mosaic" from Miletus, end of the second century CE; Digital LIMC 9623); Hatay Archaeology Museum, Antakya 10568 (mosaic from Tarsus, last quarter of the third century CE; Digital LIMC 9628; Arachne 2008173–4).

[10] E.g., Münzkabinett der Staatlichen Museen, Nomisma NDP Coin Galleries Altes Museum, Berlin, Inv. no. 18215944 (golden stater from Lamsakos, 387–334 BC; CN coin 14,915).

[11] The subject is quite common in mosaics of late antiquity (see above the references to depictions of Orpheus wearing a phrygian cap), as well as painting, e.g., Pompeii, VI.14.20, Room 18, ("House of Orpheus", mural painting, 1st c. BCE; Ditigal LIMC 88; Arachne 1105513). Two indicative examples of Orpheus as a christian symbol: Byzantine and Christian Museum, Athens, BXM 00001 (=Inv. no. T 93, table support / trapezophoron from Aegina, fourth century CE; Digital LIMC 7865; Arachne 1061680); Catacomb of SS. Peter and Marcellinus, Via Segni region, I, 6 (fresco, probably to the second half of fourth century CE; see Huskinson 1974, 87 no 13, pl. IIIb).

[12] Louvre Ma 719.b (=N 1717) and Archaeological Museum of Olympia, Λ 93 (Parian marble, ca 460 BCE; Digital LIMC 17027; Arachne 1180667).

[13] The State Hermitage Museum, ΓΡ-28190; BAPD 320368; Digital LIMC 21446.

[14] Taranto, Museo Archeologico Nazionale, inv. no. 5110 (330–310 BCE; BAPD 9003514; Digital LIMC 21456).

[15] E.g., British Museum 1800, 0712.1 (from Phrygia, 160–190 CE; Digital LIMC 17075); Archaeological Museum of Antalya, A 928 (from Perge, Pamphylia, 150–155 CE; Digital LIMC 9089).

[16] E.g., Museo Arqueológico Nacional, Madrid, room 22, no 38315 BIS (from Liria, Valencia, 201–250 CE; Digital LIMC 8890).

[17] British Museum 1958,1202.1; Digital LIMC 17366.

4 Data Processing and Interconnection

One of the main challenges of this interdisciplinary project was to find ways to combine very diverse research data from different scientific areas that have been collected within various contexts, in order to fullfil miscellaneous purposes, and that differ in content, size, scope, and internal structure. These data range from literary texts of various forms, genres, and topics, to archaeological artefacts from the Greek and Roman antiquity that relate to the mythological part of the project in the form of images, such as paintings, pottery, or coins, to contemporary information relating to different aspects of life for the visitor of the area in the form of points of interest, such as natural environment, architecture, folklore, and gastronomy, combined with diverse multimodal material (texts, audio, images). All the above represent the content of the project, which needs to be revealed to visitors while selecting their own itinerary on the map. This adds to the list of data a complex set of geographical information, which also needs to be carefully linked to all the other datasets to facilitate easy search and lead to relevant results.

4.1 Interconnecting Data

In order to integrate the aforementioned sets of diverse data that are collected from the different research teams in an effective way, there was careful planning of a detailed database schema, catering for specified allocation and documentation of each kind of data depending on their type, form, and content. To this end, the chosen content management system was carefully selected so as to provide researchers with an interoperable, flexible, mature, open-source platform that would meet all these criteria. Therefore, the back-end of the system was built on PHP (https://www.php.net/), a language commonly used to facilitate similar web platforms. Along the same lines, Leaflet JS library was used to facilitate the process of geocoding. All peripheral tools and add-ons that were incorporated in the back-end were selected to ensure ease of use, interoperability and smoothness of integration in the context of a common management mechanism. The resulting content management system guarantees the safe storage, appropriate documentation and easy access to the multimodal content of the database.

A key element that takes the system database a step further is the adoption of a suitable metadata schema that does not only ensure quick and efficient information retrieval, but is also compatible with international standards, thus allowing future cross-linking to other datasets. Therefore, each piece of data included in the database is appropriately accompanied by the respective annotations. These metadata include text indexing elements, ranging from text title, author name, date of creation, language, genre, topic, etc. for literary texts to suitable annotation for all types of multimodal material, be that sound, or image. As all parts of the information are bilingual, a dedicated part of the database contains the translations of both the content as well as the accompanying metadata. Finally, the annotation process is carried through in collaboration between experts from different fields to ensure accuracy in indexing.

4.2 State of the Art in Metadata Schemata in Relation to the Project

Research in philology and arhaeology, as in all other humanities, strongly depends on suitable classification and management through appropriate metadata schemata. Apart from their obvious use, namely, proper indexing and content retrieval, these schemata often add value to the content they characterize by allowing it to be linked to other relevant collections of data.

As far as annotation schemata are concerned, one of the most prominent and widely used standards that covers an extensive range of disciplines is the Text Encoding Initiative (TEI, https://tei-c.org/), which has been incorporated and used in several projects. Another major standard to this respect is the CIDOC Conceptual Reference Model (https://www.cidoc-crm.org/Version/version-7.2), which has been specifically used to tackle culture and heritage-related content. The metadata used to annotate the project's database is compatible with both of these standards.

As regards textual data, a wide range of relevant state-of-the-art toolsets and pipelines are available to researchers worldwide. In the more particualar case of excerpts from Ancient Greek and Latin literature, UDPipe (Straka and Straková 2019) is one of the latest tools that fascilitates annotation in language independed raw data. For the purposes of part-of-speech tagging and lemmatization of the texts in Modern Greek, a robust suit of tools that has been successfully employed in several other projects was used (Prokopidis et al. 2011).

Finally, the controlled vocabularies that have been created to annotate the contents of the database comprise a thesaurus which is used to efficiently index data and facilitate their easy retrieval by end users. Relevant state-of-the-art thesauri include the UNESCO Thesaurus (https://skos.um.es/unescothes/) concentrating on cultural content among other disciplines, which has been constantly growing in size and languages. Last, but not least the Getty Vocabularies span a wide range of culture-related subdomains with relevant schemata that also cater for several types of multimodal content. A typical example of use in the context of this project is the Getty Thesaurus of Geographic Names (TGN), used for names of geographical entities (https://www.getty.edu/research/tools/vocabularies/tgn/). Another example relevant to this project is the Art and Architecture Thesaurus (http://www.getty.edu/research/tools/vocabularies/aat/), which accommodates any object of material culture (Vacalopoulou et al. 2019a, p. 134). Such standards were used to ensure terminology control across disciplines in the content.

Metadata schemata were incorporated in the context of the project, not only to ensure proper annotation, indexing and retrieval of the data but, just as importantly, to offer interoperability with other projects, models, and databases. One of the first and most prominent applications of this feature is the incorporation of the content of the database in APOLLONIS, the Greek Infrastructure for Digital Arts, Humanities and Language Research and Innovation (https://apollonis-infrastructure.gr/), and, more specifically, in clarin- (http://hdl.handle.net/11500/CLARIN-EL-0000-0000-75F5-8).

5 Methodology

This section outlines the main methodological considerations as regards the collection, processing, and interconnection of the several types of content drawn together in the project across disciplines and modalities. After a detailed account of the methodology pertaining to the mythological and the archeological material, follows an account of the principles for handling content from current life in Eastern Macedonia and Thrace to link it with the rest of the content and showcase it via applications in tourism and educational technology.

5.1 The Mythological Material

From a total of approximately fifty (50) myths that are connected to this area, thirty six (36) myths have been selected. The selection criteria were the degree of relevance to the area of Eastern Macedonia and Thrace, the number of the texts containing references to them, and the quantity of the surviving literary and archaeological material. Each myth corresponds to an entry that contains the following:

(a) A concise introductory note, which includes the main features of each mythological person, such as biographical information and their special connection to the area.
(b) The texts (i.e., the sources) containing references to the mythological figure. Overall, two hundred and forty five (245) sources have been indexed, including eighty (80) authors (56 Greek and 24 Roman), from the Classical period and late antiquity until the Byzantine times. The exposition follows a chronological order and their length varies, from a few lines to whole paragraphs. The texts used come from critical editions, published in international series, such as the: Oxford Classical Text, Bibliotheca Scriptorum Graecorum et Romanorum Teubneriana, Les Belles Lettres, Loeb Classical Library, and others. The number of sources that are relevant to each myth vary depending on its popularity. Where the surviving evidence is meager, all sources are cited; where plenty of sources are available, only a selection. Overall, the aim of the project is not to create an exhaustive repository of texts but rather an easily accessed list of sources for the general and/or the more demanding public. A special category of sources consists of travel writings, produced by authors who visited Eastern Macedonia and Thrace in antiquity (such as Pausanias and Strabo), and in Medieval and Modern times (such as Buondelmonti and Ciriaco d'Ancona). The literary genres appearing in the sources cover almost the entire spectrum of antiquity's literary production both in prose (historiography, myth-writing, biography, rhetoric, philosophy, and scientific texts, such as geographical works, and ancient scholia) and in verse (epic, drama, elegy, epigram, lyric, bucolic, and didactic poetry).
(c) The translation of the sources in Modern Greek and in English. The Greek translations were undertaken by the authors of the entries. As for the English ones, new translations were prepared by the authors, where no published translations were available.
(d) A short note about the life and work of each ancient author.
(e) Information concerning the mythological figure (god, hero, king, nymph, etc.), and their connection to localities (city, cluster of houses, archaeological sites) and other geographical landmarks (mountain, river, lake).
(f) Key-words concerning the content of each myth (war, killings, deception, love, blinding, metamorphosis, etc.), and its connection to other mythological persons.
(g) Bibliographical references.
(h) Digital material of tourist nature that connects the ancient content with the contemporary urban and natural environment of each locality.

Tools for indexing include—among others—encyclopedic and mythological dictionaries, monographs, doctoral dissertations, bibliographical databases and digitized full-text repositories of Greek and Latin literature (TLLG, PHI Latin Texts, The Latin Library, etc.) as well as material available on the internet. Particular reference should be made to dedicated websites such as the A′ Thracian Electronic

Thesaurus (1996; www.xanthi.ilsp.gr/thraki/), and the B′ Thracian Electronic Thesaurus (2005–2008)/Guide to the Periphery of Eastern Macedonia-Thrace (www.ceti.gr/thesaurusII/index.php?lang=gr&thm=9), as well as to the projects ToposText (https://topostext.org/) and Travelogues (https://el.travelogues.gr/) of Laskaridis Foundation. Figure 1 presents a typical myth entry as viewed in the content management system.

Concerning the structure of the project, three working groups have been formed: the first group consists of nine (9) persons (5 PhD candidates and 4 graduate students), who were responsible for writing the entries and entering them in the project's dedicated platform that was set up by Athena research center. This material is complemented with extensive thematic indexes. The second group included five (5) persons (2 post-doctoral researchers, 2 PhD candidates). This group evaluated the quality/accuracy of the indexed material and approved its entry in the platform. The third group comprised the scientific board of the project (4 university professors) who coordinated the project both on academic and administrative level.

The project also features an educational application, which includes educational scenarios, worksheets, educational games, material to facilitate visits to museums and archaeological sites in Eastern Macedonia and Thrace. The educational material is aimed at both the secondary and tertiary education.

5.2 The Archaeological Material

Apart from the literary material, there is a designated form for registering data for the archaeological material. This material consists of varied ancient artefacts (Greek and Roman) that depict scenes from the selected myths. For indexing these objects our team uses some of the most prominent research tools available: *Lexicon Iconographicum Mythologiae Classicae*, Digital LIMC, LIMC France, LIMC Greece, *Beazley Archive Pottery Database* (BAPD), *Corpus NumorumOnline*, etc. For every artefact there is an entry that contains the following data (in Greek and English), as free text, as selections from dropdown lists, or as multimedia content:

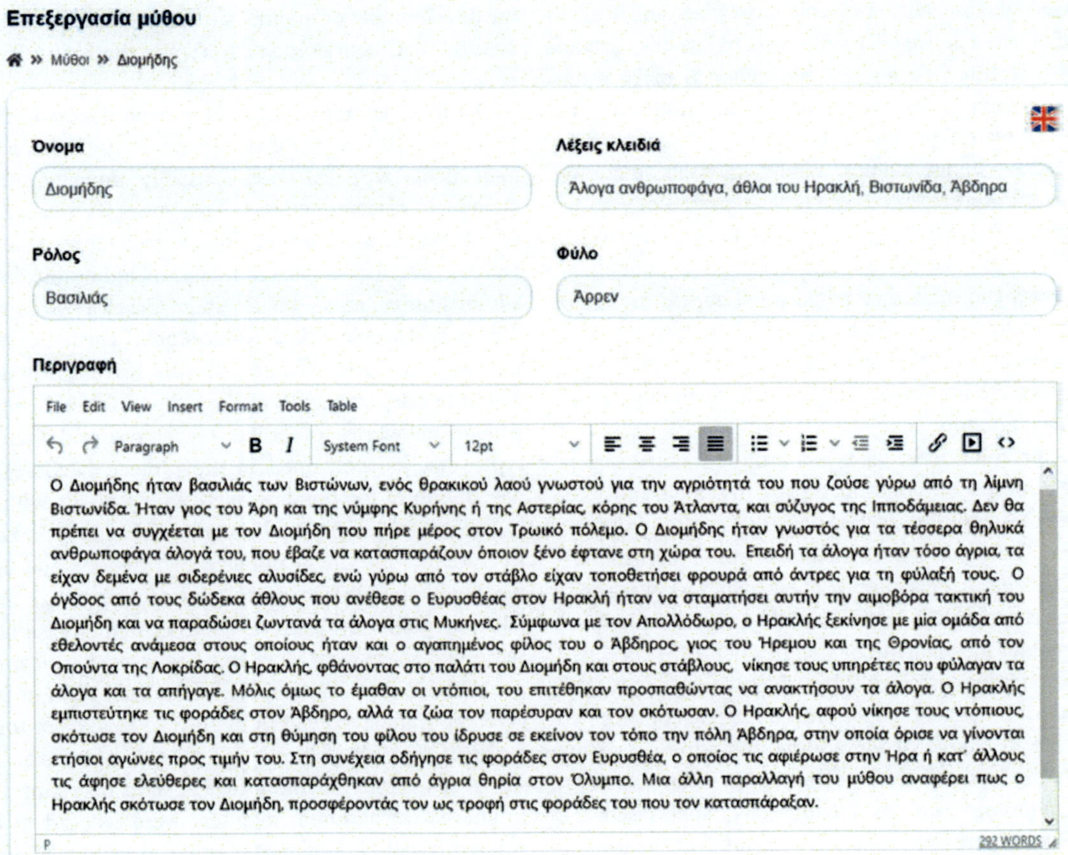

Fig. 1 Structure of a representative myth entry (Diomedes) in the content management system

(a) A general text describing the artefact, mainly the scene that is connected with the documented myth and mythological person,
(b) Information regarding the artist, the chrononology and the location of the artefact (where it was found and where it is kept today),
(c) Associations with one or more myths and mythological persons documented in the project's database, so that the artefact can help the general or specialized public to visualize these scenes and the persons involved,
(d) Associations with one or more spatial features (ancient cities, mountains, rivers, etc.), documented in the project's database, in order for the artefact to enrich visually the relevant places,
(e) Multimedia content regarding the artefact, mainly pictures, drawings, etc., and
(f) References to publications (printed or digital), in which anyone interested can find more valid archaeological documentation about the artefact itself.

In total, the archaeologists involved in the documentation or artefacts has started with gathering archaeological material from various printed or digital resources and inputting this material in the project's database. The aim was not to enter in the database all the discovered Greek and Roman artifacts that are connected to the selected myths, accompanied by scientific information, but rather to make a representative selection in order to provide users of the application an overall picture of the beauty and variety of the artefacts (vases, coins, statues, reliefs, mosaics, etc.), the artists of which were inspired by Thracian mythology within a large time range. All the gods, deities, famous musicians, heroes, or others that turned into rivers and mountains, Thracian tribes and individual persons, all appear in various artefacts of ancient Greek and Roman art. These artefacts are located in museums and collections around the world, a fact that underlines the importance of Thracian mythology in the contemporary era, too.

In the Athena RC team there is a group of archaologists, formed by four people, one director of research and three scientific associates (1 PhD in Classical Archaeology, 1 MA in Museum and Gallery Management and licensed tourist guide, 1 MA in Classical Archaeology). Their main task was registering the archaeological material in the project's platform, that accompanies the main litterary entries of the myths, based on the research tools described above. This work was layered, since one associate registered the archaeological content, two of them were in charge of the editing and approval of the entries, and the director or research, who is an expert in Thracian mythology and its iconography, supervised the whole process. Figure 2 presents a typical artefact entry in the content management system.

5.3 Showcasing Content for Applications in Tourism and Educational Technology

Alongside the mythological starting point and the accompanying archaeological material, the project content is enriched with a variety of other types of information that might be of interest to visitors of the area. The different components comrpising this content have been carefully selected to cater for a wide range of interests.

In order to make the right decisions regarding these criteria, researchers consulted a number of relevant authorities and organizations, such as the Hellenic Statistical Authority (ELSTAT, www.statistics.gr/) for matters concerning the settlements of the area; the Permanent Catalogue of the Registered Movable Cultural Goods of Greece (listedmonuments.culture.gr) for pinpointing and characterizing architectural points of interest; and Ygrotopio, the WWF portal for the dissemination of environmental information on Greek island wetlands (www.oikoskopio.gr/ygrotopio/en_US/index.php) for data concerning wetlands in the area. The wealth of pre-existing research in the context of previous projects by the research teams was also explored, the most relevant of which being the Thracian Electronic Thesaurus (www.xanthi.ilsp.gr/thraki/) and the εlectronic platform for the development of internet site aiming at the presentation of cultural and tourist content (www.xanthi.ilsp.gr/ cultureportal/).

All these are included in the project database in the form of points of interest so that they can easily be located on the map of the area and used—possibly along with other elements—to create itineraries. Apart from its name, each point of interest is accompanied by several other types of data, namely: alternative names, if any, including historical ones; administrative area—be it either Eastern Macedonia and Thrace as a whole, Eastern Macedonia, Thrace, or each of its five ρegional units (Drama, Evros, Kavala, Rhodope, and Xanthi); category, and category type (Table 1 presents the general categories of points of interest and their specific types that are contained in the project database); a short description of the point of interest; any myths with which it is linked (for example, the myth of Evros is linked to the river Evros, the mountainrange of Rhodope and the settlement which is the modern town of Orestiada); any other relevant points of interest, such as multiple sights in the same village; as well as relevant multimedia files, if any.

All the textual context offered for each point of interest is given both in Greek and in English so as to cater for a wider range of possible visitors. Special care was taken to make sure that every description text is short and simple in both languages. In addition, inputters were advised to consider different types of audience, aiming to be appealing both to the general public and to special interest groups. To this end,

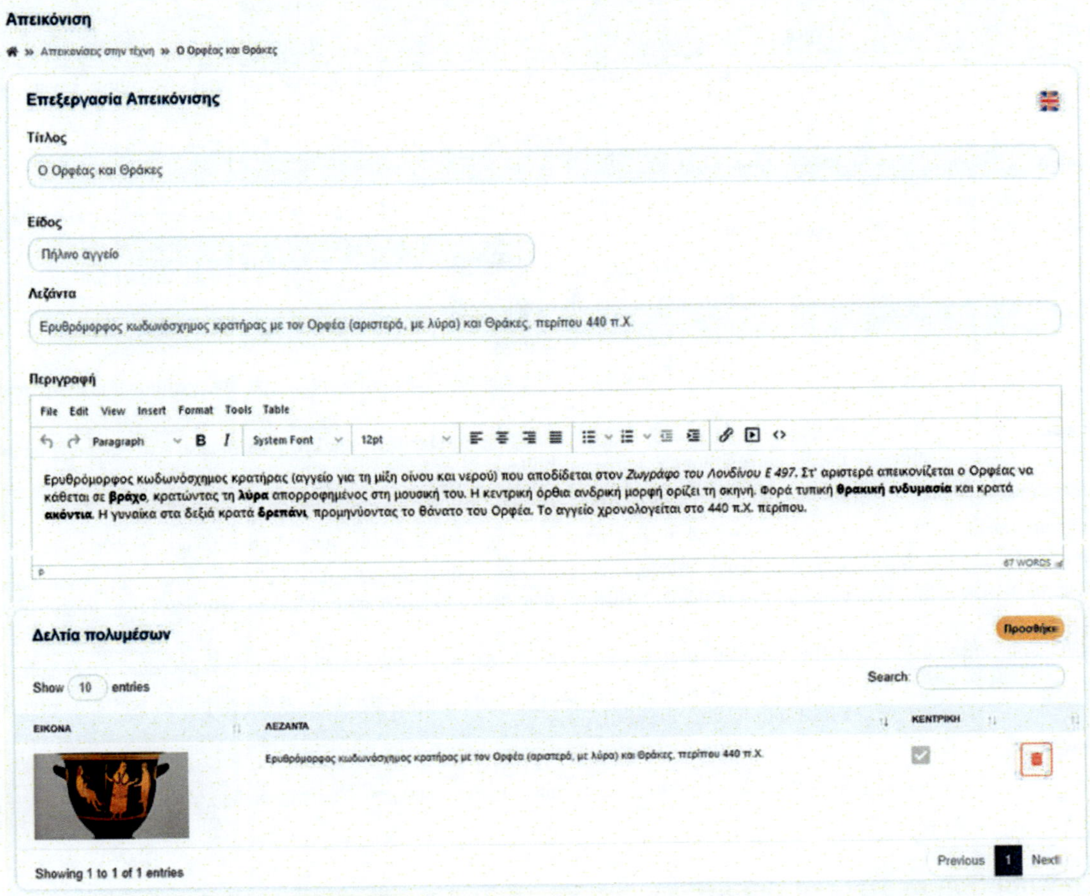

Fig. 2 Structure of a representantive artefact entry (related to the myth of Diomedes) in the content management system

all text has been designed to be both descriptive, pleasant to read, and easy to follow—using simple syntax and clear vocabulary. A typical point of interest entry in the content management system can be seen in Fig. 3.

Furthermore, selected internet links are included provided that they are alive, useful, and not tiring for users. A final critical point as far as textual material in concerned is uniformity, which is ensured by a specially designed language handbook containing details on spelling and other conventions to be followed by every member of the inputting team. Apart from written text, the content is also speech-enabled to be fully accessible in both languages by people with visual difficulties.

Apart from employment in tourism, the rich multimodal content of the database is also used in the context of an educational application. This parallel practical usage aims at the pedagogical utilization of the research outcomes by users involved in secondary and higher education. In order to produce a new environment for use as a standalone tool by this particular audience, several tools were included both for students and for educators. These comprise specially designed educational scenarios, lesson plans, exercises, as well as games. The idea behind the existence of these parellel applications is to showcase a rich and diverse multimodal content, which has come from extensive interdisciplinary research, in two distinct areas, namely, tourism and educational technology, with added value for the respective end users.

6 Research Outcomes

"Mythological Routes in Eastern Macedonia and Thrace" has resulted into a vast collection of rich and varied material as a result of extensive research in several areas of science and technology. On the one hand, this has contributed, at different levels, to the state of the art of each respective field, i.e. philology and mythology, archaeology, and metadata schemata. On the other hand, the combination of these outcomes, as showcased in the end results of the project, has led to a more general contribution to society.

Table 1 Metadata schema for types of points of interest in Eastern Macedonia and Thrace per category

Category	Types
Activity	Birdwatching, caving, climbing, cycling, diving, fishing, hiking/walking, horse riding, kayak, rafting, sightseeing, skiing, sports, swimming, water sports
Area	Historical settlement center, industrial zone, protected rural area, rural area, urban area
Gastronomy	Sweet, drink, food
Geographical element	Bay, beach, cape, cave, delta, falls, forest, gorge, hill, island, lake, lagoon, mountain, peninsula, plain, river, spring, stream
Event	Celebration, cultural event, exhibition, food/drink festival, religious festival, theater, other
Fauna	Amphibian, bird, fish, insect, mammal, reptile
Flora	Bush, herbaceous plant, tree, wildflower
Folklore	Costume, fairytale, folk dance, folk song, musical instrument, social life, traditional activity, traditional architecture
Infrastructure	Cemetery, cistern/well, dam, industrial unit, lighthouse, market, park/square, school, sports facility, theater/cinema, tunnel, university/college
Museum/Exhibition area	Archeological, art gallery, folk, historical, natural history, other
Natural environment	Area of ecological interest, area of special natural beauty, natural phenomenon, scenic route, view spot
Point of archeological/historical interest	Aqueduct, baths, bridge, cave, cemetery, fortification, industrial building, monument, production/extraction point, public building, residence, road network, sanctuary, settlement, temple/church, tomb/barrow
Point of religious interest	Bell tower, Cathedral, Christian church, Khanqah, Madrassah, Monastery, Mosque, Synagogue, Türbe
Service	ATM, bank, citizens' service centers, customs, fire department, hospital, police, port authority, post office, tourist information, townhall, other
Settlement	City, town, village
Transport	Airport, bridge, bus stop, bus terminal, marina, path, port, railway station, road, taxi rank

6.1 Contribution to State of the Art in Philology/Mythology

The main characteristic of this project is its heuristic function, as it contains not only the myths of Thrace but also the sources where these myths are mentioned, and therefore it can be used as a research tool that enables scholars to study all relevant material (literary, archaeological, and artistic) of the selected myths. Even though this material has not been thoroughly examined at this stage, we may nevertheless attempt the following initial observations regarding the contribution of the project to the state of the art, as stated above: (a) This is the first time we have a dedicated collection of myths and/or mythological references focusing exclusively on the region of Eastern Macedonia and Thrace; (b) apart from scholarly usage, the mythological references included in the project can also be used for pedagogical purposes (through the dedicated educational platform); and (c) the interconnection of mythology with cultural tourism discloses the intangible cultural heritage of Eastern Macedonia and Thrace and will thus contribute significantly towards the re-evaluation and development of the cultural potential of the region.

6.2 Contribution to State of the Art in Archaeology

As far as archaeology is concerned, a lot of fruitful preliminary outcomes can be detected. Our work on the archaeological content could delimit a research context on study of the Thracian myths in relation to ancient iconography and (ancient and modern) topography. It appears to be a wide range of materials that can be studied (sculpture, vase-painting, mosaics, metal objects, etc.) and an even wider range of dating, beginning at least from the Classical period and ending to the early Christian times. The project also shows rather emphatically that a very interesting archaeological question is who produces artefacts depicting Thracian myths, in which materials and for whom. Although there is substantial bibliography on the subject for certain types of artefacts, the combination promoted by the "Mythological routes in Eastern Macedonia and Thrace" can be considered as a fresh start.

Moreover, the close cooperation of philologists and archaeologists within the same project provides the potential to work together on ancient Thracian myths and promote the relevant research and forthcoming publications through an

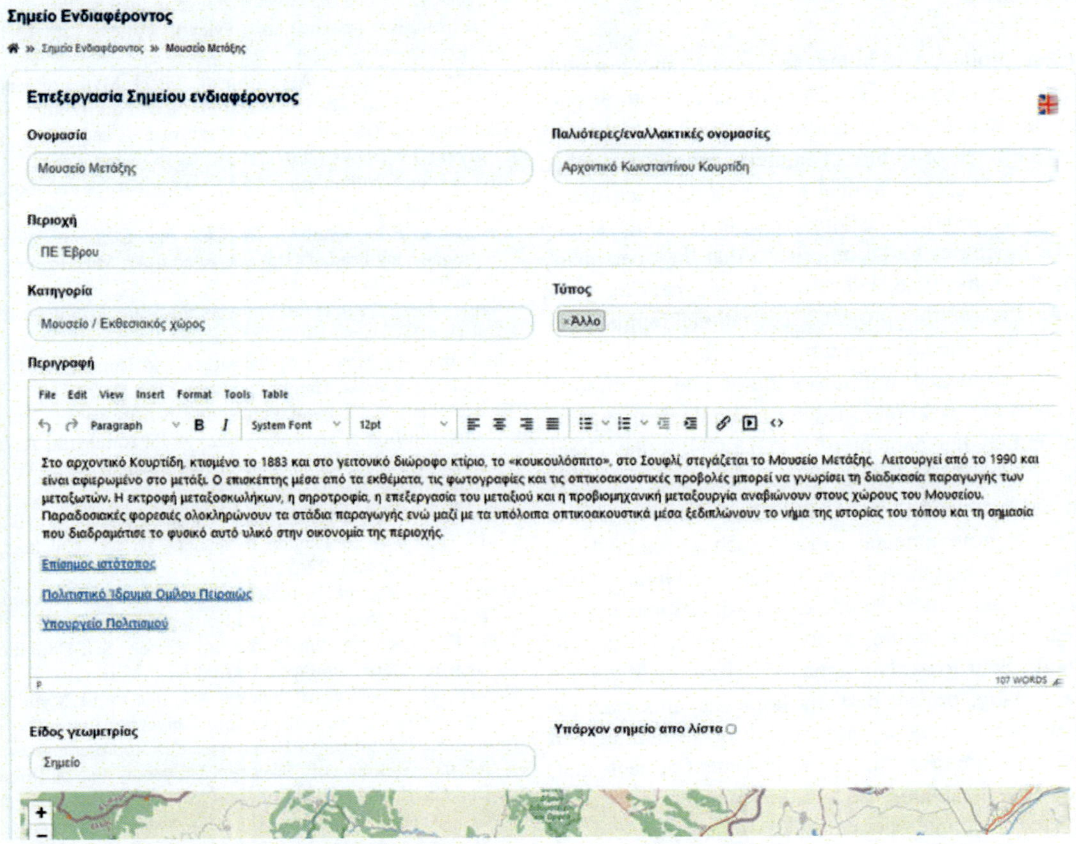

Fig. 3 Fields for a point of interest (Silk Museum) in the content management system: name (Ονομασία), Older name (Παλιότερη ονομασία), Administrative area (Περιοχή), Category (Κατηγορία), Type (Τύπος), and Description (Περιγραφή)

innovative and combined view. Finally, we shouldn't fail to mention one more observation: in our effort to relate the various mythical episodes depicted on artefacts with specific sites on the map, so as to fulfill the tourist aspect of the project, we realized that this has not always been easy, mainly because of the abstraction of many scenes, as well as the utopian nature of many mythological events. Nonetheless, the correlation of myth texts with ancient sites or even whole regions, and especially the visualization of these myths with varied original material from antiquity (coming from Thrace itself but mostly from other regions), will provide end users and visitors or the educational community with a rich, attractive, and innovative guiding application, based on the intrinsic and long lasting power of ancient art and storytelling of ancient Thracian myths.

6.3 Contribution to State of the Art in Metadata Schemata

The employment of state-of-the art schemata for data annotation has helped decisevely in the development of an efficiently indexed multimodal content for easy and fast information retrieval, which is versatile enough to be successfully used in various technological applications. As already seen in the pilot uses of two of these applications, the choice of relevant metadata schemata efficiently serves this purpose within the context of the versatile area of the humanities. Linking the project's diverse types of data to external controlled vocabularies such as thesauri was crucial to ensure consistency for the efficient management of the content as well as easy and efficient information retrieval (Vacalopoulou et al. 2019b, p. 38).

Furthermore, the project content was designed beforehand in view of enhancing the APOLLONIS national Infrastructure for Digital Arts, Humanities and Language Research and Innovation, and has successfully done that. In fact, the project is also innovative for being the first one designed to contribute to a pre-existing infrastructure and, as such, total interoperability of content has been a prerequisite. In turn, by becoming part of a national infrastructure, the content of the research process ensures maintanance and viability, and is also openly offered as such in the scientific community for further utilization and use.

7 Conclusion

The project produced a bilingual website and a mobile application in Greek and in English (https://mythotopia.eu/), a map of the area depicting different types of tourist information in the form of points of interest, and all the linked mythological and archaeological information, which lies in the core of the project. Visitors are able to easily set up customized itineraries based on criteria that they personally select. At the same time, educators and learners have the opportunity to use the project's educational application, which is based on the aforementioned material. Furthermore, the properly annotated multidisciplinary and multimodal data that form the backbone of these technological applications will enrich the content of APOLLONIS, the Greek Infrastructure for Digital Arts, Humanities and Language Research and Innovation. In the long-run, the project aspires to be of use to governmental entities, such as regional and local authorities, the Greek National Tourism Organization, archaeological and cultural authorities, educational organizations, etc. Potential future involvement with such organizations could help towards a broader utilization of project outcomes. Finally, project partners hope that this example project can be used as a case study for designing and implementing future research and development projects that, apart from their standalone use and their inherent value, can be integrated or interoperate with other services or initiatives.

Acknowledgements We acknowledge support of this work by the project "Mythological Routes in Eastern Macedonia and Thrace" (MIS 5047101) which is implemented under the Action "Regional Excellence", funded by the Operational Programme "Competitiveness, Entrepreneurship and Innovation" (NSRF 2014-2020) and co-financed by Greece and the European Union (European Regional Development Fund).

References

Ackermann, H.C.: Fondation pour le Lexicon Iconographicum Mythologiae Classicae (1981–2009). In *Lexicon iconographicum mythologiae classicae (LIMC)*. Zurich, München, Düsseldorf: Artemis & Winkler Verlag.

Arachne: iDAI.objects Arachne (n.d.). https://arachne.dainst.org/. Last accessed 8 Dec 2021

BAPD: Beazley Archive Pottery Database (n.d.). www.beazley.ox.ac.uk/carc/pottery. Last accessed 8 Dec 2021

Boardman, J.: *The Archaeology of Nostalgia. How the Greeks re-created their mythical past. (Η αρχαιολογία της νοσταλγίας. Πώς οι αρχαίοι Έλληνες αναπαρέστησαν το μυθικό παρελθόν τους)*, trans. Dimitriou, V. Athens: Patakis. In Greek (2007)

Bouzek, J., Graninger, D.: Geography. In: Valeva, J., Nankov, E., & Graninger, D. (Eds.). A Companion to Ancient Thrace. Chichester; Malden: Wiley Blackwell, 12–21 (2015)

Burgess, D.L.: Did the Greeks believe in their myths? An essay on the constitutive imagination (review). Philos. Lit. **13**(1), 184–185 (1989)

Buxton, R.: Forms of Astonishment: Greek Myths of Metamorphosis. Oxford University Press, Oxford-New York (2009)

Cartledge, P.: The Greeks: A Portrait of Self and Others. Oxford-New York: TV Books (2002/1993)

CN: Corpus Nummorum (n.d.). www.corpus-nummorum.eu. Last accessed 8 Dec 2021

De Puma, R.D., Guthrie, W.K.C.: An Etruscan mirror with the prophesying head of Orpheus. Record Art Mus. Princeton Univ. **60**, 18–29 (2001)

Digital LIMC, DaSCH (2020). weblimc.org/page/home/. Last accessed 8 Dec 21

Dunbabin, M.: Mosaics of the Greek and Roman World. Cambridge University Press, Cambridge-New York (1999)

Franceschini, M.: Iconographic series in Attic vase painting: technical simplification or semantic strategy? In: Reinhardt, A. (ed.) Strictly Economic? pp. 45–59. Universität, Ancient Serial Production and its Premises, Heidelberg (2021)

Graninger, D. Ethnicity, Ethne. In: Valeva, J., Nankov, E., Graninger, D. (Eds.). A companion to ancient Thrace. Chichester, Malden: Wiley Blackwell, 22–32 (2015)

Gruen, E.S.: Rethinking the Other in Antiquity. Princeton University Press, Princeton-Oxford (2011)

Hall, E.: Inventing the Barbarian: Greek Self-Definition through tragedy. Clarendon Press, Oxford-New York (1989)

Hansen, W.: Classical Mythology: A Guide to the Mythical World of the Greeks and Romans. Oxford University Press, Oxford (2004)

Hartog, F.: The Mirror of Herodotus: The Representation of the Other in the Writing of History. University of California Press, Berkeley-Los Angeles-London (1988)

Hawes, G. (ed.): Myths on The Map. Oxford University Press, The Storied Landscapes of Ancient Greece, Oxford (2017)

Huskinson, J.: Some pagan mythological figures and their significance in early Christian art. Papers British School Rome **42**, 68–97 (1974)

Jacoby, F. (Ed.) (1995). *Die Fragmente der griechischen Historiker (FGrHist)*. Teil 1. Genealogie und Mythographie (A. Vorrede. Text. Addenda. Konkordanz). Photomekanischer Nachdruck des vermehrten Neudruckes von 1957. Leiden: Brill

Johnston, S.I.: The Story of Myth. Harvard University Press, Cambridge MA-London (2018)

Jourdan, F.: The early Christian iconography of Orpheus and its possible meanings. Seminar of Patristic literature, London, United Kingdom. ⟨hal-00594776⟩ (2011)

Jourdan, F.: The Orphic singer in Clement of Alexandria and in the Roman catacombs: comparison between the literary and the iconographic early Christian representation of Orpheus. Sudia Patristica LXXIII, 113–127 (2014a)

Jourdan, F.: Orpheus and 'Orphism' in the Christian literature (in Greek) of the first five centuries. In: Leppin, H. (ed.) *Antike Mythologie in christlichen Kontexten der Spätantike* (Millennium-Studien *54*), pp. 193–206. De Gruyter, Berlin, Boston (2014b)

Junker, K.: Interpreting the Images of Greek Myths, trans. Künzl-Snodgrass A. & Snodgrass A., Cambridge: Cambridge University Press (2012)

Kefalidou, E.: NIKETES—The victorious athlete. a study on the iconography of ancient Greek Athetics (ΝΙΚΗΤΗΣ. Εικονογραφική μελέτη του αρχαίου ελληνικού αθλητισμού), Thessaloniki: Aristotle University of Thessaloniki. In Greek (1996)

Kefalidou, E.: The Iconography of Madness in Attic Vase Painting. In: Oakley, J., Palagia, O. (eds.) Athenian Potters and Painters II, pp. 89–98. Oxbow Books, Athens (2009)

Kondoleon, K.: Domestic and Divine: Roman mosaics in the House of Dionysos. Ithaca & London: Cornell University Press (1995)

Kostopoulou, S., Kyriakou, D., Malisiova, S., Sofianou, E., Toufengopoulou, A., Xanthopoulou-Tsitsoni, T. (eds.): The Western Silk Road in Greece. Aristotle University of Thessaloniki, Thessaloniki (2016)

Naas, V.: Pline l'Ancien a-t-il cru à ses mythes? Mythes Et Savoirs Dans Les Textes Grecs Et Latins, Pallas **78**, 133–151 (2008)

Newby, Z.: (2016). Greek Myths in Roman Art and Culture. Imagery, Values and Identity in Italy 50 BC-AD 250. Cambridge: Cambridge University Press

Prokopidis, P., Georgandopoulos, B., & Papageorgiou, C.: A suite of NLP tools for Greek. In: The 10th International Conference of Greek Linguistics. Komotini, Greece (2011)

Shapiro, H.A.: Myth into Art: Poet and Painter in Classical Greece. Routledge, London (1994)

Smith, W. (ed.): A Dictionary of Greek and Roman Geography., vol. 2. Walton & Maberly; Murray, London (1854)

Sourvinou-Inwood, C.: Myths in images: Theseus and Medea as a case study. In Lowell, E. (ed.). Approaches to Greek Myth, Baltimore: Johns Hopkins University Press, 354–406 (2014)

Straka, M., Straková, J.: ÚFAL MRPipe at MRP 2019: UDPipe goes semantic in the meaning representation parsing shared task. In: Proceedings of the Shared Task on Cross-Framework Meaning Representation Parsing at the 2019 Conference on Natural Language Learning, Hong Kong: Association for Computational Linguistics 127–137 (2019)

Thomopoulos, S., Tsimpiridis, P., Theodorou, E.-I., Maroglou, C., Georgiou, E., Christopoulou, C.: COSMOS: an innovative, graph-based application for the relational presentation and visualization of Greek mythology and the associated art. American J. Comput. Sci. Technol. **5**(1), 16–25 (2022). https://doi.org/10.11648/j.ajcst.20220501.12

Tsiafaki, D.: *Η Θράκη στην αττική εικονογραφία του 5ου αιώνα π.Χ. Προσεγγίσεις στις σχέσεις Αθήνας και Θράκης*. Μορφωτικός Όμιλος Κομοτηνής - Κέντρο Θρακικών Μελετών. In Greek (1998)

Tsiafaki, D.: Battles between Athenians and Thracians: an abstract representation or a realistic scene? In: Gicheva R., and Rabadjiev K. (eds.). Πιτύη: Studia in honorem Prof. Ivani Marazov, Sofia, 365–369 (2002)

Tsiafaki, D.: Ancient Thrace and Thracians through the Athenian eyes. Thracia **21**, 261–282 (2016)

Tsiafaki, D.: Thracians and Greeks in the North Aegean. In: Gimatzidis, S., Pieniazek, M., & Mangaloglu-Votruba, S. (eds.), Archaeology across Frontiers and Borderlands. Fragmentation and Connectivity in the North Aegean and the Central Balkans from the Bronze Age to the Iron Age (OREA—Oriental and European Archaeology 9), VÖAW, Austrian Academy of Sciences Press, 219–241 (2018)

Tsiafaki, D.: (2020). The Northern Aegean. In: Franco De Angelis (ed.). A Companion to Greeks Across the Ancient World, Wiley-Blackwell, pp 409–430

Tsiafakis, D.: The allure and repulsion of Thracians in the art of classical Athens. In: Cohen, B. (Ed.), Not the classical ideal: Athens and the construction of the other in Greek Art. Leiden: Brill, 364–389 (2000)

Vacalopoulou, A., Markantonatou, S., Toraki, K., Minos, P.: Open-Access Resour. Manage. Promot. Greek Mus. Folk Exhibits (2019a). https://doi.org/10.1007/978-3-030-12453-[illegible]_15

Vacalopoulou, A., Markantonatou, S., Toraki, K., Minos, P.: Openly available resource for the management and promotion of museum exhibits: the case of Greek museums with folk exhibits. Int. J. Comput. Methods Heritage Sci 3(1) (2019b)

Veyne, P.: Les Grecs ont-ils cru à leurs mythes? Essai sur l'imagination constituante. Editions du Seuil, Paris (1983)

Wiebe, D. Rev. of Paul Veyne. (1994). Did the Greeks Believe in Their Myths? An Essay on the Constitutive Imagination, trans. Wissing, P. Chicago: University of Chicago Press, 1983, Philosophy of Religion 35, pp. 111–112

World Tourism Organization: Western Silk Road Roadmap. UNWTO, Madrid (2018)

Xydopoulos, I. (2004). The Thracian Image in the Archaic literature. The absence of Otherness. Περί Θράκης, 4, pp. 11–22

Xydopoulos, I.: Η εικόνα των Θρακών στην κλασική ιστοριογραφία, Thessaloniki: Kyriakidis Bros. In Greek (2007a)

Xydopoulos, I.: The Thracian image in Herodotus and the rhetoric of otherness. In: Proceedings of the Mediterranean Crossroads Conference (MCC): New Trends in the Study of the Mediterranean and its history at the onset of the 21st century, 10–13 May 2005, Athens-Oxford, 594–603 (2007b)

Xydopoulos, I.: Η εικόνα των Θρακών στην ελληνιστική ιστοριογραφία, In: Αλμπανούδης Π., & Μηνούδης Γ., Πρακτικά Α' Πανελληνίου Συνεδρίου Ανατολικορωμυλιωτών, Thessaloniki, 67–90. In Greek (2009)

"The Hidden Life" of the Ceremonial Vestment: The Relic of Holy Martyr Lazar

Natalija D. Stokanović and Biljana Cincar-Kostić

Abstract

Every piece of textile has a unique journey and has a special place in different topographies. During the Middle Ages, wars, treaties and taxes shaped silk production, trade, and consumption. This paper is about one of the most important Serbian medieval relics, The Ceremonial Vestment of Holy Knez (Prince) and Martyr Lazar. The Vestment is made out of luxurious brocade silk from the Italian town of Lucca. Silk material originates from the period of the third quarter of the fourteenth century, a time before the year 1389 when Knez Lazar was beheaded in The Battle of Kosovo. Silk from the Italian town of Lucca was worn by popes and the rich, and saints were buried in silk from Lucca all over Europe, but this Vestment made out of silk of Italian origins in medieval Serbia is a unique example. Furthermore, from the execution of Prince Lazar in the year 1389, until 1733, scientists still didn't find any mention of this solemn garment. Earlier it was believed that Saint Lazar was wearing it in the Battle of Kosovo, and it has his blood stains. Until recently, it was not known where was The Vestment during The Second World War. The recent archival and field research and new digital technologies (Virtual loom program, AdaSilk project, Met Museum Open Source, Google Art, Graphic reconstructions, etc.) confirm most of the folklore and brought new scientifically proven facts around this holy object of national and religious significance, while proving the waving and embroidery technique from Lucca, time of occurrence, and explaining "the mysterious" life-path of the relic.

Keywords

Silk • Brocade • Vestment • Saint Lazar • Lucca • Museum of Serbian Orthodox Church

1 Scientific Methodology

In this paper, we are going to present the "hidden life" of a Ceremonial Vestment of Saint Lazar (Fig. 1), one of the most respected Medieval Serbian Rulers and Saint-Martyr (1329–1389). The new analysis confirmed that his incorruptible body was dressed in this garment *post-mortem* garment, that was made during the life of Saint. It is interesting that, from his execution in the year 1389, until the year 1733, scientists still didn't find any mention or visual presentation of this solemn garment. It is called the dress, a cloak, an attire, a tunic, the vestment, and most recently—the *kontuš*. Through the recent archival and on-field research, as well as new digital technologies (Virtual loom program, ADASilk project, Museums Open Source Project, Google Art, Graphic reconstructions in Adobe InDesign program, etc.) the goal of this paper is to demonstrate the communicative power of this holy object of national and religious significance, while scientifically proving the waving and embroidery technique, time and place of its occurrence (Virtual Loom Program 2017–2022; ADASilk Database 2017–2022). Today, The Ceremonial Vestment is being kept in The Museum of Serbian Orthodox Church in Belgrade, under Accession Number MSPC 1922. Explaining folklore beliefs with new scientific facts, we are going to present the inspiring life of this relic, full of richness and diversity.

N. D. Stokanović (✉) · B. Cincar-Kostić
Art Historian and Curator of the Museum of Serbian Orthodox Church, Belgrade, Republic of Serbia
e-mail: natalija.stokanovic@yahoo.com

B. Cincar-Kostić
e-mail: biljanaivangel@gmail.com

S. Kostopoulou et al. (eds.), *Cities' Vocabularies and the Sustainable Development of the Silkroads*, Advances in Science, Technology & Innovation, https://doi.org/10.1007/978-3-031-31027-0_13

1.1 The Dress of Knez Lazar (1929)

Since the rise of the national idea, especially since the middle of the nineteenth century, The Ceremonial Vestment of Knez Lazar became the subject of the first research of medieval fabrics and embroideries. The content of every scientific paper published after The World War II so far is showing that, for the basis, every researcher relies on an on-field report *The Dress of Knez Lazar*, written and published yet in the year 1929, and republished in 1937, by Lazar Mirković, one of the few Founders[1] of Museum of Serbian Orthodox Church (Mirković 1929, 48–53; Mirković 1937, 72–73). This report was made before The Second World War, while the dress was kept in the Vrdnik monastery on Fruška Gora mountain. Even in this paper, we are relying on this on-field report of Lazar Mirković published in 1929 and republished in 1937.

Professor Mirković mostly named The Ceremonial Vestment of Knez Lazar *dress* in this report, which was the safest name for this piece of clothing so far, due to the condition it was, and the inability to work with it without fear of contributing to greater damage. The expression *The Vestment of Knez Lazar* was established and accepted in the catalogs of the Museum of Serbian Orthodox Church (Mileusnić 1989, 32–33; Dušanić 2008; Mileusnić 2001; Matić 2019, 88). In the written descriptions of The Vestment, which were done many times for different purposes from the second half of the nineteenth century, we collect the names of the garment, since the shape of it was not precisely studied until now: the dress, a cloak, an attire, a tunic, the vestment, a porphyry, dalmatika, and most recently—*kontuš*.

Before The World War II, Lazar Mirković said that the monks were keeping The Vestment locked in a wooden shrine with a glass door, above a kissing icon, in the right part of a narthex of the church (Mirković 1929, 48). The Vestment was separated from the body relics of Saint Lazar. The main story the monks were telling to the believers, and to Lazar Mirković, is how Knez Lazar was wearing this golden-woven porphyry in *The Battle of Kosovo* and was beheaded wearing it. Also, some stains on the dress were considered the blood of Knez Lazar (Mirković 1929, 48).

Such information has contributed to finding statements in scientific papers as *allegedly attributed to Knez Lazar* (Nikolić 2016, 9–11) or uncertain mention, or visible bypass of this garment by many scholars whose main researches were those of the Middle Ages. The overall problem is unfounded conclusions about the history of The Vestment when there is no tangible evidence, and insufficient research, when there is tangible evidence through history. That's why Mirković asked himself, is this Vestment really a Vestment of Knez Lazar? He made a conclusion that everyone would be more than happy to have something belonging to Knez Lazar (Mirković 1929, 48).

After this folklore introduction story, he gave the basic description of *The Dress*: it is made of red silk, which has the shape of a *tunic*, narrower in the upper part, bell-shaped, and long 136 cm; it was buttoned up from the neck to the feet with many buttons; around the neck, on the sleeves and on the hem of the dress, there is a border of a silver wire. Also, in the fourteenth century, Mirković explains, Byzantine rulers and nobles wore such dresses, both tailored, and embroidered with motives of wild and fantastic animals, and decorated with pearls (Mirković 1929, 49). The concrete comparison he made with the miniature of the Great Duke of Apokaukos in *The Works of Hippocrates*, kept in The National Library in Paris (Mirković 1937, 72–73). Also, he mentioned the dresses with lions, which describes Domentian the Hilandarian (Mirković 1929, 49). Mirković described the only left button (Fig. 2), which has a symbol of the helmet with two ox horns

Fig. 1 The Ceremonial Vestment of Knez Lazar as it was exhibited in the Old Court Building for the manifestation of The Days of Belgrade (2015) (Photo-library of Museum of Serbian Orthodox Church in Belgrade, Slavomir Matejić)

[1] Considering that each of the Founders of Museum of Serbian Orthodox Church had a significant share in saving The Vestment of Knez Lazar, and each name will be significantly mentioned in this paper, their names will be listed here first, with the hope that it will make it easier for readers to follow the text, due to many new names: Lazar Mirković, Radoslav Grujić, Svetozar Dušanić, Slobodan Mileusnić. Their works and dedications make them "The League of Outstanding Gentlemen" of Serbian History and Culture.

Fig. 2 The only left button with the heraldic symbol of Holy Knez Lazar (Photo-library of Museum of Serbian Orthodox Church, Popović 1989)

made in enamel, which is the heraldic mark of Knez Lazar. The buttonholes are hammed with blue thread (Mirković 1929, 48). He dates the button back to the fourteenth century, comparing it with other heraldic symbols of that time (Mirković 1929, 48; Babić 1975, 65).

The silk is decorated with symmetrically arranged animals. Also, Mirković described the embroidery, and what he could see at that moment, mentioning the motives of lions, birds, and heart-shaped leaves, embroidered with gold and silver thread. In motives of birds, Mirković recognized herons standing on scorpions (Mirković 1929, 43, 1937, 73).

The most interesting about this report is that Lazar Mirković thought, by comparing it with the drawings from some books, that a fabric The Vestment of Knez Lazar was made from, what was *most similar* to silk produced in the Northern Italy town of Lucca in the fourteenth century (Mirković 1929, 43, 1937, 74).

Even today, these facts stated by him can only be confirmed and expanded. But for a long time, The Ceremonial Vestment wasn't researched "closely". In this paper, we are collecting most of the knowledge about the Vestment, which is available to us at this moment, but also leaving to others the opportunity to study this holy relic with admiration.

2 The Ceremonial Vestment in Popular Culture

Frequently mentioned traditions and claims maintain the respect of the cult of Knez Lazar and the attention of the wider public on The Vestment. Those traditions and claims can be summarized as follows:

1. Knez Lazar was wearing The Ceremonial Vestment during the Battle of Kosovo;
2. He was beheaded wearing it;
3. The Vestment has blood stains on itself;
4. The Vestment, allegedly, has 16 stabbing holes (Radojčić 1930);
5. The body remains of Saint Lazar were not touched for the next three years in order to complete the chemical process of decomposition;

6. The Vestment was brought to Belgrade on the bodily relics of the Saint in the year 1942;
7. In 1946, the Synod Commission determined that the Vestment on the body relics of Saint Lazar, at that moment lying in the Cathedral Church in Belgrade, is originally from the fourteenth century;
8. Since 1942, it is kept in the Museum of Serbian Orthodox Church;
9. The cut, the fabric which the Vestment was sewn from hasn't been yet completely studied and confirmed, and in scientific papers can be found different descriptions;
10. The iconography of embroidered motives is not equal in the scientific papers;
11. The time of the creation of the brocade was not completely sure, until now.

The Vestment of Knez Lazar inspired the Graduation Work of Nevenka Petrović, an applied artist, in the years 1970, when she made the replica of The Vestment for the celebration of the *Six centuries of Kruševac*, a medieval capital city of *Moravian Serbia and Holy Martyr Lazar*. For the newspaper, she said that she worked only with a magnifying glass. She wasn't allowed to come too close to the Vestment. That's why her reconstruction of the Vestment is noticeably different from the original. The loom it was used this time belongs to the Takač family in Sombor. Reconstruction of the cut also did Ana Nestorović, and Radmila Budisavljević, a sculptor, did 70 buttons needed in order to get as a faithful copy as it was possible at that moment. The silk she used was ordered from the city Titov Veles, in Macedonia, made in a fabric *Nonče Komiševa*. The golden yarn Nevenka Petrović ordered from Nurenberg (Germany) (movie *Prince Lazar's Attire* (2018)). As she said, when the curators of the Museum didn't watch, she has stolen one part (Stefanović, 2nd July 1971). Today, everyone can see this reconstruction in The National Museum in the city of Kruševac (Fig. 3).

Newspaper articles, many of them available today for online reading also, are in favor of keeping vital the cult of Knez Lazar and all the knowledge about The Ceremonial Vestment (Svečana haljina kneza Lazara 2018; Svečana haljina kneza Lazara 2018; *DOŽIVITE ISTORIJU Originalna haljina kneza Lazara izložena u Starom dvoru* 2015; *NEPROCENJIVA VREDNOST Plašt kneza Lazara pred publikom u Starom dvoru* 2015; *Plašt kneza Lazara i Pohvala monahinje Jefimije u Starom dvoru* 2015). Especially must be mentioned the manifestation held in 2015, named *Days of Belgrade* when the Ceremonial Attire was displayed in the Old Royal Court in Belgrade. It was shown in a display case, vertically attached to a red plush background, looking very representative and, indeed, royal. But, due to its importance, journalists spread the most non-objective information, mentioned above (Fig. 1). Also, today everyone can see two short documentary movies and other videos about The Vestment of Knez Lazar on YouTube: (*Muzej Srpske pravoslavne crkve: Haljina Kneza Lazara* 2017; *Odežda kneza Lazara* 2018).

Fig. 3 Replica of The Vestment of Knez Lazar in Kruševac, made by artist Nevenka Petrović (1970/1), kept in The National Museum in Kruševac (Photo-Library of Museum of Serbian Orthodox Church in Belgrade, Republic of Serbia, Svetozar Matejić)

3 Collected Knowledge

3.1 "The Hidden Life" of the Ceremonial Vestment: Medieval Sources

The poetic title of this paper, *"The Hidden Life" of The Ceremonial Vestment*, comes from the fact that the documents "are silent" in the case of the early history of this artistic object and the relic, until the year 1733. Hereof, the paper presents the Vestment in its own importance and "for now" know history, but also as a manifestation of wider geopolitical and trade circumstances in which it was made, as it served in the court of Prince Lazar and *post mortem,* but also in the wider category of Italian textiles and trading silk routs.

To explain it succinctly, due to the Orthodox "background", political alliances, and marriages, the Serbian court was always closely related to Byzantine, its imperial court and clothing culture, connected to Constantinople with the ancient Imperial Road. The usage of silk of Italian origins in medieval Serbia is, for now, in the case of The Ceremonial Vestment of Prince Lazar, a unique and exceptional example. As the thirteenth century began with the sack of Byzantine Constantinople (1204) by the armies of the Fourth Crusade, the city, which had been the seat of the Eastern Roman Empire and the center of Orthodox Christianity until that point, was occupied by the so-called Latin Empire and became subject to the Pope (Phillips 2021, pp. 28), and seems to have ended the weaving of figured compound silks there. Other parts of the eastern Mediterranean also came under Latin control, including the Morea and Thebes, both areas famed for silk weaving in the later Byzantine period. The diaspora of silk workers in Italy became principal (Phillips 2021, 28). Parallel, began the rise of silk weaving in Lucca, Venice, and other Italian cities, in diverse variations and terminology (King 1998; Punta-Ludovica Rosati 2017; Lambert and Katherine 2016). The silk from Lucca became so valued, that silk made outside this city was known as "cendata de Lucha" (Phillips 2021, 34). A possible route for the rich gold-woven silk from Lucca to Prince Lazar's court was through Dubrovnik, which sometimes paid him taxes in textiles (Mihaljčić 1984).

From the Battle of Kosovo (28th June 1389) the cult and legends about The Battle and Knez (Prince) Lazar spread very quickly and became the basis of Serbian national identity and Orthodox Christian religion. They are vital in the consciousness of the people and still important as a question related to the status of the Autonomous Province of Kosovo and Metohija. Also, Knez Lazar is a widely venerated Saint throughout Europe, in the Orthodox world, and among the Slavs. In the first place, he helped the reconciliation of unity of the Serbian and Constantinople Patriarchates in the year 1375 (Mihaljčić 1984, 76, 109). Secondly, he achieved success through diplomatic unity of the territory of the disintegrated Serbian Empire during the second half of the fourteenth century, as well as the rise of wealth and culture in *Moravian Serbia* (Mihaljčić, 116, 119). But mostly, due to the fact that he suffered death on the 28th of June 1389 in the Battle of Kosovo, while leading a Christian army against invading Ottoman Empire, led by Sultan Murat himself, who also died in The Battle of Kosovo. This battle temporarily prevented the invasion of non-Christians into predominantly Christian Europe, and the sacrifice of Knez Lazar and his soldiers has been recorded in chronicles and annals throughout Medieval Europe (Mihaljčić, 130).

Knez Lazar was beheaded. He was the first and the only one among the Serbian rulers who lost his life on the battlefield. His decision to suffer for the defense of his fatherland and people, his voluntary sacrifice, is the highest of the principles of human actions and opens the way to The Kingdom of Heaven. That's why he was immediately considered the Holy Martyr (Mihaljčić, 150; Popović 1989, 171). In Serbian epic poetry, Prince Lazar is the most commonly named Tzar (Emperor), due to his merits (Popović, 171). Since the time of his canonization, and lay the body in a *kivot* (a wooden shrine or ark) in his mausoleum, the church of Ravanica monastery, Saints relics have been attracting pilgrims and healing sicknesses (Pavlović 1965). And for centuries, traditions were rapidly growing, and are still believed and practiced, on a regular basis (Novaković 1939; Informative Service of Serbian Orthodox Church 2021).

3.2 The Death of Knez Lazar

The Battle of Kosovo shook contemporaries across Europe and quickly turned into a legend. As the younger the sources are, the less accurate they are, and through time, new motives are added to them. The fastest-spread narrative tradition shaped into epic songs, which strongly influenced generations for centuries, and contributed to the creation of the people's historical consciousness and folklore practices. The Battle of Kosovo became the basic timeline in Serbian tradition. After the battle, started the centuries of slavery in the Ottoman Empire (fourteenth–nineteenth century) (Mihaljčić, 131–132).

Ten cult writings, created between the year 1389 and the year 1419/20, also have a historical value (Mihaljčić, 132–137). Their writers are contemporaries of The Battle of Kosovo, the beheading of Knez Lazar, the transfer (translation) of the body of Lazar to the Ravanica monastery one year later (1390/1), and his canonization. In accordance with the general Christian hagiography and the celebration of the Christian sacrifice, the goal of these authors was not historical accuracy and a detailed description of the events, but rhetorical praise of Saint Lazar focused on his martyrdom (Mihaljčić, 137–138). There is no mention of The Ceremonial Vestment in the descriptions of The Battle of Kosovo, nor in the time of the beheading. The pre-knowledge of the domestic armies was that they are multiply overpowered by the Ottoman Army. And a serious military leader, as Knez Lazar was, would not go into the battle in a gold-woven court porphyry, but in full armor. It is not for sure, but it is more probable that he wasn't buried in Priština, in the church of the Ascension of Christ, in the golden-woven attire in the year 1389.

3.3 Translation of the Bodily Relics and Canonization

Knez Lazar built the Ravanica monastery. Today, it is believed he did it following the idea of the Mausoleums of the Nemanjić dynasty, as a tomb church, where a memorial service for the salvation of the soul was held by the monks. The translation of his bodily relics from Priština into the Ravanica monastery was also an act of formal canonization (Todić 1989). The translation and canonization were done by the inheritors of Knez Lazar and from the very top of the Serbian Orthodox Church (Popović 1989, 171–183).

This kind of act, which is a State event of the first order, needed, and in this case shows, a specially established order: the act of death, the appearance of miraculous signs as signs of God's presence, finding whole incorruptible relics, their raising and transfer to a new grave (elevation and translation), whereby, in this case, the relics are dressed in ceremonial attire, and a special service and biography are compiled for this occasion (Popović 1989, 171–183).

But there are no descriptions, nor a simple mention of the special attire throughout the event: from finding the relics, through the translation of his bodily relics from Priština to the Ravanica monastery, nor during lying in the prepared *kivot* in the Ravanica monastery. It is logical that the ceremonial attire had to be prepared in advance for the ceremonial transfer of the relics to the tomb church.

3.4 The Portraits of Saint Lazar on Frescoes

In a search for a visual representation of The Ceremonial Attire of Saint Martyr Lazar, we also explored the frescoes (Mirković 1929). Portraits are preserved only in the Ravanica, Veluća, and Ljubostinja monasteries, but none of them show the same attire as this one that is kept in the Museum of Serbian Orthodox Church. The portrait in Hilandar monastery is from the year 1667, and the fresco paintings from the Lazarica, Gornjak, Tismeni, Vodice, and Manasija monasteries aren't preserved.

3.5 The Description of Saint Lazars Bodily Relics in Medieval Travelogues

Scientist and diplomat from The Venetian Republic, Marco Antonio Pigafetta, who traveled with Magellan on a first journey around the world, was on a special mission for Habsburg Emperor Maximilian II when visited the Ravanica monastery. While traveling from Vienna to Constantinople in the year 1567, to negotiate peace with Sultan Selim II, he made detailed descriptions: *that day we turned off the road to see the monastery where the monks live and where lies the body of Blessed Lazar, a despot of Serbia. They showed the Saint to us, in fact, they showed us only one hand because they say that due to limited privileges, it is forbidden to show it in his entirety. So they showed us only one hand, while everything else was covered with a veil. We saw a very beautiful ruby on one finger of that hand* (Pigafetta 1567; Mihaljčić 1984, pp. 289).

David Ungdada also testifies about the relics of Saint Lazar, since he visited the monastery in the year 1572, stating that the body relics of the Martyr rested in the kivot, a wooden chest, or an ark (Popović 1989, pp. 183). He didn't give any other description of the relics, which leads to the conclusion that, indeed, no one could see the whole bodily relics, nor how the Martyr was *post-mortem* dressed. During the Middle Ages, it was only possible to see his one hand.

3.6 The Cult of Saint Lazar in the New Century

From 1389, until the year 1733 (Ruvarac 1903, 73–74), we still didn't find any mention of this ceremonial attire. During a period of the New Age, some of the thousands of immigrants run from the Balkans, from the Ottoman revenge, during the year 1690, and many times later during the eighteenth century. Migrations of Orthodox people from the whole Balkan happened often from the late Middle Ages, but *The Great Exodus of the Serbs* in the year 1690 was bigger and more massive, the most important in Serbian history, because this time, the focus of Serbian religious and political life has been moved from the Balkans to Habsburg Monarchy (Slijepčević 1966, 26; Bešlin 2014, 70). Old Serbia has been devastated. The higher and lower clergy fled, led by the Patriarch, and they left centuries-old churches and monasteries, carrying the shrines and relics of Serbian Saints with them (Slijepčević, 26, Bešlin, 70). Many people fled, led by community elders and merchants. They settled in a state whose ruler was a patron of the Roman Catholic Church (Točinac 2001, 151). Since there were no noble leaders of Serbian people in The Ottoman Empire, there was only Orthodox church clergy, who were the only ones who had some kind of education in monasteries, and in this situation were competent to stand up for the people, helping new life organization on an unknown territory. Traditionally, until the nineteenth century and the rise of the national idea, religious institutions have held together all dimensions of identity (Halpern—Ruano Borbalan 2004 \2009, 13–14). This is how the relics of Saint Lazar ended first in Saint Andrea in South Hungary, and in the year 1698, monks settled in the Vrdnik monastery, on Fruška Gora Mountain (Matić 1990; Davidov 1990, 94).[2]

[2] Because the mountain became a refuge for monks, some medieval monasteries were renovated, and many new ones were built during the New Age, so the mountain became known as *Serbian Sinai* and *Serbian Athos*.

3.7 The First Known Mention of the Ceremonial Attire in the Year 1733

The first mention of the ceremonial attire of the Holy Knez Lazar can be found in *The Inventory of Vrdnik monastery* from the year 1733 and later in the year 1753. *The Inventory* gives a few descriptions: *In the great kivot lies the whole and intact body of the holy martyr knez Lazar the Serbian, on the right side of the temple. In the hand of Knez Lazar, there is a gilded cross, chained with silver and decorated with stones.* He also mentioned *one large gold-woven porphyry of Prince Lazar, decorated with pearls* (Ruvarac 1903, 76).

3.8 Separating the Bodily Relic from the Ceremonial Vestment

Today, due to the ethnological-forensic analysis, we know that The Ceremonial Attire was removed from the saintĺs body relics on the 4th of December 1826, when Mitropolit Stefan Stratimirović, during the time of Josip Mihailović, the abbot of the Vrdnik monastery, transferred the body relics from the old *kivot* to the new one (Fig. 4), made from the cypress tree, and changed his clothes (Pavlović 1965, 123).

3.9 First Complete Description in the Year 1854

Milica Stojadinović Srpkinja, a 19th-century Serbian writer and ethnographer, lived in the village Vrdnik, next to the Monastery Vrdnik. Her father was a priest, and she left us a description of one special visitation to the Vrdnik monastery in the year 1854: *At that moment, Father Leontije opened the*

Fig. 4 *Kivot* in the Vrdnik monastery, the picture made in the year 1927 (Photo-library of Museum of Serbian Orthodox Church in Belgrade, Legacy of Radoslav Grujić, Republic of Serbia)

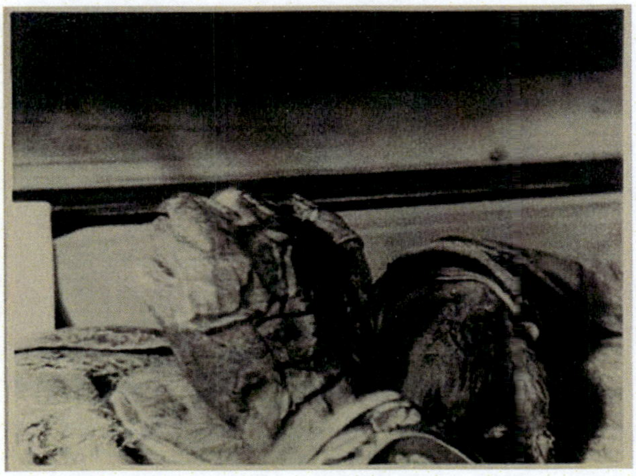

Fig. 5 Body relics of Holy Knez Lazar. (Photo-library of Museum of Serbian Orthodox Church in Belgrade, Legacy of Radoslav Grujić, Republic of Serbia)

kivot, which we all kissed. Then we kissed the holy hand that held the scepter of The Serbian Empire. On the inside of a lid, hung many sacrifices made of silver and gold. But I think that the most beloved sacrifice to his blessed soul is the one who came to him with the feeling of sadness and memory of what fell and failed for the Serbian people with him in Kosovo (Fig. 5) ... *In the narthex, the Father Leontije opened another kivot for us, in which Knez Lazars dress was, in which he was dressed in Kosovo on the last day, and —life ended! Oh, how it must be for the Serb to look at it! However, those who got that precious thing, who did not know how to evaluate it, deserve a great rebuke, as they should have preserve it better. The shape of the dress is almost unknown, the dress is torn off piece by piece, that are given from the holy fathers of the monastery to the superstitious people: one came with sick eyes, and asked for a peace to wipe his eyes, one to carry it with him always, one just like that, ele—of that important dress only the front skirts are intact!* (Stojadinović 1985, 50).

The Ceremonial Attire was kept in the special *kivot*, in the narthex of the church of Vrdnik monastery, just as Milica Stojadinović Srpkinja said in her *Description of Fruška Gora*, until The World War II, when The Ceremonial Vestment and other relics of Saint Lazar "disappeared".

3.10 *The Street of Tzar Lazar* in Belgrade

There is a famous *Report written by Professor Radoslav Grujić*, another Founder of the Museum of Serbian Orthodox Church and dedicated scientist, written on 17th April 1942, to Velibor Jonić, The Minister of Education and Religion of former Yugoslavia, about *The Mission of saving the bodily relics of Serbian Saints from Fruška Gora mountain* (Grujić 1942). In this *Report*, Radoslav Grujić explained in detail how the rescue mission took place, describing a desecration and theft by *ustaše,* the soldiers of the newly established Independent Croatian State (Grujić 1942). One of the German officers in Belgrade, was a respected acquaintance of Radoslav Grujić, while they were students in Germany. Also, in *The Rescue Report*, he mentioned the German occupier soldiers who helped him and his colleagues in an effort to transfer the relics of Saint Uroš, Saint Stefan Štiljanović, and Saint Martyr Knez Lazar to Belgrade (Grujić 1942).

But this *Report* doesn't inform what happened to The Ceremonial Vestment during World War II. This *Report* tells us that all values associated with the respect of Knez Lazar, weren't there when Radoslav Grujić came to save what ustaše left behind them, after the robbery and burning of the monasteries. Radoslav Grujić saved only the body relics of three Serbian Saints and brought them to Cathedral Church in Belgrade (Grujić 1942). Even it was known that Professor Grujić risked his life to save the most valuable Serbian relics, and kept historical documents with his body, lying over them during the bombing of Belgrade in 1944, after the war, in The Court of Honor, he was declared a traitor by The Communist authorities (Medaković 2007, 354).

The Ceremonial Vestment had a different life path than the body relics of Saint Lazar. The Vestment was brought to Belgrade in secret, by the German officer, who could safely pass by the German and ustaše guards (Mileusnić 1989, 416). The Ceremonial Vestment and other relics of Knez Lazar were kept in secret during The World War II in one private apartment in the heart of Belgrade (MSPC No. 5336/79; Glumac 1954, 2; Mileusnić 1989, 415–419). The name of this street is, no more no less, *The Street of Emperor (Tzar) Lazar 16/IV* (MSPC No. 42/1954).

It was the apartment of Professor Dušan Glumac, who investigated the Vrdnik monastery just before the war, in order to make a publication of the Serbian antiquities. During the research, he became a friend with the Archimandrite Longin, who trusted Professor Glumac enough to send him the relics, to keep them secret until the end of the war. From the year 1944, when Dušan Glumac gave back the relics kept in his apartment to Archimandrite Longin, and many years later, the ceremonial attire was kept strictly in The Chapel Treasury of Saint Simeon, in the Court of Patriarchate in Belgrade (MSPC No. 232/25. 9. 1953; MSPC No. 265/5. 11. 1953; MSPC No. 279/20.11.1953; MSPC No. 434/14. 5. 1955). After the war, in the newspaper, Professor Glumac only mentioned that The Vestment was kept in *some* private apartment (Glumac 1954, 2). Due to the newly created political climate in Yugoslavia after the war, it was wiser to keep quiet about everything. But, in the year 1979, he donated to The Museum of Serbian Orthodox Church *The Revers of The Exchange of The Relics* between him and Archimandrite Longin on the 16th June 1941, in Belgrade (Fig. 6) (MSPC No. 5336/79; Mileusnić 1989, 415–419).

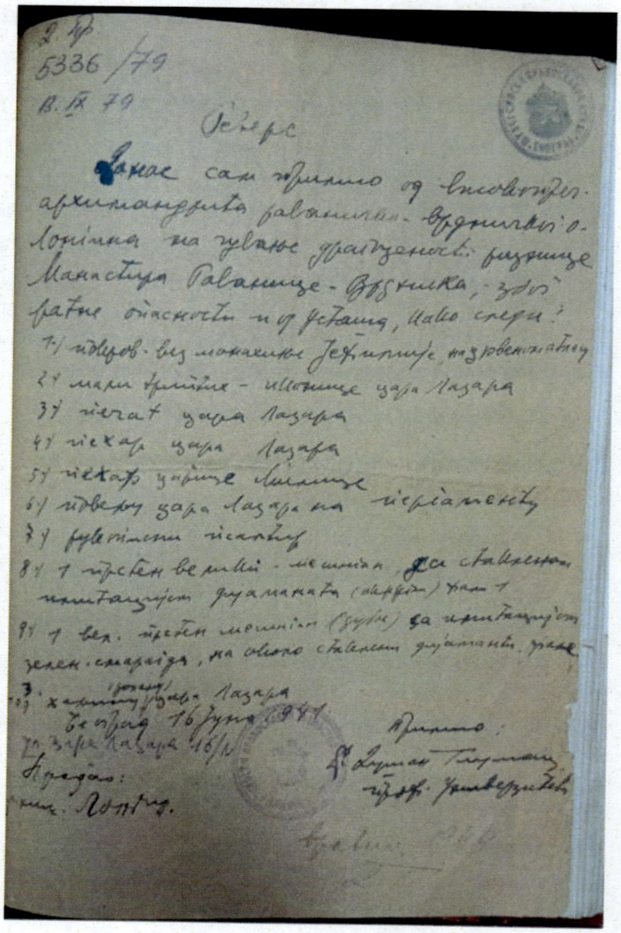

Fig. 6 Reverse made between Archimandrite Longin and Professor Dušan Glumac on 16th June 1941. (Archive of Museum of Serbian Orthodox Church in Belgrade, Republic of Serbia)

It was the first time that *The Revers of The Exchange of The Relics* (1979) the document of the Museum of Serbian Orthodox Church, was published in the book *Holy Knez Lazar: a monument to the sixth centenary of The Battle of Kosovo* published in 1989 (Mileusnić 1989), and still, according to the up mentioned movies and newspaper articles, left unnoticed. Due to the opening of the Museum, Svetozar Dušanić, the first director of The Museum of Serbian Orthodox Church after The World War II, after several written requests to Nikanor, the Bishop of Srem, finally got The Ceremonial Vestment of Knez Lazar from The Chapel Treasury of Saint Simeon, to take it to restoration and conservation (The written requests to Nikanor, the Bishop of Srem, about The Ceremonial Vestment of Knez Lazar: MSPC No. 232/25. 9. 1953; MSPC No. 265/5. 11. 1953; MSPC No. 279/20.11.1953; MSPC No. 434/14. 5. 1955). Today it is kept under special conditions, in a glass showcase.

4 Serbian Medieval Royal and Nobility Costume

Up-to-date digital reconstruction is proving that only the combination of historical research, archival and on-field results, as well as the results given by the most modern technology, can tell the story of this sensitive exhibit from the fourteenth century. It cannot be seen "with the bare eye". Due to the five centuries long and exciting history of this exhibit, as soon as the conditions for that were made, a digital reconstruction has been done in the last few years in The Museum of Serbian Orthodox Church in Belgrade, for a book *Serbian Medieval Royal and Nobility Costume* (Popović 2020, 2021).[3] Due to the great interest of the Serbian public in Medieval History and Art, for the particular topic of Medieval Costumes and Textiles, symbolic and decorative patterns, the book is already published twice by this Museum, in a record time of only several months. The first edition of this book is published on the day of the Transfiguration of Christ, on 19th August 2020, in the circulation of 800 copies. The second edition was published for the Ascension of Christ, on 26th May 2021. It is printed in the circulation of 500 copies, luxuriously equipped. Most importantly, the second edition is enriched with eighty new photographs, ten new illustrations, and digital reconstructions of clothes painted on medieval portraits, icons, and preserved items from the eleventh to the seventeenth century. Some of the digital reconstructions of the fresco portraits, quality measured with gigapixels, are presented as short animations and can be seen on the Instagram profile of The Museum of Serbian Orthodox Church, named #muzejspc. Based on the analysis of more than fifty presented frescoes and preserved items, the costumes of Serbian medieval rulers and nobility are chronologically presented, and each part of the garment is called by its name, from the time in which it was made up, as well as, what it served for. And since the book was published for the second time, The Ceremonial Vestment of Knez Lazar attracted the attention of researchers, even more, contributing to the topic with many new information, separating folklore from facts obtained through scientific research.

5 New Heritage Readings

5.1 The Cut of the Vestment of Knez Lazar

The digital reconstruction showed that its cut can be categorized as a *late outgrowth of kontuš* (pron. kontush) (Vasić 1979, 23–54; Popović 2004, 376–379; Popović 2021, 223). The term *kontuš* refers to the wide dress that was worn in Moravian Serbia, during the Lazarević and Branković dynasties (Ibid).

The Ceremonial Vestment of Saint Lazar is sewn from sixteen wedge-shaped pieces of silk brocade. It has a decorative embroidery border, which follows the lowest part of the Vestment, the opening around the neck, and the end of the sleeves. The border is embroidered with golden thread. The front of the Vestment was cut vertically. There used to be 41 buttons in front. Also, existed nine buttons per sleeve. The silk base is embroidered with gold and silk thread. The embroidered motives are lions, birds, and stylized flowers. The button is made of silver and linen thread (Popović 2021, 236, 223, 234). This is The Vestment everyone can see in the permanent exhibition of The Museum, in the room dedicated to the relics of Serbian Saints. The Vestment was porphyry originally, and it is known it had pearls sewn on, until the half of the eighteenth century (Ruvarac, 63). The pearls were torn apart, and so the color faded.

The digital reconstruction done during 2020–2021, in The Museum of Serbian Orthodox Church, has shown the real shape of The Vestment of Knez Lazar. The lower parts of the fabric are wider, so the lower part of the Vestment is bell-shaped, known also as *the gothic cut of the dress* (Popović 2021, 224). The Vestment used to have long sleeves. The sleeves narrowed to the elbows. The length of the Vestment is 142 cm, and Bojan Popović believes that it reached the ground, thus revealing information about the height of Knez Lazar (Ibid, 236).

The cut, the fabric which the Vestment was sewn from, and the iconography, haven't been yet completely studied since silk fabric is very fragile. It is an almost flat object, and with an RGB camera, in the Museum, the surface of the Vestment is documented with as much detail as possible, but the internal structure and the technique of weaving were still unknown.

In September of 2021, using the results of Project SILKNOW, helped us to look at the complex internal structure and see the model of silk fabric at the yarn level. Now we can show some more results of research on The Vestment. Comparing the Vestment with the pieces of North Italian silk brocade fabrics, we can tell for sure production place, production time span, material, technique, and iconographical motives.

The list of the Accession Numbers through different Museums materials we used for comparison consists of the fabrics from five different Museum:

I Metropolitan Museum: 07.243.4; 09.50.967; 09.50.980; 09.50.996; 09. 50. 1009; 09.50.1010; 12.55.1; 17.121.5; 17.121.6; 28.180.2; 31.69; 39.21.2; 46.156.28; 46.156.29a; 46.156.28b; 46.156.32; 46.156.36a; 46.156.36b; 46.156.38;

[3] Museum of Serbian Orthodox Church have the richest collection of Serbian medieval textiles.

46.156.38; 46.156.39; 46.156.44; 46.156.47; 46.156.55; 64.101. 1388; 1998.196; 46.156.72; etc.

II Boston Museum of Fine Arts: 35.74; 39.539; 48.378; 30.654; 39.540; 35.73; 35.78; 35.79; 41.50; 47.1039; 54.2; 1973.10a-c; 58.22; 1973.96; 31.140; 96.51b; 1973.95; 1973.94; 24.395; 96.51b; 35.75; 35.80; 35.82; 35.85; 36.163; 32.91; 31.128; 28.321; 26.793; 25.618; 24.396a-b; 22.357; 16.28; 15.844; 09.369; 07.713; 07.624; 97.564; 96.474; 96.503; 96.472; 96.468; 04.1633; 04.1637; 51.2476; 52.488; 22.83a-c; 19.717; 35.33; 35.83; 39.540; 31.13; 30.654; 28.320; 16.111; 96.610; 12.99; 04.1689; 04.1634; etc.

III LA Country Museum of Arts: M. 64.82.1.

IV Smithsonian Design Museum: 1902-1-385; 1902-1-284-a/d; 1938-57-1226-a; 1902-1-251.

V Cinquatenaire Museum, Brussels, Belgium.

5.2 Drappi d'oro et d'ariento

The most interesting about this report is that Lazar Mirković thought, by comparing it with the drawings from some books, that a fabric The Vestment of Knez Lazar was made from, what was *most similar* to silk produced in the Northern Italy town of Lucca in the fourteenth century (Mirković 1929, 43; Mirković 1937, 74). After The World War II, in 1954, we found the note, that the curators of the Museum of Serbian Orthodox Church were searching for confirmation of his assumption, that the fabric originated from Lucca in Northern Italy. At that time, it was believed it was from the end of the fourteenth century. Confirmation was sought from Fritz Volbach, director of Römish-Germanishes Zentral Museum in Mainz, who was considered an author of the best study of Italian medieval art *L'art Byzantine* (MSPC No. 23/54).

Since we had a leading belief that the attire was made from textile from Lucca, we checked all Open Source Museum Sites, and the ADASilk depository helped us to find many pieces of fabrics produced in the city of Lucca and Northern Italy, ranging from the end of the thirteenth century to the end of the fifteenth century, for making parallels and finally make conclusions about main motives, material, and technique of weaving. We used text metadata and available images, and the Virtual Loom application. Comparing the information, we got the final results, we receive confirmation that Lazar Mirković laid a foundation, which we have now supplemented with just a several new facts:

The Vestment is made from silk from Lucca. This kind of silk is called lampas, with brocade embroidery. It is weaved 2/1 twill ground and 1/3 twill pattern, which means it has three pairs of main warp threads to one binding warp thread; one pattern weft of gilt membrane thread (Fig. 7).

The textile belongs to the class of *drappi d'oro et d'ariento* (King 1998, 94–98), gold and silk woven, and gold and silk embroidery, as it was done in the middle of the fourteenth century, but here it is the more precisely third quarter of the fourteenth century, surely made before the year 1389 and The Battle of Kosovo (Punta-Ludovica 2017, 66–98; Lambert—Wilson 2016). The patterns on the fabric from Belgrade Museum, also shows today unrecognizable letters in a circular ornament (Figs. 8 and 9).

Fig. 7 The texture of lampas silk, with brocade embroidery: weaved 2/1 twill ground and 1/3 twill pattern, which means it has three pairs of main warp threads to one binding warp thread; one pattern weft of gilt and silver plated membrane thread (Photo-library of Museum of Serbian Orthodox Church in Belgrade, Republic of Serbia, Aleksandar Radosavljević 2021)

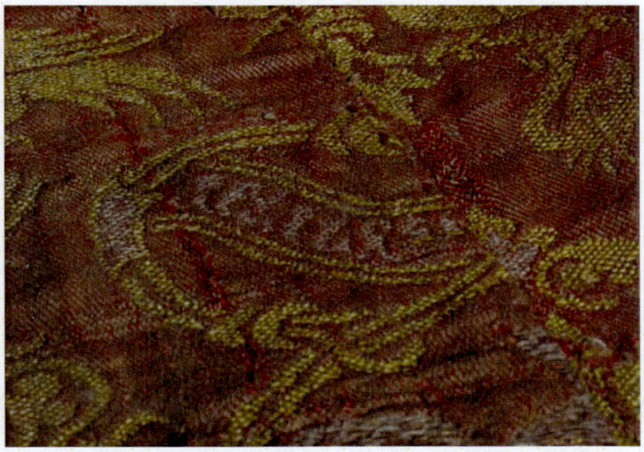

Fig. 8 Round motive with letters (Photo-library of Museum of Serbian Orthodox Church, Republic of Serbia, Aleksandar Radosavljević 2021)

Fig. 9 Two standard motives from the lampas silk from Lucca. Motives from The Vestment of Knez Lazar (Photo-library of Museum of Serbian Orthodox Church, Republic of Serbia, Aleksandar Radosavljević 2021)

The only left button from 44 of them, has parallels from Balkan. The button is showing the heraldic sign of Knez Lazar, but it can be from a later period. The only surviving button is made of twisted linen and silver threads. On the central part of the button is a silver surface with an imprinted heraldic symbol of Knez Lazar. Blue and green enamel remained around the symbol (Fig. 2). A similar button from the year 1365 has been saved (Mirković 1929, 48; Babić 1975, 65; Popović 2021, 236–234).

6 Conclusion

The textile which the Vestment of Knez Lazar was made, belongs to the class of *drappi d'oro et d'ariento* (King 1998. 94–98), gold and silk woven, and gold and silk embroidery, as it was done in the middle of the fourteenth century, but here it is the more precisely third quarter of the fourteenth century, surely made before the year 1389 and The Battle of Kosovo (Punta-Ludovica 2017, 66–98; Lambert—Wilson 2016). The poetic title of this paper, *"The Hidden Life" of The Ceremonial Vestment*, comes from the fact that the documents "are silent" in the case of the early history of this artistic object and the relic, until the year 1733. Hereof, for the first time, this paper collects in one place the "for now know" history of the Vestment and shows it in its own richness and importance. It is also a manifestation of the geopolitical and trade circumstances in which it was made, as it served in the court of Prince Lazar and *post-mortem*. The results of the paper are a supplement to the research of a wider category of historical textiles topography, and trading silk routs in medieval Serbia, Balkan, and the Mediterranean.

7 Special Mentions

We sincerely thank Father PhD Art Historian Vladimir Radovanović, the principal of the Museum of Serbian Orthodox Church, for his blessing to research The Vestment of Prince Lazar, and to publish this paper. Also, we would like to thank our colleagues from the Museum—PhD Miljana Matić, MA Marija Jović, Nataša Životić, Aleksandar Radosavljević, and Stefan Kojadinović, for helping us in many ways during the research. We owe a special thanks to PhD Mihailo Radinović and PhD Emilija Brdarić for helping us with the methodology, and MA Emilija Vuković and MA Ivana Filipović, for collecting the material during the research.

References

ADASilk database (2017–2022). The University of Valencia. https://ada.silknow.org/

Bešlin, B.: Istorija Novog Sada (History of Novi Sad). Umetnička topografija Novog Sada (Artistic topography of Novi Sad). Matica srpska. Novi Sad. (2014)

Davidov, D.: Stradanja fruškogorskih manastira 1941–1945 (Devastations of Monasteries of Fruška Gora 1941–1945). Fruškogorski manastiri (Monasteries of Fruška Gora). Galerija Srpske akademije nauka i umetnosti, Sveska 66. Beograd. 204–215 (1990)

Dušanić, S.: Muzej Srpske pravoslavne crkve (Museum of Serbian Orthodox Church). Beograd. 8 (2008)

DOŽIVITE ISTORIJU: Originalna haljina kneza Lazara izložena u Starom dvoru (Experience History: The original dress of Prince Lazar displeyed in the Old Palace). Blic. Beograd. (2015). https://www.blic.rs/vesti/beograd/dozivite-istoriju-originalna-odora-kneza-lazara-izlozena-u-starom-dvoru/vg5nlpm. Accessed 17 Dec 2021

Glumac, D.: Povodom rešenja o prenosu moštiju Kneza Lazara (Regarding the decision on the transfer of the relics of Prince Lazar). Vesnik SPC. Beograd (1954)

Grujić, R.: Izveštaj Gospodinu Veliboru Joniću o prenosu moštiju srpskih svetitelja (Report to Mr. Velibor Jonić on the transfer of relics of Serbian Saints), Museum of Serbian Orthodox Church, Belgrade (1942)

Halpern, K., Klod Ruano Borbalan, Ž.: Identitet(i): pojedinac, grupa, društvo. (Identity's: individual, group, society). (S. Džeferanović, Trans.). Clio. Beograd. (Original work published 2004) (2009)

Informative Service of Serbian Orthodox Church. (29 June 2021). http://www.spc.rs/sr/patrijarh_na_vidovdan_posetio_na_kosovo_metohiju. Accessed 17 Dec 2021

King, D.: Silk Weaves of Lucca in 1376. Collected Textile Studies. Pindar Press, London (1998)

Lambert, B., Wilson A.K.: Europe's Rich Fabric: The Consumption, Commercialisation, and Production of Luxury Textiles in Italy, the Low Countries and Neighbouring Territories (Fourteenth—Sixteenth Century), ASHGATE (2016)

Matić, V.: Arhitektura fruškogorskih manastira (Architecture of Monasteries of Fruška Gora). Fruškogorski manastiri (Monasteries of Fruška Gora). Galerija Srpske akademije nauka i umetnosti.. Sveska 66. . Beograd (1990)

Matić, M.: Osam vekova umetnosti pod okriljem Srpske Pravoslavne Crkve (1219–2019) (Eight centuries of art under the auspices of the Serbian Orthodox Church 1219–2019). Muzej Srpske Pravoslavne Crkve. Beograd (2019)

Medaković, D.: Sveta Gora—fruškogorska (Holy Mountain of Fruška Gora). Prometej. Novi Sad (2007)

Mileusnić, S.: Kud' se đede car Lazara blago (Where is the treasure of Tzar Lazar). Sveti knez Lazar: Spomenica o šestogodišnjici Kosovskog boja 1389–1989a (1989a)

Mileusnić, S.: Muzej Srpske pravoslavne crkve (Museum of Serbian Orthodox Church). Muzej Srpske Pravoslavne Crkve. Beograd (1989b)

Mileusnić, S.: Muzej Srpske pravoslavne crkve (Museum of Serbian Orthodox Church). Muzej Srpske Pravoslavne Crkve. Beograd (2001)

Mirković, L.: Iz riznice fruškogorskih manastira (From the treasury of Fruška Gora Mountain). Glasnik istorijskog društva u Novom Sadu II/2 (1929)

Mirković, L.: Haljina kneza Lazara (The dress of Knez Lazar). Umetnički pregled. Beograd (1937)

Mihaljčić, R.: Lazar Hrebeljanović: istorija, kult, predanje (Lazar Hrebeljanovic: history, cult, tradition). Nolit. Beograd (1984)

Muzej Srpske Pravoslavne Crkve: Haljina Kneza Lazara (Museum of Serbian Orthodox Church: Dress of Knez Lazar) (2017). https://www.youtube.com/watch?v=9KzTb8C4D3M Accessed 2 Dec 2021

NEPROCENJIVA VREDNOST Plašt kneza Lazara pred publikom u Starom dvoru (Invaluable: Prince Lazar's cloak in front of the audience in the Old Palace). Blic. Beograd. (2015). https://www.blic.rs/vesti/beograd/neprocenjiva-vrednost-plast-kneza-lazara-pred-publikom-u-starom-dvoru/w2ecs1e. Accessed 17 Dec 2021

Nikolić, J.: Korespodencija oslikane, sačuvane i metaforične odeće na primeru kneza Lazara i vlaškog vojvode Njegoja Basarabe (Correspodence of painted, preserved and metaphorical clothes on the example of Pince Lazar and Duke Basaraba of Wlach). Zbornik of Museum of Applied Art, Issue 16 (2020).

Novaković, K. (1939). Celebration of 550 years of The Battle of Kosovo (Бој на Косову). Novaković Movie. https://www.dailymotion.com/video/x3s2ute

Odežda kneza Lazara—kratkometražni dokumentarni film/Odežda kneza Lazara (Prince Lazar's Attire). (2018). https://www.youtube.com/watch?v=1ZoZt08n4RQ. Accessed 2 Dec 2021

Pavlović, L.: Kultovi lica kod Srba i Makedonaca (Cults among Serbs and Macedonians). Narodni muzej Smederevo. Posebno izdanje. Knjiga 1 (1965)

Plašt kneza Lazara i Pohvala monahinje Jefimije u Starom dvoru (Cloak of Prince Lazar and Praise of Nun Jefimija in the Old Palace). Politika. Beograd. (2015). https://www.politika.rs/scc/clanak/324372/Plast-kneza-Lazara-i-Pohvala-monahinje-Jefimije-u-Starom-dvoru. Accessed 17 Dec 2021

Pigafetta, A.M.: Itinerario di Marc' Antonio Pigafetta, gentil'huomo Vicentino (1567). Yale Univesity Library: https://collections.library.yale.edu/catalog/2065842. Accessed 24 Sept 2021

Popović, B.: (2020). Srpska srednjovekovna vladarska i vlasteoska odeća (Serbian Medieval royal and feudal clothing). Muzej Srpske Pravoslavne Crkve. Beograd.

Popović, D.: Ravanički grob kneza Lazara (Tomb of Prince Lazar of Ravanica), Sveti knez Lazar: Spomenica o šestogodišnjici Kosovskog boja 1389–1989 (Saint Prince Lazar: Memorial on the Sixth Anniversary of the Battle of Kosovo 1389–1989). Beograd (1989)

Popović, M.: Srpska srednjovekovna vladarska i vlasteoska odeća (Serbian Medieval royal and feudal clothing). Muzej Srpske Pravoslavne Crkve. Beograd (2021)

Phillips, A.: Sea Change: Ottoman Textiles between the Mediterranean and the Indian Ocean. University of California Press, Oakland, California (2021)

Punta, I., Rosati, M.L.: Lucca, una città di seta: produzione, commercio e diffusione dei tessuti lucchesi nel tardo medioevo. (Lucca, a city of silk: production, trade and dissemination of Lucca fabrics in the late Middle Ages). M. P. Fazzi (Ed.). Lucca (2017)

Ruvarac, D. (1903a). Opis srpskih fruškogorskih manastira 1753. godine (Description of Serbian Monasteries in Fruška Gora in 1753). Novi Sad

Ruvarac, D.: Vrdnik. Opis srpskih fruškogorskih manastira 1753. godine (Description of Serbian Monasteries in Fruška Gora in 1753), III, Srpski Sion, 3 (1903b)

Radojčić, N.: Šesnaest rana Kneza Lazara (Sixtheen wounds of Knez Lazar), Glasnik Istorijskog društva u Novom Sadu (1930)

Slijepčević, Đ. (1966): Istorija Srpske Pravoslavne Crkve (*History of Serbian Orthodox Church*). knj.. 2, BIGZ, Beograd

Stojadinović, M.: U Fruškoj gori 1854 (*In Fruška Gora Mountain 1854*). Prosveta: baština (1985)

Svečana haljina kneza Lazara (Ceremonial Dress of Knez Lazar): RTS. Beograd. https://www.rts.rs/page/magazine/sr/story/2523/nauka/3021842/svecana-haljina-kneza-lazara.html. Accessed 17th Dec 2021 (2018a)

Svečana haljina kneza Lazara (Ceremonial Dress of Knez Lazar): SPC. Beograd. http://www.spc.rs/sr/svechana_haljina_kneza_lazara. Accessed 17th Dec 2021

The written requests to Nikanor, the Bishop of Srem, about The Ceremonial Vestment of Knez Lazar, to take it to restoration and conservation, (MSPC No. 232/25. 9. 1953; MSPC No. 265/5. 11. 1953; MSPC No. 279/20. 11. 1953; MSPC No. 23/54; MSPC

No. 42/ 12.6.1954; MSPC No. 42/1954; MSPC No. 434/14. 5. 1955), Documents from The Museum of Serbian Orthodox Church, Belgrade

Todić, B.: Mitropolija u Prištini—prva grobna crkva kneza Lazara (Mitropoly in Priština: the first burial church of Knez Lazar). Sveti knez Lazar: Spomenica o šestogodišnjici Kosovskog boja 1389–1989 (Saint Prince Lazar: Memorial on the Sixth Anniversary of the Battle of Kosovo 1389–1989). Beograd (1989)

Točinac, I.: Popisi mitropolita Mojsija Petrovića za sveštenike i parohijane, (Metropolitan Mojsej Petrović's regulations for priests and parishioners). Istorijski časopis, кн. knj. XL, VIII (2001)

Virtual Loom Program (2017–2022). The University of Valencia. https://silknow.eu/index.php/virtual-loom/

Women's Intangible Heritage in Tunisia Between Profitability and Authenticity

Ons Sakji

Abstract

The pottery of Sejnane was added to UNESCO's World Heritage in 2018. It is an ancient style of pottery unique to the region created exclusively by the women there, using a special technique. This art-craft has been transmitted from mother to daughter for generations. It involves making cookware, dolls and animal figures. Today, facing changing social and economic needs, this pottery is changing and taking new forms following the demands and the modern trends of today's society. On the one hand, this can guarantee a fixed income for these women. On the other hand, this technique is losing its authenticity over time. This opinion paper shows how an intangible heritage is trying, at the same time, to survive and to preserve its authenticity.

Keywords

Intangible heritage • Profitability • Women • Contemporaneity

1 Introduction

Sejnane pottery is an ancestral know-how that has preserved its authenticity over time and has stood out comparing to other Tunisian pottery thanks to its primitive and organic appearance. At the end of 2018, this pottery has been added to the UNESCO World Heritage list. This traditional human know-how has made an essential contribution to the sustainable development and the maintenance of social cohesion and cultural diversity in the region of Sejnane. This modeling technique is a set of gestures and movements transmitted from mother to daughter from generation to generation through observation and learning. Since the end of Bronze Age, second millennium BC, this technique has stood the test of time and has been able to keep its authenticity. It has become an intangible cultural heritage that only women hold its secret. Tunisia is well known for its rich and diversified pottery in different regions, but the pottery of Sejnane stands out by the nature of its clay.

A first and only one research and investigation on this pottery, that are not scientific, were conducted by Sekik and Louhichi (2007). They made an inventory and a classification study according to its different typologies. This study was a significant contribution to the knowledge of the artistic and ethnographic heritage.

Faced with intervention on any type and form of heritage, we are always confronted with a duality of conservation and creation. However, its protection and conservation do not necessarily guarantee its permanence over time. A variety of research studies, such as that of Alexandra Georgescu Paquin (2013), have shown that updating the architectural heritage nowadays is necessary to ensure its longevity. We have already confirmed the same hypothesis in a previous work, that focused on an analysis of adaptive-reuse of existing and old buildings (Sakji and Kharrat 2019). In this study, the architectural heritage requires a contemporary architectural transformation which inserts the existing building into its current era. In this way, it gives a new image and a new symbolic meaning that will spontaneously give it a new social and economic dimension. The transformation of this heritage guarantees its survival and longevity. This hypothesis is well confirmed in the two previously mentioned works on built heritage, and we believe that it can also apply perfectly to intangible heritage.

This study tries to understand the transformation process of Sejnane pottery and to analyze the degree of authenticity of this technique in the current time. Since its registration in the UNESCO, there has been a certain enthusiasm for this

O. Sakji (✉)
Carthage School of Science and Engineering, UTC, Tunis, Tunisia
e-mail: ons.sakji@gmail.com

Research Laboratory LarPA, National School of Architecture and Urbanism, Tunis, Tunisia

S. Kostopoulou et al. (eds.), *Cities' Vocabularies and the Sustainable Development of the Silkroads*, Advances in Science, Technology & Innovation, https://doi.org/10.1007/978-3-031-31027-0_14

pottery, manifested in selling it in some shops in the capital, Tunis, and in exporting it to other countries.

The pieces that are often created and sold are not those that have been registered as World Heritage by UNESCO. New models have been developed by the women potters themselves. In this way, some believe that this technique is losing much of its authenticity in order to satisfy the tastes and the current needs of customers. In addition to that, other dangers await this expertise, such as the ease of access to the chemical products. This rural population lives in need, and many families depend on the sale of pottery. This exposes them to several dangers, most importantly the exposition to chemicals and particularly customer orders that deviate from the original pottery models. There is also a health danger from the smoke inhaled by the women potters during the firing process, which has led to the sickness of some of them.

Fig. 1 Environment of Sejnane. *Source* Author

2 An Ancestral and Authentic Expertise

2.1 The Environment of the Sejnane Region

Sejnane is a harsh and landlocked region in northeastern Tunisia in the Koumirie Mountains of the Mogod massif. It covers 60,000 hectares (43% of which is forest) with 42,000 inhabitants. This small town, which was once inhabited by a Berber population, is on the road that connects Bizerte to Tabarka. This road crosses a series of fairly high hills covered with myrtles and mastic trees and pines, eucalyptus and wild olive trees (Fig. 1).

The composition of Sejnane soil makes it much more suitable for pottery rather than agriculture because of the nature of the clay that covers it. The soil is characterized by its richness in good quality clay veins of various colors, ocher, red, white or brown.[1]

2.2 Sexual Division of Tasks

In Tunisia men often emigrate from this region, to other regions like the city of Bizerte, Tunis or to Europe to look for work, or they work in agriculture. Otherwise, they spend most of their time playing cards at local coffee shops.

However, the very active women stay to take care of the children and the livestock. They do daily housework, collect wood for the fire, bring water, milk the cows and maintain the backyard. In addition to this, they make the pottery for which this region has been known since antiquity. Women do not only have the know-how of pottery making but much more, such as spinning, weaving, grinding grain and cooking.

2.3 Transmission of the Pottery Making

The art of pottery making is a skill of women, who ensure its transmission in a spontaneous and random way to the little girls of their families. Despite their schooling, this technique persists and is still transmitted, in a disorganized way, through community social relations based on kinship and neighborhood ties. Today, it is the National Crafts Office that carries out on integrated development programs and training cycles (Fig. 2) for young women in Sejnane who want to devote themselves to this activity.

Fig. 2 Training center in Sejnane realised by raw materials and traditional technique of build. *Source* Author

[1] "*........cette couche argileuse qui couvre plus de 70% des sols de Sejnane*". Sekik and Louhichi (2007). *Les potières de Sejnane: des femmes et un savor-faire*. Editions Finzi, p. 15.

Fig. 3 Pottery of Sejnane for sale exposed on the road in Sejnane. *Source* Author

2.4 The Functional Nature of Pottery

Pottery in Tunisia differs from one region to another according to distinctive rounded shapes or dyes or decorations used. There are some types of utensils and cookware that are still used such as the «kanoun», the «tagine» or the «tabouna» in certain rural or mountainous regions. Until the middle of 1950, the pottery of Sejnane had been used as cookware for the kitchen, wall decoration and sometimes some furniture, which the women renew annually.

Cookware that is used in ovens or on stoves that are hot do not have ornamentation like cooking containers and plates. Women add some decoration on storage jars and extensive decoration on glasses, plates, bowls and tableware (Fig. 3).

2.5 The Know-How

Unlike other potters, the women of Sejnane do their models without lathes or molds. These are works that they shape and adorn with their own hands without any mechanical tools while using local and natural materials (Sejnane 2005). Each piece is unique, and the potters cannot reproduce the same models or the same dimensions. It is an archaic technique that lends itself to several techniques called coil modeling, "*The oldest and most used technique in North Africa for the manufacture of domestic tableware ... It proceeds by adding coils*"[2] (Camps 1961, p. 230). The coil modeling technique

Fig. 4 Methred (fouted bowl). *Source* Sekik and Louhichi (2007) p. 18

dates back to the Bronze Age, to the Neolithic period around the middle of the 4th millennium BC, specific to North Africa (Fig. 4). According to the Larousse dictionary, coil technique is "a sausage of soft paste used to shape ceramics without using the lathe".

Comparisons have been made between pottery from the medieval period and that from the Middle Ages. The technique from the Middle Ages resembles contemporary pottery, according to Adnene Louhichi (Sekik and Louhichi 2007). "It thus appears indisputable that survivals in forms, techniques and decoration are confirmed by specialists between ancient and medieval ceramics and current pottery, in particular those from the Sejnane region. The indigenous fund is very tenacious and despite the various successive cultural contributions, it has been able to cross the ages to continue until today".

The clay is extracted in the neighborhoods of the dwellings, on the side of the slope or on the side of the wadi. The women potters get rid of impurities and pebbles. Then they put it in clumps in the courtyard of the house and put it in water for one or two days. It decomposes following the fermentation of organic matter, which gives it a plasticity effect. Then the potter takes a lump of clay and adds an equal amount of "tafoun": it is a powder obtained after grinding worn or damaged pottery, which then goes through a sieve. After mixing, this powder turns into a homogeneous paste (Fig. 5) which preserves the pottery from bursting and cracking during drying and firing.

The first tool that the potters use is the «r'leg» (Fig. 6). It is a support or a work platform that allows the elevation and rotation of the object. The potter begins to put degreasing powder on this base to prevent the clay from sticking to it. Then she places a lump of clay on it, she flattens and regularizes it with her fingers. She always starts from the bottom of the object, and she tries to round off the first clump (Fig. 6). In the same way, other coils are superimposed and shaped until the side of the object rises. Then they are smoothed by hands with a mixture called «talya» using a squeegee called «machât». The potter smoothes the piece and shapes it at the same time. This facilitates the cohesion

[2] «*la technique la plus ancienne et la plus employée en Afrique du Nord pour la fabrication de la vaisselle domestique....elle procède par ajout de colombins*» Camps (1961). *Aux originess de la Berbérie, Monuments et rites funéraires protohistoriques*. Délégation générale en Algérie.

Fig. 5 Powder of "tafoun", a sieve step. *Source* Author

Fig. 6 Hadda regulating and shaping a bowl with her fingers, on the support of work "rleg". *Source* Author

and the assembly of the coils. Then, she puts the pieces out for drying according to the ambient temperature and the season, in the sun or in the shadow. This allows the evaporation of the water from the clay to be slow. The objective here is to avoid rapid evaporation during the firing process, which will certainly causes cracks in the pottery. The next step is to polish the exterior and interior using a shell called «baboucha». A second layer of «talya» is applied to give a beautiful coloring.

The final step of this process is firing, which is done in the open air. The pieces to be fired are placed on an organic fuel made of branches or cow dung mixed with straw. The fuel is arranged in a circular shape (Fig. 7a, b) and directed towards the center of the hearth to receive a maximum of heat.

Once fired, the objects are decorated. For cookware intended for the fire or heating, they are devoid of any ornamentation or are decorated in a simple manner on the exterior sides. For the others, like tableware, they are decorated with geometric designs organized vertically. These designs are traced with red ocher or lentisk leaves juice, a juice that turns into a black color after the second firing (Fig. 8).

3 The Capability Between Authenticity and Subsistence

3.1 The Degree of the Pottery's Authenticity

The technique of making this pottery remains identical to that of the past. The women potters still respect the stages of preparation and production. Given the continuous development of our lifestyle today, this modeling technique has certainly kept much of its authenticity, but it has also evolved and has become more skillful and confident. Even if we find a striking resemblance of the current Sejnane pottery with the models made in the Bronze Age, we can observe a

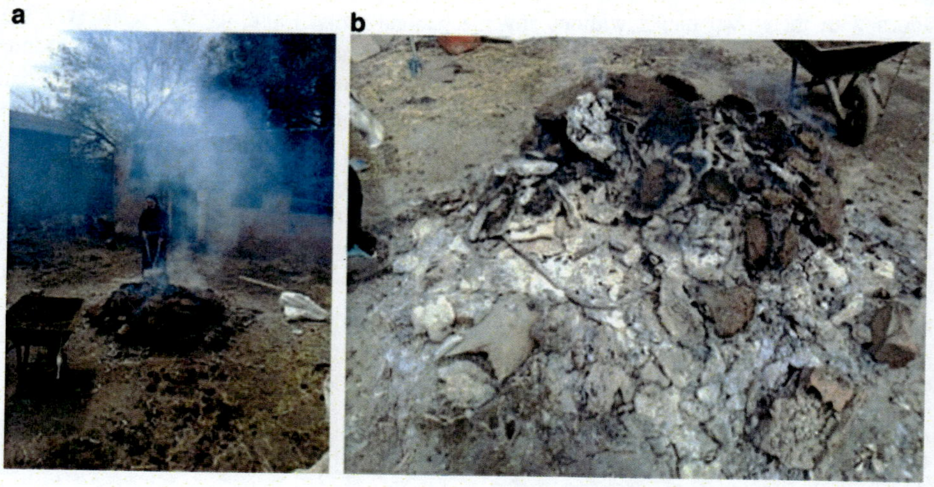

Fig. 7 **a** Organic fuel arranged in a circular shape. **b** Fuel turned into ashes, pottery is cooked *Source* Author

Fig. 8 Pottery decorated with both: lentisk juice and red ochre. *Source* Author

certain maturity and precision in the current products (Figs. 9 and 10).

However, with the industrialization and the arrival of glass and plastic tableware with attractive colors, women were pushed to change their cookware and to buy new tableware. This led to a change in the function of this pottery, as it became more decorative. Consequently, this evolution reduced the overall production.

Women's perspective and vision have been evolving over time. Through a social survey we have found that some went to school and stopped their studies at the primary level as they must help their mothers to take care of the house and make pottery. This certainly has both a negative and positive

Fig. 9 Sahfa (jatte), Nécropolis of Smirat, The Sahel of Tunisia III–II. B.C. National Museum of Bardo, Tunisia. *Source* Sekik and Louhichi (2007)

Fig. 10 Sahfa (jatte), The beginning of twentieth century, Ksar Said, Tunisia. *Source* Sekik and Louhichi (2007)

impacts that are manifested in the creations produced, and in particular the method of creation.

Then arriving to the era of technology, television, photography, and telephones that facilitated the access to panoply of images and contributed in leading these women to imitate and mimic industrial pottery. This certainly encouraged these women to create new models that are easily sold and to be in line with the market and the current requirements of the buyer.

3.2 Pottery Livelihood of a Rural Population

These women potters want to earn money to pay for their children's education and weddings, as well as for their hoe extension. They are ready to adopt any style or design without any requirements. From a photo, they can study the piece and determine its dimensions and proportions. They can enlarge or shrink it without using a measuring device. These women have a good visual memory, it is an innate gift. The design remains in their memories, and they no longer need the photo to reproduce it.

We believe that this pottery has always been more successful than others for its primitive and natural look. It is quickly distinguished from the pottery of other regions by a clear, visible, pure, natural and organic effect. To sell the pottery, these women participate in fairs and their men help them.

The price of this pottery has significantly increased since its addition to the UNESCO World Heritage art form. This important step made these women prouder and more confident. The profession of the women potters has been transferred for generations, from mother to daughter. Also, they train their daughters-in-law if they are not from the region, as Fatma, daughter-in-law of Sallouha, said "*I am from Ain Draham where we also do pottery, but it's different. After my marriage, when I moved to Sejnane, I observed all these women who master this technique from the youngest to the eldest and more over they do it with household chores. I learned this technique, which has now become my livelihood. Mastering in this technique was not easy*".[3]

Fig. 11 Figurines of animals and dolls decorated. *Source* Author

3.3 The Contemporaneity of Pottery for Its Durability

More modern pottery produced by the women of Sajnene has emerged. These are decorative objects and trinkets such as figurines inspired by their familiar environment of animals, lucky fish, cats, turtles, birds and dolls which make this pottery famous and specific (Fig. 11). The women have also developed more elaborated items such as sauce boats with compartments, incense burners, and jewelry boxes. When they try to innovate or modernize the pottery, generally the objects are no longer functional, and the lines are not well drawn. They start from the usual forms, then they try to change.

Today, the production of pottery has become a daily activity. It is their livelihood, and they rarely produce domestic pottery only, but rather decorative, which is the fruit of their imagination and tradition. They make different designs for some shops with added colors or materials (Figs. 12, 13 and 14).

Fig. 12 Pottery of Sejnane with new material added vegetable fibre. *Source* Author

4 Conclusion

Conserving the technique is much more difficult than conserving the cultural property, no matter its dimensions. Considered as a human heritage, its preservation depends on the people who maintain it. These people who are living in a particular context and conditions are sometimes pushed to neglect the original technique in order to be able to afford their vital needs. This technique lives till today thanks to the sales, efforts and interest of young designers who are aware of the heritage and aesthetic value of this pottery. However, to be able to make their projects profitable, the women potters must follow certain designs with proportions, shapes and sometimes colors that are different from the old pottery. Sometimes these products are exported to other countries, and it costs ten times more than the purchase price.

The canons of beauty today favor simple and minimalist shapes with no ornamentation. Women potters must accept current demands in order to survive. These brave women need to earn money to be able to continue making pottery and preserve this technique. They are very attached to their craft and art form. Moreover, they perfectly respect all the stages of this manufacturing process as described above. In addition to that, sometimes because of the lack of time, they use Chinese ink instead of tree juice and varnish instead of glue. But for those who are familiar with this, it is very noticeable because Chinese ink gives a uniform effect and a particular smell with a speckled and heterogeneous appearance.

The women potters do not try to replace the natural kiln with an artificial one. They always try to respect the process from the local raw materials to the manufacturing tools.

The idea of revisiting the pottery of Sejnane is the only alternative for its preservation and its durability over time. Of course, some graphic and ornamental aspects of this technique can be lost along the way. However, we ensure the perpetuity of the technique, which will allow the

[3] Personal investigation in Sejnane, 5 October 2021. Interview with one of the women Potters called Fatma.

Fig. 13 Pottery of Sejnane with painting. *Source* Author

Fig. 14 Pottery of Sejnane with natural colors. *Source* Author

descendants of the potters to study the theme and to develop this technique and the traditional ornamentation.

Introducing the contemporaneity to this technique is necessary for the development of this form of art and its profitability, so the women of Sejnane will be able to have a comfortable and decent life. In this way, we can guarantee its survival and durability. We think that contemporaneity styles and pieces maybe the future of this rich heritage.

References

Camps, G.: Aux originess de la Berbérie,Monuments et sites funéraires protohistoriques. Délégation générale en Algérie (1961)

Sejnane. Au pays d'une tradition millénaire. La poterie modelée des Femmes de Sejnane. Institut national du patrimoine. Badisches Landesmuseum (2005)

Paquin, A.P.: L'actualisation du patrimoine par la médiation de l'architecture contemporaine. Sciences de l'information et de la communication. Université d'Avignon, 2013. Français. ⟨NNT: 2013AVIG1132⟩ (2013)

Sakji, O., Kharrat, F.: Adaptive re-use in Tunisia between remembrance and contemporaneity. In: Hawkes, D. et al. (eds.) Conservation of Architectural Heritage. Advances in Science, Technology and Innovation (IEREK Interdisciplinary Series for Sustainable Development). Springer, Cham (2019). https://doi.org/10.1007/978-3-030-10871-7_10

Sekik, N., Louhichi, A.: Les potières de Sejnane: des femmes et un savor-faire. Editions Finzi (2007)

The Lustre Pottery Techniques Continuum Through the Silk Roads

Stratos Karakitsos

Abstract

In this paper we will attempt a macrohistorical examination of the Silk Roads and the resulting dissemination of ceramic techniques. Our research will concentrate on lustre technique in the production of ceramics. *Lustre Pottery* comes into existence at the beginning of the 9th c. in Abbasid Iraq and remained in use during the Fatimid, Seljuk, Ilkhanid and Safavid eras. Thence, it spread westwards to central and western Mediterranean, to Italy and Andalusia. *Lustre Pottery* technique was later adopted in Western Europe, finding its way in the workshop of internationally acclaimed British potter and ceramic scholar Alan Caiger-Smith (1930–2020). Our paper will concentrate on the enduring interest in the art of *Lustre Pottery* and the enchanting qualities of the surfaces it creates, irrespective of the cultural background of their provenance as well as on the strong continuity in the exchange of ideas and commercial goods between Europe and Asia which builds on the Silk Roads.

Keywords

Macrohistory • Silk roads • Ceramics • Lustre • Techniques

1 Introduction

The term *Silk Roads* was coined by the geographer von Richthofen at the end of the nineteenth century referring to trade routes linking east and west or China, India and the Mediterranean world, through Central Asia (Karakitsos 2021). Nowadays, the term *Silk Roads* or the *Great Silk Roads*, refers to the geophysical and historical socially determined transcontinental commercial and cultural routes —primarily in Afro-Eurasia-. The *Silk Roads* are continental and maritime providing commercial, cultural and technological exchange between the countries. In regards with the merchandise trade, high value goods like silk, paper, metals, horses, lower-cost bulkier goods like olive oil grain and manufactured goods such as ceramics, furniture etc. are being exchanged. Cultural exchanges include religious and philosophical systems, arts, literature and also scientific and technical knowledge e.g. paper, printing technology and gunpowder (Barisitz 2017).

The continental Roads interconnect the geographically demarcated areas with surplus resources, high population density and ongoing long-running wars, which are based on water complexes: the central ones, Indus Valley, Tiger and Euphrates (Mesopotamia), Nile and Danube with a second pole, Yangtze and Yellow rivers—through Amu Darya and Syr Darya basin-. At the same time, all of them, besides the latter, are coastal, so they act as potent maritime *Silk Road* hubs. All that results in the connection of the Pacific Ocean, first of all, with the Mediterranean Sea through the Indian Ocean and secondly with the Atlantic Ocean, through Mediterranean Sea or South West Indian Ocean (Peponis 2021).

Silk Roads exist for millennia. Their traces can be found to the trading activities of the Andronovo cultures in the Central Asia around 1500 BC (Perdue 2003). We see the Sogdian merchant networks and later the Achaemenid Empire road network which connected the Black and the Aegean Sea with the Central Asia and beyond (Vaissière 2005). Trade and cultural exchanges continue flourishing in the Diadochi period and Romans after their domination in Mediterranean Sea will fight with the Persians for the heart of the world, Mesopotamia (Frankopan 2015). In this first half of the first millennium AD, the maritime relations between the eastern countries and the Mediterranean and the Gulf started. Following the monsoonal climate Roman Egypt

S. Karakitsos (✉)
Faculty of Theology, Aristotle University of Thessaloniki, Thessaloniki, Greece
e-mail: stratos@karak.gr

S. Kostopoulou et al. (eds.), *Cities' Vocabularies and the Sustainable Development of the Silkroads*, Advances in Science, Technology & Innovation, https://doi.org/10.1007/978-3-031-31027-0_15

connected with southern India in the first century AD and in the fifth century AD the maritime trade between Arabs, Persians and Chinese begins (Shen 2017).

In the seventh century the core of Islamic rule lies in Mesopotamia, which is located in the crossroads of Afro-Eurasia and has direct contact with the Mediterranean Sea and the Indian Ocean through its ports. Islam soon secured the second important river complex, Egypt, uniting the economic heartlands of the former Sassanian and Roman Empire, which were stretching from west Mediterranean to Central Asia. The early Muslim scholars consolidated this spatial supremacy in maps and treatises, while the merchants travelled to every direction, importing and exporting goods (Frankopan 2015). Some early geographic treatises are: al-Khwārizmī (d. 850), *Shape of the Earth* (Sūrat al-arḍ) and Ibn Khurradādhbih (d. 912), The *Book of Routes and Realms* (Kitāb al-Masālik wa-l-Mamālik) (Park 2012). It is attested that even earlier, around 750 AD, İbāḍī seafarers, as Abū ʿUbayd al-Qāsim and al-Najjār b. Majmūn, have travelled to al-Ṣīn (China) (Lewicki 1935). The main termini of the eastern trade were the ports Siraf, Suhar, Baṣra and the new cities Samarra and Baghdad. All of them flourished economically and demographically as centers of the market extending from west Mediterranean to Far East (George 2015). It's not strange that Chinese ceramics are bestowed as gifts to caliph Harūn al-Rashīd (r. 786–809) through the continental roads (George 2015). Chinese ceramics are excavated in Samarra, as well in the ports of Siraf and Suhar, coming from the Eastern Seas. Thousands of Chinese ceramics were found in the ninth century *Belitung wreck* south-east from Sumatra. The ship was directing from China to Siraf (Wood and Priestman 2016). Via continental and maritime *Silk Roads*, goods and specifically ceramics crafted in Islamic lands are circulated in Eastern and Western Mediterranean reaching the northern lands. Furthermore, after the conquest of Iberia and Sicily in eighth and ninth century respectively, the craftsmen of these areas started using the Muslim production techniques. (Messina et al. 2018; Amorós and Gutiérrez 2020).

2 Lustre Pottery

In the first part, we will follow the spread of lusterware in the eastern and western Muslim lands and in the second part we will see its adoption and use beyond the Muslim territory (Fig. 1). It is important to note that the Islamic or early Abbasid glazed pottery appeared and rised quite suddenly. By the tenth century glazed and lustre ceramics could be found from Spain to Central Asia. Especially in this early period we meet three interrelated ways of circulation of glazed pottery, as O. Watson observes:

(1) Trading ceramics as physical goods within and out of the Islamic world
(2) Ceramics travel as ideas and habits following the arrival of the objects and the immigrants
(3) Potters travel for sundry reasons and set up their workshops in new places which leads to the dispersal of the techniques, technology and styles (Watson 2017).

It is said that the main impetus for this sudden glazed pottery emergence in the Islamic world is the arrival of Chinese stonewares in the Abbasid court in the ninth

Fig. 1 Main *Lustre-Pottery* production centers

century. However, as O. Watson mentions, that may not be the case. There is clearly a Chinese influence on the potters and customers but that does not explain

> the revolution in the making and consumption of glazed fine wares, a practice which had no precedent in Iraq. It fails to address the widespread use of glazed pottery in Egypt and Syria before the ninth century and does not explain the complete independence of many early Islamic wares from Chinese types. (Watson 2017)

On top of that, there is already a glazing tradition in Mesopotamia since Parthian times. Most of the production is consisted of large jars and vessels with limited decoration. The technology is simple: The clay is earthenware and the glaze is alkaline with copper giving a turquoise color. With the advent of Islam new social needs have to be satisfied. A new type of kiln is designed and better glazes, impermeable, easy to clean and brilliant are being produced. The first fine Islamic glazed ware, the *Coptic* or *Yellow glazed*, is found in Alexandria, Egypt, before the fall of the Umayyads in 750 AD. This type is mainly circulating eastwards toward Levant and Iraq. In the next century more types of clays and glazes will emerge in Mesopotamia. The opaque white glazed ware will replace the Yellow glazed, which is clearly a Chinese impact (Watson 2017). At that time we see the emergence of lusterware pottery, a unique technique which is totally different than any other category of pottery, glazed, unglazed or decorated with slips. But obviously, the potters, the market and the technology had reach such a point of sophistication that made this innovation possible.

The first applications of the lustre technique are first observed in glass, possibly in Egypt, in the Ptolemaic period if not in the Eighteenth period (Martin 1929; Whitehouse 2001). Al-Kindī (d. 873) writes treatises on industrial chemistry and refers to luster-painted glass in the *Kitāb Kimiyā' al-'itr wa al-taṣ'īdāt* (Book of the Chemistry of Perfume and Distillations). Jābir ibn Ḥayyān (d. 815) mentions among others, recipes on the staining of glass in *Kitāb al-durra al-maknūna* (The Book of the Hidden Pearl) and *Kitāb al-Khawāṣṣ al-Kabir* (the Great Book of Properties). The alchemical performance and transubstantiation of the Aristotelian philosophy, from the Hellenistic era onwards, led to experiments that contributed to the transmutation of one element into another and their application in crafts. *The Art of Alchemy* ('Ilm al-San'a) refers to a theoretical and a practical science, although here, it seems that we are dealing with practical industrial chemistry (al-Hassan 2009).

On the surface of the wares, luster pigment grants a fine decoration thin film of metal oxides obtained by a complex technical process. Al-Neyshapuri, in his treatise on precious stones, minerals and metals, *Javaher-name-ye Nezami*, in 1196 AD and Abu'l-Qāsem, a potter of Kashan, in 1301 AD, mention lustreware recipes according to which the suspension, more or less, includes powdered copper compounds, iron pyrites, silver sulphide and other ingredients dissolved in grape juice or vinegar (Roqué et al. 2008). That pigment was painted or brushed on top of pre-glazed pottery, composed of lead and/or tin oxides. The painted pottery was then given a second firing in a low-temperature smoky kiln, which produced an iridescent luster surface due to the metal reduction, which once polished, "reflects like red gold and shines like the light of the sun", as Abu'l-Qāsem wrote in the aforementioned treatise (Caiger-Smith 1985; Allan 1973).

In 836 AD, the new capital of the 'Abbāsid caliphate will become Samarra under Caliph al-Mu'taṣim. The first lusterwares excavated at Samarra dated at the beginning of the ninth century, represent the earliest examples of lusterware pottery (Sarre et al. 1925; Kervran 1977). At the same time, from ninth to late tenth centuries we see that Baṣra is the earliest center of production of the lusterware period (Mason and Keall 1991). The early Abbasid ceramics are polychrome and monochrome with a variety of decorations. We can see Roman designs, such us the Peacock-Eye, Sassanian symmetrical wings and palmettes and generally vegetal and figural, abstract and naturalistic decorations. The most important is that all of them share the common surface lustrous effect which is increased by the natural light. As M. Saba writes, the categorisation of these early lustre ceramics may supersede the abstract or naturalistic and vegetal or figural dichotomy (Saba 2012). Lusterware pottery transferred to Nishapur where beyond bright monochrome and polychrome lusterware we see items bearing inscriptions and at the same time, during the ninth century we observe that the Great mosque of Qayrawan, the capital of the Aghlabid Empire, is decorated with Başra lusterware tiles and a lusterware industry set up in Qayrawan. In the late tenth century artisans move to Fatimid al-Qāhira, where the lustre decorations and colors are enriched and reflect the cultural traditions of the new dynasty (Caiger-Smith 1985). Also, we observe the transition from earthenware to proto-fritware. Fritware is hardwearing white clay and is the precursor of the famous and so-called *Faience*. It seems that fritware or stonepaste technology is introduced to Syria and Iran after 1075 AD and will become the primary body for all fine ceramics including lustre wares (Mason and Tite 2007). Ayyubid Raqqa industry will flourish at this point (Mason 1997). We see similar type of pottery in Qal'at, Ja'bar and Balis Meskene (Blackman and Redford 1997). The Seljuks use the luster-fritware technology and produce exuberant arabesques and foliated kufic designs. They enrich the visual culture of Iran with a new figurative imagery inspired by Central Asia and the Buddhist arts. At the same time, during the twelfth–thirteenth century, Seljuks establish a stunning lusterware industry in Kashan, where high quality of luster wares with golden colors can be seen (Caiger-Smith 1985). From Kashan the technique spread to the eastern cities Nishapur and Samarqand (Watson 2017). The ceramics technology and innovation developed rapidly and new styles around the thirteenth century emerged as the famous *Mina'i* overglazed wares with enamel painting (Watson 1975a, b, 2004). Kashan potters continued in the service of the Mongol İl-Khanid dynasty till the fourteenth century. The cobalt borders and the luster animal illustrations on tiles produced for İl-Khanid patrons display high aesthetic value. During the next two centuries the lusterware falls out of fashion in the region of Iran and the depressed industry is revived around 1650 by the Safavids (Watson 1975a, b). The Safavid style of luster glazed pottery becomes distinct as new colors are being produced such us the coppery luster of pinkish hue (Golombek et al. 2014). Beyond the Islamic world towards east, *Lustre Pottery* travelled via the sea routes to western India and Southeast Asia but as far as we know not in China (Watson 2017).

On the other hand, from east to west, the *Serçe Limanı shipwreck* near Rhodes island in 1025 show us the level of interaction between the Muslim and the Christian spheres in Mediterranean (Jenkins 1992). Among the objects found were byzantine coins, metalwork, jewelry, different types of ceramics etc. In regards with the ceramics we see Byzantine type amphorae and Fatimid glazed ceramics coming from Egypt and Syria, attesting the circulation of the crafts in the Mediterranean (Hoffman and Redford 2017). The *Serçe Limanı shipwreck*

> provides evidence both dramatic and mundane of the extent to which Islamic art and material culture were enmeshed within the medieval Mediterranean sphere. The process of transculturation[1] in the particular context of art and culture of the Mediterranean during the eleventh to thirteenth centuries offers important insights into the nature of localization and identity between centers in the Maghrib, Africa, Egypt, Sicily, Venice, and al-Andalus, and between Islamic, Norman, and Byzantine realms. (Hoffman and Redford 2017)

Fatimid and Ayyubid iconographic and stylistic influences appear already in the pre Norman Sicily and it is said that at some point a *Siculo-Arabic* artistic court style emerged (Kapitaikin 2017). Also, it is not well known that the commercial and holy spaces of the north and south sides of the Mediterranean intersected thanks to the network of Crusader circulation and exchange. Fatimid crafts were a "recognizable signifier for the memory and presence of the Holy Land" (Hoffman and Redford 2017). So, these exchanges and affinities are based on substantial diplomatic, mercantile and cultural links (Kapitaikin 2017).

We should not forget the intellectual sphere exchanges: Already from the eighth century we witness the study, translation and commentary of the Greek and Hellenistic scientific and learned works into Arabic (Gutas 1998). In fact, the denizens of the Islamic world continued the inherited traditions of their homelands by transforming this corpus fit to their own needs. Representatives of the Islamic Aristotelianism, such us Ibn Sina, Ibn Rushd, Maimonides converse with the Latin and ecclesiastical wise men, such us Albert the Great and Thomas Aquinas, acting as a catalyst in the formation of Scholasticism and later Nominalism (Ziakas 2007). In regards with the Arabic chemistry, the first alchemical treatise that is translated in Latin occurs in 1144 by Robert of Chester. More Arabic alchemical treatises are translated in the twelfth and thirteenth centuries, such as the *Summa Perfectionis* of the aforementioned Jābir ibn Ḥayyān under the Latin name Geber (al-Hassan 2009). That is the way the foundations of Latin alchemy were found not by the Eurocentric framed allegations that Julius Ruska and Berthelot used in order to disconnect the Latin alchemy from its Arab roots in the nineteenth and twentieth centuries, as Ahmad Y. al-Hassan proves (al-Hassan 2009).

The application of the early Islamic chemistry, the glazes, passed through Başra and Mesopotamia in North Africa and then moved to the northern side of the Mediterranean. Craftsmen that dispersed from Fustat and maybe Kashan, reached Malaga at the end of the twelfth century, establishing the first lusterware production center (Caiger-Smith 1985). In the well-known Alhambra palace located in Granada, the largest lustre pots made in Malaga were found, intended to occupy the wall-niches. The luster production thrived under the protection of Nasrid dynasty and lasted for two centuries till 1450. At that time, the production moved to Manises. The luster changes and is based only in tin, in order to highlight the glaze more brightly. The designs take on their own local character here as well. Thus, we observe Christian motifs like that of St. George and the dragon. These items were made for export to Europe (Frothingham 1951; Caiger-Smith 1985). Italy imported ceramics from Spain. The term *Maiolica*, which describes the later polychrome Italian ceramics, seems to be connected with Spain. Italians imported wares from the Islamic world, primarily from North Africa for centuries prior to the Spanish imports. By the fourteenth century, this trade had dried up as Italians learned to make things by themselves. The main *Lustre Pottery* centers have been established in Deruta and Gubbio and flourished till the mid-sixteenth century. The painter Bernandino Pintoricchio (1454–1513) and the potter Cipriano Piccolpasso (1524–1579) are the prominent figures (Lightbown and Caiger-Smith 1980; Caiger-Smith 1985). The reduced-lustre production with iridescent effects didn't really continue in the late sixteenth and seventeenth centuries and never transferred into northern Europe. The art survived in Spain till the eighteenth century. At that time, Italians started experimenting with the effects of real gold

[1] We use this term within the specific context of the Mediterranean sphere and its Islamic material culture and not in a colonial or post-colonial frame eg Latin American Studies or Neoliberalism (Hoffman and Redford 2017; Flood 2009).

applications on their pieces prior to firing. It was at this very point that European fashion began to develop an interest for the next two centuries in tableware with gilded surfaces, in a style very different from the old, traditional lustre techniques (Hunt 1979; Caiger-Smith 1985).

A revival of interest took place in late nineteenth century Europe, particularly as part of the Aesthetic and Arts and Crafts movements in France and Britain. Luminaries such as Joseph Theodore Deck or William de Morgan researched Islamic pottery and experimented with techniques to produce ranges of sophisticated designs that were fashionable until the outbreak of the First World War. The artisans recreated Nasrid Alhabra ceramic vases and were inspired by *Hispano-Moresque Pottery*. They were also inspired by Persian, Syrian and Iznik ceramic styles. At the same time, they produced their own types of clays and colors of luster glazes. In areas beyond the northwestern countries of the European continent, the monuments and ceramics made by innovative techniques and style that been found in them were preserved. These objects were considered archaeological, studied by scholars, categorized and to this day are exhibited in museums. Also, treatises on historical, philosophical and aesthetics were written concerning their forms and motifs-ornaments. This age is referred by Necipoğlu as dillentantism,

> which encompasses the enthusiasm for Islamic art that was heightened with the taste for romanticism, exoticism, and eclecticism among 'amateurs,' including artists, architects, collectors, art dealers and travellers. (Necipoğlu 2012)

The pottery workshops that produced luster pottery continued in the second part of the twentieth century (Caiger-Smith 1985). The greatest example of fine craftsmanship is without saying Alan Caiger-Smith (1930–2020) a British studio potter and writer. He established the Aldermaston Potter in 1955 and by trial and error reconstructed the Islamic lusterware technique as well as wrote treatises in Islamic ceramic arts and lusterware. His book *Lustre Pottery: Technique, Tradition and Innovation in Islam and the Western World* is undoubtetly the bible of lustres, which undoubtly inspired in multiple ways the modern ceramics. Nowadays we can find more artisans worldwide using the various techniques of luster e.g.: Joan R. Carillo from Spain, Greg Daly and Johan Demaine from Australia, Danny Moohead from New Zealand, Sutton Taylor and Pauline Monkcom from UK, Ken Turner and Cathy Keen from US. In the Islamic lands we find Sevim Çizer (see Çizer 2010), Mehmet Tuzum Kizilcan, Emre Feyzoğlu from Turkey, Abbas Akbari from Iran, Kashan (see Akbari 2013), Ossama Eman from Egypt, all of them applying the traditional techniques of luster.

Generally, the modern ceramic artisans recreate the lustre pigments with modern materials, build kilns specific for lustre firing and expand the possibilities of lustre applications by experimenting and mastering new recipes in glazes, pigments and clays. In some occasions there is a new meaning of their symbolism in the aesthetic and philosophical frame, following the cultural environment in which they are born. Finally, it is interesting to mention that while at the beginnings of the luster pottery, the secrets of the techniques were kept well hidden by a small number of potters, nowadays we can attend classes in art universities, train ourselves in a potter's atelier, read articles and books and even watch videos in the internet. But it is also true that the lustres still hold a very small niche of the wide range of the clay applications, as they require quite a sophisticated level of experience.

3 Further Studies and Development

Lustre Pottery field has the potential to grow further and people can be acquaintant with it. Among others, tourism could be a way of such an acquaintance. For example, nowadays we observe that travel experiences include visits to places where specific types of traditional pottery are carried out as well as seminars by contemporary ceramists regarding their techniques. This venture may be further developed in the above mentioned *Lustre Pottery* production centers of *Silk Roads*.

Besides this type of cultural and art education, it is worth noting that we are currently working on developing an innovative product named Innovative Cultural Experience (ICE), T2EΔK-02564, which combines the following cutting-edge technologies: 1. Transparent Window—touch screen, interactive technology, 2. Augmented Reality, monuments-exhibits, 3D immersive video, intangible heritage narration. The product will provide a unique atomic or mass-touring experience, utilizing information based on material and intangible cultural heritage, through narrative scenarios. ICE will be a Transparent Window's advanced state of the art, in which any exhibit that will enter (e.g. ceramic object) it will be able to display visually enhanced and augmented Reality information to the user/visitor (admin enabled). In addition, visitors can, using the built-in 3D Camera, upload their own Video with their views or impressions from the exhibit. The advantage of ICE is based on the collection of alternative multimedia sources of information (material–immaterial), and it can adopt alternative forms of presentation of the aggregated information. For more information see https://ice.web.auth.gr/.

4 Conclusion

In this paper we attempted a macrohistorical examination of the *Silk Roads* and the resulting dissemination of ceramic techniques. We found out that already from the ancient times, East Asia and Europe were in fact connected thanks to the continental and maritime *Silk Roads*. By virtue of this geographic and social reality it is that civilizational process came to life. At the times of the Islamic command, we saw the birth of lusterware in Mesopotamia a basic artery of the *Silk Roads*, an innovative technique of ceramic wares application of a decorative film made of metal oxides. This technique turned out that was the result of a multilevel and age-old continuum of scientific and artistic achievements in which many peoples contributed. The *Lustre Pottery* circulated towards east and west in Muslim lands and beyond. Iraq, Egypt, Syria, Iran, Andalusia and Italy were the most important centers of lustre manufacture and distribution. Later a revival of interest took place in Western Europe which continued to this day influencing modern ceramic art and artisans worldwide.

In conclusion, we can say that the lusterware was born in the middle of the *Silk Roads* and has survived to this day, receiving special hues, techniques, individual elements and meanings depending on the region, the craftsmen, the market demand and the time. We observe that there is a dynamic in the circulation of this technique, which transcends established religious, cultural and linguistic identities and endures over time. Today, *Lustre Pottery* is being studied thoroughly by scholars, lustre objects are exhibited in museums and the technology and materials industry is developing rapidly. In general, the multiple sides of the lustres have a lot of room for cultural, trading and touristic development.

References

Akbari, A.: *The Vessel*. Peykareh Publications (2013)

al-Hassan, A.Y.: *Studies in al-Kimya': Critical Issues in Latin and Arabic Alchemy and Chemistry*. Olms (2009)

Allan, J.W.: Abū'l-Qāsim's treatise on ceramics. Iran **11**, 111–120 (1973)

Amorós, R.V., Gutiérrez, L.S.: Ceramics in transition: ceramics from the first Islamic period in the western Mediterranean—the example of al-Andalus. Libyan Stud. **51**, 99–125 (2020)

Barisitz, S.: *Central Asia and the Silk Road: Economic Rise and Decline over Several Millennia*. Springer (2017)

Blackman, M.J., Redford, S.: Lustre and fritware production and distribution in medieval Syria. J. Field Archaeol. **24**, 233–247 (1997)

Caiger-Smith, A.: *Lustre pottery: Technique, Tradition, and Innovation in Islam and the Western world*. Faber and Faber (1985)

Çizer, S.: *Lüster. Tarihi, Tekniği, Sanatı*. İzmir (2010)

de la Vaissière, É.: *Sogdian Traders: A History*. Leiden (2005)

Flood, F.B.: Objects of Translation: Material Culture and Medieval "Hindu-Muslim" Encounter. Princeton University Press (2009)

Frankopan, P.: *The Silk Roads. A New History of the World*. Bloomsbury (2015)

Frothingham, A.W.: *Lustreware of Spain: Printed by Order of the Trustees*. Hispanic Society of America (1951)

George, A.: Direct Sea Trade Between Early Islamic Iraq and Tang China: From the Exchange of Goods to the Transmission of Ideas. J. R. Asiat. Soc. **25**(4), 579–624 (2015)

Golombek, L., Mason, R.B., Proctor, S.P., Reilly, E.: *Persian Pottery in the First Global Age: The Sixteenth and Seventeenth Centuries*. Brill (2014)

Gutas, D.: Greek Thought, Arabic Culture: the Graeco-Arabic Translation Movement in Baghdad and Early 'Abbāsid Society (2nd–4th/8th–10th centuries). Routledge (1998)

Hoffman, E.R., Redford, S.: Transculturation in the Eastern Mediterranean. In: Necipoğlu, G., Flood, F.B. (eds.) *Blackwell Companion to Islamic Art and Architecture*, vol. 1, pp. 405–430. Wiley-Blackwell (2017)

Hunt, L.B.: Gold in the pottery industry. The history and technology of gilding processes. Gold Bull. **12**, 116–127 (1979)

Jenkins, M.: Early medieval Islamic pottery: the eleventh century reconsidered. Muqarnas **9**, 56–66 (1992)

Kapitaikin, L.A.: Sicily and the staging of multiculturalism. In: Necipoğlu, G., Flood, F.B. (eds.) *Blackwell Companion to Islamic Art and Architecture*, vol. 1, pp. 378–404. Wiley-Blackwell (2017)

Karakitsos, S.: Chinese porcelains embedded in Omani Miḥrābs. International trade and cultural relations. In: Fu, Z., Ziaka, A. (eds.) *The Silk Roads Between China and Oman. Networks of Communication and Transmission of Ibadi Knowledge*, pp. 121–144. Olms (2021)

Kervran, M.: Les niveaux islamiques du secteur oriental du Tépé de l'Apadana. Cahiers De La Délé Gation Archéologique Frangaise En Iran **7**, 75–161 (1977)

Lewicki, T.: Les premiers commerçants Arabes en Chine. Rocznik Orientalistyczny **11**, 173–186 (1935)

Lightbown, W.R., Caiger-Smith, A.: *Cipriano Piccolpasso. The Three Books of the Potter's Art = I Tre Libri Dell'arte Del Vasaio: A Facsimile of the Manuscript in the Victoria and Albert Museum*. Scholar Press (1980)

Martin, F.R.: Lustre on glass and pottery in Egypt from the period of Hadrian to Saladin. *Collana di Studi D'arte Ceramic*, 3rd edn. Lega (1929)

Mason, R.B., Keall, E.J.: The 'Abbāsid glazed wares of Sīrāf and the Baṣra connection: petrographic analysis. Iran **29**, 51–66 (1991)

Mason, R.B., Tite, M.S.: The beginnings of Islamic stonepaste technology. Archaeometry **36**, 77–91 (2007)

Mason, R.B.: Medieval Syrian lustre-painted and associated wares: typology in a multidisciplinary study. Levant **29**, 169–200 (1997)

Messina, M., Arcifa, L., Barone, G., Finocchiaro, C., Mazzoleni, P.: Islamic pottery production in eastern Sicily (10th-11th centuries): preliminary archaeometric data on local and imported products from Paternò (Sicily). Mediter. Archaeol. Archaeom. **18**(5), 207–223 (2018)

Necipoğlu, G.: The concept of Islamic art: ınherited discourses and new approaches. In: Benoit, J., Georges, K., Stefan, W., Gerhard, W. (eds.) *Art and the Museum*. Saqi Books (2012)

Park, H.: *Mapping the Chinese and Islamic Worlds. Cross-Cultural Exchange in Pre-modern Asia*. Cambridge (2012)

Peponis, D.V.: Peri tis Dimiourgias tou Kratous (on the creation of the state). *Cosmoidioglossia*. https://cosmoidioglossia.blogspot.com/2021/02/blog-post_7.html (2021). Accessed Nov 2021

Perdue, P.: Silk Road. In: Mokyr J. (ed.) *The Oxford Encyxlopedia of Economic History*. Oxford (2003)

Roqué, J., Molera, J., Cepriá, G., Vendrell, M., Pérez-Arantegui, J.: Analytical study of the behaviour of some ingredients used in lustre ceramic decorations following different recipes. Phase Transitions **18**, 267–282 (2008)

Saba, M.D.: Abbasid Lusterware and the Aestetics of ʿAjab. Muqarnas **29**, 187–212 (2012)

Sarre, F., Herzfeld, E., Arnold, H.: *Die Keramik von Samarra*. Reimer (1925)

Shen, H.M.: The China–Abbasid Ceramics Trade during the Ninth and Tenth Centuries. In: Necipoğlu, G., Flood, F.B. (eds.) Blackwell Companion to Islamic Art and Architecture, vol. 1, pp. 197–217. Wiley-Blackwell (2017)

Watson, O.: Ceramics and circulation. In: Necipoğlu, G., Flood, F.B. (eds.) Blackwell Companion to Islamic Art and Architecture, pp. 479–500. Wiley-Blackwell (2017)

Watson, O.: Persian lustre-painted pottery, the Rayy and Kashan styles. Trans. Orient. Ceram. Soc. **40**, 1–19 (1975a)

Watson, O.: Persian lustre ware, from the 14th to the 19th centuries. In: *Le Monde Iranien et l'Islam*, 3rd edn, pp. 63–80 (1975b)

Watson, O.: *Ceramics from Islamic Lands: The Al-Sabah Collection*. Thames & Hudson (2004)

Whitehouse, D.: The growth of interest in Islamic glass. In: Carboni, S., Whitehouse, D. (eds.) Glass of the Sultans, pp. 8–13. MET (2001)

Wood, N., Priestman, S.: New light on Chinese tang dynasty and Iraqi blue and white. Bull. Chin. Ceram. Art Archaeol. **7**, 47–60 (2016)

Ziakas, G.: *Ta Ellinika Grammata kai o Aristotelis stin Araviki Paradosi* [The Greek Letters and Aristotle in the Arabic Tradition]. Agra (2007)

Silk Road Tourism Development

New Directions for the Development of Sustainable Tourism Using the Potential of the River

Giorgi Giorgadze

Abstract

The article explores the possibility of maximizing the use of the river by the countries that use communication corridors. The new possibilities for the development of the Mtkvari river are considered as the carriers of great historical-informational, and recreational meaning. The identification of new, previously unused potential of the river is associated with the multidisciplinary targeted study of the Mtkvari channel and adjacent territories along its entire length. At this stage of comprehending the general goal and priority tasks of the upcoming project, the main areas of the research can be determined as following: The geographical description of the river; the landscape and climatic characteristics; the historical facts about the Mtkvari river; the attractive background of the spatial structure of the channel and the banks; etc. The successful achievement of the regional mission of the "Silk Road" depends on a comprehensive understanding of the great idea—connecting the people of the region. The formula for the success of the idea is to live in peace and use this road for the prosperity of its surrounding countries. The reality of the twenty-first century and especially the events of recent years have reshaped many parameters of this idea, including sharing, logistics, recreation, and entertainment. These changes create the need to search for new ways of spatial and functional expansion of the "Caravan Road". Integration of the Mtkvari river into the regional "Silk" system is an important part of this research.

Keywords

Communication corridor • Regional mission • Caravan road

1 Introduction

The importance of the Great Silk Road has been appreciated throughout the history of world civilization. This road was a connecting bridge between different empires: the caravans had been moving through the continuous path, which was a place for trade between different parties; the people from diverse cultural backgrounds had been sharing an experience with each other; and the relentless battles had occurred for seizing the road.

The main objectives of this study are:

1. Determining the role of Georgia in the formation of the Transcaucasian fork of the historical Silk Road
2. Analyzing the space-planning possibilities for the Mtkvari river as the transport and communication corridor
3. Finding out the promising directions for the development of sustainable tourism using modern technologies.

The study and the research methodology for the topic are based on several factors including: The features of the existing development in the historical environment; characteristics of the river, and the orography of the area; hospitality and traditions of local culture; specifics of building conditions in the riverbed, etc.

Historical documents, along with online research used for collecting the necessary data, provided the most accurate and authentic information. This type of approach has been valid at the given stage of the project however, more detailed exploration will be needed in the next phases of the project's development. (3D scanning, etc.)

G. Giorgadze (✉)
East European University (EEU), Tbilisi, Georgia
e-mail: giagiorgadze888@gmail.com

S. Kostopoulou et al. (eds.), *Cities' Vocabularies and the Sustainable Development of the Silkroads*, Advances in Science, Technology & Innovation, https://doi.org/10.1007/978-3-031-31027-0_16

The significance of this study is determined, first of all, by the main context of the topic—the restoration of the Great Caravan Road. The most important factors are also: project scale, real possibilities for Georgia to be integrated into the international program, the need to prepare project bases for urban and regional planning, etc.

The reliable historical sources describe the routes of the Silk Road that crossed Georgia:

> …The Great Silk Roads, that connected China with Mediterranean world, developed since second century BC. Its main route went through Iran and Syria. At the same time there exist assumptions, arguing that several branches of this road crossed the territory of Georgia as well… (Baker 2016)

Modern researchers find the evidence of these path's existence in the descriptions of ancient authors:

> …According to Strabo, the Oxus River was navigable, and many Indian goods were transported and traded over the Hyrcanian Sea (also known as the Caspian Sea). Subsequently, the goods were ferried to Albania or Azerbaijan, through the Cyrus (that is, the Kura River), to the territories of the ancient Georgian kingdoms, Iberia and Colchis, and until they reached the Euxeinos Pontos (Black Sea) (Strabo, XI, VII, 3). Items and artifacts found in the Khwarezm settlement, Kalalygyr II, in Transcaucasia, confirmed that the northeastern route was in use from the fourth to the second century BC. Subsequently, goods were transported from the Euxeinos Pontos to the Roman Empire, which in the first century AD ruled almost all of the regions around the Black Sea… (Alimova et al. 2009)

1.1 The Caravan Has Passed Through Georgia

The first caravan passed through Georgia in 568. The new road was much more difficult than the route taken—via Iran and Syria, but politically and militarily it turned out to be more convenient and reliable, as disagreements between Iran and the Byzantine Empire questioned the safety of the old road. At the end of the sixth century and the first half of the seventh century, a significant amount of Chinese silk was supplied to Byzantium via the Caucasus.

From the fourteenth century on, the Caucasian part gradually lost its significance, as nomads threatened the merchants in that region: Europeans preferred to bypass India—a new navigable route for safety (Fig. 1).

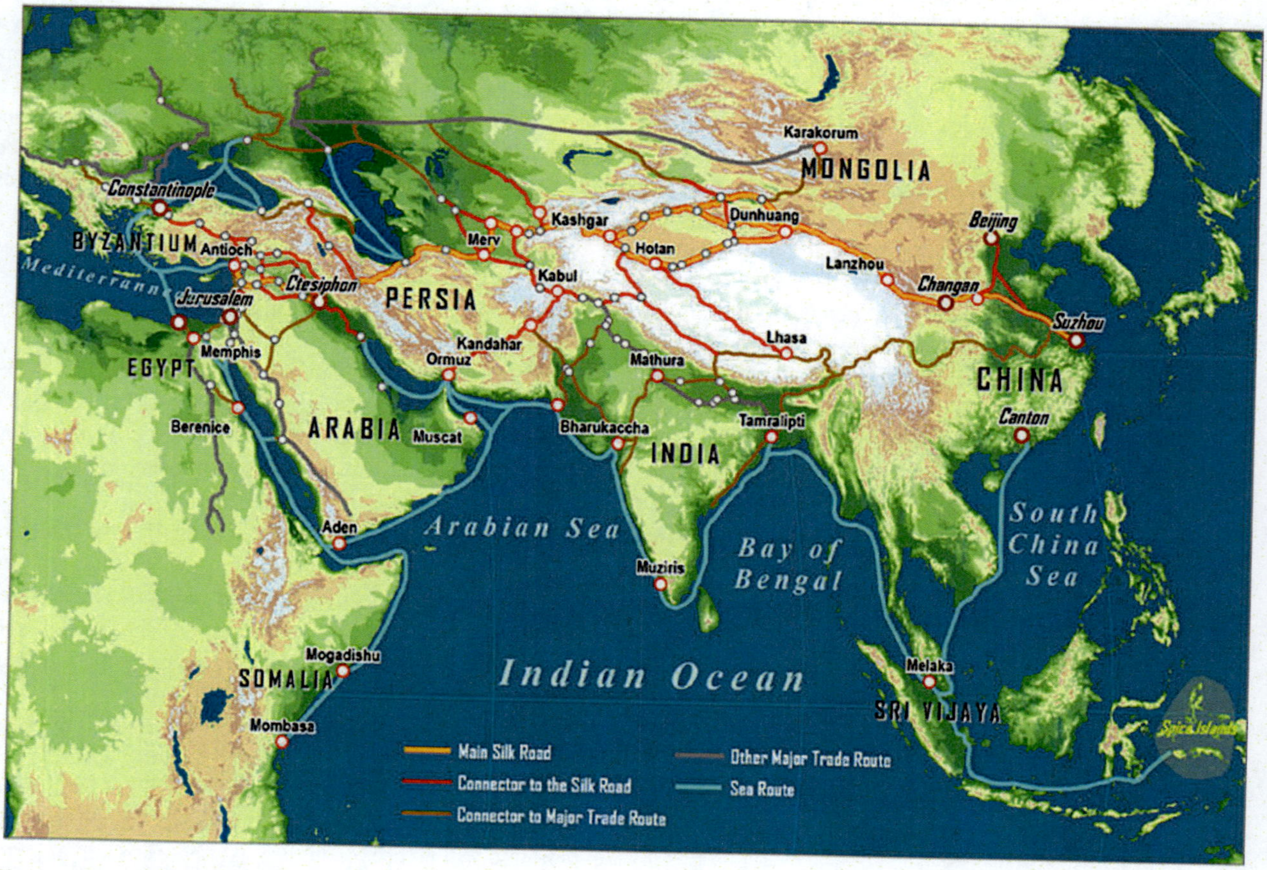

Fig. 1 Historical route of the Silk Road in Georgia. *Source* https://thedustbinblog.wordpress.com/2014/09/16/chinas-maritime-silk-route/

1.2 The New Silk Road

The second life of the Silk Road dates back to the 1990s. By the end of the twentieth century, the geopolitical redistribution had come to an end. The new global policy has encouraged restoring the old route. The EU, along with China, were the initiators of the historic Silk Road revival; It was acceptable for the EU to develop the concept of a relatively small-scale transport corridor—from Europe to Asia via the Caucasus and the Caspian Sea. The European Union, dependent on Russian oil, was looking for an alternate energy source. The main goal in this scenario was to bypass Russia and gain access to a region rich in oil and other resources. At the same time, Georgia, which had just gained independence, was looking to obtain an international function that would increase its role and regional importance. That is why the country supported the EU-East-West Corridor project called "Transport Corridor-Europe-South Caucasus-Central Asia" (TRACECA), which was informally the new Silk Road (Miodrag 2017).

The initiative gained widespread support: In 1998 in the United States. The Senate passed a resolution by supporting the new Silk Road. Which also was supported by international organizations and approximately 150 countries. The project covered several areas, including free trade, joint infrastructural development, political coordination, and the strengthening of interpersonal ties. The Silk Road Fund, with a starting capital of $40 billion, was set up to support the development of the project, and major investment projects were planned in the participating countries. The main actors in the project—the EU and China—aimed to develop the transportation corridor and reduce the cost of transportable products. To achieve this, the development of the roads between these two points was initiated. Georgia joined the initiative in 2015.

Caucasus has the minor importance under the project, however, all of the countries on the route can benefit from this initiative. (Historical Silk Road and TRACECA 2021).

The analysts predict, that Georgia will become a more competitive transit route if it harmonizes transportation regulations and legislation with other members of the corridor and simplifies the operational procedures (Fig. 2).

Several actions have taken place inside the Caucasus to improve the technical equipment and increase the load flow in the TRACECA transport corridor. The TRACECA program and the restoration of the Great Silk Road are two of the greatest challenges of the twenty-first century. Sooner the comprehensive development of this program is ensured—faster the integration of the South Caucasus, Central Asian countries, and Europe into a single transport-trade-cultural system will happen. Georgia is intensively and permanently carrying out its share of activities for formation of this important global network.

1.3 New Challenges in the Tourism Industry

One of the first and most fruitful results of the 'Great Caravan Road" historical route restoration and the intensification of traffic between the countries will be the development of international tourist routes. The effects of the global pandemic and the constraints associated with new realities have led to the search for entirely new types of approaches to form the spatial-organizational structure of the tourism industry. It became necessary to limit the physical contact, increase the distances between tourist groups, change the dynamics of their movement for these groups in areas with a high concentration of visitor flows. In response to these restrictions, new physical standardizations have been introduced indoors (and even in open spaces): the distances between the visitors, exposures, evacuation exits platforms, waiting areas, etc. Arranging a large-scaled performance using digital technologies on existing building facades has become one of the prioritized activities; holding the laser shows "on top" of the entertainment centers and historical monuments; creation of the special information terminals and holograms; involvement of air balloons for placing remote ads, etc.

1.4 River and Tourist Infrastructure

The role of the river in the development of tourism infrastructure is quite important. Since ancient times the main roads have been located along the riverbeds. Cities, villages, and various important urban development elements were found in the river valleys. The Mtkvari river is one of the main components of such urban order in Georgia.

Almost all of the important settlements, fortifications, and monastery complexes in eastern and southern Georgia, where the Mtkvari river flows is located in the immediate vicinity of its channel. The Mtkvari river has been making a significant contribution to the adaptation of the tourism industry to new realities. For example, in the capital of Georgia—Tbilisi the most important architectural, historical, and landscape monuments like Metekhi, Narikala, Sioni, Anchiskhati, Krtsanisi Valley, and others are located along the river Mtkvari. The inclusion of the river as a transportation and communication artery in terms of historical development, panoramic perception of the city, and underwater cultural heritage—is covered in a separate article (Giorgadze 2020). The urban-aesthetic functions of the Mtkvari river are described in the European Union project, which was developed in order to improve the management of the Mtkvari river within the administrative boundaries of the city of Tbilisi: "...A good balance" between natural elements and the built-up environment is very important for

©1994 Magellan Geographix SM Santa Barbara, updated in December 1999 by D.Feldheim & M.Graille for the TRACECA PROGRAMME of the European Commission

Fig. 2 TRACECA transport corridor. *Source* https://reliefweb.int/sites/reliefweb.int/files/resources/E5BFD302D8B1378BC1256F2D0047FF6F-traceca_geo1299

urban development in many cities today. The Mtkvari river flows 34 km through the city from its northern to its southern border. The same is the case with the metropolitan territory, therefore many opportunities exist to experience the relationship between the river and its urban riverfront. From the moment the river was channelled and the roads were constructed on both sides of the river—the distance between the river and the riverfront has increased, causing the riverfront to lose its importance.

In the past, the riverfront was on the banks of the river. To see the evidence, one has to look at one of the massive new buildings nowadays, such as the House of Justice, a building of 44,000 m^2. The building's only facade faces the road in front and not the river… The conservation of historical buildings facing the river becomes an important element to consider.

The City Hall program for the historic city could, include some of the buildings indicated by the consultants, with the support of UNESCO, the European Union, and with the participation of the New Architectural Faculty of the Orthodox Church Patriarchate, in order to preserve them, as they are examples of ancient wooden structures and part of the city's history… (Bradley et al. 2012).

The expansion of this topic and the search for new opportunities to maximize the use of the city's tourism resources lead to design of new touristic facilities and the

Fig. 3 Project proposal of the night show "Silk Road and Georgia" of Metekhi temple on the adjacent vertical rock wall in Tbilisi. Author's collage—G. Giorgadze

involvement of the existing natural environment in the unified informative and spectacular installations. The following places can be considered examples of such installations: the left bank of the river Mtkvari—a vertical rock mass near the Metekhi temple (Fig. 3), where the slide show can be performed; the surface of the air balloon at Rike Park (Fig. 4) with the advertising slogan of the Silk Road; etc.

The historical and cultural heritage located along the river will be reflected in the special tourist information terminals, which will be located on the main transport-communication artery and will be uploaded to the navigation system (mobile application). Monuments of historical Georgian architecture from Eastern and Southern Georgia (Fig. 5) will be uploaded to this unified information system in the form of multimedia files.

The list of these monuments from Eastern Georgia includes: historical monuments in Tbilisi; historical-cultural heritage of Mtskheta; Uplistsikhe rock-cut town; Gori Fortress and from Southern Georgia; Tmogvi Castle, Khertvisi Fortress, Atskuri Fortress, Rabat Castle and the rock-cut city of Vardzia.

1.5 Preparatory Work for Investment Projects

The twenty-first century in the history of Georgia was marked by the prospects of large-scale investment by Chinese capital. Georgia met the Chinese initiative to restore the historic Silk Road, which was positively reflected in the economy of the country.

Work on large-scale investment projects started in Georgia. In anticipation of the "Big Caravan", Georgian companies began to think about creating the profile enterprises that would meet the new challenges of the East-West partnership. The planners focused on the design of service facilities in the areas of traffic transit and logistics.

Eastern Europe University (EEU) Project Research Bureau is working on an in-university grant project: "River in a historic city as a definition of tourism infrastructure. New realities" (Giorgadze 2021).

The project is about the impact of the Mtkvari river as the most important urban development factor on the restoration and renovation. The historic districts of the city are the main areas for attracting visitors and serve as profiled spaces for

Fig. 4 Silk Road advertising slogan on an air balloon in Rike Park, Tbilisi. Author's collage—G. Giorgadze

the realization of potential within the tourism infrastructure. The proposed project addresses urban management on the banks of the Mtkvari river. The need for large-scale, profile scientific research work is on the agenda; Summarizing the results of this study will provide the basis for creating a holistic model with a specific urban concept, project proposals, and appropriate investment planning (Fig. 6). It is planned to present a holistic model for the urban and architectural services of Tbilisi Municipality, the presentation material will be developed for discussion at the City Government meeting. A new vision of connecting the city and the river will be prepared, this vision will be firmly reasoned, scientifically validated, and equipped with practical design and estimation documentation. The municipality will gain an effective tool for attracting strategic investors; Tbilisi's tourism infrastructure will have a real opportunity to use an alternative investment mechanism to compensate the damage caused by pandemics, which itself is a good prospect for the integration of this field into the policy of the circular economy. Historical heritage has always been considered a priority for the development of the tourism industry. The centuries old history of Georgian art, contemporary culture, and epic poetry is very interesting for the locals and especially for foreign visitors.

> …The river plays quite an important role in the perception of the country's history and, particularly the capital city, Tbilisi. Many dramatic twists and turns are associated with the river Mtkvari; Tbilisi and Mtskheta are a continuous chain of distant and recent memories, bound by: the Svetitskhoveli and Sioni cathedrals, Pompeus Bridge, and the self-sacrifice of one hundred thousand martyrs at Metekhi, Krtsanisi Valley, Island of Madatovi, Ortachala Gardens, etc. Until recently, some tour providers organized river trips with improvised rafts, stylized ferries, sightseeing boats, etc. Rike Park and Abanotubani received numerous visitors during the traditional Tbilisoba festival on entertainment events. The subject of Mtkvari remains one of the most attractive factors in the tourism industry; Understanding this topic in a new way and discovering its hidden reserves will most likely be a guarantee of a new life for the sector… (Giorgadze 2009)

Fig. 5 Cultural heritage along the Mtkvari River in the eastern and southern regions of Georgia. Author's collage cf the tourist terminal—G. Giorgadze

Fig. 6 The role of the river in a historic city—Eastern European University (EEU) ongoing grant project (supervisor—G. Giorgadze). Situation and 3D modeling for investment planning with tourist routes and accentuated locations

References

Alimova, D., Rtveladze, E., Abdurasulov, U.: Central Asia-Transcaucasia-Rome: the significance of the Amu Darya water route via the Caspian Sea to Transcaucasia. https://en.unesco.org/silkroad/sites/default/files/knowledge-bank-article/route%20via%20the%20caspian%20sea.pdf (2009)

Baker, E.: Georgia and the Silk Roads. Advance online Publication, Tbilisi (2016)

Bradley, G., Adeishvili, M., Chikviladze, K., Bello, C., Giorgadze, G.: Ensuring the improvement of the Mtkvari river management in Tbilisi Area, final report (draft). Framework Contract Beneficiaries 2009, Lot 6 Letter of Contract N° 2012/294630—Version 1. Advance Online Publication, Tbilisi (2012)

Giorgadze, G.: Design of tourism infrastructural objects in conditions of circular economy. East European University, Tbilisi (2009)

Giorgadze, G.: Historical heritage in the city below and above the river level. In: *Urban Planning and Architectural Design for Sustainable*

Development, 5th edn (online conference). The Academic Research Community Publication, Florence (2020)
Giorgadze, G.: The river in the historic city as a determinant of tourism infrastructure. New realities. Ongoing Research Project. East European University (EEU), Architecture Research and Design Bureau. Advance Online Publication (2021)
Historical Silk Road and TRACECA: Advance Online Publication, Baku. https://www.azerbaijans.com/content_790_en.html (2021)
Miodrag, S.: China's 'New Silk Road' goes straight through the Caucasus. https://www.dw.com/en/chinas-new-silk-road-goes-straight-through-the-caucasus/a-41930469 (2017)

The Role of Art-Marketing in the Sustainable Development of International Cultural Heritage Tourism

Veronica Bulat and Maria Hamuraru

Abstract

Research and protection of cultural heritage is a particularly relevant field for international tourism. The problem offers multiple topics for scientific investigation, which still expects researchers to be tempted to examine changes in culture as a whole, in various historical periods, and especially, for the Republic of Moldova's case, in the Soviet period, as well as their effects on the entire contemporary society. The interdependence of sustainable marketing with other sciences, the heterogeneity of economic and social processes and phenomens that shape its area of research is the framework for this young economic science, to manifest itself as a multidisciplinary science, with an open character, which also ensures its universality. By appealing to the field of arts, marketing generates a chain of reactions for visualizing the strategies for promoting the national cultural heritage. The aims of the study are to present how art is known in our time, to emphasize not only its embodiment of beauty, which performs both educational and aesthetic functions, but also the process of integrating a brand into a work of art. This problem is important according to the management process in the field of public relations because it creates not just a trend in society, it also contributes to the sale of goods at a more favorable price for the customer.

Keywords

Art-marketing • Cultural heritage • Sustainable tourism • Marketing strategies • Ethnological tourism

V. Bulat (✉) · M. Hamuraru
Moldova State University, Chisinau, Republic of Moldova
e-mail: veronica.bulat@usm.md

M. Hamuraru
e-mail: maria.hamuraru@usm.md

1 Introduction

Cultural heritage stores the historical memory of human societies. It contributes to the national identity, scientific and cultural value, as well as it gives prestige to the country, improving its image (Swanson and Davis 2012, pp. 264–265). The concept of cultural development should pay particular attention to the symbiosis between culture and tourism (Crick 1989, pp. 307–344). The development of cultural tourism is an important factor for sustainable economic development, it is a profit and a job generator, and it is a fundamental factor for creating the image of any tourist destination (Santagata 2002, pp. 9–23). The interaction of tourism and the anthropogenic environment is versatile and has different dimensions. Its positive socio-cultural effects are the local production of goods and souvenirs, the promotion and revival of local art forms, such as folklore and gastronomy.

The cultural vestiges preserved and passed on by previous generations are conclusive proofs of the values they had and preserved. The attitude we have towards this legacy today can ensure a prosperous or uncertain future. Marketing, by its intrinsic nature, appears as a sustainable complex system of the organization that expresses a way of thinking, an attitude that is reflected in a set of coordinated and organized activities with the help of appropriate tools. The objectives of the research consist in: presenting art-marketing as a form of sustainable tourism in promoting the values of the cultural heritage of the Republic of Moldova; characteristic of the basic activities carried out by the young generation in the field of cultural heritage protection; analysis of the attitude of the young generation regarding the level of education in the field of protection of the national cultural heritage; arguing the opportunities of Europeans to capitalize on the potential to promote cultural values through art-marketing strategies.

As a concept, sustainable tourism was originally set out in the Brundtland Report (1987), which states that in order to

S. Kostopoulou et al. (eds.), *Cities' Vocabularies and the Sustainable Development of the Silkroads*,
Advances in Science, Technology & Innovation, https://doi.org/10.1007/978-3-031-31027-0_17

function at optimal levels, a balance must be ensured between economic, social and environmental aspects (McNulty and Koff 2014, pp. 50–56). Then, later, in 1995, as a result of the first World Conference on Sustainable Tourism, in Lanzarote (Spain), there were laid the foundations for a Charter for Sustainable Tourism, making reference to these three components, which generate sustainability (Candrea and Nechita 2015, pp. 8–10). In addition to this document, comes the Environmentally Responsible Tourism Guide (2005), published by the United Nations Environment Program (UNEP). The European Commission (2002) has also developed an information material, entitled "Basic Orientations for the Sustainability of European Tourism", whose purpose was to guide tourism agents in the development process and in 2012, the EC puts forward "European Charter for Sustainability and Responsible Tourism", in order to encourage the sustainable and responsible tourism policies and actions that are required to be implemented in Europe, so that afterwards would be promoted worldwide (Colbert and St-James 2014, pp. 566–575). The most widespread definition of sustainable tourism belongs to the World Tourism Organization (WTO), which defines it as the component that meets both the current needs of tourists and host regions, meanwhile protecting increasing opportunities for the future.

Trans-positioning the concept of sustainable development in the tourism business manifests some major features, resulting from the forms and types of businesses found in this area of activity, as well as the various influence companies' actions have upon the environment.

Sustainability in marketing strategy not only helps to gain a competitive advantage, but also opens the door to cost savings and innovation (Carillat et al. 2014, pp. 314–320). It is truly believed that travel agencies cannot continuously resist on the market if they do not become truly sustainable (Juslin 2013, pp. 235–266). However, there is only a limited number of agencies that are trying to adopt sustainability in their marketing strategy. Moreover, one of the objectives of the marketing strategy for sustainability should focus on segmentation, targeting and customer positioning based on sustainability criteria, along with designing the sustainable marketing mix for higher quality tourism products and services, better prices, a more effective distribution and a more attractive promotion.

Sustainable international tourism is playing a much more important role in the economic development and growth, especially in pandemic conditions. Accordingly, a greater visibility of this sector is needed in terms of promotion strategies. For this matter, tourism meets the need of an ambitious political framework, which should include specific actions designed to support the conservation and consolidation of national and cultural heritage, as well as the global promotion of the country's image as a unique destination (Swanson and Davis 2012, pp. 263–278). A strategic and integrated approach to sustainable tourism would include all the effects that tourism manifests, including the social consequences, related to the environment. The universal principles of sustainable tourism are widely recognized and can be used as a global framework for creating detailed reference points and indicators for the sustainable development of tourism.

The study analyzes the efficiency of art marketing strategies in promoting the cultural heritage of the Republic of Moldova, following international experience in this way. The presented data are based on the concrete aspects taken from official sources and specialized bibliography. For research, there were used the methods of analysis, synthesis and comparison.

2 Results and Discussions

All at the macro level, sustainable development is associated with an increase in the well-being of the population and a reduction of poverty and inequality (Nasser 2003, pp. 467–479). The authentic historical and cultural heritage, the traditions and the particularities of the host communities must be respected, their preservation and development needing to be promoted (Evans et al. 2012, pp. 309–341). Tourism is a sector that uses the cultural heritage of humanity and contributes to its enrichment (van den Berghe 1980, pp. 345–391). One of the main problems of strategic tourism management is related to the sustainable development of anthropogenic resources that are irrecoverable (Carillat et al. 2014, pp. 314–335). It requires the development and use of specialized approaches, conservation methods and techniques, their interpretation and use for tourism purposes. The protection of cultural heritage is a systematic process of discovery, research, identification of cultural assets and archaeological sites, as well as their promotion (François et al. 2018, pp. 114–118). It also includes actions of documentation, registration, conservation, restoration and adaptation.

The protection of cultural heritage is a function and a responsibility of the authorities and institutions organized in a national system at national, regional and local level.

Achieving cognitive objectives depends largely on the promotion of cultural heritage and the development of cultural tourism—its marketing, permanent advertising and specialized commercial publications; popular educational programs and multimedia products available for building knowledge and trips to tourist sites—cultural values.

Ethnological tourism is a specialized type of cultural tourism, which is associated with the ethnographic components of the visited destination (Akathleen 2006, pp. 112–130). It recreates and transfers people's memory over generations, being available both for local visitors and for international tourists.

The resources of this type of cultural tourism are ethnographic evidence of life, crafts, skills, and beliefs that are inseparable elements of the spatial environment of which they are a part of, and are ethnologically significant (Raymond 2008, pp. 8–10). They are real and mobile, mainly architectural and historical buildings, and objects of cultural value. Most ethnographic anthropic tourist resources belong to the intangible cultural heritage. It consists of traditions, crafts, shows, social practices, rituals and festive events.

The protection of the intangible cultural heritage aims at the protection, conservation, and respect for:

- linguistic traditions and expressions, including language as a carrier of intangible cultural heritage;
- performing arts;
- social practices, rituals, and festive events;
- knowledge and practices related to nature and the Universe;
- knowledge and skills associated with traditional crafts.

Preservation and display of local cultural, historical, and natural heritage can be included in the company's marketing strategy and aimed at improving the quality of life in the region by developing and promoting small businesses (hotels and small restaurants) as a sustainable local economy (Ngondo et al. 2020, pp. 14–21). The model of investment activities can be directed towards:

- supporting existing groups in presenting popular customs, for example, by purchasing costumes, musical instruments, and audio-visual equipment;
- creating a permanent and itinerant exhibition of the local ethnographic, natural and cultural-historical heritage, with the application of audio-visual clips.

Cultural and natural heritage is among the strategic factors for the sustainable development of tourism, including hotel and restaurant business, as part of the tourism product (Lapum et al. 2012 pp. 100–115). They gradually transform from a conservation object into a tool for development. The integrated approach to the protection and enhancement of the natural and cultural heritage is an important precondition for increasing the attractiveness of regions, both for investment and for the development of the quality of the environment and the preservation of regional identity.

In this paper we present the authors' vision regarding the principles of market segmentation of cultural heritage values. According to Fig. 1, the right side is aimed at protecting the authenticity of the product, preserving the aesthetic spirit and the cultural value of the cultural heritage. Quadrant 1 is oriented towards the production of a prototype with original socio-economic value. Quadrant 2 is aimed at the active involvement of marketing strategies that can capitalize on a cultural product in mass production, so as to gain added value for the market.

The left side of the figure represents the market orientation and obtaining the maximum profits, the obtained commercial effect being the main evaluation criterion. Quadrant 3 is aimed at intensively promoting marketing strategies for mass production, neglecting the cultural values of national heritage, while Quadrant 4 focuses on building prototype products or services, capitalized intensely by market-oriented marketing tools but making little use of the values of authentic cultural heritage.

At the level of promoting culture for a stronger Moldova, a series of activities were carried out, such as:

1. Development of support policies—promoting the role of culture and creativity in social and economic development.
2. "Creative Europe" program—promoting cultural cooperation through the EU's emblematic cultural program. In 2015, the Republic of Moldova joined the Creative Europe program, the EU's emblematic program to support the cultural, creative, and audiovisual sectors. Following the agreement, Moldovan cultural organizations now have the opportunity to participate fully in the Creative Europe Cultural Program, which supports projects in the following areas: cultural cooperation; literary translations; cultural networks; platforms for cultural operators, which promote up-and-coming artists, etc.
3. Culture and Creativity Program—for three years, from February 2015 to January 2018, the Eastern Partnership Culture and Creativity Program supported the role of culture and creativity in the social and economic development of the six Eastern partner countries, including the Republic of Moldova.
4. Cross-border culture—As part of the EU-funded cross-border cooperation project between Romania, Moldova, and Ukraine—"The Medieval Jewels: Hotin, Soroca, and Suceava"—the great fortress of Soroca on the Dniester was restored as part of a development project of tourism by uniting the common heritage of the three medieval fortresses in the three countries. As a result of the extraordinary restoration, tourism in the cross-border region has increased, and 60,000 tourists recently participated in a medieval festival organized simultaneously in Soroca, Suceava, and Hotin. Soroca also benefited from EU's support through the COMUS project for community-led urban strategies in historic cities, which supported the city in undergoing a heritage development and rehabilitation plan.

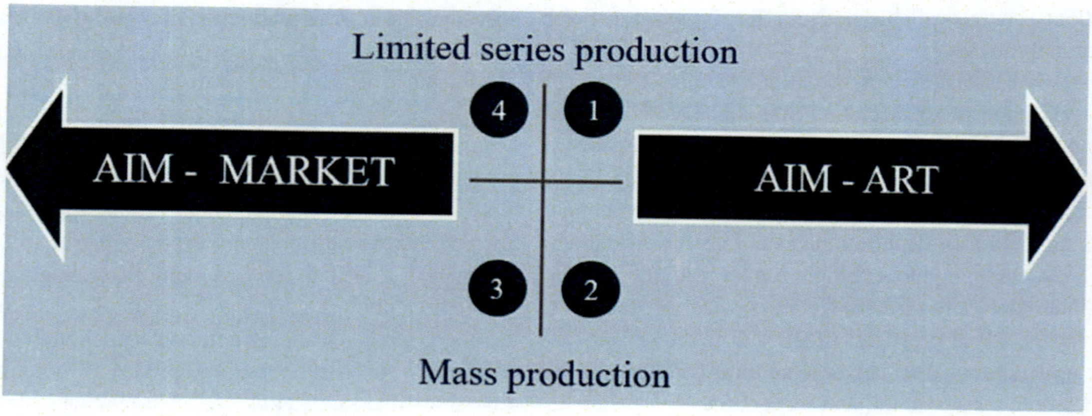

Fig. 1 Principles of segmentation in cultural heritage market

Art marketing is a systematized process of creating and increasing the degree of recognition of interest in the company, product, or service, reflecting the purpose of attracting buyers to purchase goods or services. According to Juslin (2013), art-marketing is used to communicate offers, promote brands, identify new development perspectives and strengthen the connection with the target market. In the field of arts, marketing generates a chain of reactions from potential customers, using visual art in strategies to promote on Social Media, StoryTelling, and advertising "live", resulting in increased recognition (brand awareness) of the artist's work.

In art, the term "Brand Awareness" is associated with the process by which strategies are developed to promote works of art among potential consumers, not necessarily collectors or specialists in the field (Bradshaw 2010, pp. 8–19). Through the effective use of various marketing methods, Art-marketers keep alive the interest of the public and the degree of recognition—after which the actual sales opportunities appear. Although, there are cases when a buyer goes through all the stages: awareness, interest, desire, purchase decision, characteristics of art fairs, still a large part of sales are made after a while, as a result of repeated viewing display of works of art (Candrea and Nechita 2015, pp. 1–16). In the field of art marketing, the following promotion techniques are recommended:

Development of cultural consumption practices and habits. The habits of public cultural consumption of young adolescents are divided into three categories: cultural activities, entertainment, and relaxation activities, and community activities.

According to Moisei (2013), cultural activities include leisure activities that are based on participation in various cultural events in the public space and on the consumption of products/services of cultural institutions. These activities have, in principle, an artistic or educational character. Participation in cultural activities is also a source of social differentiation. The degree of participation in these types of activities is the lowest, concerning the entertainment and relaxation activities, as well as the community ones. This is not only specific to young people, but also to adults. The authors of a study believe that the low level of participation can be explained by the fact that artistic activities are generally attended by people who are frequent consumers of culture and not just occasional consumers (François et al. 2018, pp. 48–60). Another cause of the low degree of participation may be the competition in which cultural activities are with other types of recreational activities.

Entertainment and relaxation activities are characterized by the little effort required to practice them; they are pleasant, short-lived, and do not require previous experience to be practiced. Unlike cultural or community activities, these activities are based on the production of a sense of inner well-being and not on the accumulation of new experiences (cultural, educational). These types of activities fall into two categories depending on the effort required to practice them: activities that involve passive consumption, which do not require concentration on the part of participants (cinema, attending sports competitions, participating in entertainment or music shows), and activities that involve active consumption, which require a certain degree of direct involvement (walking in parks, trips outside town, walking in malls, discos/clubs and restaurants/bars/cafes). These types of activities have the highest degree of consumption in the public space among young adolescents. Of these, walking in parks or green areas, walking in the mall (for shopping, games, movies, restaurants, etc.), and walking in cafes/restaurants/bars (other than those in the mall) is most often done by young people. Community activities are characterized by the fact that participation in them creates a sense of social inclusion through the interactions generated between members of a community. At the base of community activities is the assertion of a collective identity, an identity that is most visible in these activities, compared to cultural, entertainment or relaxation.

Education through culture (values of culture) is done in several school subjects (history, civic education, technological education, music education). The purpose of this article is to present the ways to capitalize on cultural heritage through art education and music education classes. Subsequently, artistic education helps to manifest the human rights to education and cultural participation, the development of individual abilities, the improvement of the quality of education, and the promotion of the expression of cultural diversity. Education prepares people for both the need of art and beauty, and for their possibility of using art and beauty in their own existence. The importance of education through and for beauty is vigorously asserted. This means involvement in education, firstly, as artistic beauty, and secondly, as beauty beyond art, as a bearer of traditional values. Education must increase receptivity to the ways of beauty in nature, in social life, the beauty of interpersonal relationships, the beauty of industry, craftsmanship, beauty in clothing, in the arrangement of interior and exterior architectural spaces, design, beauty with moral connotations. Beyond the fact that education involves culturalization and, implicitly, the transmission of identity values, to speak in schools about the preservation and transmission of traditional values would mean to speak briefly about a concern that appears too little in the curricula, and often not between extracurricular activities.

Children and young people learn to reflect on their own culture, the culture of others, and culture in general. This capacity of reflection can be formed through several ways: studies, cultural events, and arts. The art is one of the main manifestation of cultural self-awareness. In the Republic of Moldova, the values of cultural heritage are promoted through educational disciplines, taught in schools (art education, music education, civic education), specific authentic laboratories (for example, the Moldova State University has the Methodical Center "CASA MARE" for students of Hotel Services, Tourism and Leisure specialty), as well as the organization of events to promote traditions and customs (holiday "MARTISOR" and "SINZIENE"), see Fig. 2.

Therefore, in this field, the study of popular traditions, knowledge, and preservation of identity values, stimulating young people's interest in these values, cultivating their artistic or craft skills would contribute to the harmonious development of personality in young democracies subjected to continuous challenges. Promoting traditional values through artistic and cultural education can contribute to the spiritualisation and socialization of young people, to the orientation towards a beneficial meaning of life, towards a morally superior existential motivation, which young people are often deprived of in schools, in an increasingly obvious manner. The concept of traditional value in contemporary society has undergone various interpretations over time through changes in the cultural perception of society. Value itself has different meanings depending on the discipline. For example, in aesthetics and philosophy, the concept has a normative connotation, it refers to criteria based on which what is beautiful is distinguished from what is ugly or what is good from what is bad. The concept of traditional value, on the other hand, is often restricted to the specific meaning of customs or the accumulation of artistic or cultural practices, passed down from generation to generation, preserved as an identity over time in the form of cultural heritage. But what needs to be clarified from the beginning is that traditional values in a society refer to much more. Traditional values refer to the ideational substratum that has endured over time in a society and dictates social and individual behaviors and attitudes, which trace the identity of a nation, its specificity compared to other nations.

When we say "cultural heritage", we think of all forms of artistic expressions, forms that bear witness to the creativity of our people throughout its history. Whether we are talking about literature, music, painting or theater, I believe that art, traditions, and customs are the treasure that defines our people. The authority of the Republic of Moldova has the noble task to support in all possible ways the preservation and development of these elements that define us not only as a people but also as a nation.

This study aimed to focus more attention on peripheral areas, areas with a potential of unique cultural heritage. Results show that cultural heritage has a universal value for us as individuals, communities, and societies. The study helps to form a better cooperation in the field of culture, enriches relations between societies and promotes cultural diversity and the intercultural dialogue.

3 Conclusions

In order to achieve more sustainable forms of tourism, it is necessary to focus more on peripheral areas, areas with a potential of unique cultural heritage. Therefore, it can be realized by increasing the involvement of competent authorities, various professional categories, local population from a certain tourist destination, in finding the weak points and, consequently, turn them into possible advantages that will attract tourists.

Cultural heritage has a universal value for us as individuals, communities, and societies. It is important to preserve and pass it on to future generations. Even if we think of heritage as "from the past", it evolves as we interact with it and plays a key role in building the future.

The practices and habits of cultural consumption of adolescents mainly include entertainment and relaxation activities. Walking in parks or green areas, walking in the mall (for shopping, games, movies, restaurants, etc.), and walking in cafes/restaurants/bars (other than those in the

Fig. 2 Examples of cultural heritage promoted through studies, cultural events, and laboratories (**a**—"MARTISOR" holiday; **b**—"SINZIENE" holiday; **c**—methodical center "CASA MARE" at Moldova State University)

mall) is most often done by young people. A possible explanation for these preferences is that there is much easier access to these places (proximity, free of charge, permissiveness) compared to other places.

Cultural heritage is among the top cultural activities preferred by young people, visiting museums or art galleries, a historical monument or an archaeological site, represent the occasional activities with the highest percentage of attendance among teenagers. The attitude of young people towards cultural heritage is generally one of protection, pride, and responsibility.

Cooperation in the field of culture enriches relations between societies and promotes cultural diversity and intercultural dialogue. At the same time, cultural and creative industries, based on crafts and tourism, can help stimulate economic growth. In line with the EU's strategy on international and cultural relations, the EU supports a dynamic cultural sector in the Republic of Moldova, involving citizens, public authorities, and cultural operators, strengthening civil society and providing greater economic opportunities.

The art-marketing strategies in promoting the cultural heritage of the Republic of Moldova focuses on the sustainable development of anthropogenic resources; the protection of the intangible cultural heritage; preservation and display of local cultural, historical, and natural heritage and protection of the authentic historical and cultural heritage. The perspective for further research is directed to analyze the activity of art marketers who are faced with the task of not only acting as a generator of ideas dressed in elements of culture, but also having the ability to correctly select the most commercially effective types of art. Art marketers are not just people with an education in the field of culture, but also the ones who understand the mechanisms of marketing, people who satisfy the business objectives of brands. Therefore, the success of any product directly depends on the professionalism of specialists working in this field.

Acknowledgements Research and Innovation Project: "Strengthening sustainable public procurement in the Republic of Moldova" 20.80009.7007.15, financed from the state budget within the State program 2020–2023.

References

Akathleen, M.: Art as Politics: Re-crafting Identities, Tourism, and Power in Tana Toraja. University of Hawaii Press, Indonesia (2006)

Bradshaw, A.: Before method: axiomatic review of arts marketing. Int. J. Cult. Tourism Hospitality Res. **4**(1), 8–19 (2010). https://doi.org/10.1108/17506181011024724

Brundtland, G.H.: Our Common Future: Report of the World Commission on Environment and Development. Geneva, UN-Dokument A/42/427 (1987). http://www.un-documents.net/ocf-ov.htm

Candrea, A., Nechita F.: Interpretation and promotion of cultural heritage in museums. Transilvania University (2015)

Carrillat, F.A., Colbert, F., Feigné, M.: Weapons of mass intrusion: the leveraging of ambush marketing strategies. Eur. J. Mark. **48**(1–2), 314–335 (2014). https://doi.org/10.1108/EJM-11-2011-0641

Colbert, F., St-James, Y.: Research in arts marketing: evolution and future directions. Psychol. Mark. **38**(8), 566–575 (2014). https://doi.org/10.1002/mar.20718

Crick, M.: Representations of international tourism in the social sciences: sun, sex, sights, savings, and servility. Annu. Rev. Anthropol. **18**, 307–344 (1989). http://www.jstor.org/stable/2155895

European Commission.: Basic Orientations for the Sustainability of European Tourism (2003)

Evans, J., Bridson, K., Rentschler, R.: Drivers, impediments and manifestations of brand orientation: an international museum study. Eur. J. Mark. **46**(11/12), 1457–1475 (2012). https://doi.org/10.1108/03090561211259934

François, C., Ravanas, P., Brunet, J., Restuccia, M., St-James, Y., de gestion des arts Carmelle et Rémi-Marcoux, C.: Marketing culture and the arts, 5th edn. Carmelle and Rémi Marcoux Chair in Arts Management, HEC Montréal (2018)

Juslin, P.N.: From everyday emotions to aesthetic emotions: towards a unified theory of musical emotions. Phys. Life Rev. **10**(3), 235–266 (2013). https://doi.org/10.1016/j.plrev.2013.05.008

Lapum, J., Ruttonsha, P., Church, K., Yau, T., David, A.M.: Employing the arts in research as an analytical tool and dissemination method: interpreting experience through the aesthetic. Qual. Inq. **18**(1), 100–115 (2012). https://doi.org/10.1177/1077800411427852

Making Tourism More Sustainable, UNWTO and UNEP.: Basic Orientations for the Sustainability of European Tourism (2005). https://eur-lex.europa.eu/LexUriServ/LexUriServ.do?uri=CELEX:52003DC0716:EN:HTML

McNulty, R., Koff, R.: Cultural Heritage Tourism, Partners for Livable Communities, Washington, DC (2014)

Moisei, L.: Ways to promote cultural heritage in the young generation. New Trends Prot. Promot. Natl Eur Cult. Heritage **9**, 202–209 (2013)

Nasser, N.: Planning for Urban Heritage Places: Reconciling Conservation, Tourism, and Sustainable Development. J. Plann. Lit. **17**(4), 467–479 (2003). https://doi.org/10.1177/0885412203017004001

Ngondo, E., Hermann, U., Boshoff, L.: An exploration of tourism experiences at a world heritage site in Namibia. Auckland University of Technology (2020). https://search.informit.org/doi/10.3316/informit.046433383001886

Raymond, A.: Cultural and heritage tourism. Eastern Michigan University (2008). https://www.researchgate.net/publication/237461371_CULTURAL_AND_HERITAGE_TOURISM

Santagata, W.: Cultural districts, property rights and sustainable economic growth. Int. J. Urban Reg. Res. **26**(1), 9–23 (2002). https://doi.org/10.1111/1468-2427.00360

Swanson, S.R., Davis, J.C.: Delight and outrage in the performing arts: a critical incidence analysis. J. Mark. Theory Pract. **20**(3), 263–278 (2012). http://www.jstor.org/stable/23243705

van den Berghe, P.L.: Tourism as ethnic relations: a case study of Cuzco. Peru Ethn. Racial Stud. **3**(4), 375–392 (1980). https://doi.org/10.1080/01419870.1980.9993312

Virtual Tourism and Silk Routes

Ioanna C. Chatzopoulou and Theodoros N. Mitakos

Abstract

Silk Routes were and still are a renowned network of trade routes that connects east and west. This network interconnects different lands that consist of versatile landscapes and are inhabited by diverse people with long history and rich culture. Not only trade goods but also ideas travel over these routes. In this paper, we aim to present the high potential of virtual tourism over the silk routes. In addition, we want to present the peculiarities and uniqueness of the silk routes as a tourist destination and to explain why virtual reality can play a vital role in the tourist industry over the silk routes. We review existing paper and online literature on virtual reality. We also examine numerous internet sites related to the silk routes. These sites offer virtual reality experiences to their visitors. We classify the sites analogous to their digital content and their primary operational function. Additionally we examine several virtual reality applications that give users not only information about silk routes but also invest on their experience. The conducted research shows that protection of natural and cultural environments that the silk routes encompass can be achieved through virtual tourism. The main objective of this article is to show that new technologies can overcome obstacles that have to do with space and time barriers, war, pandemic outbreaks such as COVID-19, human rights discrimination and support social inclusion. Finally, yet importantly, our research shows that virtual tourism on the silk routes can be combined with ecommerce. This way the silk routes acquire a new form of modern digital trade routes. Along the benefits that virtual reality offers to a silk route virtual traveler, we mention the main challenges and the problems that virtual reality faces.

Keywords

Virtual tourism • Virtual reality • Augmented reality • Sustainable development • E-commerce

1 Introduction

For many economies and human societies today, tourism is a driving force (Guojun 2015). Tourism can be an important source of income and employment (Manzoor et al. 2019). It can also boost the development of infrastructures such as roads, harbors or airports. It also promotes international relations and mutual understanding between nations. It also offers educational and recreational experiences to tourists. On the other hand, the tourism industry is vulnerable for a number of reasons, such as global economic instability, political unrest or the spread of a pandemic. According to the United Nations World Tourism Organization (UNWTO 2021), in 2020 international tourism recorded its worst year ever on record; international tourism declined by 74 percent, Middle East by 75 percent and Asia and the Pacific 84 percent since pre covid-19 levels, see Fig. 1.

There are also some other drawbacks such as putting in risk sensitive natural ecosystems, monuments or historical places due to the large number of visitors negatively affecting the cultural identity of ethnic groups or unilateral development of the economy.

Virtual tourism emerges as a new and promising paradigm in tourism industry. It combines virtual reality and travel. Computer and information technology advances offer innovative solutions to risks and problems faced by the tourist industry. It offers immerse user experience at low cost and many alternative ways for tourists to explore the places

I. C. Chatzopoulou (✉)
Department of Tourism Management, University of Patras, Patra, Greece
e-mail: ichatzop@upatras.gr

T. N. Mitakos
Software and Computer Engineer, Hellenic Electricity Distribution Network Operator S.A, Athens, Greece
e-mail: t.mitakos@deddie.gr

S. Kostopoulou et al. (eds.), *Cities' Vocabularies and the Sustainable Development of the Silkroads*, Advances in Science, Technology & Innovation, https://doi.org/10.1007/978-3-031-31027-0_18

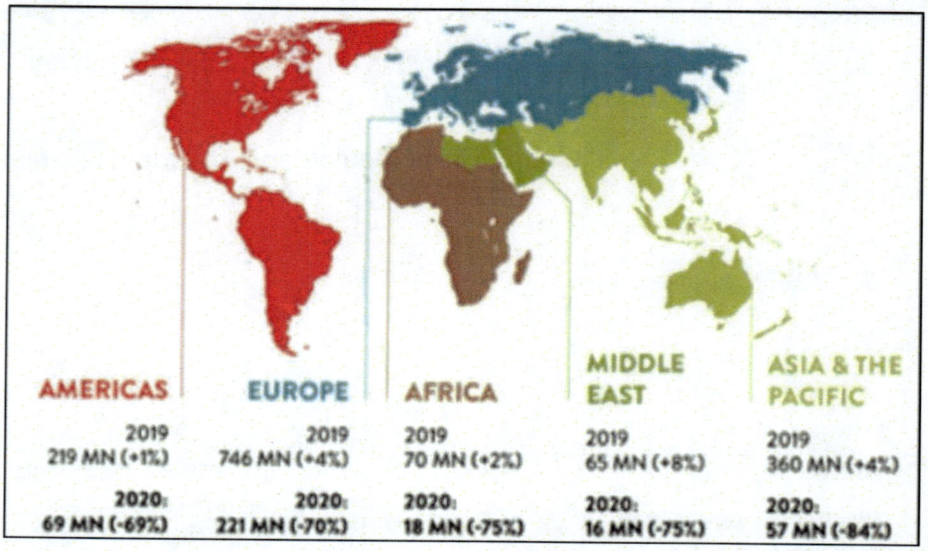

Fig. 1 Unprecedented fall of international tourism (UNWTO 2021)

they plan to visit without being there in person. Research conducted in China (Talafubieke et al. 2021) and Africa (El-Said and Aziz 2021) shows that virtual reality emerges as an important dimension in the tourism business. One the one hand, virtual reality can be proved not only an alternative product during times of crisis but also a new tool for the tourism business. On the other hand virtual tourism may be a strong analogue to a real-world experience but it cannot completely replace real tourism (Wagler and Hanus 2018).

In this paper we advocate that virtual tourism can be used as a new vehicle to boost tourism over Silk Routes. In the past the Silk Routes were the prime highway of international trade (Çinar et al. 2015). The fingerprint of civilizations that flourished over the Silk Routes still exists. It excites the imagination of people and increases their desire for a visit. Nevertheless, a real travel is difficult to be accomplished even nowadays. Obstacle is not only the cost and the enormous distance a traveler has to cover, but some parts of the Silk Routes are closed because of war conditions, or difficult to access because of the COVID-19 pandemic. On the other hand virtual tourism overcomes these obstacles and gives new opportunities for economic growth. At the same time the environment is protected and the cultural heritage is preserved.

2 Silk Routes—Silk Road

Although, there is not a single Silk Road but a series of Silk Roads that form the Silk Routes both terms are used interchangeably. Silk Road is called a network of overland and maritime trade routes that was and still is used to interconnect Europe with China, see Fig. 2. The maritime Silk Road interconnects Asia with the east coast of Africa. That is Silk Routes interconnect three continents, that is Europe, Asia and Africa. The overland Silk Road crosses a large number of lands including nowadays countries such as Greece, Turkey, Jordan, Iraq, Iran, Afghanistan, India, and China. The Han dynasty is considered the initiator of the Silk Road in the second century BCE. Many civilizations developed and flourished due to the Silk Road trade routes. Luxurious and precious goods were traded offering wealth and power to traders. Moreover ideas were exchanged offering new knowledge and enlightment to the people along the Silk Road. It is also worth noting that the trip from West to East and vice versa was long and dangerous. The Silk Road was most successful during times of peace, whereas in times of war there was little or no trade.

The Silk Road consists of three major land routes and a major one maritime route. These main routes have many secondary branches. In the following we mention only the major Silk Routes.

The northern route started at Chang'an (https://en.wikipedia.org/wiki/Chang%27an) and travelled northwest. It bypassed the Taklamakan Desert (https://en.wikipedia.org/wiki/Taklamakan_Desert) to reach the ancient kingdoms (https://en.wikipedia.org/wiki/Monarchy) of Parthia (https://en.wikipedia.org/wiki/Parthia), Bactria (https://en.wikipedia.org/wiki/Bactria) and eventually Persia (https://en.wikipedia.org/wiki/Persia) and Rome (https://en.wikipedia.org/wiki/Rome).

The southern route or Karakoram route was mainly a single route from China through the Karakoram Mountains (https://en.wikipedia.org/wiki/Karakoram). West of the Karakoram Mountains, the southern route had many routes heading south to the sea, since many travelers wished to continue by ship instead of going overland.

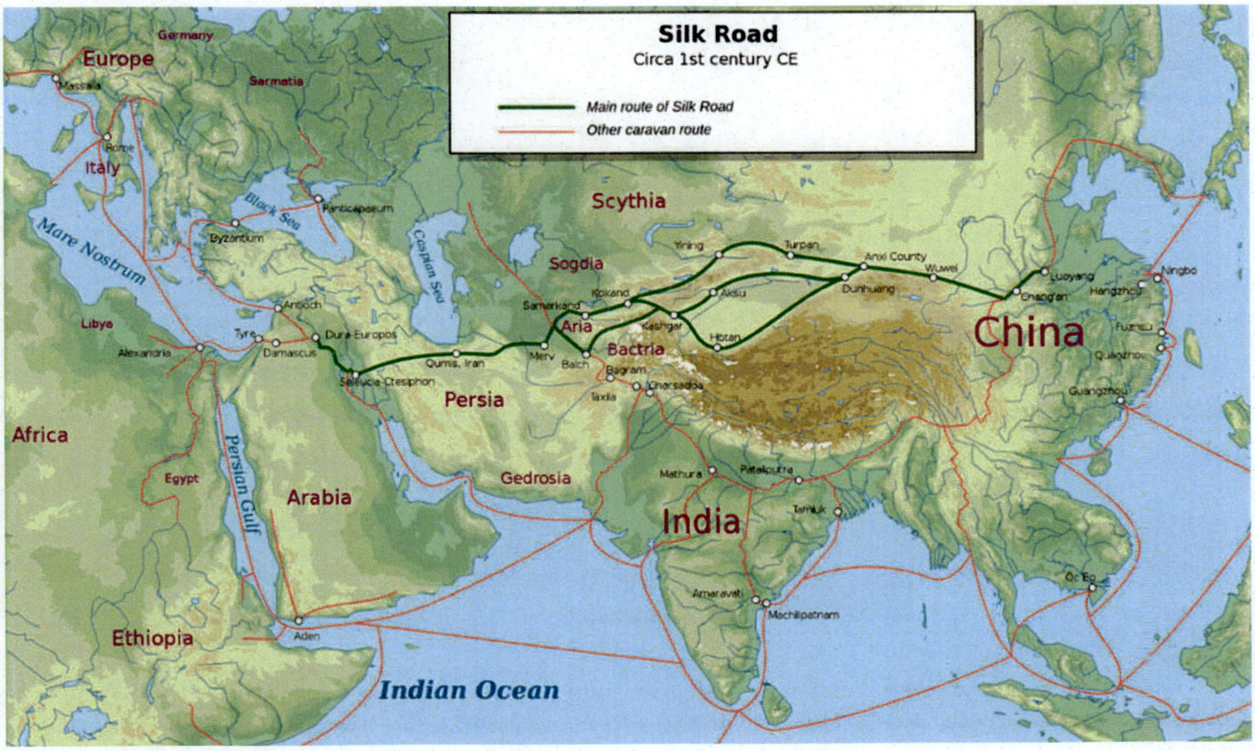

Fig. 2 The Silk Road in the first century (Kaidor 2018)

The southwestern route is considered to be the Ganges (https://en.wikipedia.org/wiki/Ganges)/Brahmaputra (https://en.wikipedia.org/wiki/Brahmaputra) Delta route. This region was an important trading hub. Archeological excavations have found relics of goods from various parts of the world. Strabo a 1st-century CE Greek (https://en.wikipedia.org/wiki/Ancient_Greece) geographer (https://en.wikipedia.org/wiki/Geographer), philosopher (https://en.wikipedia.org/wiki/Philosopher), and historian (https://en.wikipedia.org/wiki/Historian) has mentioned the Ganges region, showing a connection between Rome and China.

The use of maritime routes can be traced back thousands of years in the past till the present time. The maritime links among the Arabian Peninsula, the Mesopotamia and the Indus valley were in use long before the Silk Road. The maritime Silk Road or maritime Silk Route refers to the maritime (https://en.wiktionary.org/wiki/maritime) section of the historic Silk Road. It links China to Southeast Asia, the Indonesian archipelago (https://en.wikipedia.org/wiki/Indonesia), the Indian subcontinent (https://en.wikipedia.org/wiki/Indian_subcontinent), and the Arabian Peninsula (https://en.wikipedia.org/wiki/Arabian_peninsula), all the way to Egypt and finally Europe.

In the present day there are airways, railways and shipways that interconnect Europe, Africa and Asia. Still travelling and trading goods is a challenge for the demands of our times. The Suez Canal has shortened by a lot the voyage from the Chinese harbors to Europe by ship. On the other hand the recent accident of the Evergreen cargo ship blocked the traffic through the Suez Canal disrupting global trade. This accident has demonstrated how dangerous is such a voyage and even in the present day there is no cheap, easy and quick way to interconnect Europe and far Asia. It also brought to the fore the discussion on a more active use of other Silk Routes. As both Europe and Asia, being product producers, major trade centers and homelands of great civilizations both benefit from their close interconnection.

Along the Silk Road a vast number of commodities traded, just to mention some, included gold, jade, tea, spices, textiles, metal work, fragrances, jewels, ivory, sugar, wine, furs, carpets, glass, various artifacts, weapons, and slaves (Advantour n.d.). Judging by the road's name, silk was the main and lucrative good that was exported from China. In the sixth century CE silkworm eggs were smuggled out of Central Asia by two byzantine monks. Until the thirteenth century, the Byzantine Empire had the monopoly of the trading silk with Western Europe (Laiou and Morrisson 2007). In the same era a Venetian merchant, explorer and writer called Marco Polo was the first who described the trip from Europe to China along the Silk Road. The term Silk

Fig. 3 Karakoram Mountains, view of Hunza Valley, Ganish Bridge on Silk Road (Rizwan 2017)

Road was coined by the German researcher Ferdinand Richthofen in his fundamental work, "China", (von Richthofen 1877).

Along the Silk Routes not only trade goods were carried but also ideas were exchanged. We mention advances by religions, such as Islam, Christianity and Buddhist. Nestorian Christianity reached China at least in the seventh century CE (Bays 2011). Also, Islam reached China in the seventh century CE. On the other hand Europeans conducted Buddhism as early as the third century BCE through Alexander's conquests. Zoroastrianism, Manichaeism Confucianism, Daoism, syncretic philosophies, and various other religions and philosophical schools moved along the Silk Road. Moreover missionaries and merchants not only carried their faith and their beliefs, but they also built churches, monasteries, temples and shrines along the Silk Road.

The Silk Road was also used as a means for interchange of scientific discoveries and technologies between East and West. We mention paper, gunpowder and the compass. Papermaking was invented in China circa 179 BCE. During the eighth century CE it was spread to the Islamic world and finally it was brought to Europe in the eleventh century CE. The Chinese were first to use paper, not only to write a text but also to print currency. Gunpowder is a Chinese alchemists' invention from the ninth century CE. Originally, it was used to make fireworks for festivals and major events. Later it was utilized as an explosive substance in cannons, fire-arrows, and other military weapons. The compass is an instrument equipped with a needle that always points to the north. This feature has numerous applications in navigation, field exploration, trade, war and other fields. The compass's origins may be traced back to the fourth century BCE. In the tenth century CE this instrument was brought to the Arab world and Europe.

The landscape along the Silk Route is diverse, see Figs. 3 and 4. The trip starts from the Mediterranean coasts, it continues through the mountains of Asia Minor, then it goes through the plains, mountains and deserts of central Asia to end in mainland of China or to China's Pacific Ocean Coast. The climate conditions vary accordingly. It can be Mediterranean, desert, or continental monsoon.

Along the Silk Road not only constructions for religious purposes were built but also numerous buildings that served as accommodation of travelers. Among them caravanserais stand out, see Fig. 5. A caravanserai was a place where both caravans and single travelers could find shelter and food. It was also a place where merchants could trade their goods and also meet and interact with other people from distance and diverse places that the Silk Road crossed.

Thus, the Silk Road served not only as a route of economic value but also as a route of great cultural value for the civilizations that used it.

We conclude that a trip over the Silk Road was and still is an amazing experience. The Silk Road is an ancient road. Though it has lost its economic importance it is still in use. It has a long history and was important for the development of human civilization. Its history is connected with various people, ethnic groups and various cultures. Furthermore, it has overland and maritime paths. It covers vast areas with versatile landscapes. Sometimes these landscapes are inhospitable and the distances a traveler has to cover from one end to the other are very long.

Fig. 4 Gobi desert landscape along the Silk Road near Turfan, Xinjiang, China, Asia (Owen n.d.)

Fig. 5 Pasangan caravanserai on the silk road from Tehran to Isfahan (Hojjat ataei 2017)

Modern technology can overcome many of the obstacles mentioned above.

3 Virtual Reality

Advances in computer and information technology have given rise to new concepts such as augmented reality, mixed reality and virtual reality. These technological advances will yield new opportunities and application possibilities for the tourism industry (Beck et al. 2019).

Augmented reality is an immerse user experience where digital information such as audio, images, text and animation augment or enhance an individual's perception of the real world. For example, an augmented reality application installed in a mobile device can give information about tourist attractions during a tour.

Mixed reality is a concept that combines elements of digital world and real world. These elements are combined and interact. For example, medical students and doctors can use mixed reality holograms to practice various complex procedures.

Virtual reality or VR is a cutting-edge and an emerging computer technology that combines simulated environments with real user experience. The use of VR technology alters the user's perception of reality. A digital environment can be like a real one or completely different from it. A user can interact with a computer-generated environment using special devices that affect the senses of user according to his actions.

A user can have the illusion of being in a three-dimensional environment. Head-motion-tracking sensors identify the movements of a person's head. A software changes the field of vision as a person turns his head this or that way, he can view different perspectives of an artificial environment. He can also have the illusion of existence and move in a computer-generated environment different from the natural one. There are also eye-tracking sensors that make it even more realistic. He can also have the sensation of touch using special equipment.

A VR helmet is a head-worn device that gives the user a 360-degree view of a simulated environment. Stereoscopic vision can be achieved, creating a sense of depth and dimension, giving each eye the same image taken from a different perspective. The VR audio is binaural, in other words a different audio signal is fed to each ear to create the perception of a three-dimensional sound field. Alternative to VR helmet is VR headset. A virtual reality headset is depicted in Fig. 6. A VR headset does not cover the whole head but covers the eyes completely so that the user has lost the ability to see any aspect of the real world. We also mention the existence of special gloves that not only are designed for capturing full hand and finger action, but they create the feeling of touch as well. The most advanced VR gloves give the wearer sensations that mimic the weight and feel of real objects. There also exist exoskeleton suits that give the user the ability to walk freely through a virtual world. Aroma capsules can be incorporated in a VR helmet making user experience more engaging and effective. A fully immersive technology experience can happen when all elements of sight, sound, touch and smell come together.

Fig. 6 A virtual reality headset (Evan-Amos 2017)

A virtual reality world is created by specialized software. A lot of software developers need to work in order to produce a virtual reality application. A virtual reality application most of the time uses computer graphics, special algorithms, artificial intelligence, databases and network technologies. A user can also have virtual reality experiences on the internet. Even without special VR equipment a user can interact with internet cyberspace. Cyberspace is the "space" that exists only as data inside a computer.

In the following we mention a range of fields where virtual reality technology can be applied. VR can be used as a powerful tool in education. Students benefit from simulated worlds or situations e.g. a virtual tour or a virtual experiment. VR can be used in medical training. Medical students can practice surgeries and medical procedures, avoiding the risk of inflicting harm or making a mistake while practicing on real patients. VR can be used in training of pilots thus eliminating the cost and the risk of flying a real plane. VR can also help handicapped people. It gives them the opportunity to extend their physical abilities e.g. to visit places that it would be difficult if not impossible to visit. VR is also used in business to simulate business situations and help decision makers to take decisions. We also mention the use of VR in computer games industry. The applications of VR are numerous. In the following we elaborate on the applications of VR on the field of tourism and we also discuss the advantages and disadvantages of virtual tourism.

4 Virtual Tourism

Virtual tourism is a concept that was created by the application of augmented reality, mixed reality and virtual reality technology in tourism. It is a way to delve in the experience of visiting places, destinations, monuments, accommodations and other travel attractions or venues without being there in person. A tourist can also be given guided information or view a historic place how it was in its time and not only how it is in its present time e.g. how was Hagia Sophia in Istanbul (Constantinople) the time it was built or how psalms were heard when it functioned as a church, see Fig. 7. Since a person does not have to move anywhere this type of travelling stimulates the interest of tourists and is growing rapidly. This is especially true during the COVID-19 pandemic period because it eases anxieties and overcomes travel restrictions. There are many different forms of virtual tourism. This mainly has to do with the technological capabilities that are given to users.

The simplest type of virtual tourism is screening a set of photos, listening to an audio clip or watching a video. Virtual tourism allows virtual tours of galleries and museums around the world. A culture and art lover can explore via internet numerous galleries and museums (Carvajal et al.

Fig. 7 Hagia Sophia, interior (Julien 2019)

2020). A person who loves animals and wildlife can visit zoos and thematic parks without leaving the comfort of his home. Also, a traveler can organize better his travel before it actually happens. He can virtually visit the accommodation places before he decides to book a room. Moreover, travels that are impossible to take place can happen in a digital environment. Such travels include space travels or travels to the past.

Virtual tourism can also be used by tourist operators as a marketing tool. It can be used as a means to promote real travels. According to (digitalintheround 2021):

- "Millennials and Generation Z are 130% more likely to book a place if there is a virtual tour.
- 50% of adult users on the internet rely on virtual tours in their research and decision-making process.
- Customers spend 5 to 10 times more on websites with virtual tours.
- 67% of people want more businesses to offer virtual tours.
- There are more than 5 million visits daily on worldwide virtual tours."

YouGov announced at the Arabian Travel Market 2016 the results of its research in the MENASA market (Middle East, North Africa and South Asia). According to this research (YouGov n.d.):

- "51% of the 10,000 respondents from 21 countries consider VR as the most appealing technological concept to aid their travel booking process.
- 64% would be willing to visit a travel store with the technology and 31% could imagine booking a trip after experiencing VR in a store.
- 71% would download travel-specific VR content."

Google pioneers with its virtual reality application that is called "Street View". This application supports virtual tours in the streets all over the world. Google also claims that "360 images and tours drive results" and can lead from looking to booking (Google n.d.). Moreover, "Amazon Explore" is a platform launched in 2020 by Amazon. Users of this platform can enjoy individual virtual tourism experiences. The virtual travel happens in real time with the assistance of a local travel guide (Amazon n.d.).

Virtual tourism can preserve the memory of destroyed monuments. For example the giant Buddha's statues of Bamiyan in central Afghanistan (https://en.wikipedia.org/wiki/Afghanistan) were two 6th-century monumental statues that were blown up due to war in 2001 (BBC n.d.). There are many pictures and videos that depict the statues. Moreover through 3D projection a virtual representation can be put in the exact position of the original statues and tourists have the illusion that the statues still exist in their place. Palmyra is an ancient city in Syria that was strategically placed on the Silk Road. The monuments of Palmyra have suffered severe damages due to war and was closed for tourism for a long period. There are virtual tours that allow tourists to visit Palmyra without leaving their home. Moreover important monuments such as the Temple of Bel in Palmyra, Syria, are digitally reborn. The Temple of Bel was leveled due to war but it is digitally preserved by modern technology (LDS Living 2014), (Denker 2017), (Arck-project Sketchfab (https://sketchfab.com/3d-models/temple-of-bel-interactive-3d-experience-7a82cb38ebf042b28735339379d911e5)), see Figs. 8 and 9.

Virtual tourism can help sustainable development of ecosystems. Along the Silk Road a huge diversity of heterogeneous ecosystems exist. Some of them are fragile, vulnerable to human activities and some of them are difficult

Fig. 8 Front view of 3D reconstruction of Bel Temple (Arck-project Sketchfab)

Fig. 9 Side view of 3D reconstruction of Bel Temple (Arck-project Sketchfab)

to be accessed by a typical tourist. Virtual tours ease the access to tourists and protect ecosystems (Suocheng et al. 2017).

Technological infrastructure is vital for the development of virtual tourism. Networks should not only exist but need to be fast and have large bandwidth as well. Databases should not only keep data but also analyze them in real time. Smart hardware devices should be used together with software that uses advanced algorithms and artificial intelligence.

Criticism on virtual tourism advocates that a person will never have the experience of traditional tourism since technology has limitations. A real tourist not only gets a feel for the atmosphere of a place, but he will also interact with the locals and other tourists. Virtual tourism may be profitable for the persons that use it but not for the people who work in traditional tourism business. Restaurants, taverns, hotels and airlines need to serve real customers in order to make a profit.

5 Silk Road and Virtual Tourism

The easiest way to explore virtual tourism in the Silk Road is through the use of internet. Although, a web surfer cannot find a single site that offers all the information related to the Silk Route there are plenty of sites devoted to it. There are numerous sites devoted to monuments, history, environment, arts, culture, accommodation, e-commerce and almost everything a tourist would like to explore. Moreover there are sites that offer innovative services to willing customers. We classify the sites analogous to their digital content and their primary function.

There are sites where their digital content is primarily in text form. They usually give general information about the history of the Silk Road e.g. Wikipedia (UNESCO n.d.).

There are sites that offer unique audio experiences. For example “auralizations of the Byzantine cathedral of Hagia

Sophia chant are imprinted with the acoustics of the current condition of the great churches' interior". This was made possible taking advantage of accurate computer-enabled simulation of Hagia Sophia's unique acoustics. Traditional and modern music in all the countries that the Silk Route crosses can also be found in many sites.

There are sites where their primary content is photos. These photos depict, graphics, pictures, maps and visual content related to the Silk Road and its various trade routes. The visitors of these sites acquire visual information about various Silk Road components such as landscapes, monuments, and artifacts.

There are also sites that offer video clips. These video clips give a more realistic and complete view to the user. They usually combine audio and visual effects together. Their content, quality and their duration varies. Youtube is a prime source of videos related to Silk Routes. It offers not only regular videos but three dimensional videos as well, that is videos that give the viewer the feeling of depth in space. Additionally, there are 360-degree videos that a view in every direction is recorded and can be put in sight, see Fig. 10. Moreover there are VR videos that a visitor has to wear virtual reality glasses to have the full experience. Videos are mainly used to present landscapes, museums, galleries and accommodation places to potential future visitors. Some videos present fiction environments where reconstruction of imaginary places or situations is presented e.g. a Silk Road market or trading. The videos can be free of charge to be accessed or the visitor has to pay a fee for the video tour.

Live streaming, that is video sent over the internet in real time, without first being recorded and stored, extends the user experience. An innovated service is booking of real time virtual guided tours. A tour guide performs a ground tour but the customers of the service attend the tour online from their place (Culture Questtours n.d.). The customers of the service can also communicate with the guide before, during and after the tour. The combination of modern technologies such as internet and products such as zoom have made this kind of service feasible. During the tour certain events can happen e.g. visit to local markets, discussions with local cooks that reveal secrets of traditional cooking etc. After the virtual tour the customers can have after sales services such as the order of products or souvenirs they would like to acquire.

Moreover, due to the COVID-19 pandemic China's live performance industry has moved online. This was a necessity since many artists were left without an income. According to the China Association of Performing Arts, the loss of revenue from January 2020 to March 2020 amounted to an aggregate box-office loss of over 2 billion yuan (Xuejiao 2020). Moving music and theater online gives audiences from all over the world the opportunity to attend performances virtually. This way revenue is made, and loss of jobs and money is prevented.

Home entertainment is extended with gaming. There are many computer games that draw inspiration from the Silk Routes. For example "Silkroad Online" is a fantasy MMORPG that is massively multiplayer online role-playing video game. It simulates a traders' trip along the Silk Road between China and Europe in the seventh century CE. This video game is a role-playing video game where many players play online at the same time.

There are many VR applications that approach Silk Routes from an educational point of view (UNIVERSITY OF CAMBRIDGE n.d.). A lot of free educational material can be found on the internet. Moreover, there are paid online courses and paid material offered by various educational institutions. E-learning courses can be of general interest, e.g. giving knowledge for Silk Routes or of special interest e.g. Silk Road is used as a test case for training qualified tour guides (EUROPEAN FEDERATION OF TOURIST GUIDE ASSOCIATIONS n.d.).

The countries that exist in the Silk Routes regions have understood the significance of the old trade routes and they are trying to revive them. There is an ambitious vision for the creation of a "New Silk Road" (BRI n.d.). The "Belt and Road Initiative" (BRI) is a transcontinental long-term policy and investment program that aims to the economic integration of countries along the route of the historic Silk Road. It combines two initiatives, that is the (land based) Silk Road Economic Belt, and the 21st Century Maritime Silk Road (BRI n.d.). There is also the technology dimension of China's Belt and Road Initiative (https://reconasia.csis.org/category/chinas-belt-road/) that is called Digital Silk Road (Hillman 2021). Chinese politicians were the first that envisioned and took steps to revive the Silk Road. China is moving with fast pace in cutting edge technologies such as informatics and space. Fast progress in areas such as

Fig. 10 Three screenshots taken from the same scene of a 360 degree video (Singapore Tourism Board 2021)

telecommunications, databases and artificial intelligence gives China the opportunity to reap commercial and strategic rewards, especially as emerging economies grow. China competes with other technologically advanced countries for sharing global economy. Virtual reality and virtual tourism can certainly help this effort since China is rich in history, culture and human resources.

We have to mention that virtual tourism has some prerequisites. That is, it depends heavily on technology. Internet lines must me fast enough to support multimedia. Computers have to be powerful enough to run the VR applications. The VR applications must be friendly and easily accessible by the users. The hardware equipment should be cheap and easy to buy.

6 Conclusions

Although the Silk Routes ceased to exist as a main international trade and culture exchange route in the fourteenth century CE, their fingerprint in human civilization still exists. Modern technology gives unprecedented possibilities for West and East interconnection since innovative services emerge and new opportunities for wealth and prosperity are given. Virtual tourism can open a new trade and culture interchange route for the countries of the Silk Routes. It can also give reliable solutions to new and old problems of trade, mobility, and tourism. E-commerce trading of goods and services is perfectly combined with virtual tourism. We also emphasize that virtual tourism is a product of tourism and emerging information technology. Thus it depends on technology that is not still mature and sometimes it is expensive. The future of virtual tourism seems promising. We conclude by advocating that people living along the Silk Routes and people exploring them can benefit from a mixture of new and traditional forms of tourism.

References

Advantour: Goods Traded on the Silk Road (n.d.). https://www.advantour.com/silkroad/goods.htm
Amazon: Amazon Explore. https://www.amazon.com/b?node=19424628011&ref=ae_stfrt_hro_lrn. Accessed 8 Apr 2022
Arck-project Sketchfab: Retrieved from https://sketchfab.com/3d-models/temple-of-bel-interactive-3d-experience-7a82cb38ebf042b28735339379d911e5
Bays, D.H.: A New History of Christianity in China. Wiley, pp. 7–15, 18–21 (2011). ISBN 978–1–4443–4284–0
BBC: In pictures: 3D return for Bamiyan Buddha destroyed by Taliban (Buddha statues) (n.d.). https://www.bbc.com/news/world-asia-56337042
Beck, J., Rainoldi, M., Egger, R.: Virtual reality in tourism: a state-of-the-art review. Tour. Rev. **74**, 586–612 (2019)
BRI: Belt and Road Initiative (n.d.). https://www.beltroad-initiative.com/belt-and-road/. Accessed 2 May 2021
Carvajal, D., Morita, M., Bilmes, G.: Virtual museums. Captured reality and 3D modeling. J. Cult. Heritage **45**, 234–239 (2020)
Çinar, E.M., Geusz, K., Johnson, J.: Historical perspectives on trade and risk on the Silk Road, Middle East and China. Top. Middle Eastern North African Econ. **17**(2), 156–191 (2015)
Culture Questtours: Experience Any Special Destination You Like! (n.d.). https://culturequesttours.com.au/
Denker, A.: Rebuilding Palmyra virtually: recreation of its former glory in digital space. Virtual Archaeol. Rev. **8**(17), 20–30 (2017)
digitalintheround: Retrieved from https://digitalintheround.com/virtual-tour-statistics/ (2021). Accessed 5 Apr 2022
El-Said, O., Aziz, H.: Virtual tours a means to an end: an analysis of virtual tours' role in tourism recovery post COVID-19. J. Travel Res. **61**(2), 1–21 (2021). https://doi.org/10.1177/0047287521997567
European Federation of Tourist Guide Associations: Western Silk Road Experience Online Course (n.d.). https://www.feg-touristguides.com/western-silk-road-experience-online-course.php
Evan-Amos: An Oculus headset released in 2016. In: Wikipedia (2017). https://en.wikipedia.org/wiki/Virtual_reality_headset#/media/File:Oculus-Rift-CV1-Headset-Back.jpg
Google: Street View: Retrieved from https://www.google.com/streetview/business/ (n.d.). Accessed 7 Apr 2022
Guojun, Z.: Tourism and Hospitality Development Between China and EU (2015). Springer Berlin Heidelberg. ISBN: 978-3-642-35909-5
Hillman, E.J. Mapping China's digital Silk Road. In: Reconnecting Asia (2021). https://reconasia.csis.org/mapping-chinas-digital-silk-road/
Hojjat ataei: Retrieved from https://commons.wikimedia.org/wiki/Category:Pasangan_Caravanserai#/media/File:Panorama1_copy.jpg (2017)
Kaidor: The Silk Road in the 1st century. In: Wikipedia (2018). https://en.wikipedia.org/wiki/Silk_Road#/media/File:Silk_Road_in_the_I_century_AD_-_en.svg
Laiou, AE, Morrisson, C.: The Byzantine Economy. Cambridge Medieval Textbooks, Cambridge (2007)
LDS Living: New Virtual Tour Lets You Visit Palmyra Without Leaving Your House (2014). https://www.ldsliving.com/new-virtual-tour-lets-you-visit-palmyra-without-leaving-your-house/s/76949
Manzoor, F., Wei, L., Asif, M., Haq, M.Z., Rehman, H.: The contribution of sustainable tourism to economic growth and employment in Pakistan. Int. J. Environ. Res. Public Health **16**(19), 3785 (2019). https://doi.org/10.3390/ijerph16193785
Julien, M.: Retrieved from https://commons.wikimedia.org/wiki/File:Hagia_Sophia_(48985225283).jpg (2019)
Owen, A.: Retrieved from https://www.westend61.de/en/imageView/RHPLF14729/buildings-dwarfed-by-vast-gobi-desert-landscape-along-the-silk-road-near-turfan-xinjiang-china-asia (n.d.)
Rizwan, S.: View of Hunza Valley from Baltit Fort. In: Wikimedia (2017)
Singapore Tourism Board: Virtual Experiences: Immerse in 360° Experiences from Your Home (2021). https://www.visitsingapore.com/virtual-experiences
Suocheng, D., Kolosov, V., Li, Yu, Li, Z., Li, F., Zhao, M., Shi, G., Yu, H., Cheng, H., Guo, P.: Green development modes of the belt and road. Geogr. Environ. Sustain. **10**(1), 53–69 (2017)
Talafubieke, M., et al.: Evaluation of the virtual economic effect of tourism product emotional marketing based on virtual reality. Front. Psychol. **12**(759268), 1–11 (2021). https://doi.org/10.3389/fpsyg.2021.759268
UNESCO: Silk Roads Programme (n.d). https://en.unesco.org/silkroad/about-silk-roads

University of Cambridge: The Silk Route (n.d). https://www.ice.cam.ac.uk/course/silk-route
UNWTO (United Nations World Tourism Organization): UNWTO World Tourism Barometer. COVID-19 and tourism 2020: A year in review (2021). https://webunwto.s3.eu-west-1.amazonaws.com/s3fs-public/2020-12/2020_Year_in_Review_0.pdf. Accessed 2 Apr 2022
von Richthofen, F.: Über die zentralasiatischen Seidenstrassen bis zum 2. Jh. n. Chr. Verhandlungen Der Gesellschaft Für Erdkunde Zu Berlin (in German). **4**, 96–122 (1877)
Wagler, A., Hanus, M.D.: Comparing Virtual Reality Tourism to Real-Life Experience: Effects of Presence and Engagement on Attitude and Enjoyment. Faculty Publications, College of Journalism and Mass Communications (2018). https://digitalcommons.unl.edu/journalismfacpub/121
Xuejiao, C.: China's live performances go virtual amid COVID-19 epidemic. In: The Theatre Times (2020). https://thetheatretimes.com/chinas-live-performances-go-virtual-amid-covid-19-epidemic/
YouGov: The Future's Looking Virtual for Travel Booking in MENASA (n.d.). https://mena.yougov.com/en/news/2016/04/26/futures-looking-virtual-travel-booking-menasa/. Accessed 4 Apr 2022

Sustainable Tourism Development Two Types of Leverage: Innovative Tradition and Tourism for the Future

Andreas Andreopoulos

Abstract

Modern Civilization is in jeopardy due to global problems (biodiversity loss, climate crisis, financial crisis, diseases). What is needed are global frameworks and innovative tools for keeping body and soul together. The desideratum, therefore, is a collective contribution from the three basic achievements of humanity (Science, Spirituality, Fine Arts); and last but not least, from the six fundamental resources (people, ideas, material reality, the planet, finance.) To these must be added three temporal parameters (past, present, future.) Cultural Heritage, Sustainable tourism, UNESCO World Cultural and Natural Heritage, One Belt-One Road, China's century, 4th Industrial Revolution, Sustainable Development, Ecological Civilization are among the insightful ingredients for compounding the formula for the future ("The Recipe For The Future"). Current research enhances the added value of the above ingredients, delivering frameworks and tools such as Innovative Tradition, karma, Tourism for the Future, etc., based on reality checks, original monitoring compilations (human activity sectors; modern societal priorities; humanistic awareness parameters; sustainable indicators; perennial skills), paradigms and tangible models via for the most part uncommon methodologies. What is eventually suggested are practical steps towards implementation—with Sustainable tourism becoming the global "soapbox" for humanity's well-being, moving ordinary travelers towards Sustainable Development, making them instigators of Ecological Civilization, cultural pilgrims, active citizens, re-baptizing them netizens/earthlings on a PEACEFUL planet.

Keywords

Karma • UNESCO world cultural and natural heritage • One belt • One road • Ecological civilization • Cultural heritage • Sustainable tourism

Nomenclature: Multi-disciplinary

BRI	Belt and road initiative
CH	Cultural heritage
EC	Ecological civilization
GSC	Global short circuit
HC	Human civilization
ICH	Intangible cultural heritage
IT	Innovative tradition
OC	Original compilation
SD	Sustainable development
SDHA	Sustainable development holistic accelerator
St	Sustainable tourism
TF	Tourism for the future

***Contents**—in logical sequence: (1) The current state of human civilization - What is the problem? The big picture. (2) A reality check of modern mega-trends: factors exerting a positive or negative influence on the problem. (3a) The research methodology. (3b) Research findings. (3c) Two great opportunities. (3d) "Playing for time" - example of a disruptive way of thinking about research. (4) Innovative Tradition: Transforming Cultural Heritage into a handy tool for left-handed people: **Innovative Tradition**; leverage solution #1 Paradigm: **Parampara** and more examples. (5) Sustainable Development and Ecological Civilization paradigm: **The New Silk Road/BRI**. (6) Tourism for the Future: leverage solution #2; example: **Sustainable Development Holistic Accelerator**. (7) Intercultural and inter-religious meeting point: **Karma** as core spiritual paradigm. (8) Conclusions.*

A. Andreopoulos (✉)
Metis, Global Awareness Network, Accredited Observer to the UNFCCC, Tristeno, Zagori, 44014, Greece
e-mail: mhtiswan@gmail.com

S. Kostopoulou et al. (eds.), *Cities' Vocabularies and the Sustainable Development of the Silkroads*, Advances in Science, Technology & Innovation, https://doi.org/10.1007/978-3-031-31027-0_19

Sustainable Tourism Development two types of leverage

1. **INNOVATIVE TRADITION**
2. **TOURISM FOR THE FUTURE**

Aim: Enhance Sustainable Tourism (ST) through additional types of leverage (tools) so as to become more effective within two frameworks; (a) the traditional one (UNESCO World Cultural and Natural Heritage global network); (b) the modern one (The New Silk Road/BRI (Apostolopoulou 2022) global construction of infrastructure) taking advantage of intercultural and inter-religious elements and emerging technologies.

Objectives: (1) Promote added value elements of *cultural heritage* (CH) by collecting them and integrating them into the practical novel toolkit named "*Innovative Tradition*" (1st leverage) based on ancient accumulated experience. (2) Promote the novel concept and content (2nd leverage) of "*Tourism for the Future*" by integrating the BRI spirit in concrete touristic packages. (3) Intertwine and interweave the two types of leverage in BRI infrastructure in ongoing concrete projects. (4) Shift and promote "Karma" (Kaefer 2022) concept of a supportive core spiritual principle of sustainable tourism.

1 Current State of Human Civilization—What Is the Problem? The Big Picture

Three components combined in innovative ways for the next qualitative cultural leap. (a) Shortage and abundance of resources; (b) The scarcity of time and enlarging it; (c) Narrowing and widening the original content of Civilization.

Human civilization is a journey *back and forth* in time. A journey lasting 10–15 thousand years so far. Until two centuries ago, the human footprint on the planet was positive, or, at least not negative. In the last one hundred years, humans have grown exponentially and exorbitantly in number and become overburdened with things they produced by spending, without planning, resources created over 4.5 billion years—on a planet that has grown not a single square meter since then; with ideas primarily oriented towards material production and consumption without moderation, and with an extraordinary absence of harmony and happiness (the main characteristics of civilization). This has resulted in a multi-level *global short circuit*. Despite COVID-19 crisis Tourism sector remains an exceptional paradigm because it covers people's core needs for communication, recreation, and leisure.) This is, in short, the panorama of human civilization based on the six (6) basics (people; ideas; material reality; time; place; economy) as Andreopoulos constructs his material (Andreopoulos 1996).

This unbalanced conjunction of factors had the expected result of civilization breaking sharply and being in danger of sliding. According to research findings and suggestions there were prospects of it discovering its lost heritage… just in the nick of time. It would focus on traditionally tried and tested principles: sustainable development, innovation, human solidarity. Realizing the heaviness of being burdened by so many 'things'—most of them useless, it would decide to put its house in order. Efficient and effective management of the remaining resources would make possible the rehabilitation of a dangerously disturbed environment. ***Scarcity would give birth to maturity***. This process would be accelerated by rediscovering the power of cultural heritage and harnessing the exponential achievements of the human mind, with the additional support of strategic plans on a planetary scale.

2 A Reality Check of Modern Mega-trends: Factors Exerting a Positive or Negative Influence on the Problem

Global human civilization has been hit hard over the last decades. Where significant levels of achievement were once recorded, in areas such as poverty elimination and the combating of epidemic diseases, present-day performance levels have receded. It has not been possible to utilize the many resources available to solve problems, although they were there ready to be used.

1. The ***negative*** mega-trends caused by local or international short circuits are: biodiversity loss; the climate crisis, the financial crisis, the COVID-19 outbreak. Google search results verify their importance (3/11/2021: COVID-19 outbreak 1,020,000,000 results 3/11/21; biodiversity loss 89,100,000; climate change 10,050,000,000; Climate crisis 983,000,000; Financial crisis 1,050,000,000. The ***spillover effects from these trends are,*** in short: Climate change became critical due to loss of biodiversity. Global development has stalled primarily because of the priority necessarily accorded to disaster restoration. COVID-19 came to give a decisive final blow both to the well-being and to the financial status of the population (1 billion fewer international tourist arrivals; loss of US$ 1.3 trillion in total export revenues from international tourism.) (Figs. 1 and 2).
2. Fortunately, ***positive*** mega-trends have also emerged and grown recently. Mega-trends with ***spillover benefits include***: Global Tourism; The rise of China; The Chinese century; The New Silk Road/BRI: an infrastructure plan with unprecedented potential (Google search results 3/11/2021: Global Tourism 947,000,000; China's rise 114,000,000; Chinese Century 913,000,000; new silk road /BRI 671,000,000). The ***spillover benefits from***

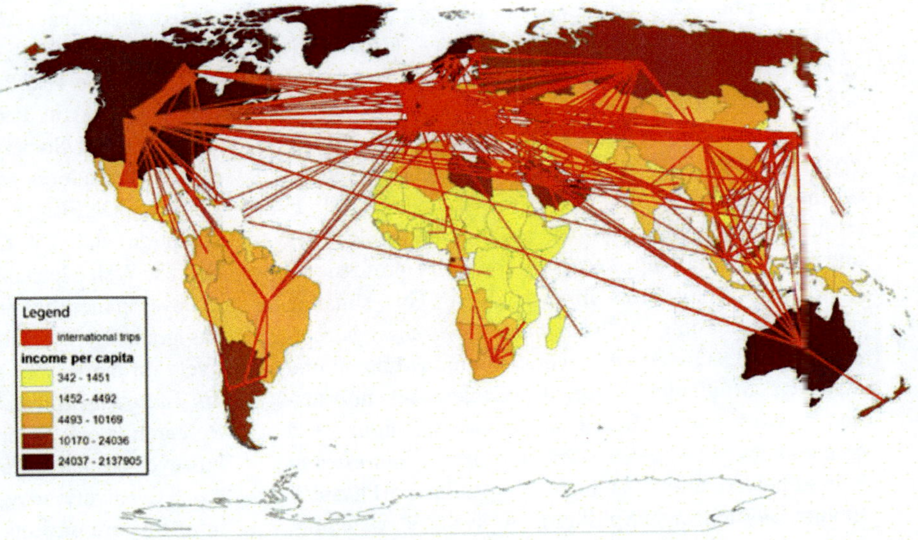

Fig. 1 International travel world map

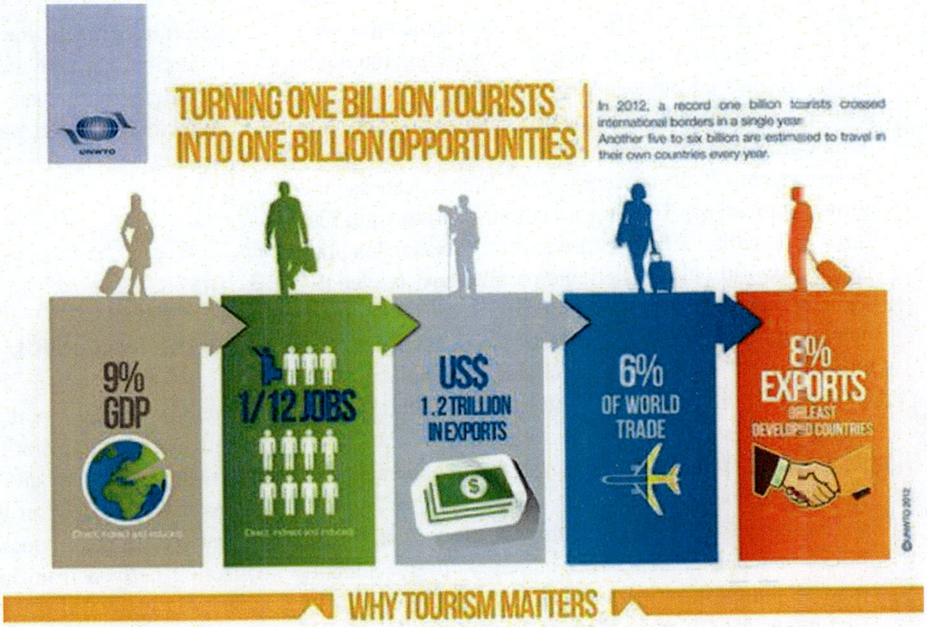

Fig. 2 Data on tourism industry

***these trends**, in short: Global tourism is a lever for international communication, understanding and co-operation. The rise of China promises a global platform for peace and prosperity. The New Silk Road/*BRI has started to implement a global infrastructure development beneficial to all mankind.

3. These mega-trends interact on many levels. The *negative* ones prevent the positives from growing according to modern needs. There is a *global short circuit* on development, employment, professional training, health, communication due to the end of the era of *Industrial Civilization*. These challenges are having a crucial effect on factors indispensable for the continuation of civilization. The three original compilations are NOT 'numerus clausus'. They are open to become more. First of all, the 35 areas of human activity (1st ORIGINAL compilation: (1) Organization—Planning (2) Land (3) Population (4) Legislative Power (5) Judiciary Power (6) Executive Power (7) Public Administration (8) Local Municipality (9) National and Internal Security (10) Health (11) Social Welfare (12) Communications (13) Publications (14) Mass Media (15) Labor, Employment, Business (16) Education (17) Church (18) Science and Technology Research (19) Culture (20) Sports (21) Tourism. Travel

and Recreation (22) Transport and Storage (23) Public Transport (24) Economy (25) Energy (26) Trade (27) Industry, Small Industry, Products Processing (28) Mining (29) Environment (30) Physical Planning—Public Works (31) Shipping (32) Agriculture (33) Fishery (34) Cattle (35) Forestry), already mentioned as part of the all-inclusive lists above—if not almost all. In particular employment, public health, lifelong learning and fast track training, necessary for every professional to follow developments in every area. This situation makes it impossible for "*Humanistic Awareness parameters*" (2nd ORIGINAL compilation: Multicultural diversity; Multicultural acceptance/tolerance; Information network circuit; Blood and Organ donation; Dead cremation/burning ashes for fertilizer; Self-esteem/Self-empowerment; Human potential (Physical/Emotional/Intellectual/Spiritual); Active TV watching; Positive Internet; Human interconnections/global brain; No excess consumerism/materialism; Love; Whole-systems/Chaordic thinking; New communication technologies intelligent user; Social networking intelligent user; Species-consciousness; Empathy for other life forms; Responsible Production and Consumption; Employment with social benefit; Culturally sustainable development; Consciousness of consciousness/inner growth processes; Tourism as cultural pilgrimage; Cultural and natural heritage proactive; National language usage as world cultural heritage part; Nature as sacred—Gaia (earth self-regulating organism); Reestablishment of communities natural infrastructure; Recognition and affirmation of soulful nature; Global consciousness/planet—global village; Proper nutrition; Holistic health/alternative medicine/Health-orientation; Functional literacy; Opinion for the future; Cosmic perspective; Against Social exploitation; Media/Audiovisual/TV/Advertisement literacy (grammar and syntax); CSR active member; Alternative currency user; Socially responsible investor) to emerge (part of all-inclusive lists)—let alone for "*Modern Societal Priorities*" (3rd ORIGINAL compilation: Sustainable Development, Regional Awareness and County Development; Employment and Education; Lifelong Learning; Digital Literacy; Knowledge Society; Creative Occupation; Natural and Cultural Environment Preservation; Multicultural Diversity and Tolerance; Anthropocentric Technology and Consumption; Alternative Energy Sources; Biotechnology; Thematic Tourism (AGROTOURISM, Cultural Tourism, Cultural Itinaireries + 60 More); E-Governance; E-Business/Commerce; International Cooperation For Development; Volunteerism; Humanitarian Awareness; Roadmap To Sustainability 6. Six Core Ingredients (A) Human Potential; (B) Ideas; (C) Homeland; (D) Concrete Reality; (E) Time Period; (F) Financial Situation The Above Ingredients Are "numerus clausus" For All Living Or NonLiving Stuff: There Exists Not Even One Additional One To Use From The Creation Of The Universe To A Successful Party; Innovation and Industrial Production; Civil Society; Corporate Social Responsibility; Health Promotion and Quality Of Life; Change and Crisis (CONFLICT) Management; Responsible Consumption and TV Literacy; Social Networking; Web Search, Web Mechanics) (also part of all-inclusive lists) to develop their full potential. However, a positive course has gradually been developed and challenges are effectively on the way to being tackled.

4. ***4th Industrial Revolution and BRI*** (Anonymous 2019): Research is only indirectly related to the issue. It should however be pointed out that "*One Belt, One Road/BRI*" has extensively integrated emerging technologies into infrastructure projects. Its main ingredient for the future, between present and future, is the extension of mega-projects along the traditional silk routes such that the future meets the past. UNESCO monuments and intangible heritage elements all along the BRI road and including more than 140 countries could easily find a collectively agreed place for integrating past and future features.

3 .

3.1 The Research Methodology

The purpose of the research is to support civilization's quest to find a correct trajectory. Its two most disruptive findings, "innovative tradition" and "tourism for the future" could not be developed simply on the basis of conventional methodologies. Quite the opposite: it relied on a method of data collection derived from incorporation of the following elements: (1) Insightful penetrative observation; (2) A qualitative approach; (3) Descartes' universal method of inductive reasoning (Richard 2002) comprising four rules: a. accepts nothing as true that is not self-evident; b. divide problems into their simplest parts; c. solve problems by proceeding from simple to complex; d. recheck the reasoning; (4) Visual semantics (Giunchiglia et al. 2021). (5) Mesoscience innovative method (Huang et al. 2018) (6) All-inclusive lists (above as "ORIGINAL") originally compiled on the basis of modes of thinking which include oblique thinking (Brian 2009), lateral thinking (De Bono 1970), creatively destructive thinking, chaordic thinking (Dee 2000), holistic thinking, antifragile thinking (Nicholas 2014).

This *hybrid methodology* has already produced ORIGINAL *findings* and *compilations above* supporting the results of this research in concrete ways: for instance, Parampara has collected material from several domains of human activity; research has identified a need for further investigation on the basis of a comprehensive list of human activities. Particular highlights include the ease of identifying unknown gaps between the known (based on Mesoscience), spillover effects and benefits of regions with exponential dynamics (list below) based on non-linear thinking. Finally, one important core element is the "Time Game"—another means of introducing disruption—not directly derived from quantum theory. The methodological approach is at once, interdisciplinary (in scientific terms) and inter-artistic (in cultural heritage terms and orientation towards the fine arts) but also multicultural (embracing karma).

Methodological procedure—in summary:

a. ***As regards the leverage of "Innovative Tradition"***: Long-term observation of (empirical research on) the static and underutilized nature of the valuable cultural heritage material still lying neglected in museum showcases. Systematic monitoring and identification of sustainable methods and material from the past that can be utilized today.
b. ***As regards the leverage of "Tourism for the Future".*** *The sustainable development holistic accelerator* (SDHA) proposal, presented as a model, is the product of research in, and synthesis of, three areas: (1) The complex selection procedure and content nomination of UNESCO Cultural and Natural Heritage sites; (2) The content and spirit of The New Silk Road infrastructure projects. (3) The creative synthesis of the following multilevel factors: Location of the monument; location of the BRI development project; concentration of thousands of visitors at the site; visitors' positive disposition towards the experience, signaling willingness to assimilate its value; employment of immersive technologies (Pierre (Pete) Routhier 2022) enhancing visitor's perceptual capacity; production content combining innovative traditional elements (***past***) integrated into the BRI project (***future***) at the specific ***present*** moment when the visitor is in situ, with the anticipated result of elevating the visitor's consciousness to a higher level (e.g. from run-of-the-mill visitor to cultural pilgrim).

The above elements, taken together, are not explicable as the outcome of research, selection and collection of data from just one level of cognitive sources. They stem from meta-analysis and inspired *meta-synthesis* (**Research synthesis** is the process of combining the results of multiple primary studies aimed at testing the same conceptual hypothesis) of prior research at various levels, including inputs from science (semiotics, emerging technologies), fine arts (the cultural heritage) and spirituality (karma). The final result is not just an idea but an unprecedentedly serviceable product of high added value.

The educational component of the SDHA product (the "Social Matrix Revealed") based on secondary data analysis ([13] 3859 activities [14] (1976–2019) are mostly of technical and scientific nature; not holistic course like **Social Matrix Revealed** (SMR) exist; the reason is simple: SD and, more specifically, Ecological Civilization are considered ways of thinking, holistic ones. They are approached within a circular framework producing causes, effects and solutions. UNFCCC (2019) Organization site Activities from UNFCCC capacity building portal: (a) Institutional capacity building, including the strengthening or establishment, as appropriate, of national climate change secretariats or national focal points: 284 (11.5%); (b) Enhancement and/or creation of an enabling environment: 189 (7.6%); (c) National communications: 64 (2.6%); (d) National climate change programs: 95 (3.8%); (e) Greenhouse gas inventories, emission database management, and systems for collecting, managing and utilizing activity data and emission factors: 122 (4.9%); (f) Vulnerability and adaptation assessment: 94 (3.8%); (g) Capacity building for implementation of adaptation measures: 265 (10.7%); (h) Assessment for implementation of mitigation options: 233 (9.4%); (i) Research and systematic observation, including meteorological, hydrological and climatological services: 151 (6.1%); (j) Development and transfer of technology: 148 (6%); (k) Improved decision-making, including assistance for participation in international negotiations: 143 (5.8%); (l) Clean development mechanism: 107 (4.3%); (m) Needs arising out of the implementation of Article 4, paragraphs 8 and 9, of the convention: 45 (1.8%); (n) Education, training and public awareness: 419 (16.9%); (o) Information and networking, including the establishment of databases: 117 (4.7%). **Activities by region** (March 1, 2019): Africa 138 (45.8%); Latin America and the Caribbean 44 (14.6%); Asia and the Pacific 114 (37.9%); Eastern Europe 5 (1.7%). **Examples**: Canada 3; USA 5; Mexico 76; Colombia 128; China 140; south Africa 113; Greece 1 (2016 Forum of Marine Protected Areas in the Mediterranean); Germany 7; Saudi Arabia 25; Cabo Verde 76; Uganda 108; Côte d' Ivoire 108; etc. In total: from 1976–2019: 3859 total activities. Education, training and public awareness; 419 (16.9%) (UNFCCC 2019) UNFCCC capacity building is already being promoted by the UNFCCC (UNFCCC 2019). The construction component is proceeding as part of the UN sustainable development campaign (UNSDGS 2016). UNESCO monuments and ongoing New Silk Road projects overlap at certain points and comparative observation suggests that wherever creative coupling of the two occurs, this is the result of semiotic research. Maps and diagrams are selected following macroscopic consideration.

3.2 Research Findings

1. ***From the past***: Cultural heritage, the legacy of tangible and intangible heritage assets of a group or society that is inherited from past generations. Included are valuable elements (such as the Parampara collection) which have solved everyday problems for thousands of years.
2. ***From the past and for the future***: The silk road was and still is a network of trade routes connecting east and west. The New Silk Road (or One Belt, One Road, is a global infrastructure development strategy adopted by the Chinese government in 2013 for investment in nearly 70 countries and international organizations. Many of the infrastructure works are located near UNESCO sites. (Figs. 3 and 4).
3. ***From the past for the present***: Karma (In the Epic, free will has the upper hand. Only when a man's effort is frustrated or when he is overcome with grief does he become a predestinarian (believer in destiny); "Quote—This association of success with the doctrine of free will or human effort (purusakara) was felt so clearly that among the ways of bringing about a king's downfall is given the following simple advice: 'Belittle free will to him, and emphasize destiny'". (Mahabharata 12.106.20) (Daniel and Ingalls 1957) has been among the handful of spiritual values that most religions share.
4. ***From the present***: Tourism; *Sustainable Tourism.* Resource scarcity demands leverage and sustainable management. It is therefore necessary to invent new frameworks and highlight novel tools. For this, the contribution of all the achievements of the human intellect is required. To cite them in alphabetical order: The fine arts [cultural heritage]; science [rationalization, emerging technologies]; and spirituality: among the main concepts of spirituality is karma [action]). Each of these would contribute at many levels through time (past, present and future).
5. ***Connecting the dots:*** An innovative product or service is not composed exclusively of novel elements. Innovation also intertwines and interweaves several already known ingredients into a ***novel service product***. This is the case for the two types of leverage in this research: "Innovative Tradition" (1) and "Tourism for the Future" [with its concrete exhibition centers] (2) These two types of leverage are integrated in the two already established networks (UNESCO locations (3) and BRI works (4) The attractive element insightfully inserted into the program is karma (5) In the diagram below elements #1, 2, 3, 4, and 5 are represented with explanatory notes. In part A elements #1, 3, 4. In part B the procedural steps.

Novel Service Product Ingredients and Steps Towards Integration: A UNESCO Location Near a BRI Project/Work

See Fig. 5.

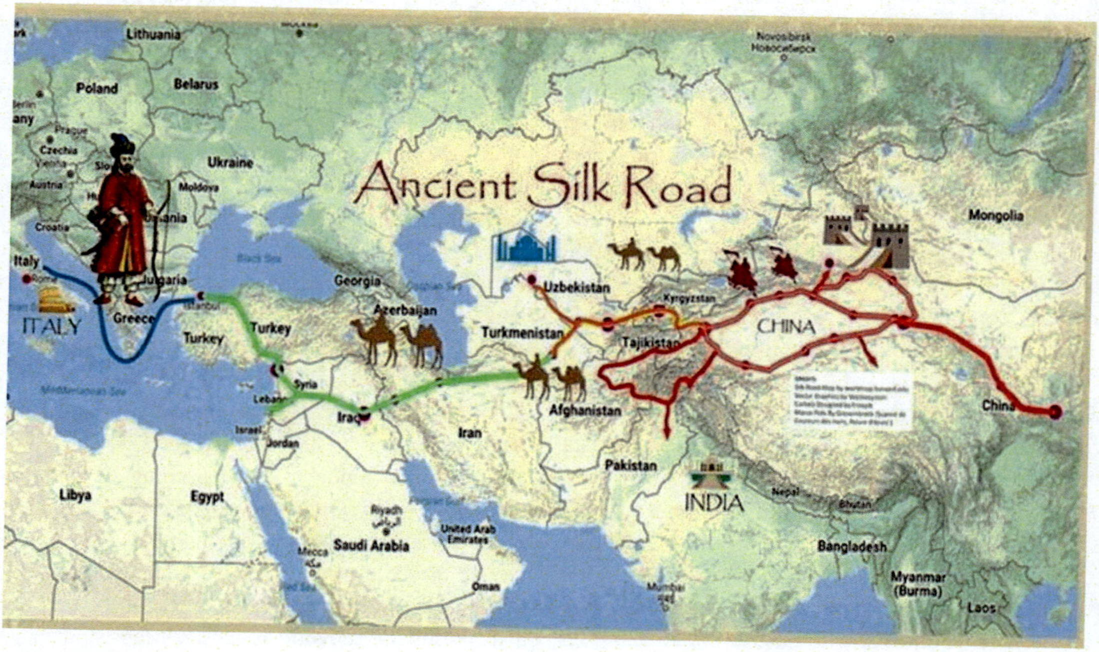

Fig. 3 Ancient Silk Road map

Fig. 4 BRI mega projects in the eastern hemisphere

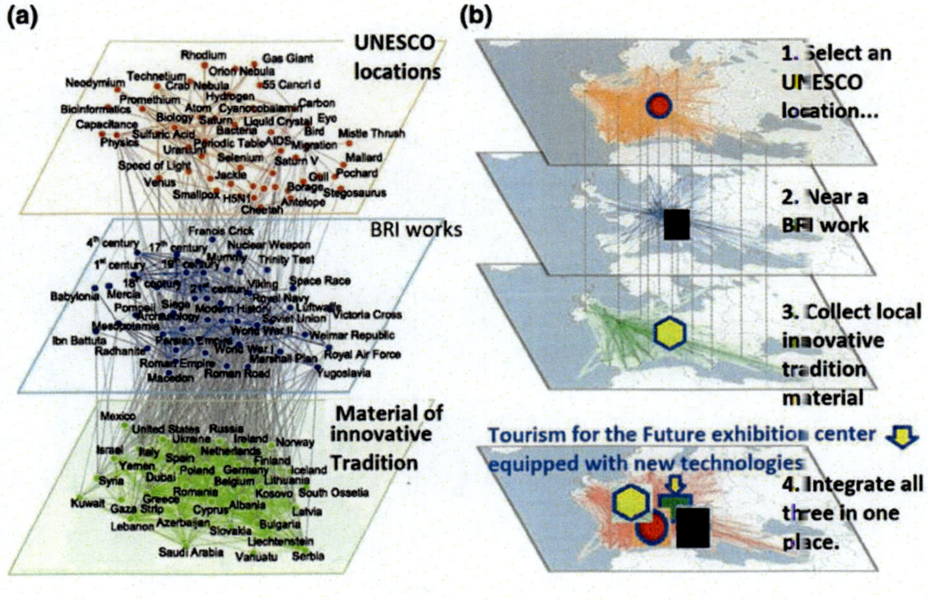

Fig. 5 Gradual intertwining and interweaving step by step

3.3 The Above Findings Provide Two Great Opportunities for Developing Tools and Promoting Platforms

(A) ***Innovative Tradition (IT) is*** the ***first leverage tool***: Evaluated elements derived from intangible cultural heritage would be integrated into implementation of Sustainable Development (SD) and Ecological Civilization's (EC). (B) ***BRI*** (*The New Silk Road*) is the ***platform***, the panorama of the Future. To accelerate civilization, more and more development projects and countries are included in the BRI strategy. ***Tourism of the Future*** is the ***2nd leverage tool***. The future would now be presented to the global visitor in a context that combines past and future: the UNESCO World Cultural and Natural Heritage, in conjunction with the projects under the surveillance of the New Silk Road (Tim 2016a, b, Figs. 6 and 7).

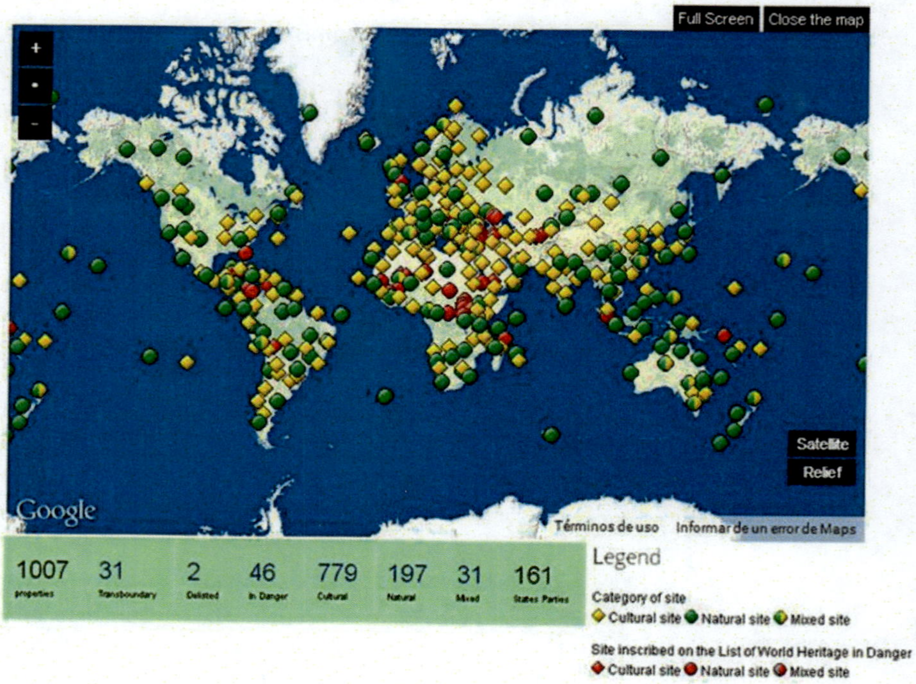

Fig. 6 UNESCO cultural and natural heritage global network

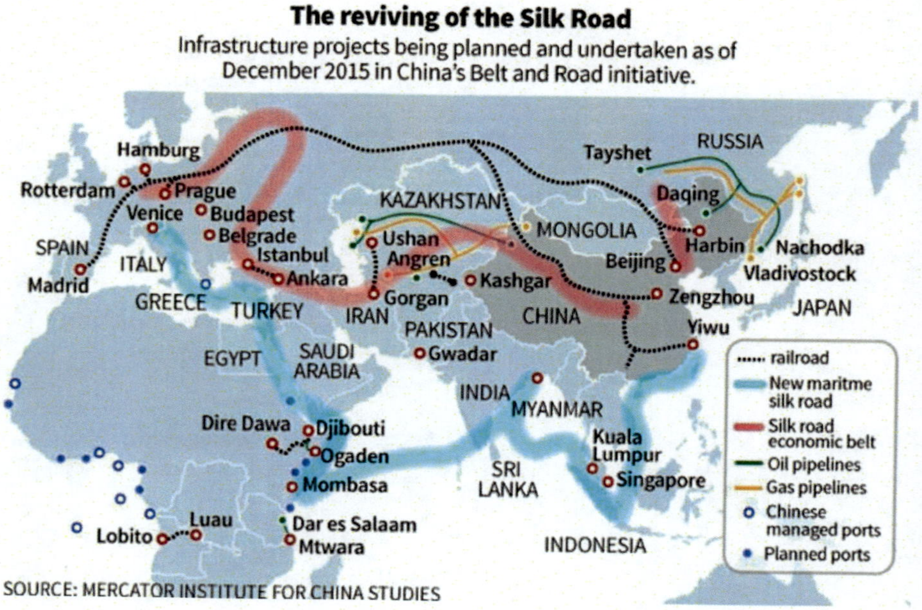

Fig. 7 BRI countries and kinds of works

3.4 Playing With Time (In the Appropriate Style)—Example of a Disruptive Way of Thinking About Research

Crises, Time and Solutions. What is the relation between time, ***biodiversity loss, climate change, financial crisis and COVID-19*** (in short), the negative mega-trends mentioned above and their spillover effects?

1. **Past = Innovative Tradition** Tangible knowledge and experience from the near past (one or two generations back) to thousands of years back. The more its data are practical and usable, the more innovative tradition works as a ***motive force***. (1) *Biodiversity*: The gradual enrichment of species, or sometimes imposed restriction, has become the incubator for a rather notable species… humans. In my personal opinion, China is the most

characteristic paradigm of modern civilization. After a century of humiliation China comes on the stage again as the Modern civilization paradigm with twenty-first century to be charactetized Chinese Century. If asked to select one Chinese proverb to illustrate, in generalized terms, the uniqueness of Chinese culture and how Chinese people behave in various situations I would select this one: ***"The old man lost his horse, but who knows if this is a bad luck"*** ("塞翁失马, 焉知非福": "Sai weng shi ma, yan zhi fei fu"). "Once upon a time there was an old man who lived with his son in a small remote mountainous village. The family was poor but they had a lovely horse and a peaceful life. One day the horse unfortunately strayed off. The neighbors all came and said "Sorry, old man; what bad luck you've got". But the old man said: "What makes you believe this is a bad luck?" Several months later, the disappeared horse all of a sudden came home together with a number of beautiful female wild horses. Then, the neighbors all came to congratulate on the old man's good luck. But the old man said: "What makes you believe this is a good luck?" Now the family had got a number of horses, the son took to riding one of the female wild horses and as a result broke his leg; he became handicapped because of this accident. Once again all the neighbors came to show their sympathy with the family's bad luck. But once again the old man was asking why his neighbors would believe this is a bad luck. A year later, a war broke out; the emperor's army came to recruit all the young males in the village to go to war. Because the son was lame he was except from going to war while all other young males left the village and never returned. The son who girls earlier didn't want to go out with turned out to be the only young man available in the village. He was now able to go out, got married, and become father to many lovely children. This small society survived and the family s name carried on from generation to generation. (Tony 2012) comprise a really distinguished branch of the tree of life,—the very branch we treat with tremendous zeal nowadays… (2) *Climate Change*: This has been a natural cyclical event, a mechanism for the maintenance of planetary equilibrium and dating back 4.5 billion years. In the current Anthropocene Era humans have behaved like sorcerer's apprentices, dramatically reversing the situation in just 200 years. (3) *Financial crisis*: In the olden days the crisis was local and usually pertained to food provisioning. The main reason for it was the potential greed of one or two hunters who killed more animals than they could consume. Farmers, wisely trained by nature, have never been greedy… (4) *COVID-19*: There is no relevant case from antiquity. The corona virus was originally detected in 1965. Although it sounds unbelievable since 2000 vaccine patents have been registered.

2. **Present = Objective Reality**. The more it is based on reliable data, the more real and tangible it becomes - in the absence of reliable data we are guided by daydreams or even worse seek to profit from selling dreams. The relation with time: apparently ephemeral. Transitioning from the past to the future we traverse the anti-chamber of the present—the present, therefore, does not exist at all theoretically but acquires substance in the context of given time-frames e.g. life stages (puberty, adulthood, old age). The more reliable the data, the more the present becomes a dynamic ***springboard for the future***. (1) *Biodiversity*: A unique solution is to treat as a personal affair the loss of any species, wherever it lives, just as we would mourn for the loss of a close and dear relative. It is an issue of conscience and awareness. (2) *Climate Change*: What is happening here and now is happening, more or less, all over the planet; it is disastrous for human civilization but a side issue for the planet, which has more than once faced similar situations (the crises of bacteria, and/or of the dinosaurs) effectively. There is only one remaining solution: to tackle it collectively today! (3) *Financial crisis*: Western world's multilevel crisis is being transmitted around the globe in the form of a pandemic—not only metaphorically as in the past, but today literally. The only solution is to do away with the over-production, over-consumption model. (4) *COVID-19*: The disease appears to be an accidental side effect; but is rather an opportunity for those GREEDY FOR WEALTH and POWER to restrict civil rights and democratic freedoms. The only solution is drug awareness and informed consent.
3. **Future = Virtual Reality**. Our projected wishes are based on predictions (virtual time of unlimited duration from… tomorrow to whenever we want). The more reliable the data these predictions are based on the more this virtual reality shifts to become a tangible **force of attraction**. (1) *Biodiversity*: The only solution: Let nature do her job with her customary efficiency as she has done for 4.5 billion years and will do for another 4.5 billion years—with humans not necessarily included. The planet is unique, unprecedented, with dozens of specific and probably unrepeatable features. Stop the loss in biodiversity, a persistent crime against the universe! (2) *Climate change*: The only solution is for it to be tackled collectively, since… yesterday. This is the only way for human civilization to survive. Otherwise the human species will be registered in nature's archives simply as homo (from humus, soil) with the two terms (sapiens sapiens) erased forever… (3) *Financial Crisis*: The vocabulary of the future will indicate that industrial civilization collapsed due to over-production and over-consumption against a background of overpopulation, before ecological civilization took over.

(4) *COVID-19*: The disease will be eliminated through a combination of help from nature (natural immunity), human solidarity, reliable medication and treatment, enhancement of the immune system, local community support, strengthening of the public health system, working and living in a safe environment—and not, exclusively through chemical based vaccines with unknown or suspicious side effects, supplemented by global governmental, administrative and police measures.

4 Innovative Tradition: Transforming Cultural Heritage into a Handy Tool for Left-Handed People

The real wealth of a planet is in its landscape, how we take part in that basic source of civilization—agriculture—Dune (Frank 1966)

1. We usually 'worship' '*tradition*', '*heritage*', '*cultural heritage*', etc., as values of the *past*—although most of them still live. They are often praised but rarely utilized in practice—and they are almost never taken to be an exponential asset for the *future,* which is what they are. Since the dawn of human civilization, tradition has integrated innovative features into everyday life. The reason for this is simple. People could not go 'back to the basics' because that era was the era of the basics. There was no scarcity of resources because most resources had not yet been discovered or invented.
2. Today, in an era of scarcity it is urgently necessary for "'*ol' but good tricks*" to be rediscovered, along with *methods, tactics', techniques, procedures, practices, formulas, processes, courses of action, recipes,* etc., and all relevant nuances in between because each different *word* has a reason to exist (Dictionary.com 2010); methods, tools, ideas, with smart management and economy of resources. The concept of 'waste' does not exist in the traditional ancient lifestyle. The same applies for the principle of "reduction, recycling and reuse". All this content becomes comprehensible in the light of innovative tradition, a framework for advancing an understanding of climate change, adaptation and mitigation in the context of culture and traditional knowledge. Such a rediscovery must focus rigorously on exposing what would be useful for implementation of sustainable development. Among the most successful global efforts in the field is UNESCO's "*Intangible cultural heritage*" initiative.
3. Cultural heritage does not end at monuments and collections of objects. It also includes traditions or living expressions inherited from our ancestors and passed on to our descendants. These may be oral traditions, performing arts, social practices, rituals, festivals, knowledge and practices pertaining to nature and the universe or the knowledge and skills to produce traditional crafts, all of which are collectively designated "Intangible cultural heritage" (ICH). ICH, or what some call "living heritage" also includes the stories we tell, the family events we celebrate, our community gatherings, the languages we speak, the songs we sing, knowledge of our natural spaces, our healing traditions, the foods we eat, our holidays, beliefs and cultural practices.
4. "*Innovative Tradition*" focuses on the systematic rediscovery, evaluation and use of methods, tools and ideas. These 'tools' are taken out of the showcases of museums and are available to solve modern problems. The traditional knowledge can complement broader-scale scientific research with local precision and consciousness of nuances. For decades, and with strict criteria, UNESCO has been collecting and certifying cultural and natural world heritage (tangible and intangible). It is content which can be used for two purposes.
   Through hosting exhibition presentations of works in the region of 1300 + monuments of the past, these monuments can be connected with, and can coexist with, the achievements of the New Silk Road.
5. *Intangible heritage* can be utilized as a rich pool of innovative traditional artefacts. There are many examples of compilations. Parampara, India's culture of climate friendly sustainable practices is one such (Centre for Environment Education 2015). It could be a model for all regions—some countries are too big for uniform methods to be applied on all of their territory. Among sustainability's most important features is therefore the *regional footprint*.
   Examples: (1) *Parampara*; (2) *Ancient and abandoned* (Lighting Top Channel 2020) *Technologies*; (3) *Ancient Origins*.
6. **Innovative tradition paradigm: Parampara, a continuation of tradition without interruption**
   The fact that we eat rice and daal together could be attributed to it being such an old practice that it has become a habit for most people involved. Scientific studies have indicated that consuming these two together increases a person's protein intake (Centre for Environment Education 2015)
   India's vast cultural diversity is in itself an essential component of its intangible cultural heritage. The intangible cultural heritage of India includes forms of popular and traditional expression such as: oral literature; music; dance; games; mythology;

rituals; costumes; craftwork know-how; cultural spaces, where popular and traditional cultural activities take place (sites for story-telling, rituals, marketplaces, festivals etc.). Parampara's contents:

Solar and wind energy: drying in the sun for long-term food preservation passive solar building design; drying clothes in the open; sun dried spices; traditional air conditioning methods; windmills and wind pumps; sailing; etc.

Agriculture: aqua-culture; farming below sea level; rice fish culture; mixed farming to counter rain dependency; rain water harvesting and soil management; a non-mechanized system for irrigation in hilly terrains; seasonal vegetable cultivation for food security; traditional weather management; homestead development; chemical free preservation techniques; plant products as alternative to chemical pesticides; eradication of insect pests using ash; use of bio-pesticides to store grains; sustainable farming through traditional seed preservation; etc.

Biodiversity: Creating carbon sinks with sacred groves; the Agaria technique; bio-indicators for weather prediction; local products from local material; biodegradable leaf plates; etc.

Food: naturally fermented food for healthy living; an energy-saving milk preservation technique; winter food for warming; preservation by salting; a process of non-mechanized butter making; the traditional knowledge behind eating food with hands; etc.

Health: staying healthy with Neem in a changing climatic world; ginger as a preventive for winter ailments; etc.

Shelter: earthquake-resistant construction technology; designs for disaster risk reduction; etc.

Textile, clothing and lifestyle products: the craft of recycling waste paper; an elegant alternative to plastic mats; production without using fossil fuels: weaving looms; doing more with less; etc.

Water: the earthen pot as a water cooler; flood water harvesting for irrigation; groundwater recharge through water storage; etc. Additional innovative tradition systematic research would be enriched by equivalent climate-friendly sustainable practices material on Labor; *Employment; Business; Education; Sports; Transport and Storage; Economy; Trade; Small Industry, Products Processing; Mining; Physical Planning—Public Works; Shipping; Fisheries; Cattle; Forestry; Science and Technology*—for this latter, interesting attempts to collect material below.

7. **Ancient and abandoned technologies** An attempt to compile data on ancient and forgotten technology: "Not everything our ancestors knew was passed on to their children. All over the world there are examples of buildings or objects that appear to have been made using technology impossibly complicated for the people who are said to have created them. In some cases it is a type of technology that we would struggle to create today even with our computers and all of our modern equipment and knowledge; from tiny devices that would fit into the palm of your hand to enormous buildings".
8. **Ancient Origins** seeks to uncover what we believe is one of the most important pieces of knowledge we can acquire as human beings: the knowledge of our origins (…); everything concerning human origins, myths and legends, archaeological news, latest discoveries, ancient places, mysterious artefacts and unexplained phenomena (Fig. 8).

Fig. 8 Parampara edition

5 Sustainable Development and Ecological Civilization Paradigm: The New Silk Road/BRI

1. *Ecological Civilization* seems to be the next phase of civilization after the era of industrial civilization. Sustainable development as a model of development based on sustainable principles and insightful resource management to cover all human activities has been proved a very complicated concept, providing so far a number of effective but fragmentary examples; this is the main reason it still faces various difficulties of large-scale implementation. In contrast, "*Ecological Civilization construction*" has seen a number of successful achievements in China. This led the Chinese government (https://en.wikipedia.org/wiki/Government_of_China) in 2013 to design a model (The New Silk Road or One Belt, One Road) for international implementation, prioritizing the development of infrastructure based on humanity's successful past experience (*The Silk Road (*https://en.wikipedia.org/wiki/Silk_Road*))*.
2. "The Silk Road" (https://en.wikipedia.org/wiki/Silk_Road) (Chinese (https://en.wikipedia.org/wiki/Simplified_Chinese_characters): 丝绸之路) was and is a network of trade routes (https://en.wikipedia.org/wiki/Trade_route) connecting the East (https://en.wikipedia.org/wiki/Eastern_world) and West (https://en.wikipedia.org/wiki/Western_culture), from the second century BCE to the eighteenth century CE. It was central to the economic, cultural, political, and religious interactions between these regions. "The Silk Road" usually refers to certain land routes, but it may also refer to sea routes that connect East Asia (https://en.wikipedia.org/wiki/East_Asia) and Southeast Asia (https://en.wikipedia.org/wiki/Southeast_Asia) with South Asia (https://en.wikipedia.org/wiki/South_Asia), Persia (https://en.wikipedia.org/wiki/Persia), the Arabian Peninsula (https://en.wikipedia.org/wiki/Arabian_Peninsula), the Near East (https://en.wikipedia.org/wiki/Near_East), East Africa (https://en.wikipedia.org/wiki/East_Africa) and Southern Europe (https://en.wikipedia.org/wiki/Southern_Europe) (Fig. 9).
3. The New Silk Road (or One Belt, One Road, or the Belt and Road Initiative (https://en.wikipedia.org/wiki/Belt_and_Road_Initiative) BRI or BandR), known in Chinese and formerly in English as **One Belt One Road** (Chinese (https://en.wikipedia.org/wiki/Chinese_language): 一带一) or **OBOR** for short, is a global infrastructure development (https://en.wikipedia.org/wiki/Economic_development) strategy adopted by the Chinese government (https://en.wikipedia.org/wiki/Government_of_China) in 2013 for investment in nearly 70 countries and international organizations. This strategy will in its broad sense re-engineer (**Business process re-engineering (BPR**) is a business management strategy (https://en.wikipedia.org/wiki/Strategic_management), originally pioneered in the early 1990s, focusing on the analysis and design of workflows (https://en.wikipedia.org/wiki/Workflow) and business processes (https://en.wikipedia.org/wiki/Business_process) within an organization. BPR aimed to help organizations (https://en.wikipedia.org/wiki/Organization) fundamentally rethink how they do their work in order to improve customer service (https://en.wikipedia.org/wiki/Customer_service), cut operational costs (https://en.wikipedia.org/wiki/Operational_costs), and become world-class competitors (https://en.wikipedia.org/wiki/Competitor) global development in the direction of sustainability. ("Maintaining stability in China's neighborhood is the key objective of peripheral diplomacy. We must encourage and participate in the process of regional economic integration, speed up the process of building up infrastructure and connectivity. We must build the Silk Road Economic Belt and 21st Century Maritime Silk Road, creating a new regional economic order". [Xi Jinping's Important Speech at the Peripheral Diplomacy Work Conference] (Xinhua News Agency 2013). "The Belt and Road Initiative aims to promote policy coordination, facilities connectivity, unimpeded trade, financial integration and people-to-people exchange among the countries along the Belt and Road (hereinafter referred to as "countries involved"), promote orderly and free flows of economic factors, efficient allocation of resources and deep integration of markets, and jointly create an open, inclusive, and balanced regional economic cooperation framework that benefits all. Financial Integration is an important underpinning for implementing the Belt and Road Initiative" (Tim 2016a, b).

6 Tourism for the Future: Leverage Solution #2; Example: Sustainable Development Holistic Accelerator

Tourism, in general, is travel for pleasure or business. With one billion tourists a year (WHS 2016), tourism has become a global carrier. Tourism also bears considerable responsibility for the climate crisis. *Sustainable tourism* (St) is a concept that covers the overall tourism experience, including concern for economic, social and environmental issues as well as attention to improving tourists' experiences and addressing the needs of host communities.

Silk routes: valuable past concepts. BRI: present and future framework. Tourism and, more specifically,

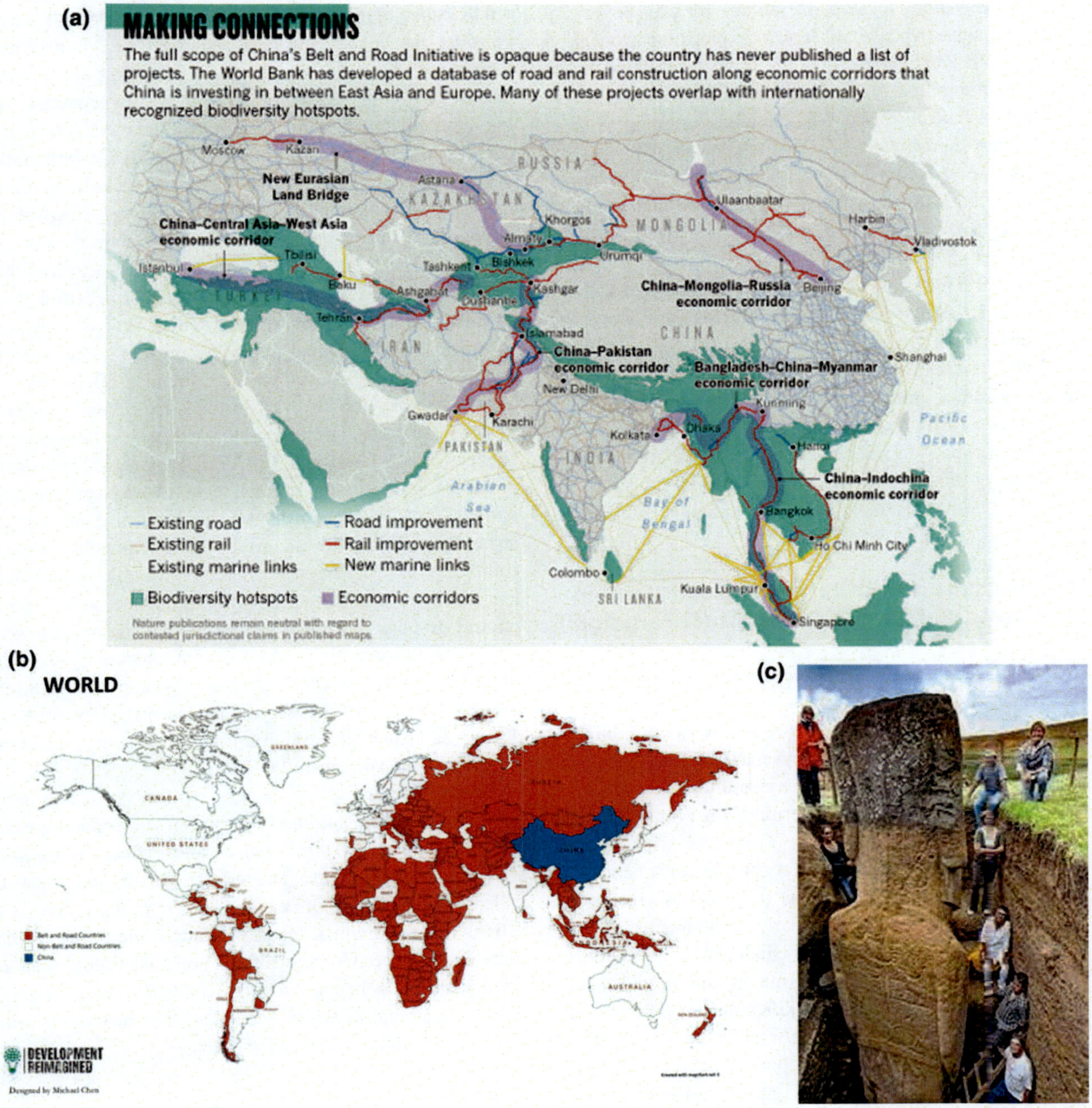

Fig. 9 BRI works near biodiversity hotspots. In BRI works high quality science has been integrated (Masoud 2019) Source: https://www.nature.com/immersive/d41586-019-01124-7/index.html

sustainable tourism, a rapidly growing trend, is the ideal highway for the promotion and publicizing of cultural heritage, innovative tradition and BRI large-scale construction projects such as the sustainable development and ecological civilization panorama. Sustainable tourism could be a global vehicle for Sustainable Development. Numerous projects could fall into this category. The pre-eminent paradigm is the *Sustainable Development Holistic Accelerator* (SDHA). The *SDHA integrates the two frameworks into a third one;* the *BRI* (global strategy for SD and Ecological Civilization (EC) in 140 countries with a rich cultural heritage and requirements for the FUTURE) and UNESCO World Cultural and Natural Heritage. The new framework is *Tourism for the Future* (TFF). It would be developed by selecting UNESCO monuments neighboring BRI constructions and proposing *TANGIBLE MODELS* of SD and EC next to or near ***UNESCO Visitor Centers*** for hundreds of millions (https://www.worldheritagesite.org/connection/One+million

+visitors+or+more) (Geoparks attract globally over 250 million visitors who would like to explore the wonders of nature (UNEP and UNWTO 2005) of visitors. These models would promote Innovative Tradition methods integrated into modern sustainable works in the region.

A short description of the Sustainable Development Holistic Accelerator (SDHA): A project within a project (BRI) within a project (Human Civilization advancement): There are two parts to the SDHA:

1. A generic model of sustainable development presented in various forms (from virtual reality to more straightforward guises) to be hosted next to UNESCO World Heritage sites.
2. A fast track training program (500 + item syllabus) to make it easier for millions of people to enhance their formal education and re-orient themselves professionally securing jobs related to sustainable development.

7 Intercultural and Inter-Religious Meeting Points; Karma as Core Spiritual Paradigm

Intercultural Meeting Point, Innovative Tradition Has Been Described Above

Karma (ACTION) is an inter-religious meeting point, a cognitive/ethical element. "Karma meaning deed or action; in addition, it also has philosophical and technical meaning, denoting a person's deeds as determining his future lot". The philosophy of karma is closely associated with the idea of rebirth (https://en.wikipedia.org/wiki/Reincarnation) in many schools of Indian religions (https://en.wikipedia.org/wiki/Indian_religions) (particularly Hinduism (https://en.wikipedia.org/wiki/Karma_in_Hinduism), Buddhism (https://en.wikipedia.org/wiki/Karma_in_Buddhism), Jainism (https://en.wikipedia.org/wiki/Karma_in_Jainism) and Sikhism (https://en.wikipedia.org/wiki/Sikhism), as well as Taoism. This concept has also been adopted in Western popular culture (https://en.wikipedia.org/wiki/Popular_culture), in which the events which happen after a person's actions may be considered natural consequences. Christianity: Mary Jo Meadow suggests karma is akin to "Christian notions of sin (https://en.wikipedia.org/wiki/Christian_views_on_sin) and its effects". She states that the Christian teaching on a Last Judgment (https://en.wikipedia.org/wiki/Last_Judgment) according to one's charity is a teaching on karma. Christianity also teaches morals such as one reaps what one sows (https://en.wiktionary.org/wiki/reap_what_one_sows) (Galatians (https://en.wikipedia.org/wiki/Epistle_to_the_Galatians) 6:7) and live by the sword, die by the sword (https://en.wikipedia.org/wiki/Live_by_the_sword,_die_by_the_sword) (Matthew (https://en.wikipedia.org/wiki/Gospel_of_Matthew) 26:52). Most scholars, however, consider the concept of Last Judgment as different from karma, with karma as an ongoing process that occurs every day in one's life, while Last Judgment, by contrast, is a one-time review at the end of life. Judaism: There is a concept in Judaism called in Hebrew midah k'neged midah, which literally translates to 'value against value', but carries the same connotation as the English phrase 'measure for measure'. The concept is used not so much in matters of law, but rather, in matters of ethics, i.e. how one's actions affects the world will eventually come back to that person in ways one might not necessarily expect. David Wolpe (https://en.wikipedia.org/wiki/David_Wolpe) compared midah k'neged midah to karma. (Encyclopædia Britannica 2010). IT/**TFF** and karma develop St; visitors become cultural pilgrims, SD/EC instigators, and active citizens, and are rebaptized as netizens /earth- lings on a PEACEFUL planet. This research promotes karma in its original sense (deed, work, or conscious action). It is considered the key spiritual point of departure for the forward drive. Current practice is most commonly the very opposite. The great problems and challenges of the time are dealt with through procrastination, through grand words rather than tangible work, or spasmodic and fragmentary motions which do not, obviously, bring the expected results. It is very interesting that karma is a concept common to most of the world's religions (Hinduism; Buddhism; Jainism; Sikhism; Taoism; Christianity; Judaism. (Encyclopædia Britannica 2010).

Karma means action, work, or the deed. The term also denotes the spiritual (https://en.wikipedia.org/wiki/Spirituality) principle of cause and effect (https://en.wikipedia.org/wiki/Cause_and_effect), often descriptively called the **principle of karma**, wherein the intent and actions of an individual (cause) influence the future of that individual (effect): Good intent and good deeds contribute to good karma and happier rebirths (https://en.wikipedia.org/wiki/Reincarnation), while bad intent and bad deeds contribute to bad karma and bad rebirths.

8 Conclusions

1. ***Opportunities for development are also opportunities for peace***. National differences are diminished through promotion of the elements common to different religions, i.e. karma. Political differences are also minimized through rediscovery of traditional principles common to neighboring countries.
2. Cultural heritage upgraded to ***innovative tradition*** gets sustainable principles and capacities out of museum showcases thereby extracting for present-day implementation material elements of the past.

3. IT would develop a ***multidisciplinary toolkit embracing*** all sectors of human activity; societal priorities; humanistic awareness parameters; six fundamental elements; sustainable indicators; timeless skills, so as to gather, evaluate, classify, and disseminate methods for adaptation to modern needs.
4. ***Karma*** would be a notion to be disseminated in global sustainable tourism programs for the purpose of motivating local populations and visitors to adopt the same perspective of well-being.
5. ***Tourism for the future*** would enhance the generic notion of sustainable tourism, introducing a concrete series of tourism packages for paradigmatic monuments, thus integrating past and future in the present.
6. ***BRI*** would enhance its infrastructural identity with cultural high added value.
7. ***UNESCO World Heritage*** would enhance regional monuments Exhibition/Visitor Centers by technological high added value at the same time solving permanent financial issues of conservation.
8. ***Sustainable Development Holistic Accelerator (SDHA)*** and/or similar models would be used for integration of UNESCO Monuments and BRI projects.
9. **Paradigm usage** (Parampara, Karma, BRI, tourism of the future, *sustainable development holistic accelerator)* would exponentially facilitate the next steps in human civilization. … if applied in an antifragile way; but this is another story/research… Years of research *"gave me insight as to how to shepherd innovation through 'the system", regardless of where I happen to be and how many colleagues are telling me that "it can't be done"*. (Turrin 2019).

Acknowledgements Many more bits of information than meet the mind have formed this research; some of them recognizable and citable but the majority lost in the chaotic ways the human mind works.The whole research is inspired by

1. "A book entitled a short History of the World by H.B. Wells (1933)…" 1933, Greek edition by Neos Kosmos journal edition. The book firmly focuses on the victory of the hunting nomads over the agriculturists/migratory cattle farmers/settlers.

2. Nassim Nicholas Taleb "Antifragile: Things That Gain from Disorder", 2012

Continuous back and forth from past to present and future, from science to fine arts and spirituality, from figures to songs (still haunted by "The Green hills of Earth" https://www.youtube.com/watch?v=iRAG16mrkxA), from global maps to tiny relics of culture hidden in the sand, eventually crafted this research. All this information has been offered by people, books, and rags of ideas, fractions of time, places and opinions. To all these I owe significant gratitude—at least.Last but not least, I owe eternal and multilevel acknowledgements to my muse for life, Nausicaa obviously my being Odysseus.

References

Andreopoulos, A.: Life-surfer's 365 adventures in 365 picture-thoughts. Strategy Multimedia—Verettas (1996)

Anonymous: The fourth industrial revolution and BRI practices: RDI annual report 2018. In: Amazon (2019)

Apostolopoulou, E.: China's Belt and Road Initiative Sustainability in the New Silk Road. The University of Cambridge Institute for Sustainability Leadership (2022)

Brian, E.: Oblique Strategies—the Ultimate Music Production Tool. Productionadvice.co.uk (2009)

Centre for Environment Education (CEE): Parampara, India's culture of climate friendly sustainable practices. In: Ministry of Environment, Forest and Climate Change, Government of India (2015)

Daniel, H., Ingalls, H.: Dharma and Moksa. Philosophy East and West (1957)

De Bono, E.: Lateral Thinking: Creativity Step by Step. Harper and Row (1970)

Dee, H.: The Art of Chaordic Leadership. Leader to Leader (2000)

Dictionary.com: Method definition. In: Dictionary.com (2010)

Encyclopædia Britannica: Karma. In: Encyclopædia Britannica (2010)

Frank, H.: Dune. New English Library (1966)

Giunchiglia, F., Erculiani, L., Passerini, A.: Towards Visual Semantics. Cornell University (2021)

https://www.ancient-origins.net/; https://www.minds.com/Ancient-Origins https://www.facebook.com/ancientoriginsweb/?ti=as

Huang, W., Li, J., Edwards, P.P.: Mesoscience: exploring the common principle at mesoscales. Natl. Sci. Rev. **5**(3), 321–326 (2018). https://doi.org/10.1093/nsr/nwx083

Kaefer, F.: Karma Tshering on Sustainable Tourism in Bhutan Sustainability Leadership in Tourism. Springer (2022)

Lighting Top Channel: 12 most ancient and abandoned technologies. In: Lighting Top Channel (2020)

Nicholas, T.: Antifragile: Things That Gain From Disorder (Incerto). Amazon (2014)

Pierre (Pete) Routhier: Immersive Technologies: Virtual Reality, Augmented Reality, Mixed Reality and the Immersive Ecosystem. Amazon (2022)

Richard, W.: Descartes, rules for the direction of the mind. In: Britannica Encyclopedia (2002). https://www.britannica.com/biography/Rene-Descartes#ref478297

The Diplomat. Retrieved from https://www.stuff.co.nz/business/world/92542916/new-silk-road-the-146-trillion-push-to-reshape-the-world-in-chinas-image

Tim, W.: Heritage diplomacy along the One Belt One Road. In: International Institute of Asian Studies Newsletter, vol. 74, pp. 8–10 (2016a). https://whc.unesco.org/en/list/

Tim, W.: One Belt, One Road, One Heritage: Cultural Diplomacy and the Silk Road (2016b)

Tony, F.: Understanding Chinese Culture and Communication. Stockholm University (2012)

Turrin, R.:Innovation Lab Excellence Digital Transformation from within. Gold River (2019)

UNEP and UNWTO: Making tourism more sustainable—a guide for policy makers. In: UNEP (2005)

UNESCO: The role of visitor centres in UNESCO designated sites: report of the first regional workshop for Europe. In: UNESCO (2019)

UNFCCC: Social Matrix Revealed. In: UNFCCC (2017)

UNFCCC: UNFCCC capacity building portal. In: UNFCCC (2019)

UNSDGS: EcoVirtualLands. In: UNSDGS Platform (2016). https://sdgs.un.org/partnerships/ecovirtualands-global-investment-initiative-social-cause

WHS: One million visitors or more. In: WHS (2016)

Xinhua News Agency: Speech at symposium on neighboring diplomatic work. In: Xinhua News Agency (2013)

Geoparks and Sustainable Tourism Development. The Role of Internet and Social Media

Theodora Lanara, Valia Xanthopoulou-Tsitsoni, Stella Kostopoulou, and Thekla K. Tsitsoni

Abstract

A Geopark is defined as a geologically significant area, with clearly defined boundaries. The most essential parts of planet history, culture, and biodiversity include geological and geomorphological legacy. A Geopark's establishment strives to capture what constitutes a world-class outstanding geological heritage, while also exploring, developing and highlighting the relationship between geological heritage and all other issues and aspects of a region's natural and cultural heritage. The primary concept of establishing Geoparks is to combine conservation with sustainable development and active participation of local communities in the region's economic development. Geopark, through geotourism activities, are seen to be the most suitable for sustainable tourism development of the region, aiming at stimulating the local economy and community. Highlighting all those elements that make up the Geopark, such as geology, natural ecosystems, cultural elements—archaeology, promotion of local identity, and sustainable development, is an important development keystone that has a clear impact on the preservation of geological heritage and cultural values. The use of new technology applications such as social media, smartphone applications, augmented reality (AR), virtual tours can serve as additional tools for promoting the multifaceted role of a Geopark, highlighting the geopark's multifunctional role, and providing information (images, videos, texts) about points of interest in specific geographical sites where the user/visitor is located and aims with the camera of the mobile phone. In this paper the use of digital technologies, such as social media, on marketing activities of Geoparks for sustainable tourism development is explored. The study investigates the Greek Geoparks policies in relation to the use of digital technologies and especially social media.

Keywords

Geological territories • Local communities • Digital applications

1 Introduction

People across the world are becoming more interested in protecting nature and the landscapes in which they live, which has led to increased support for environmental efforts in national, regional, and worldwide organizations and networks. The globalised efforts for protecting important ecosystems and areas has led to the dynamic development of the UNESCO Global Geoparks (UGGs) initiatives. The definition of Geoparks given by UNESCO (2021) refers to "*single, unified geographical areas where sites and landscapes of international geological significance are managed with a holistic concept of protection, education, and sustainable development*". A Geopark through its geological heritage, in combination with all the other aspects of the area's natural and cultural heritage, could raise awareness of the local communities, mitigate the effects of climate change, environmental pollution, deforestation, and contribute to the reduction of the impacts of natural disasters. A Geopark is not just a geologically significant area, but also a rural area that is globally recognized for its geological legacy.

T. Lanara (✉) · T. K. Tsitsoni
School of Forestry and Natural Environment, Aristotle University of Thessaloniki, Thessaloniki, Greece
e-mail: theolana@for.auth.gr

T. K. Tsitsoni
e-mail: tsitsoni@for.auth.gr

V. Xanthopoulou-Tsitsoni · S. Kostopoulou
School of Economics, Aristotle University of Thessaloniki, Thessaloniki, Greece
e-mail: xanthopv@econ.auth.gr

S. Kostopoulou
e-mail: kostos@econ.auth.gr

S. Kostopoulou et al. (eds.), *Cities' Vocabularies and the Sustainable Development of the Silkroads*, Advances in Science, Technology & Innovation, https://doi.org/10.1007/978-3-031-31027-0_20

Geological territories are widely acknowledged for their geological characteristics but also their archaeological sites, historical events, culture and natural environment (McKeever et al. 2010). The basic idea of Geoparks' establishment is the combination of nature conservation and sustainable development along with the collaboration of local communities and stakeholders, who are involved in the economic development of the region. According to Zouros and McKeever (2008:2), a Geopark "*must have clearly defined boundaries and sufficient surface area for true territorial economic development, primarily through tourism*".

Worldwide, the organizational and institutional development of Geoparks began in 2001. In 2004, 17 European and 8 Chinese Geoparks formed the Global Geoparks Network (GGN) (http://globalgeoparksnetwork.org/?page_id=5), under the auspices of UNESCO. In 2015, the 195 Member States of UNESCO ratified the UNESCO Global Geoparks as a new institution. This holistic approach expressed the governmental admission of the importance of managing outstanding geological localities. At present 169 UNESCO Global Geoparks have been legislated in 44 countries (UNESCO: Global Geoparks 2021).

This paper explores the use of digital technologies, such as social media networks, on the marketing activities of Geoparks for sustainable tourism development. The study investigates the Greek Geoparks policies about the use of digital technologies and especially social media. The methodological approach examines the digital profile of Geoparks by the use of online content analysis, which is based on structural characteristics of digital technologies and social media, with the aim to promote a national geotourism brand name. The following is a breakdown of the paper's structure. The definition of Geoparks and their active participation as sustainability territories are discussed in Sect. 2. The contribution of Geoparks to the promotion of sustainable tourism development is described in Sect. 3. Section 4 discusses the use of digital communication, such as social media as a powerful tool for promoting Geoparks' diverse role. In Sect. 5 the methodology used to examine the digital profile of Greek geoparks is briefly outlined. In Sect. 6 the research findings are analyzed and the role of social media in the geotourism sustainable development in Greece is documented. Conclusion and suggestions are included in Sect. 7.

2 Geoparks: Sustainability Territories

Geoparks are made up of various geosites, that are defined by their territory (EGN 2021). Geosites are natural formations with distinct characteristics such as groups of rocks and minerals, geoformation of fossils, ground formation or geological structure as an outcome of the creation or evolution of the Earth's layer. Geosites provide information about the geological history of the territory they are taking part in, allowing geologists to make forecasts for the coming years based on scientific analyses (Çiftçi and Güngör 2016; Özgeriş and Karahan 2021).

The European Geoparks Network was founded in June 2000 by four European countries, France, Germany, Spain, and Greece (Zouros and McKeever 2008). The objective of this network was to foster cooperation across regions with similar natural and socioeconomic characteristics, for the protection of the geological heritage while also ensuring the success of their areas through sustainable development. (Zouros 2004, 2013). At present, the European Geopark Network includes 88 Geoparks from 26 Countries around the European Union (EGN n.d.) (Fig. 1). According to the European Geoparks Network (2021) charter "*A European Geopark is a territory which combines the protection and promotion of geological heritage with sustainable local development, supported by a European program. It must have clearly defined boundaries and sufficient surface area for true territorial economic development*".

A Geopark must innovate novel methods and tools for the sustainable conservation of geological heritage, as well as conduct scientific research in Earth Sciences (Zouros 2014). Education and training of the public and local communities in geological sciences, natural environment and cultural heritage, is the keystone for promoting the knowledge and value of geological heritage, biodiversity, and the geodiversity of the region (Zouros and McKeever 2008).

Furthermore, a Geopark must have an active role in the region's socio-economic development by boosting the promotion of the geological heritage and the development of geotourism as a new concept (Zouros 2004). Geoparks have to cooperate with the people who live within them for the promotion and sustainable development of responsible tourism in rural areas by developing innovative local development strategies. Geotourim is an alternative mode of tourism, that provides new experiences to visitors to discover rural areas while obtaining valuable knowledge of the Earth's history through geosites and natural environment (Martínez-Graña et al. 2017).

3 Geoparks and Sustainable Tourism Development

According to the report of the World Commission on Environment and Development (1987), the definition of sustainable development is "*a process to meet the needs of the present without compromising the ability of future generations to meet their own needs*". United Nations World Tourism Organization (UNWTO) defines sustainable tourism as "*tourism that takes full account of its current and future economic, social, and environmental impacts,*

Fig. 1 European geoparks network (EGN n.d.)

addressing the needs of visitors, the industry, the environment and host communities". All stakeholders involved must be included in the development of sustainable tourism, while strong political leadership is necessary to ensure that all parties participate. Sustainable tourism development is a continuous process and it is required to criticize the impacts constantly, introducing the appropriate preventive measures whenever necessary (UNEP and WTO 2005).

The main characteristics of sustainable tourism emphasize three primary dimensions, economic, environmental, and socio-cultural (Farsani 2012). The economic dimension refers to citizens' needs and implies that the economy of the region supports employment and quality of life. The environmental dimension characterizes the need to manage the environment within ecological limits. The social-cultural dimension is related to the individuals' needs for a dignified and healthy life (Farsani 2012). Sustainable tourism entails making the appropriate use of resources, such as biodiversity, by minimizing the environmental, cultural, and social negative impacts, and maximizing the benefits of natural and cultural heritage preservation, as well as economic benefits for local communities. The principles of sustainable tourism development are in line with the main goals of Geoparks (Özgeriş and Karahan 2021). Geoparks are regarded as ideal susntainable tourism destinations in terms of maintaining resource values, building up socio-economic benefits for inhabitants, and supporting economic development in the territory (Özgeriş and Karahan 2021; McKeever et al. 2010).

Geoparks are established under international terms, although they are managed by local governments and agencies. One of the most important challenges for a Geopark is to encourage actions involving local and indigenous peoples in preserving and celebrating their culture (Global Geoparks Network 2020; Zouros 2004). Local communities are encouraged to take part in activities that promote sustainable tourism development. Geoparks can also help local economies by promoting geotourism development and environmental initiatives related to the geological heritage (Farsani 2012). Geoparks contribute by enhancing and promoting a certain image connected with geological heritage and sustainable tourism development programs. This could have a significant impact on the rural area, altering local communities' daily living conditions and rural environment. The main objective is to help residents to reclaim the values of their territory's heritage and actively participate in the cultural revitalization of their entire land (Global Geoparks Network 2020; McKeever et al. 2010).

The formulation of integrated sustainable tourism development programs must be based on cooperation among local and regional authorities, on systematic promotion, publicity and advertising activities for the region serving as tools, as well as continuous search for new tourism markets and target groups, and finally, cooperation among local authorities, local enterprises and communities throughout the territory (Kostopoulou et al. 2017). Through sustainable geotourism activities, new employment opportunities, particularly for young people, are created by the manufacture of unique handmade goods from local businesses, as well as high-quality training courses incorporating the territory's geological resources (Global Geoparks Network 2020). As a result, digital communication tools widely used to promote tourism destinations, are considered to be most suitable for geotourism development in Geoparks territories, with the goal of boosting the local economy and community.

4 Digital Communication and Promotion of Geoparks: The Role of Social Media

Digital communication is widely acknoweledged to become the future language of tourism, a new form of communicating with guests that allows for two-way dialogue. It refers to the use of the internet based on technology that includes email, video conference, and modern marketing initiativies such as social media to convey informmation to a wider audience and support sustainable tourism development (Fiori and Burlando 2017). The definition of information communication and internet technologies relates to mobile communications that enable people to move around and generate, transfer, and receive various types of data. Digital users in tourism can interact and share information with other tourists/friends using various platforms (blogs, forums, wikis, video and photo sharing to social networks, virtual communities, chat rooms and podcast) (Živković et al. 2014). Tourism and hospitality businesses may now stay in touch with their customers in real time thanks to the introduction of new, less expensive communication tools such as instant messaging apps like WhatsApp, or Messenger (Lo Presti et al. 2022). Geoparks use digital technologies to manage geoheritage, such as geolocation and digital mapping, imaging and modeling, and hybrid virtual environments. Digital tools are adaptable, easily modified and updated, user-friendly and accessible to all, utilized both online and in situ, combine a variety of resources and means and are supported by a variety of devices, rapidly developing with expanding target groups, widely embraced by younger generations (Fasoulas et al. 2022; Cayla 2014). Nowadays, when tourists visit a Geopark, they want to have a prime role in the promotion of the park, by writing reviews and posting on blog spots and social media networks, with digital users' comments serving as an independent and objective critique of the park's attractions, where interaction is the cornerstone (Hysa et al. 2021; Beier and Wagner 2016). Geoparks should improve and modernize their communication methods in order to reach new potential visitors, particularly younger people (Fiori and Burlando 2017). The promotion and marketing of the Geopark needs to be based on a strong communication-mix strategy that includes television and radio, information points, leaflets, signage, interpretation panels, as well as a variety of technology advanced digital platforms such as the internet, in a sustainable way (Martínez-Graña et al. 2017). Various organizations have changed their way of interacting with the media and stakeholders, as well as the way they communicate with citizens and apply marketing strategies. New technology applications (QR codes, videogames, Augmented Reality AR, 360° Virtual Tours, WebGIS applications, digital maps) are important tools for the promotion of Geoparks and the interaction between visitors and geological features through scientific, educational, and environmental activities (Martínez-Graña et al. 2017).

One of the main difficulties for Geoparks is to increase their appeal as a destination with unique geological and geomorphological features, natural ecosystems to improve the protection, conservation and promotion, of their territories (Zouros 2014). Geoparks management requires a holistic approach of to protection, education and sustainable development, as well as increased community awareness of the critical role of protecting and conserving the Earth's unique geological legacy (Zouros 2004). In addition, Geoparks have an active role in the organization and participation of educational and training projects at all education levels in Earth Sciences, as well as in the improvement of natural environment and sustainable development policies (Zouros 2004, 2013).

It is critical for the successful implementation of a continuous communication strategy for the promotion of the territory, education, local products and stakeholders, particularly those involved in the sustainable economic development of the local community through activities such as geotourism. Thus, tourists, as well as locals should be able to find information about the Geopark through visibility, such as panoramic panels, info—kiosks in geologically significant sites and areas, communication through regular presence in media and social networks, information transmission, such as the production of a promotional video, leaflets, digital maps, online applications, special publications for younger visitors and for education awareness (Nunes et al. 2017). The key challenge for Geoparks is to provide geological information that is appropriate for a wide range of visitors

(Lansigu and Desbois 2017). As mentionned by Lansigu and Desbois (2017:104) "*the information must interest the visitors in people's relationship with the Earth's history and the environment in a very attractive way, avoiding the trap of being too complex or academic*". Digital images, interactive maps, and 360° virtual tours are just a few of the innovative technologies tools used to hold public's interest in geology, the natural environment, and sustainable development of rural areas (Lansigu and Desbois 2017). Nowadays, in the era of social media, digital images have great influence on the promotion of Geoparks (Cimermanová et al. 2017). Because of the development of digital technologies, social media networks will have a significant impact on marketing activities in sustainable tourism development (Hysa et al. 2021). Social media networks have already been utilized to raise awareness of the Geoparks's values, geoheritage, and local communities, as well as to establish communication channels for information sharing with visitors before, during, and after their visit to the Geopark (Hysa et al. 2021).

In recent years, the Internet has become the most popular means to instantly disseminating knowledge and expanding information globally (Li et al. 2015). As a result, offering Geopark information and digital interactive applications on the Internet may aid to promote geotourism initiative (Li et al. 2015). New strategies for promoting geoconservation and geotourism are still being encouraged. Social media, for example, is a set of digital-based technologies that facilitate the success of new ideas, concepts, and information spread via virtual platforms and digital communication channels. Social media refer to a variety of mobile phones and online applications that allow individuals and communities to produce, distribute, and modify information through interactive digital platforms (Beier and Wagner 2016).

By applying modern technologies that are prevalent in everyday activities, Geoparks could develop educational programs and interactive digital applications that establish new perceptions, enhancement, protection, and dissemination of the geological heritage (Martínez-Graña et al. 2017). The promotion of geological heritage via social media is a powerful tool for communicating Geoparks activities and an easy way to collaborate with UNESCO Global Geoparks from countries all over the world and share ideas for new projects (Caballero et al. 2017). Tourists and visitors, nowadays, are increasingly using social media at various stages of their journey, including planning, realizing, and sharing trip experiences, because the postings provided by digital influencers have a direct impact on making decisions about choosing a sustainable tourist destination such a Geopark (Hysa et al. 2021). Because social media was founded on the technological peak of Web 2.0, all users can share information content on web-based and mobile platforms by constantly generating, modifying, and communicating their opinions (Beier and Wagner 2016). Platforms and social media networks offer digital tools for creating profiles and disseminating files and digital material. As a result, all digital users can utilize social media platforms to share their thoughts and experiences from visiting a Geopark by writing a simple text or uploading content, such as documents, pictures, videos, or recording files with nature sounds (Beier and Wagner 2016).

5 Methodology

The primary goal of this study is to examine the dynamics of Geoparks in Greece for geotourism development, and to investigate the digital presence and promotion techniques applied by the geoparks' managing agencies, with a focus on social media. The ultimate goal of the study is to support building Geoparks sustainable tourism networks based on their unique natural amenities as explored through their digital presence (online and social media) in order to promote a national geotourism brand name. To better explore Geoparks and geotourism development perspectives in Greece, research analysis is organized in the following sections:

a. First, the geomorphological profile of the geoparks in Greece is examined, which includes data on their administrative, geomorphological and tourism information aspects.
b. In the following section, the geoparks' digital profile is examined via online tracking (website, social media, Twitter, Youtube videos etc.) sorted in categories.
c. Then, the research findings are analysed and a database is built as a methodological tool in order to identify networking opportunities and suggest new geotourism networks based on local natural assets.
d. Finally, concluding remarks and recommendations for geotourism development are presented.

The methodological steps of the research include the elaboration of data collected in order to formulate a nationwide inventory of Geoparks' presence in the social media. Extensive secondary research (studies, published articles, official data sources, reports, promotional material related to Geoparks, as well as relevant information obtained through extensive online research) is the basis for the identification of the sustainable tourism "image" of Geoparks in Greece.

Following the research process, a database of Geoparks in Greece was elaborated. This database included information on the Geoparks′ location, organizational structure, and marketing digital profile (Internet/Social Media), in order to demonstrate their tourism dynamics. Geoparks were identified by conducting a systematic search of relevant websites and social media, as well as internet search engines.

The data basis about Geoparks in Greece, can be used as a tool to provide organized information, addressed to a wide range of people including tourists, decision-makers, tourism industry professionals, researchers and others. Information in the inventory of Greek Geoparks' identified so far is organized based on a number of criteria: Name of the Geopark, location (Region), Geosites within the Geopark area, Geopark Area Surface (km^2), UNESCO Global Geopark Designated Year, Information Centre, Website, Facebook, Instagram, Twitter, Youtube. In order to check the visibility of the Geoparks, a qualitative evaluation of their presence on social media, including Facebook, Instagram, Twitter and Youtube was elaborated, based upon the content of information provided. The Facebook platform was used to uncover Geoparks' accounts while, Instagram accounts were found using the application. The same approach, through the application, was used to locate the Twitter accounts of each Geopark.

Geoparks digital profile (elements).

1. Website
2. Website translation in English or other languages
3. Web search
4. Blog
5. Facebook page/group/profile of the Geopark/Geopark management agency
6. Facebook followers
7. Facebook likes
8. Twitter account of the Geopark/Geopark management agency
9. Instagram account
10. Instagram number of posts
11. Instagram number of followers
12. Youtube—videos uploaded by the organizers and number of view
13. Youtube—videos uploaded by others apart from the organizers, i.e. the locals or local authorities, visitors, media etc. and number of views.

6 Research Results: Geoparks and Social Media in Greece

6.1 Greek Geoparks

Six areas in Greece have been designated as Geoparks (Table 1, Fig. 2), with the goal of promoting exceptional geological and geomorphological geosites through educational programs, projects involving local stakeholders, and rural development based on geotourism promotion. In 2011, the Hellenic Geoparks Forum was established. One of the main priorities was the coordination of Greek Geoparks on topics related to the promotion of geological heritage and the

Table 1 The Greek Geoparks: Geomorphological Profile

Geopark name	Region	Geosites	Geopark area surface (km^2)	UNESCO global geopark designated year	Information centre
Lesvos Island Geopark (Fig. 4)	North Aegean	Sigri petrified forest, lava domes in Molyvos, Tria Marmaria—Kryfti gorge—Nappe, Gulf of Gera hot spring	1636	2015	Natural history museum of the Lesvos Petrified Forest
Psiloritis Natural Park (Fig. 5)	Crete	Talea ori section: "a 250 million-year journey in the past", Psiloritis karstic landscape, Amari cliffs	1200	2015	Plsiloritis information centre, Anogeia, Rethimno
Chelmos—Vouraikos Geopark (Fig. 7)	Peloponnese	The steep and imposing gorge of Vouraikos river, Cave of the lakes, Aroanios river springs, Tsivlos lake (landslide)	647	2015	Chelmos—Vouraikos information centre, Kalavrita
Vikos—Aoos Geopark (Fig. 3)	Epirus	Vikos gorge, Aoos Gorge, "Stone forest", Vradeto karstic landscape, "Drakolimni"	1200	2015	Vikos—Aoos information centre, Megalo Papigo—Voidomatis (Kleidonia)
Sitia Nature Park (Fig. 6)	Crete	Mavro Mouri and Poplou monastery Tafoni, Meteorakia, karstic geotopes, Kato Zakros gorge, Fangromouro hippopotami	517	2015	Sitia Nature Park information centre, Karidi village
Grevena—Kozani Geopark (Fig. 8)	West Macedonia	Tsouriakas—geowonder site, Parorio geo—view of Orliakas, "Onion skin" exfoliation of sediments, tectonic rock arch "Dragon eye"	2486	2021	Grevena—Kozani Geopark information centre, Grevena

sustainable tourism development of rural and natural areas. The activities of the Hellenic Geoparks Forum include actions for the recognition of Geoparks at the national and global level, as well as assistance for the European Geoparks Network and the Global Geoparks Network through the organization of special issues (Museum of Lesvos Petrified Forest: Geoparks Network n.d.) The Greek Geoparks are members of the European Geopark and UNESCO Global Geoparks Network (Figs. 2, 3, 4, 5, 6, 7 and 8).

6.2 The Role of Internet, Social Media and Communication Technologies in Geotourism Development in Greece

Making use of today's dominant digital technologies, Geoparks generate educational content and innovative multimedia applications that provide promotion and protection of the geological, biological, ecological and cultural heritage of their territory. Geoparks often develop initiatives to train residents and visitors about the need to protect and manage the area. They also implement actions to raise their visibility and promote their territory both domestically and internationally and introduce new experiences and tourism services (Fasoulas et al. 2022; Zouros 2017). These initiatives are facilitated through the Geoparks websites and socia media tools such as Facebook, Instagram and Youtube. According to the survey results, Greece has a rich natural and gelogical heritage, as six Geoparks have been identifed so far (Table 2). All Greek Geoparks have a website and Facebook address, only one is without an Instagram address, and three are without a YouTube channel.

However, the majority do not use exceedingly the potential of digital tools to promote sustainable tourism development. According to more detailed information on the Geoparks's digital profile (Table 3), with regard to social media tools, starting with Facebook, Greek Geoparks have an account on this network, while only two of the Geoparks have a Twitter account. As for Instagram, it has been confirmed that not all Geoparks have an account on this social network. Greece is endowed with abundant Geoparks natural resources, which can be boosted by the use of new technologies and the dynamics of social media and to make more active use of social networks.

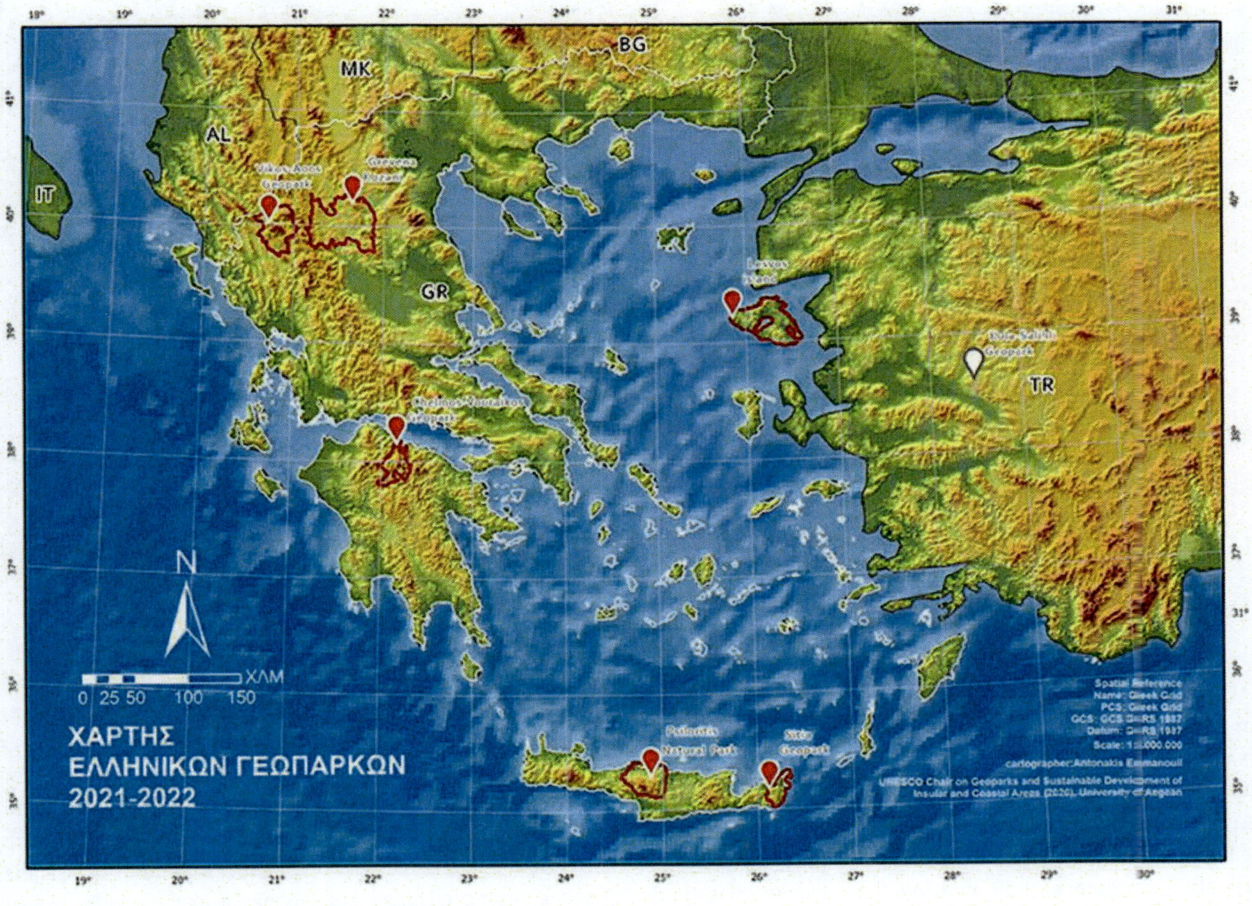

Fig. 2 Map of Greek Geoparks (Hellenic Geopark Forum 2021)

Fig. 3 Vikos gorge, Vikos—Aoos Geopark (©Theodora Lanara)

Fig. 4 Petrified Forest park "Bali alonia", Lesvos Island Geopark (https://www.lesvosmuseum.gr/)

Approaching Geoparks has revealed disparities in using the digital and social media tools and online promotion. According to the research findings, there are Geoparks rich in geosites such as Chelmos—Vouraikos Geopark, Sitia Nature Park and Grevena—Kozani Geopark, however, they are not fully promoted online and lack digital tools. According to Table 3, the digital profiles of the Geoparks show a variation on digital presence, with only one having a low digital presence, three with medium and two with high presence by effectively using all of the social media platforms included in the study (Lesvos Island Geopark and Psiloritis Natural Park). It is evident that all Geopark managers understand the necessity of having a website and also translation of it in order to attract tourists who search for what Geoparks provide. By keeping an updated profile on all available social media, it is easier to rank higher on a web search and again attract visitors. Psiloritis Natural Park is the pioneer among all the other Geoparks in the proper use of

Fig. 5 Nida plateau, Psiloritis Natural Park (https://tours.nhmc.uoc.gr/geoparks/map/idi/el/#)

Fig. 6 Geological geotopes "Rouso spasma", Sitia Nature Park (https://www.sitia-geopark.gr/)

social media tools and platforms since it is the only Geopark that keeps its platforms updated constantly with interesting information in order to attract a wider audience.

Geoparks managers need to get much more information, be more active on social media, reach a greater extension in wider range of park posts, create more events, receive more "likes", and interact with followers and visitors more often. In brief, Greek Geoparks need to be more dynamic and proactive on social networks. A notable finding is the lack of Geoparks' social media activity, implying a missed opportunity of focusing on visitors, especially young visitors, as well as local stakeholders in terms of establishing a national geotourism brand image.

The Geoparks identified all with websites and social media, while some with only Facebook page, implying that Instagram and Twitter are not used to promote and direct

Fig. 7 Ydata Stygos, Chelmos—Vouraikos Geopark (©NECCA, Management Unit of Chelmos—Vouraikos National Park and Protected Areas of the Northern Peloponnese, https://fdchelmos.gr/)

Fig. 8 Portitsa gorge, Grevena—Kozani Geopark (©Theodora Lanara)

tourists to Geoparks. Five of the Greek Geoparks have an Instagram account, with the exeption of Grevena—Kozani Geopark, and only two Geoparks have a Twitter account, Vikos—Aoos Geopark and Lesvos Island Geopark. According to certain survey results, social media facilitate online interaction and demonstrate visitors' participation. There are videos on Youtube from the six Geoparks which are promoting the geosites attractions, most of which are submitted by visitors and local media, illustrating the popularity of the Geopark, while there are also videos uploaded by the Geoparks management bodies via their Youtube channels, such Lesvos Island Geopark, Vikos-Aoos Geopark and Psiloritis Natural Park (Table 4).

7 Conclusions

The Geopark territory, its management authority, local communities, the public including tourists, educational institutions and local businesses emphasize the promotion

Table 2 The Greek Geoparks: Digital profile

Geopark name	Website	Facebook	Instagram	Youtube
Lesvos Island Geopark	https://www.lesvosgeopark.gr/	https://www.facebook.com/lesvosgeopark	https://www.instagram.com/lesvosgeoparkunescogg/	Petrified forest lesvos https://www.youtube.com/channel/UCpw2rJKDE80DkDpIRKsM5ZQ
Psiloritis Natural Park	https://www.psiloritisgeopark.gr/	https://www.facebook.com/PsiloritisGeopark	https://www.instagram.com/psiloritis_geopark/?fbclid=IwAR0yNw74nFrZaX10SVgfTOgbkx6ObFCgdgpwj26SISEf5nT5zrRwcJ53l9c	https://www.youtube.com/channel/UCWnBgExuHP-GsLFk0YQdg6w
Chelmos—Vouraikos Geopark	https://fdchelmos.gr/el/	https://www.facebook.com/chelmosvouraikosgeopark	https://www.instagram.com/chelmosvouraikospark/	
Vikos-Aoos Geopark	https://vikosaoosgeopark.com/	https://www.facebook.com/vikosaoosgeopark	https://www.instagram.com/vikosaoosunescogeopark/	https://www.youtube.com/channel/UCqAsM9p2lmyxtJdkaXyy0uw
Sitia Nature Park	https://www.sitia-geopark.gr/	https://www.facebook.com/sitiageopark	https://www.instagram.com/sitiageopark/	
Grevena—Kozani Geopark	https://www.geoparkgrevenakozani.com/	https://www.facebook.com/grevenakozani		

Table 3 Content evaluation of Geoparks digital profile

Geoparks digital profile													
Geopark name	Website	Website translation	Web search	Blog	Facebook page	Facebook followers	Facebook likes	Twitter account	Instagram account	Instagram number of posts	Instagram followers	Youtube—videos by geopark	Youtube—videos by others
Lesvos Island Geopark	Yes	Yes	Yes	No	Yes	3373	3203	Yes	Yes	78	402	Yes	Yes
Psiloritis Natural Park	Yes	Yes	Yes	No	Yes	3400	–	No	Yes	245	450	Yes	Yes
Chelmos—Vouraikos Geopark	Yes	Yes	Yes	No	Yes	729	680	No	Yes	54	561	Yes	Yes
Vikos-Aoos Geopark	Yes	Yes	Yes	No	Yes	1.922	1.864	Yes	Yes	35	525	Yes	Yes
Sitia Nature Park	Yes	Yes	Yes	No	Yes	2.073	1.889	No	Yes	53	416	No	Yes
Grevena-Kozani Geopark	Yes	Yes	Yes	No	Yes	1.042	1.023	No	No	–	–	No	Yes

Table 4 Geoparks overviews digital profiles

Digital technologies—social media network	Number of geoparks
Interactive maps	3
Digital maps and leaflets	6
Applications for smartphone—Ipad	2
Virtual tours	3
Facebook	6
Instagram	5
Twitter	2
Youtube channel	3

strategy through communication and information actions utilizing a range of marketing techniques. Geoparks, thanks to their unique natural resources as well as educational and trainining activities, are proving their potential to become geotourism development poles of attraction especially for rural areas hosting significant geological heritage, natural ecosystems, often also cultural elements and archaeological sites. The digital promotion of Geoparks (online and social media activation) is expected to attract new tourism flows of travelers seeking to experience the study area's unique natural environment, while also participating in alternative forms of tourism (i.e. geotourism, mountain, ecotourism).

Geoparks appear not to have successfully attempted to fully exploit the potential of digital tools for forming partnerships and communicating with visitors and residents. Maintaining a website, a Facebook page and/or Instagram account, as well as remaining active throughout the year, are examples of efforts to keep in touch with prior or future visitors. On a local and regional scale, these resources have the ability to enhance alternative and distinctive tourism destinations with a national geotourism brand name. The findings of the study clearly demonstrate the Geoparks' networking and tourist development potential based on the geological natural heritage.

The use of new technology applications as an additional tool to promote sustainable tourism development should be included in the Geoparks' digital communication strategy. Geopark managers should be encouraged to plan and design multilingual apps, virtual tours, interactive maps, Youtube channels for enhancing the values of the park, including its geodiversity, biodiversity, historical and cultural heritage. The proposed methodology is important for Geoparks that may have previously gone unreported, due to a lack of promotion. The purpose of this study is to integrate Geoparks that are not sufficiently digitally promoted into networks in order to emphasize a more comprehensive and exclusive list of Geoparks and boost synergies. Geoparks that are active on the internet and social media may serve as best practices and their management authorities could share their technology and management expertise to help in the promotion of other geoparks. Furthermore, the identification, classification, and networking of geoparks on the basis of content analysis and digital presence, and the database created highlight the resources recorded.

The promotion of a Geopark through social media networks, national and regional websites, and among the large number of national and international visitors who share photographs and activities via social media such as Facebook, Instagram, Twitter, and other platforms, allow Geopark management authorities to accelerate Geopark science popularization and education activities, while also strengtheing interactive communication between Geoparks and their visitors. Geoparks already use social media to build the brand name of the region, and the geosites' attractions, as well as create relationships with visitors before, during, and after their visit. The database introduced by this paper as a methodological tool, may facilitate the development of effective geotourism branding strategies.

References

Beier, M., Wagner, K.: Social media adoption: barriers to the strategic use of social media in SMEs. In: 24th European Conference on Information Systems, ECIS 2016, Istanbul, Turkey (2016). Retrieved from https://www.researchgate.net/publication/301357654_0. Accessed on 26 Nov 2021

Caballero, J.L., Sánchez, I.C., Barrea, J. M.: Social media connect us!. In: Lima, E.A., Nunes, J.C., Meirinho, P., Machado, M. (eds.) 14th European Geoparks Conference—Geoparks:Pathways of Sustainable Tourism of Development—Abstracts Book, p. 122 (2017). Retrieved from http://www.europeangeoparks.org/?page_id=4926. Accessed on 21 Nov 2021

Cayla, N.: An overview of new technologies applied to the management of geoheritage. Geoheritage **6**(2), 91–102 (2014). https://doi.org/10.1007/s12371-014-0113-0

Ciftci, Y., Güngör, Y.: Proposals for the standard presentation of elements of natural and cultural heritage within the scope of Geopark project. Bull. Miner. Res. Explor. **153**, 223–238 (2016). Retrieved from https://bulletin.mta.gov.tr/article/show/670.html. Accessed on 30 Nov 2021

Cimermanová, I., Drexler, S., Szarvas, I.: The power of photography. In: Lima E.A., Nunes J.C., Meirinho, P., Machado, M. (eds.) 14th European Geoparks Conference—Geoparks:Pathways of

Sustainable Tourism of Development—Abstracts Book, p. 93 (2017). Retrieved from http://www.europeangeoparks.org/?page_id=4926. Accessed on 21 Nov 2021

European Geoparks Network (EGN): What is a Geopark (n.d.). Retrieved from http://www.europeangeoparks.org/?page_id=165. Accessed on 29 Nov 2021

Farsani, N.T.: Sustainable Tourism in Geoparks through Geotourism and Networking, Doctoral dissertation, Universidade de Aveiro, Portugal (2012). ProQuest Dissertations Publishing. Retrieved from http://hdl.handle.net/10773/10597

Fasoulas, C., Nikolakakis, E., Staridas, S.: Digital tools to serve geotourism and sustainable development at psiloritis UNESCO global geopark in COVID times and beyond. Geosciences **12**(2), 78 (2022). https://doi.org/10.3390/geosciences12020078

Fiori, C., Burlando, M.: Beigua app: digital innovation to promote tourism. Eur. Geoparks Netw. Mag. **14**, 26 (2017). Retrieved from http://www.europeangeoparks.org/?page_id=395. Accessed on 21 Nov 2021

Global Geoparks Network: UNESCO Global Geoparks 2019/2020, 2004–2019 15 years of collaboration for geo-conservation and sustainable development. In: UNESCO (2020). Retrieved from http://globalgeoparksnetwork.org/wp-content/uploads/2021/03/GGN-2019-2020.pdf. Accessed on 29 Nov 2021

Hellenic Geoparks Forum: Hellenic Geoparks Map (n.d.). Retrieved from http://www.hellenicgeoparks.gr. Accessed on 30 Nov 2021

Hysa, B., Karasek, A., Zdonek, I.: Social media usage by different generations as a tool for sustainable tourism marketing in society 5.0 idea. Sustainability **13**(3), 1018 (2021). https://doi.org/10.3390/su13031018

Kostopoulou, S., Tsitsoni, T., Lanara, T.: Sustainable tourism local cluster development in protected areas: the case of Parnassos. In: Proceedings of 18th Hellenic Forestry Congress and International Workshop, pp. 1345–1353 (2017). Hellenic Forestry Society. Retrieved from https://www.forestry.gr/en/announcement/confere-preoceedings/2017

Lansigu, C., Desbois, J.L.: Visual media for sharing geoheritage with both partners and visitors: the example of the massif des bauges ugg. In: Lima, E.A., Nunes, J.C., Meirinho, P., Machado, M. (eds.) 14th European Geoparks Conference—Geoparks: Pathways of Sustainable Tourism of Development—Abstracts Book, p. 104 (2017). Retrieved from http://www.europeangeoparks.org/?page_id=4926. Accessed on 21 Nov 2021

Li, Q., Tian, M., Li, X., Shi, Y., Zhou, X.: Toward smartphone applications for geoparks information and interpretation systems in China. Open Geosci. **7**(1), 663–677 (2015). https://doi.org/10.1515/geo-2015-0060

Lo Presti, L., Maggiore, G., Marino, V.: Digital communication tools and tourism engagement: instant CRM strategies through instant messaging apps. In: Gursoy D., Kaurav, R.P.S. (eds.) Handbook on Tourism and Social Media, pp. 242–255 (2022). ISBN: 9781800371408. https://doi.org/10.4337/9781800371415

Martínez-Graña, A.M., Serrano, L.J., González-Delgado, A., Dabrio, C. J., Legoinha, P.: Sustainable geotourism using digital technologies along a rural georoute in Monsagro (Salamanca, Spain). Int. J. Digital Earth **10**(2), 121–138 (2017). https://doi.org/10.1080/17538947.2016.1209582

McKeever, P.J., Zouros, N., Patzak, M.: The UNESCO global network of national geoparks. George Wright Forum **27**(1), 14–18 (2010). Retrieved from https://www.jstor.org/stable/43598130. Accessed on 26 Nov 2021

Natural History Museum of the Lesvos Petrified Forest: Geoparks Network (n.d.). Retrieved from https://www.lesvosmuseum.gr/en/geoparks-network. Accessed on 29 Nov 2021

Nunes, J.C., Machado, M., Lima, E., Costa, M.P.: Visibility, communication and dissemination of information in the Azores UNESCO global geopark: a powerful trilogy aimed for public awareness. In: Lima, E.A., Nunes, J.C., Meirinho, P., Machado, M. (eds.) 14th European Geoparks Conference—Geoparks: Pathways of Sustainable Tourism of Development—Abstracts Book, p. 94 (2017). Retrieved from http://www.europeangeoparks.org/?page_id=4926. Accessed on 21 Nov 2021

Özgeriş, M., Karahan, F.: Use of geopark resource values for a sustainable tourism: a case study from Turkey (Cittaslow Uzundere). Environ. Dev. Sustain. **23**, 4270–4284 (2021). https://doi.org/10.1007/s10668-020-00773-3

UNEP and WTO: Making Tourism More Sustainable—A Guide for Policy Makers (eBook version) (2005). https://doi.org/10.18111/9789284408214. Accessed on 21 Nov 2021

UNESCO: UNESCO Global Geoparks (2021). Retrieved from https://en.unesco.org/global-geoparks. Accessed on 29 Nov 2021

United Nations: Report of the World Commission on Environment and Development (Annex to document A/42/427) (1987). Retrieved from http://www.un-documents.net/wced-ocf.htm. Accessed on 30 Nov 2021

Živković, R., Gajić, J., Brda, I.: The impact of social media on tourism. In: Proceedings of the 1st International Scientific Conference—Sinteza-2014. E-Business in Tourism and Hospitality Industry, Serbia, pp. 758–761 (2014). https://doi.org/10.15308/sinteza-2014-758-761

Zouros, N.: The European Geoparks network. Geological heritage protection and local development. Episodes—J. Int. Geosci. **27**(3), 165–171 (2004). Retrieved from https://www.researchgate.net/publication/258100233. Accessed on 26 Nov 2021

Zouros, N., Fasoulas, C.: Geoconservation and Geoparks in Greece. World Geoparks Newsl. **4**, 8–14 (2007). Retrieved from https://www.psiloritisgeopark.gr/en/the-park/publications/documents/. Accessed on 22 Nov 2021

Zouros, N., McKeever, P.: European Geoparks: geoconservation and sustainable local development. In: Proceedings of the International Conference "Studying, Modeling and Sense Making of Planet Earth", Lesvos, Greece, pp.1–6 (2008). Retrieved from http://www1.aegean.gr/geography/earth-conference2008/papers/papers/A10ID195.pdf. Accessed on 23 Nov 2021

Zouros, N.: European Geoparks: new challenges and innovative tools towards earth heritage management and sustainable local development. In: Proceedings of the 12th European Geoparks Conference National Park of Cilento, Vallo di Diano and Alburni Geopark, Italy, pp. 5–9 (2013). Retrieved from https://www.researchgate.net/publication/258763580. Accessed on 22 Nov 2021

Zouros, N.: Global Geoparks—a tool for outreach in earth sciences, purpose and reality. In: Proceedings of the Conference GeoFrankfurt—Earth System Dynamics, Frankfurt, Germany, p. 644 (2014). Retrieved from https://www.researchgate.net/publication/282366033. Accessed on 22 Nov 2021

Zouros, N.: Global Geoparks network and the new UNESCO global geoparks programme. Bull. Geol. Soc. Greece **50**(1), 284–292 (2017). Retrieved from https://ejournals.epublishing.ekt.gr/index.php/geosociety/article/view/11729. Accessed on 22 Nov 2021

Environmental Awareness and Sustainable Practices of the Tourism Companies of Epirus in the Context of Corporate Social Responsibility

Maria Papavasileiou, Alexios-Patapios Kontis, Efthimia Sarantakou and Paris Tsartas

Abstract

Despite the temporary effects of the coronavirus pandemic, tourism is still considered the largest industry in the world and is expected to continue to grow even more in the twenty-first century. The recent health crisis has forced tourism businesses to adopt strict procedures with strong social responsibility characteristics, at an unprecedented level. However, even before the recent health crisis, there was a rapidly growing trend of consumers preferring tourism companies to adopt socially responsible practices. In the context of Corporate Social Responsibility, tourism companies aim not only at the diversification and enrichment of their tourism services, but also at the strengthening of the values provided in three main dimensions: economic, social, and environmental. The paper attempts to present the current state of environmental awareness and sustainable practices implemented by the hospitality companies of the Epirus Region, such as hotels and guesthouses in the context of Corporate Social Responsibility. At the core of the Corporate Social Responsibility strategy is the shared value which highlights the positive relationship between running a business that benefits society and the environment and achieving high financial returns. The results of the research show that there is a significant deviation between the stated intentions for the adoption of sustainable practices and their actual level of implementation.

Keywords

Sustainable development • Corporate social responsibility • Tourism business • Environmental awareness

1 Introduction

The socio-economic changes of recent years have influenced consumer perceptions and behaviors, with the result that, among other things, a trend has emerged of consumers becoming more aware of the impact of their purchasing choices on sustainability issues in the areas of environment, society and governance. Modern consumers are now more empowered and have an expanded role (Prahalad and Ramaswamy 2004), and through social networks they collaborate and interconnect with other consumers so that they are well informed, dynamic, and proactive in their relationship with companies (Kotler et al. 2010). In response to the above shift in consumer interests, there is a proliferation of businesses that, realizing the power of both the swelling consumer wave that is approaching and the influence they exert on society, are moving away from the narrow logic of profit maximization and undertaking a series of socially responsible initiatives. Kotler (2003) points out that “companies can no longer operate as self-contained, fully capable units without dedicated partners”. Therefore, as the marketing function recognises its vital interconnection with society, the areas of business ethics, social responsibility and stakeholder care become crucial for the successful and sustainable development of businesses (Parmar et al. 2010). The international literature records the interest of researchers in understanding the interrelated business issues connected to areas such as the development of commercial value,

M. Papavasileiou
Hellenic Open University, Patras, Greece
e-mail: mariapap@otenet.gr

A.-P. Kontis (✉)
University of the Aegean, Mytilene, Greece
e-mail: apkontis@aegean.gr

E. Sarantakou
University of West Attica, Aigaleo, Greece
e-mail: esarantakou@uniwa.gr

P. Tsartas
Harokopio University, Kallithea, Greece
e-mail: tsartas@hua.gr

S. Kostopoulou et al. (eds.), *Cities' Vocabularies and the Sustainable Development of the Silkroads*, Advances in Science, Technology & Innovation, https://doi.org/10.1007/978-3-031-31027-0_21

balancing the relationship between ethics and capitalism, and the formulation of appropriate management models in relation to the social responsibilities and obligations of business (Parmar et al. 2010). The relevant research has shown that one of the key tenets of CSR is that managers must accept their social responsibilities (Andriof and Waddock 2002). According to McIntosh (2007), the business entity is seen as an entity that has rights, duties and obligations to society. It is involved and participates in the governance of a society like a normal citizen in various ways but of course with more power and influence.

Corporate Social Responsibility is based on extending the relationship between society and business to include understanding, and therefore assuming, the social, environmental, and political responsibilities of business (McIntosh 2007). Interestingly, Swift and Zadek (2012) distinguish between first and second generation CSR. Companies belonging to the first generation of CSR focus very little on traditional forms of philanthropy and follow specific CSR practices, while companies belonging to the second generation give a more strategic role to CSR initiatives as part of new management and corporate governance models. Based on the above approach, there are companies that establish their entire operations on the principles of Corporate Social Responsibility, while on the opposite side there are companies that limit the implementation of CSR policies to specific functions or departments (e.g. Human Resources Department, Technical Services, etc.).

At the core of the CSR strategy is the shared value that emphasises the positive relationship between running a business that benefits society and the environment and achieving high economic returns. Thus, nowadays, references to CSR are placed in the same context as sustainable development and corporate governance. In this paper, a first attempt is made to assess the current situation in the hospitality businesses of the Region of Epirus in relation to environmental awareness and sustainable practices.

2 Literature Review

Studying the historical development of the conceptual definition of Corporate Social Responsibility (CSR) reveals the dynamism of the concept, which, passing through some key points presented below, is constantly improving, trying to cover the rapid changes in both supply and demand (Haski-Leventhal 2018). In the early 1950s, Bowen (2013) conceived of corporations as centers of power, authority and decision-making and observed that their operation affects individuals and the societies in which they function. In relation to the Social Responsibility of entrepreneurs, he referred to the obligation they have to choose policies and decide according to what is desirable to the values and goals of society. Similarly, Frederick (1960) recognizes elements of Social Responsibility in business when its financial and human resources are used more broadly to achieve wider social goals, moving beyond the narrow logic of pursuing individual and corporate profit. Carroll (1979, p. 500) claimed that "the social responsibility of business encompasses the economic, legal, ethical, and discretionary expectations that society has of organizations at a given point in time". According to the above definition, the economic component implies that society expects businesses to produce goods and services and sell them at a profit. The legal dimension represents the framework of rules within which the enterprise is required to fulfill its economic mission. The ethical responsibility dimension represents the attitudes and practices that society expects businesses to follow, which go beyond mere obedience to the law. Finally, the fourth dimension, discretionary expectations, is driven by the desire of companies to assume social roles that are not imposed or required by legislation and are not expected in the moral sense, but are left to the discretion of senior managers and have the character of strategic interventions (e.g. charitable offers, programs to remove social exclusions, etc.).

An equally pivotal point in the conceptual definition of Corporate Social Responsibility (CSR) was the contribution of Freeman (1984), who introduced the "Stakeholders Theory", arguing that companies are not only responsible to their shareholders, but also to a wider range of stakeholders who have a direct or indirect relationship with the company. Freeman (1984) defined stakeholders as 'any individual, group or organization that is affected, positively or negatively, by the achievement of the organization's purpose'. Stakeholders therefore include the company's employees, partners and suppliers, customers, members of local communities, governmental and non-governmental bodies, and the wider community. Stakeholder theory is one of the most applied theories in CSR research, helping to broaden the perspective of corporate responsibility (Haski-Leventhal 2018). When stakeholders' interests conflict, company managements should address the emerging issues, to the extent possible, in a way that creates even more value for all stakeholders (Harrison et al. 2018). According to Smith (2003), stakeholder theory is based on the principle that companies should take seriously the effects of their actions on all those directly and indirectly involved with them, even if it affects their profitability. In the past, the value of a business was based solely on its financial performance, but now, in a globalised market with informed, active, and aware consumers, this is not enough. Businesses are obliged to be accountable for their actions and their impact on the wider society. Thus, socially responsible businesses must be managed following Stakeholders Theory (Freeman 1984).

The literature review for understanding the incentives that lead firms to adopt CSR applications distinguishes two different but not necessarily conflicting approaches. According to the first, companies take social problems into account as soon as they are integrated and interact with society (Granovetter 1985), while in the second, they perceive the business benefits of adopting CSR. These benefits are related to improving the image of the business, increasing customer loyalty and increasing sales, retaining/attracting competent employees, increasing capital value, attracting capital and investors, etc. (Turban and Greening 1997; Brown and Dacin 1997). It should be noted that the corporate benefits of implementing CSR depend directly on the way it is implemented (e.g. first or second generation CSR). When corporate social initiatives are the result of the culture and strategy of the company (second generation CSR) the benefits are substantial and long term. On the contrary, the superficial approach and implementation of CSR actions as a means of promoting public relations, when not accompanied by substantial commitment and sincerity, can only benefit companies in the short term or even harm them (Kotler et al. 2010).

Lee and Kotler (2019) identify the six most typical CSR policies, which are: (1) Cause Promotions (supporting social causes through promotional sponsorships), (2) Cause-related Marketing (making a contribution or donating a percentage of revenues to a specific cause based on product sales or usage), (3) Corporate Social Marketing (supporting behaviour change campaigns), (4) Corporate Philanthropy (making direct contributions to a charity or cause), (5) Community Volunteering (providing volunteer services in the community), (6) Socially Responsible Business Practices (adopting and conducting discretionary business practices and investments that support social causes).

The World Business Council for Sustainable Development (WBCSD), linking the achievement of sustainable development with Corporate Social Responsibility (CSR), describes it as "continuing commitment by a company to behave ethically and contribute to economic development while improving the quality of life of its workforce and family members, as well as the local community and society at large" (WBCSD 1999). Emphasizing long-term social welfare, several scholars (Freeman and Velamuri 2006; Jamali 2008; Ketola 2006) acknowledge that in the context of CSR even more attention is gradually being paid to environmental issues, due to the emerging debate on the effects of climate change. The apparent lack of reference to the environmental dimension in the definition of CSR led to the formulation and addition of new concepts such as "Corporate Social and Environmental Responsibility" (CSER), "Corporate Sustainability Management" and "Corporate Sustainability and Responsibility" (Salzmann et al. 2005; Lund-Thomsen 2004; McIntosh 2003).

Focusing on tourism, CSR development has become one of the latest trends in the global tourism industry (Jones et al. 2016; Murphy and Schlegelmilch 2013), due to the strong social orientation of tourism activity and its interdependence on improving people's quality of life (Kumar and Kumar 2018). Thus, CSR for tourism businesses focuses on taking initiatives to effectively manage the positive and negative impacts of their activities in the areas of environmental protection, and economic and social development (Meyer and Meyer 2016; Muafi 2017). CSR is consistent with the concept of sustainable tourism development, which focuses on the rational use of available tourism resources through the long-term achievement of a balance in the realization of economic, environmental, social and cultural development objectives for all those directly and indirectly involved (e.g. tourists, sending and receiving destinations, local population).

The spatial dimension of tourism activities creates strong interdependencies with local communities and the qualitative environmental characteristics of destinations. Therefore, the integration of CSR strategies by tourism enterprises is becoming increasingly important, with particular emphasis on environmental issues. Bohdanowicz and Zientara (2012, p. 96) highlight the practical feasibility of linking CSR with environmental practices, because "in most cases (at least in hospitality) corporate social responsibility and environmentalism are intertwined with each other". Thus, in the context of CSR, it seems that environmental actions have an advantage over socio-economic actions either because they can be more easily implemented within companies or because they have a direct recognition by and positive impact on consumers (Sheldon and Park 2011; Jones et al. 2016).

Previous research has shown that there is a high recognition of the need to adopt CSR practices in the tourism sector, but there are relatively few companies implementing specific actions either due to a lack of resources or a lack of understanding of the concepts involved (Sheldon and Park 2011). In a survey by de Grosbois (2012), the positive intention to implement CSR actions is particularly high (73%), but significantly lower is the formulation of specific objectives (36%) and even lower the reporting of results from CSR implementation. Research on CSR in tourism shows that the environmental dimension includes several aspects such as waste management (Font et al. 2016), certification schemes (Karlsson and Dolnicar 2015), environmental sustainability (Esparon et al. 2014), energy saving actions (de Grosbois 2015) and efficient use of resources (Cvelbar and Dwyer 2013).

3 Methodology

The general objective of this study on the environmental awareness of tourism enterprises in the context of Corporate Social Responsibility (CSR) was to determine the degree of awareness and sensitization of hospitality tourism enterprises on sustainability issues and whether they implement or intend to implement sustainable development practices within the framework of Corporate Social Responsibility. This general objective was made up of individual specific objectives which relate to:

1. identifying the characteristics of the hotels and accommodation participating in the survey, such as category, capacity, seasonality, and services provided.
2. drawing conclusions on the attitude of hospitality companies towards CSR principles.
3. determining the degree of implementation of practical energy saving measures, reduction of water waste, rational use of waste, etc.
4. the identification of trends in the hospitality sector regarding the implementation of Environmental Management Systems, the acquisition of eco-labels and certificates and, in general, the environmental awareness of businesses on sustainable development issues.

The factors examined were the management's philosophy on environmental protection issues, the use of human resources (participation, training, awareness raising), the use of new technologies and information from relevant organizations and institutions as well as the motivation for continuous improvements and the acquisition of green certification.

The target population of the research concerned the accommodation businesses of the Region of Epirus and the island of Lefkada, due to the proximity and interest of the area to be studied. The survey was conducted between May and June 2020 (10/05/20–30/06/20) and was based on the voluntary participation of businesses, and their participation was anonymous. From the initial population census and from the information collected during the survey, a total of 66 hotels and accommodation establishments were identified. The aim when conducting the survey was to achieve as wide a spread as possible in the representation of accommodation and hotels, which appears to have been achieved.

The survey was mainly conducted electronically, and the data was initially collected by sending the questionnaires to the accommodation businesses through the local Chambers and the Hotel Associations and Federations. Subsequently, as the number of responses was not satisfactory, due to the problems created by the COVID-19 pandemic in the hospitality industry, actions were taken to communicate directly with the informants for confirmation and reminders, as well as sending again by e-mail to each company individually and not through Federations or Associations. After the collection of primary data from the informants was completed, the statistical program SPSS was used for the analysis of the data, as well as for the extraction of statistical results and conclusions.

Some important issues encountered in the conduct of the research should be mentioned. The biggest difficulty in the research was primarily the reversal of the original planning of the survey, in that the timeline set by the researcher towards the beginning of March 2020 coincided with the outbreak of the COVID-19 pandemic, so given the circumstances, the survey has deviated from the planned timetable. The response of businesses was unexpectedly low because most businesses did not operate in the summer due to the pandemic and also did not have the willingness to bother with completing the questionnaire.

Finally, in several cases, either owners or employees contacted by telephone by the researchers were quite reluctant to participate in the survey, claiming that they were busy or were unwilling to disclose business information and data through the survey, although it was clear that the responses were anonymous.

4 Results and Discussion

The survey performed among Epirus and Lefkada hoteliers casts some light on the perception of environmental issues in the context of CSR within the hotel sector and on factors affecting relevant attitudes. The hoteliers participating in the study show that results of the survey on the familiarity and attitude of businesses towards the practice of Corporate Social Responsibility, which has emerged as a new way of doing business that has been gaining ground in recent years, are discouraging.

More specifically, 77.3% of respondents said they were not at all, slightly or moderately familiar with the term, while only 12.7% said they were familiar with the term. Following on from the previous results, most respondents were either undecided (19.7%) or agreed to a small extent (37.9%) with the proposition that CSR concerns all businesses (regardless of their potential or size) in order to benefit their economic, environmental and social effectiveness in the short and medium term. The lack of a clear answer may be related to the low degree of familiarity with the term Corporate Social Responsibility. Only 42.4% strongly agreed with the proposal.

When the hoteliers were asked to report the incentives they considered important for the adoption and implementation of CSR strategies, 42.3% of respondents included in their answers the respect for environmental and social commitments, making this aspect the most important. At the same time, 36.3% of respondents included in their answers the improvement of the company's efficiency, while 18% looked forward to increasing customer satisfaction. Only 7.5% responded that they wanted to increase employee satisfaction through CSR. Overall, the main motivations for adopting CSR strategies related to external corporate image, business performance and customer satisfaction.

The findings of this study about the importance of Corporate Social Responsibility for each respondent's hotel showed that 60.6% of respondents considered Corporate Social Responsibility to be fairly or very important to their hotel, while 37.9% responded that Corporate Social Responsibility is somewhat or less important. Only 1.5% said they considered Corporate Social Responsibility completely unimportant.

In any business challenge, such as the adoption of Corporate Social Responsibility practices, public policy plays an important role. Fiscal incentives such as capital subsidies, operating grants and tax reductions would increase CSR activities for 87.8% of respondents (they considered it quite likely or very likely), while 12.2% considered it somewhat likely or only slightly likely that government interventions would increase actions. Thus, the above results emphasize the hoteliers' willingness to implement practices if they have the support of and information from the government.

Hoteliers also reported that the non-implementation of CSR strategies is more related to lack of information and extra (high) costs than to the suitability of staff, lack of time or not being seen as necessary (Table 1). Given the limited knowledge of CSR of the participating enterprises, as found in the survey, the bodies responsible for tourism, associations, and federations, should be more active and informed on CSR issues, and disseminate this information to their members. The State should provide direct information and cooperate with the Chambers of Commerce, primarily of the Prefectures, as most hotels and accommodation establishments belong to them, and inform businesses about the multiple benefits and advantages of CSR.

The survey also points to several issues that need to be further investigated in relation to the implementation of measures to protect the environment (energy and water saving, recycling, responsible waste management, use of renewable energy sources). In general, it appears that companies do not give so much importance to the use of renewable energy sources, while on the contrary there are several that give importance to the recycling process, with a percentage (21.2%) fully implementing any measures required in this direction (Table 2). Energy-, water-, and waste -related issues, recycling and use of renewable energy sources are the areas targeted by the majority of pro-ecological solutions. The types of measures are additionally influenced by corporate policy, the level of environmental awareness at management level and the financial state of the establishment Hotel and tourism associations, including hotel corporations, are believed to have an important role in promoting environmental awareness and advocating more sustainable practices among the operators of the hospitality industry. This study illustrates the still low level of corporate hotel market development in Epirus and Lefkada.

The hoteliers of Epirus Prefecture and Lefkada participating in this survey were also asked to rank the incentives capable of encouraging them to get involved in environmentally oriented initiatives according to their importance and using the 5-point Likert type scale. For the purpose of analysis, average values were calculated (Table 3).

An overwhelming majority (69.7%) considered reducing operating costs most important, with the remaining respondents attaching importance to it (answers 1–3 cover 93.9%). This result coincided with the findings of our research where 87% of respondents stated that fiscal incentives such as capital subsidies, etc., would increase CSR activities for them. Few considered it of minor importance as a motivator (6%).

The reduction of the environmental impact of the business seems to be considered an equally important factor as 40.9% considered it important. On the other hand, the assets of a commercial nature were considered somewhat less important compared to the previous ones. Comparing the assets of a commercial nature to the availability of professional advice, the research found that the brand name and the trademark seem to be more important as incentives than professional opinions, since for the former 72.7% responded that they are quite or very important, while for professional opinions the corresponding figure was 62.2%.

Table 1 Main reasons for non-implementation of CSR policies by hospitality businesses in the study area (% of respondents on 5-point Likert scale)

	Very important 1 (%)	2 (%)	3 (%)	4 (%)	Not important 5 (%)
High costs	51.5	18.2	16.7	7.6	6.1
Lack of information	56.1	19.7	12.1	9.1	3.0
Lack of appropriate staff	39.4	21.2	22.7	12.1	4.5
Lack of time	34.8	18.2	22.7	13.6	10.6
Not considered necessary	6.1	10.6	42.4	6.1	34.8

Table 2 Main areas of environmental activities undertaken by hoteliers (% of respondents on 5-point Likert scale)

	Full implementation 1 (%)	2 (%)	3 (%)	4 (%)	No implementation 5 (%)
Energy conservation	16.7	47.0	22.7	9.1	4.5
Water conservation	13.6	39.4	31.8	12.1	3.0
Recycling	21.2	33.3	22.7	16.7	6.1
Responsible waste management	18.2	39.4	24.2	12.1	6.1
Use of renewable energy sources	7.6	31.8	22.7	12.1	25.8

Table 3 Incentives to introduce more environment-friendly initiatives (% of respondents on 5-point Likert scale)

	Very important 1 (%)	2 (%)	3 (%)	4 (%)	Not important 5 (%)
Reduction of operating costs	69.7	13.6	10.6	3.0	3.0
Reduction of the environmental impact of the business	40.9	39.4	13.6	1.5	4.5
Assets of a commercial nature (brand name, trademark)	30.3	42.4	16.7	6.1	4.5
Availability of professional advice	25.8	36.4	27.3	6.1	4.5
Satisfaction of customer needs	47.0	34.8	10.6	4.5	3.0
Improving the image of the business	48.5	42.4	1.5	6.1	1.5

Customer satisfaction is an important motivator for environmentally friendly initiatives according to 47% of the respondents who answered that this factor is very important. Improving the company's image as an incentive to adopt environmentally friendly measures is similar to the results obtained from the answers to the previous question, which is to be expected since meeting customers' needs contributes to improving the image of the company.

The availability of professionals advice seems to have the least weight as a motivator compared to the previous ones, as only 25.8% considered it to be very important.

As it is known, environmental aspects are one of the pillars of Corporate Social Responsibility. The participants' assessments of the trend in demand for green hospitality in the near future showed that 39.4% were confident that there will be a large increase in the near future, while 33.3% thought it is very likely. Just 9.1% believed it will not increase at all or very little. The creation of an environmentally-sustainable "green hotel market" is one of the challenges today and the respondents believe that there will be benefits through environmental protection initiatives, with 71.2% having respond that they strongly or almost strongly agreed that there are sufficient benefits through environmental initiatives and only 1.5% strongly disagreed. The trend for demand for greener hotels is stronger nowadays and the companies that take the lead by undertaking environmental actions and efforts will reap most of the early benefits.

5 Conclusions

Several studies highlight the growing interest of consumers in products and services that minimise negative impacts, or better still, make a positive contribution to areas such as the environment and society. Similarly, previous studies show that business leaders recognise their social and environmental responsibilities, resulting in a proliferation of companies that, realising the shift in consumer attitudes and influence, are taking initiatives that have a positive impact on society and the environment. The present study shows that although hoteliers in the study area are not familiar with the relevant terminology, in reality the majority of them understand and recognise the need to undertake CSR business initiatives. Since the region of Epirus and Lefkada are characterized by their rich natural and cultural wealth, and as it is widely accepted that the environmental and cultural resources of a region can be the driving force for its development through appropriate planning based on the promotion of innovative and flexible measures, tourism businesses are willing to choose a model of sustainable development that will enhance their image and reputation as environmentally responsible hotels which, especially today, is a factor in choice for customers.

At the same time the overwhelming majority considers that the reduction of operating costs, the satisfaction of customer needs, the improving of the image of the business, and the reduction of the environmental impact of the business are the most important benefits of CRS. Thus, the main motivations for adopting CSR strategies are related to external corporate image, business performance and customer satisfaction, showing limited interest for increasing employee satisfaction. Based on the above, the profile of tourism entrepreneurs participating in the survey seems to adopt first generation CSR practices.

Participants indicate that the non-implementation of CSR practices is due to high costs, lack of information and lack of qualified staff, while arguing that the provision of additional information by relevant local and national sectoral bodies as well as financial incentives would significantly increase the adoption of CSR practices.

The period during which the survey was conducted is likely to have influenced not only the response rate of hoteliers in the Epirus and Lefkada region but also the results of the survey. It would be of research interest to revisit the questions in a more positive business context in the post-COVID period as well as in destinations with a diversified tourism profile and level of development.

References

Andriof, J., Waddock, S.: Unfolding stakeholder engagement. In: Andriof, J., Waddock, S., Husted, B., Rahman, S.S. (eds.) Unfolding Stakeholder Thinking, pp. 19–43. Routledge (2002). https://doi.org/10.4324/9781351281881

Bohdanowicz, P., Zientara, P.: CSR-inspired environmental initiatives in top hotel chains. In: Leslie, D. (ed.) Tourism enterprises and the sustainability agenda across Europe, pp. 93–121. Routledge (2012). https://doi.org/10.4324/9781315550725

Bowen, H.R.: Social Responsibilities of the Businessman. University of Iowa Press (2013). https://doi.org/10.2307/j.ctt20q1w8f

Brown, T.J., Dacin, P.A.: The company and the product: corporate associations and consumer product responses. J. Mark. **61**(1), 68–84 (1997). https://doi.org/10.2307/1252190

Carroll, A.B.: A three-dimensional conceptual model of corporate social performance. Acad. Manag. Rev. **4**(4), 497–505 (1979). https://doi.org/10.2307/257850

Cvelbar, L.K., Dwyer, L.: An importance-performance analysis of sustainability factors for long-term strategy planning in Slovenian hotels. J. Sustain. Tour. **21**(3), 487–504 (2013). https://doi.org/10.1080/09669582.2012.713965

De Grosbois, D.: Corporate social responsibility reporting by the global hotel industry: commitment, initiatives and performance. J. Hosp. Manage. **31**, 896–905 (2012). https://doi.org/10.1016/j.ijhm.2011.10.008

De Grosbois, D.: Corporate social responsibility reporting in the cruise tourism industry: a performance evaluation using a new institutional theory based model. J. Sustain. Tour. **24**(2), 245–269 (2015). https://doi.org/10.1080/09669582.2015.1076827

Esparon, M., Gyuris, E., Stoeckl, N.: Does ECO certification deliver benefits? An empirical investigation of visitors' perceptions of the importance of ECO certification's attributes and of operators' performance. J. Sustain. Tour. **22**(1), 148–169 (2014). https://doi.org/10.1080/09669582.2013.802325

Font, X., Guix, M., Bonilla-Priego, M.J.: Corporate social responsibility in cruising: Using materiality analysis to create shared value. Tour. Manage. **53**, 175–186 (2016). https://doi.org/10.1016/j.tourman.2015.10.007

Frederick, W.C.: The growing concern over business responsibility. Calif. Manage. Rev. **2**, 54–61 (1960). https://doi.org/10.2307/41165405

Freeman, R.E., Velamuri, S.R.: A new approach to CSR: company stakeholder responsibility. In: Kakabadse, A., Morsing, M. (eds.) Corporate Social Responsibility. Palgrave Macmillan (2006). https://doi.org/10.1057/9780230599574_2

Freeman, R.E.: Strategic management: a stakeholder approach. Pitman (1984)

Granovetter, M.: Economic action and the problem of embeddedness. Am. J. Sociol. **91**(3), 481–510 (1985). https://www.jstor.org/stable/2780199

Harrison, J.S., Bosse, D.A., Phillips, R.A.: Managing for stakeholders, stakeholder utility functions & competitive advantage. Strat. Manage. J. **31**(1), 58–74 (2018). https://doi.org/10.1002/smj.801

Haski-Leventhal, D.: Strategic Corporate Social Responsibility: Tools & Theories for Responsible Management. SAGE Publications (2018)

Jamali, D.: A stakeholder approach to corporate social responsibility: a fresh perspective into theory and practice. J. Bus. Ethics **82**(1), 213–231 (2008). https://doi.org/10.1007/s10551-007-9572-4

Jones, P., Hillier, D., Comfort, D.: Sustainability in the hospitality industry: some personal reflections on corporate challenges and research agendas. Int. J. Contemp. Hosp. Manag. **28**(1), 36–67 (2016). https://doi.org/10.1108/IJCHM-10-2012-0180

Karlsson, L., Dolnicar, S.: Does eco certification sell tourism services? Evidence from a quasi-experimental observation study in Iceland. J. Sustain. Tour. **24**(5), 694–714 (2015). https://doi.org/10.1080/09669582.2015.1088859

Ketola, T.: From CR-psychopaths to responsible corporations: waking up the inner sleeping beauty of companies. Corp. Soc. Responsib. Environ. Manag. **13**, 98–107 (2006). https://doi.org/10.1002/csr.113

Kotler, P.: Marketing Management. Prentice-Hall (2003)

Kotler, P., Kartajaya, H., Setiawan, I.: Marketing 3.0. From Products to Customers to the Human Spirit. Wiley (2010)

Kumar, R.B., Kumar, M.: Corporate social responsibility in Indian tour operation industry. Int. J .hosp. Tour. Syst. **11**(2), 82–88 (2018)

Lee, N.R., Kotler, P.: Social marketing: Behavior change for social good (6th ed.). SAGE Publications,Inc. (2019)

Lund-Thomsen, P.: Towards a critical framework on corporate social and environmental responsibility in the South: the case of Pakistan. Development **47**(3), 106–113 (2004). https://doi.org/10.1057/palgrave.development.1100059

McIntosh, M.: Raising a Ladder to the Moon: The Complexities of Corporate Social and Environmental Responsibility. Palgrave Macmillan (2003)

McIntosh, M.: Corporate citizenship. In: Visser, W., Matten, D., Pohl, M., Tolhurst, N. (eds.) The A to Z of Corporate Social Responsibility, pp. 97–102. Wiley (2007)

Meyer, D.F., Meyer, N.: The relationship between the tourism sector and local economic development (LED): the case of the Vaal Triangle Region, South Africa. J. Environ. Manage. Tour. **7**(3), 466–472 (2016). https://doi.org/10.14505/jemt.v7.3(15).12

Muafi, M.: From Company reputation to environmental performance: the context of corporate social responsibility port manager in Indonesia. J. Environ. Manage. Tour. **8**(7), 1386–1398 (2017). https://doi.org/10.14505//jemt.v8.7(23).08

Murphy, P.E., Schlegelmilch, B.B.: Corporate social responsibility and corporate social irresponsibility: Introduction to a special topic section. J. Bus. Res. **66**(10), 1807–1813 (2013). https://doi.org/10.1016/j.jbusres.2013.02.001

Parmar, B.L., Freeman, R.E., Harrison, J.S., Wicks, A.C., Purnell, P., de Colle, S.: Stakeholder theory: The state of the art. Acad. Manage. Ann. **4**, 403–445 (2010). https://scholarship.richmond.edu/management-faculty-publications/99

Prahalad, C.K., Ramaswamy, V.: Co-creation experiences: the next practice in value creation. J. Interact. Mark. **18**, 5–14 (2004). https://doi.org/10.1002/dir.20015

Salzmann, O., Ionescu-Somers, A., Steger, U.: The business case for corporate sustainability: literature review and research options. Eur. Manag. J. **23**(1), 27–36 (2005). https://doi.org/10.1016/j.emj.2004.12.007

Sheldon, P., Park, S.Y.: An exploratory study of corporate social responsibility in the US travel industry. J. Travel Res. **50**(4), 392–407 (2011). https://doi.org/10.1177/0047287510371230

Smith, N.C.: Corporate social responsibility: whether or how? Calif. Manage. Rev. **45**, 52–76 (2003). https://doi.org/10.2307/41166188

Swift, T., Zadek, S.: Corporate Responsibility and the Competitive Advantage of Nations. The Copenhagen Center & Account Ability (2012)

Turban, D.B., Greening, W.D.: Corporate social performance and organizational attractiveness to prospective employees. Acad. Manag. J. **40**(3), 658–672 (1997). https://doi.org/10.2307/257057

World Business Council for Sustainable Development: Corporate Social Responsibility: Meeting Changing Expectations. WBCSD Publications (1999)

The Role of the Belt and Road Initiative in Driving Tourism in the Post COVID-19 Era

Jacob Wood, Haejin Jang, Zahra Pourabedin, Redeem Sabacan, Benedict Atkinson, Hong Hanh Nguyen, and K. Thirumaran

Abstract

As the Central Asian region emerges from the grips of the COVID-19 pandemic, it can look ahead with a sense of optimism. Nestled within its confines, lie many unique tourist attractions that provide a variety of year-round quality experiences, which can form the catalyst for sustainable economic growth over the long-term. This study examines the important role the Tourism sector can play in driving economic growth along in the Central Asian corridor. It also details the key policy initiatives that Governments can look to implement and in particular, the role the Belt and Road Initiative (BRI) plays in driving key infrastructural projects that enhance the level of connectivity to tourism markets in the region. In addition to this, our study assesses what the potential green growth developmental agenda might look like for the tourism sector. Our research showed that the BRI will potentially prove to be the largest infrastructure project in history, enormously affecting the growth of tourist industries in the regions including and surrounding the Silk Road In these regions, institutional growth and social change, both affected by technological developments, will likely drive the growth of tourism industries. The BRI provides the backbone to these developments.

Keywords

Tourism • COVID-19 • Economic development • Sustainability

J. Wood (✉) · B. Atkinson · H. H. Nguyen · K. Thirumaran
JCU Singapore Business School, James Cook University Singapore, 149 Sims Drive, Singapore, Singapore
e-mail: jacob.wood@jcu.edu.au

B. Atkinson
e-mail: benedict.atkinson@jcu.edu.au

H. H. Nguyen
e-mail: honghanh.nguyen@my.jcu.edu.au

K. Thirumaran
e-mail: k.thirumaran@jcu.edu.au

H. Jang
JCU Singapore Learning Centre, James Cook University Singapore, 149 Sims Drive, Singapore, Singapore
e-mail: haejin.jang@jcu.edu.au

Z. Pourabedin
College of Business, Shepherd University, Shepherdstown, WV 25443, USA
e-mail: zpourabe@shepherd.edu

R. Sabacan
Central Luzon, Philippines

1 Introduction

Since the beginning of the COVID-19 pandemic in early 2020, the global economy has been dealt a significant economic shock (Wood et al. 2022a). While the impact has been felt across supply chains (Wu et al. 2021a, b, c), cross-border e-commerce (Du et al. 2021), and intra-industry trade (Wood et al. 2021)—the economic and social impacts on the tourism and hospitality sectors has been profound (Qiu et al. 2020; Thirumaran et al. 2021, 2022). Exacerbating these problems, particularly in Asia, has been the China—US trade war, which has negatively impacted the two countries specifically involved but also key trading partners (Wood et al. 2020, 2022b; Wu et al. 2021c). Work has also examined the way that infrastructure has mitigated the impact on the tourist sector of natural disasters in the Asian region (Azzali et al. 2020) so that more resilient communities could be developed (Chaiechi and Wood 2022).

The inward looking perspectives that have generated as a result of COVID-19 (Atkinson et al. 2022), and a growing focus on ensuring economic development is achieved in a more environmentally sustainable manner (Sikder et al. 2022), will mean that the new growth path forged by the tourism sector may be slightly different than what had previously been prescribed. Despite the significant impacts to the tourism sector, the Central Asian region is very well

S. Kostopoulou et al. (eds.), *Cities' Vocabularies and the Sustainable Development of the Silkroads*, Advances in Science, Technology & Innovation, https://doi.org/10.1007/978-3-031-31027-0_22

placed to benefit from the longer-term tourism trends that were in existence before the pandemic once international borders fully reopen and international travel rebounds. One of the key opportunities is taking advantage of the demand for sustainable tourism practices and authentic, experiential travel. In order to capitalize on such opportunities, countries in the region must work to further develop. A key catalyst behind these developments is the Chinese led Belt and Road Initiative (BRI).

An important developmental project in the Central Asian region is the Silk Road, which is steeped in tradition, culture, and trade in the most precious commodities of ancient times: silk, spices, and gems (ADB 2021) and the BRI, which was established by the Chinese Government in 2013 (Wang et al. 2020b, c, d). The main objective of the BRI project, is to "promote connectivity between the Asian, European and African continents and their adjacent seas, establish and strengthen partnerships among the countries along the Belt and Road, set up all-dimensional, multi-tiered and composite connectivity networks, and realize diversified, independent, balanced and sustainable development in these countries" (Guoqiang 2015, p. 154). It seeks to build upon on the rich trade opportunities and negotiated trade agreements that exist in the region and beyond (Wood and Khan 2015; Wood 2017)Encapsulated within this broad initiative lie two specific initiatives, these are firstly, the land-based Silk Road Economic Belt, comprising six development corridors, and secondly, the sea-based 21st Century Maritime Silk Road.

The Silk Road Economic Belt's primary focus is the development of six significant infrastructural projects that improve the level of connectivity and economic cooperation that exists in the Eurasian region. These six corridors are: (1) the New Eurasian Land Bridge Economic Corridor (NELBEC); (2) the China—Mongolia—Russia Economic Corridor (CMREC); (3) the China—Central Asia—West Asia Economic Corridor (CCWAEC); (4) the China—Indochina Peninsula Economic Corridor (CICPEC); (5) the Bangladesh—China—India—Myanmar Economic Corridor (BCIMEC); and (6) the China—Pakistan Economic Corridor (CPEC) (BRI 2022).

Meanwhile, the 21st Century Maritime Silk Road seeks to better connect a series of important sea routes, namely China to Southeast Asia, Indonesia, India, the Arabian Peninsula, Somalia, Egypt and Europe, encompassing the South China Sea, Strait of Malacca, Indian Ocean, Gulf of Bengal, Arabian Sea, Persian Gulf and the Red Sea (BRI 2022). Overall, these two projects involve more than 71 countries and almost 65% of the global population globally (Rauf et al. 2020). Moreover, it is forecast that the BRI will, over the coming decades, grow to become the biggest infrastructure project in human history, with more than $8 trillion USD invested (Hughes 2019). The BRI initiative is expected to continue to play an indispensable role in fostering the sustainable development of the tourism sector across the Central Asian region. From a tourism perspective, the flow of tourists from these BRI infrastructural developments, combined with the vast array of natural and man-made historical attractions means that the sector will play a pivotal role in enhancing trade and sustainable development in the region (Santana-Gallego et al. 2016).

Previous research on the important role the tourism sector plays in the Eurasian region has examined a range of areas including: the role the BRI plays in driving sustainable tourism development, infrastructure development, and residents' standard of living (Mamirkulova et al. 2020; Li et al. 2020). For instance, the infrastructure initiative of New Silk Road leads to the improvement in tourist experiences, residents' standard of living, employment opportunities, the conservation of cultural and historical values, and natural landscapes (Bi et al. 2020; Kanwal et al. 2020; Shafiee et al. 2019; Wang et al. 2020a).

Based on these empirical works, and the key areas of academic focus, we have developed several research questions (RQ):

RQ1. To what extent does the BRI contribute to sustainable development in the region?

RQ2. How does the BRI benefit the countries in the BRI highway benefit?

RQ3. Does the BRI actually contribute to improvements in the lives of the people?

RQ4. How does the BRI bring China, Europe and Africa closer in international trade and people to people exchanges?

To answer the research questions, our research utilized a qualitative systematic review approach, also referred to as qualitative evidence synthesis, which is a method for integrating or comparing findings from qualitative studies. This method looks for 'themes' or 'constructs' that lie in or across individual qualitative studies. This method is an intentional and coherent approach to analyzing data across qualitative studies. It is a process that enabled the researchers to identify specific research questions and then will search for, appraise, summarize, and combine qualitative evidence to address the research questions. This process used rigorous qualitative methods to synthesize existing qualitative studies to construct greater meaning through an interpretative process. (Erwin et al. 2011). Sandelowski and Barroso (2007) described qualitative synthesis as an interpretation of qualitative findings that are themselves interpretive syntheses of data including phenomenologies, ethnographies, grounded theories, and other integrated and coherent descriptions or explanations of phenomena, events, or cases that were the hallmarks of qualitative research.

The rest of the study is as follows. Section 2 examines the significant empirical studies around the role that tourism will play in driving economic growth in the post-COVID-19 era and in how the Silk Road and BRI may act as important developmental catalysts for the sector. Section 3 outlines some of the key policy initiatives that should be implemented to further enhance Tourism sector growth and economic development in the Eurasian region. Finally, Sect. 4 then draws a series of relevant conclusions and areas of research that should form key areas of focus in the future.

2 Literature Review

The tourism sector has been and will continue to play an important role in the development of the Eurasian region. By providing a detailed assessment of the extant empirical literature, we provide a detailed assessment of the tourism sector to the region's development and in particular, the key part both the Silk Road and the BRI will play in driving further tourism development in the post-COVID-19 era. In addition to this, we also examine leading scholars' works on how a potential green growth developmental agenda might also enhance the tourism sector moving forward.

Nowadays, tourism is a central part of a country's development strategies (Arghutashvili 2018) as it contributes to the GDP, creating jobs, as well as many service sectors such as transportation and communication (UNWTO 2017). The World Travel and Tourism Council (WTTC) succinctly explained how tourism contribute to the development and growth of the global economy as follows:

In 2013, Travel and Tourism's total contribution to the global economy rose to 9.5% of global GDP (US $7 trillion), not only outpacing the wider economy, but also growing faster than other significant sectors such as financial and business services, transport and manufacturing. In total, nearly 266 million jobs were supported by Travel and Tourism in 2013—1 in 11 of all jobs in the world. The sustained demand for Travel and Tourism, together with its ability to generate high levels of employment continues to prove the importance and value of the sector as a tool for economic development and job creation (WTTC 2014, Foreword). However, the COVID-19 pandemic has unfortunately put the fragile tourism sector under unprecedented pressure, triggering significant concerns for the future development and survival of this vulnerable industry (Kaushal and Srivastava 2021).The beneficial relationship between New Silk Road infrastructure and demand in the tourism sector has been widely accepted (Mamirkulova et al. 2020). In addition, previous studies revealed that community members agreed that the development of tourism infrastructure leads to growing economic stability and a more sustainable tourism sector over the long-term (Almeida-García et al. 2016; Látková and Vogt 2012). Likewise, other studies have proposed that investment in infrastructure positively contributes to achieving economic, social, cultural, and environmentally sustainable development in areas attracting tourists (Fu et al. 2020; Greiner 2010; Kavaliausk and Kocyt 2014). Tourism development programs not only bring in revenue but create a multiplier economic effect of creating new jobs, businesses and industries along the New Silk Road (Financial Times 2015). As a remarkable strategy of the initiative to revive the Silk Road, the bilateral relations of the Pakistani and Chinese governments have marked a significant milestone, represented by the launch of their own mega-project, titled The China-Pakistan Economic Corridor, which is valued at more than $54 billion USD. This significant joint-venture was developed in an attempt to better connect Pakistan with other infrastructural projects in the region. This in turn provides a key catalyst for helping create innovative tourist experiences and exceptional service quality for tourists visiting these new destinations (Ali et al. 2017).

The Eurasia region has immense potential in terms of tourism development. However, this region traditionally faces several issues in the areas of transportation, security, infrastructure, and economic development (Koh and Kwok 2017). For instance, the main reason for the low number of Chinese tourists in Central Asia is "shortage of tourism infrastructure, unprofessional tourism workers, and low service level" (Li 2016, p. 128.). In spite of the importance of the tourism industry for economic development, Eurasia is still in the early stages in using its resources for promoting tourism destinations. Currently there are two programmes which are working on tourism development along the Silk Road. The first one is the UNWTO Silk Road tourism program (World Tourism Organization and Global Tourism Economy Research Centre 2014) which aims to develop a sustainable tourism programme along the Silk Road. The second one is The Shanghai Cooperation Organization (SCO), a Central Asian focused intergovernmental organization founded by China, Russia, Kazakhstan, Tajikistan, Kyrgyz Republic and Uzbekistan in 2001, which aims to develop a common tourism program for Central Asia. However, according to Koh and Kwok (2017) SCO has shown little progress in terms of promoting tourism. Costa et al. (2013) suggested that SCO can learn from the EU's success story in regional tourism development to promote Eurasia as a single tourism destination.

Liu and Suk (2021) explored tourism cooperation between Azerbaijan and China after BRI. The authors found that the BRI initiative has led to rapid tourism development between Azerbaijan and China to promote cultural exchange and economic cooperation (Liu and Suk, 2021). One of the main benefits of BRI for Azerbaijan was the infrastructural development which is one of the main requirements of sustainable tourism development. Mamirkulova et al. (2020) divided the New Silk Road tourism infrastructure to three categories of physical infrastructure, social infrastructure, and environmental infrastructure. In 2017, the

Baku-Tbilisi-Kars railway officially entered into service, offering high potential for the comprehensive advancement of the Belt and Road construction in Azerbaijan (Hajizade 2015). Modern and new tourism infrastructure influence the tourists' satisfaction, increasing the number of loyal tourists (Bi et al. 2020; Mathew and Sreejesh 2017; Shafiee et al. 2019).

In 2019, China and Azerbaijan signed a Memorandum of Understanding on Tourism Implementation Plan. By holding B2B tourism exhibitions and opening direct flights with Beijing, Azerbaijan could see the rise in the number of inbound Chinese tourists (Zeybek 2021). However, Azerbaijan still needs the support of local entrepreneurs and institutions to benefit from increasing BRI related tourism demand (Of routes and corridors: Challenges and opportunities for Silk Road destinations in the southern Caucasus). According to Liu and Dunford (2016, p. 335), tourism is a key service sector representing symbolic projects to 'increase support and capacity for the BRI' with the aim to deepen cooperation between government and the business sectors across participating countries. Another study (Daye et al. 2020) examined the attitudes of local tourism stakeholders in Kazakhstan toward BRI. Finding showed that tourism, civic and business stakeholders in Kazakhstan strongly support the effective role of BRI in tourism development. BRI modernization projects in Kazakhstan have helped the tourism sector to improve the tourism infrastructure and amenities such as restaurants, attractions, and hotels to expand international tourism. The number of foreign tourists' arriving at Kazakhstan in 2017 reached 4,559,500, which amounts to 3077.5 million US dollars or 1.6 percent of the world GDP (Crotti and Misrahi 2017) and created 150,585 jobs. Mamirkulova et al. (2020) revealed the positive role of tourism opportunities created by the New Silk Road Initiative. The residents' quality of life is improved through benefits such as higher income, new job opportunities, and better leisure and spiritual life in Azerbaijan. The four indicators of quality of life are Material well-being, community well-being, health and safety well-being, and emotional well-being. Although tourism infrastructure is a powerful tool for improving the quality of life. Economic, social, cultural, and environmental factors are important as well (Mamirkulova 2020).

In addition, BRI via logistics development will improve tourist accessibility. For instance, the China US$2 billion China–Kyrgyz–Uzbekistan railway project opens the Asia–Pacific market for Uzbekistan which can facilitate tourism flow. Within the Eurasia region, for example, Georgia is one of the countries which has an increasing number of inbound tourists. The country expects long term economic benefits resulting from tourism (Arghutashvili 2018). However, Georgia sees some barriers such as lack of development in infrastructure, low level of social development as well as expensive hotels and air tickets. But according to Pechlaner et al. (2021), the BRI can led to local service industry development in Georgia due to its position as a transit hub. Pechlaner et al. (2021) did a case study to investigate the linkage between BRI and tourism in Georgia. The results of the 19 semi-structured interviews with local stakeholders, national authorities and international experts in regional development showed positive statements regarding the BRI benefits and economic development. In addition, the results showed that BRI can foster tourism destination development via opportunities for farm tourism and small hotels. On the other hand, tourism is seen as a beneficial local service that can be an extra feature of an increased mobility and attract income flows. This claim is in line with the EU initiative in Guria which has been helping and educating local stakeholders to promote and develop their region as tourism destination. BRI can enable countries to enhance their destination image. An evidence which supports the positive role of BRI in destination image creation is the increasing number Chinese medical tourists in Malaysia with the establishment of BRI (Cham et al. 2021), helping Malaysia to improve its image as a medical tourism destination. BRI related initiatives such as E-visa and collaboration among hospitals in China and Malaysia led to increasing number of Chines medical tourists (The Star 2018).

3 Policy Development and Discussion

3.1 Institutional Quality

Institutional quality has long been seen to play a key role in the field of economic development (Heshmati et al. 2019; Kim and Wood 2020; Zhang et al. 2022). Mushtaq et al. (2020) opined that institutional quality signifies the quality of institutions that govern laws, government property rights, constitution, and traditions necessary for personal relationships amongst stakeholders. Iftikhar et al. (2022) assumed that institutional quality is a determining factor in a nation's tourism development and sustainability. Using the ARDL model in the Indian context, for instance, Mushtaq et al. (2020) noted institutional quality's positive influence on international tourism demand from 1955 to 2016. Similarly, with the aim to determine the effect of institutional quality in Malaysia's tourism and financial development, Lee (2020) utilized the country's data from 1996 to 2015 and found that government effectiveness and corruption control contributed a pivotal role on tourism and financial development.

A vast array of literature (see, Mushtaq et al. 2020; Iftikhar et al. 2022; Ghalia et al. 2019; Atkinson and Hamilton 2003; Hyndman 2015; Khan et al. 2020; Balli et al. 2018; Demir and Gozgor 2018; Kim et al. 2018; Tang 2018; Lai

2020; Lee 2020; Stoever 2012; Meyer et al. 2017; Bekhet and Latif 2018) has explored institutional quality's role in tourism and sustainable development. Of these, Iftikhar et al. (2022) observed that higher institutional quality can affect the tourists' flow. Atkinson and Hamilton (2003) suggested that institutional quality can play a crucial role in the economy's sustainability in natural resource-abundant countries. For Hyndman (2015), both political risk and poor governance can be detrimental in tourism undertakings. Moreover, Ghalia et al. (2019) explored institutional quality's impact related to power, socio-economic factors, and political risks on tourism and noted that institutional quality is pivotal in determining factor for tourist flows. Balli et al. (2018), Demir and Gozgoz (2018), and Kim et al. (2018) observed that recent studies investigate the combined facets of political influences on the tourism sector (e.g., nation's regional integrity, safety, security, social stability, institutions, and peace) that affect tourist flows.

In 2015, the BRI witnessed significant Chinese willingness to assist with development and trading frameworks with Central and Eastern European countries (Ministry of Foreign Affairs of PRC 2015). The Suzhou Guidelines which emerged from the 4th Summit between China and the Central and European countries clearly outlined not only infrastructure and connectivity development funded by China, but also an attempt by China to increase economic and cultural exchanges. While trade related agricultural and forestry cooperation has been a cornerstone of the Suzhou guidelines, China has also managed to imprint its own science and technology exchanges with the Eastern European countries along with increased people to people exchange. In International relations and trade, these types of summit proclamation and outcomes are set to significantly increase the long arm of friendship and cooperation across the Asian European continent.

Hyndman (2015) and Khalid et al. (2019) surmised that the presence of political instability may impede tourist operators and service providers to sustain their undertakings due to the lack of security and peace. This is related with what Tang (2018) revealed that compared with voice and accountability, global tourists are more concerned on governmental effectiveness, political stability, regulations, laws, and corruption. Khan et al. (2020) reported that terrorism is detrimental to the growth of the tourism industry as it can coerce tourists to change their travel destination since individual safety has become a priority. Corruption, on the other hand, has varied effect on tourism. Corruption may damage the growth of tourism sector as it affects the nation's image, or it can raise business transactions by criminal and illegal activities such as gambling and prostitution.

3.2 Promoting and Communicating the Benefits of BRI

Globally, cities and towns face an array of challenges—some of these are unique, and a number of them are common. Hence, more than ever, the demand for effectual cooperation and knowledge sharing among cities and local governments has become essential to deal with complex and emergent urban issues. In the early years of city-to-city communication and collaboration connections, most peer-to-peer connections were established in developed countries. Later, many developing countries too joined the bandwagon and benefitted from decentralized cooperation between foreign cities. Also contributing key role in developing connections between cities for mutual benefit are the Local Government Associations. Specifically, the United Cities and Local Governments in Asia Pacific (UCLG ASPAC) has consistently improved communication among local governments and enabled collaboration between local governments to leverage collective knowledge and resources of local governments in the region. Further, UCLG ASPAC works on augmenting cooperation and strengthening communication among the member cities by constituting Belt and Road Local Cooperation, a committee composed of over 80 member cities connected in the Belt and Road Region.

The COVID-19 pandemic has accelerated the pace and frequency of communication among cities. Today's generation witness extensive paradigm shift, brought about by digital technologies, in the field of intercity-community and decentralized cooperation vis-à-vis immediate benefits of such connections when cities with standing relationship with foreign local governments helped their peer in a time of need. Due to individual city's preferences on deciding communication channels, language constraints, machineries at local government's disposal, government officials' capacity among others, developing a formal city-to-city cooperation and communication mechanism framework has become a complicated endeavor. As such, the current pressing need is for member cities to develop communication and international cooperation initiatives and based on commonality, all these have to be integrated in development cooperation programs. What is perceptible at the moment is that every local government establishes its priorities based on the citizenry's demands, challenges they face, available resources, workforce competence, and long-term plans. Hence, assessment of individual city's joint economic, social, ecological development goals can set the foundation for building a communication strategy for decentralized cooperation among local governments. To access the best resources and ideas available, innovative tools, sustainable practices, and insights from local governments' experiences in specialized domains, strategic communication and collaboration between cities have to be established.

Finally, Liow et al. (2021) concluded that, at least for now, the prospect of elevating the BRI to the global scale seems remote since great power rivalry, retreat of liberalism and COVID-19 pandemic strengthens the notion of the sovereign state and changes the notion of connectivity to one of self-reliance and isolation.

3.3 Engaging Local Stakeholders

Engaging with local stakeholders is a very important component when looking to develop any industry, particular Tourism (Azzali et al. 2020). Based on the collective reports from recent polls and research trend presented by the European Commission in (2021), consumer behavior has been greatly affected by the COVID-19 health crisis. Short-term changes have been noted (e.g., during lockdown and confinement, online transactions and services significantly improved) and are projected to persist in the medium and long term. Likewise, EU foreign travel is also projected to be less intense in the post-COVID-19 crisis than precrisis levels (Orîndaru et al. 2021).

In their study on government measures and tourism industry performance during COVID-19, Aharon et al. (2021) recommended that owners and businesses operators in the hospitality industry need to prepare for future activity along with the post COVID-19 tourism environment. Further, stakeholders need to modify their businesses for both the short and long terms and minimize their enterprises' sensitivity to the tendency of future government restrictions even after the resumption of tourism economy. They also have to consider innovations in their businesses with regard to social distancing and maintaining a sanitized environment. There is also a need for owners and hospitality service-related vendors (e.g., airline companies, cruise companies and multinational hotel chains) to collaborate with each other, and enforce strict protocols for food handling and preparation. Through offering flexible rebooking options and free health insurance that includes COVID-19, travel agencies can motivate people to travel. Likewise, efficient COVID-19 tests prior to departure and post arrival will help airline companies and airport authorities to restore confidence in the safety of travel, which will encourage tourism and prevent infection in public areas.

To provide better financial and technological support to the various BRI initiatives, a multi-dimension cooperation network needs to be established among partner countries, multilateral development banks and the private sector. Cities and local governments are the main actors for they can assure the realization of a sustainable development agenda and establish an efficient system of sharing knowledge, learning and resources in the post-Pandemic period. Such associations require smooth communication among cities in real-time. To face new challenges, it is needed that the capacities of LGs, especially those in the developing countries, are improved with partnerships and collaborations. It is sure that the experiences and know-how accumulated through peer-to-peer exchanges not only benefit cities and associations directly but can provide valuable inputs to solve global issues.

The pandemic was a huge wake-up call for cities globally, making them realize they cannot 'live like an island' and for them to survive and thrive, collaboration is the key. The current society taking shape cannot afford local governments working silos and needs coordinated, planned action to overcome global issues like the COVID-19, future imminent pandemics, global warming, climate change, and economic recession. Finally, cities must keep the people at the center of any collaboration. This may facilitate in formulating projects and policies that are congruent with the people's needs and not the administration or local government's ideals or visions.

One important consideration that should not be neglected when considering how local stakeholders can engage with BRI-related development is the possibility that such development may result in adverse consequences. BRI development, although potentially offering the means to greatly enlarge tourism in the regions to which the BRI applies, should not be regarded uncritically by tourism providers. Loans made available to them under the BRI could be subject to restrictions controlling how services are provided and onerous repayment conditions. As noted above, people are at the centre of any collaboration. Policy on BRI-related tourism should focus on the needs of the people likely to be affected by agreements reached under the auspices of the BRI.

4 Concluding Remarks and Future Areas of Study

Recognizing the significant contribution of the tourism industry to the growth of national prosperity, governments of developing countries have widely prioritized tourism development in their long-term strategy. For the sustainable development of tourist attractions, research has shown that investment in infrastructure is one of the fundamental factors that positively influence the flow of tourists coming to the destinations (Wang et al. 2020e, 2022). The beneficial relationship between tourism infrastructure and tourism demand not only help boosting the local economy but it also contributes to the cultural, social and environmental aspects of the tourist attractions. In this regard, Eurasia region, with abundance of natural and cultural resources, has not yet optimized their potentials for tourism development. The growth of this sector in the region has been hindered by numerous issues, including: safety and security, transportation and accessibility, as well as general infrastructure and economic development.

However, with the current initiative to develop BRI modernization projects, tourism in the Eurasia region starts witnessing remarkable improvements in various tourism evaluation index. Firstly, the launch of BRI is crucial for tourist accessibility via logistic development. Secondly, BRI increases the opportunity for the region to attract foreign visitors thanks to the expansion of tourism facilities and amenities, such as hotels, restaurants and tourist attractions. Thirdly, BRI opens new rooms for the development of other tourism-related concepts, including but not limited to farm tourism or cultural-focused tourism. Last but not least, the development of BRI helps enhancing the image of Eurasia region as an attractive tourist destination in the global tourism market. To conclude, all the above-mentioned benefits of BRI would only be achievable if relevant local tourism stakeholders of every country in the region fully understand and strongly support the role of BRI for the tourism development of their region.

In terms of policy recommendations, there are several key areas that countries in Eurasia region are suggested to focus on, consisting of: improving institutional quality, promoting and communicating the benefits of BRI, and engaging local stakeholders. For the tourism industry of a destination to develop sustainably, it is essential that sufficient attention is paid to the destination's institutional quality. This is due to the direct impact of this determining factor on the power, socio-economic and political risks of the destination, which may eventually influence the tourist flows to the region (Ghalia et al. 2020). Amongst all push and pull factors to attract visitors to a destination, global tourists are most concerned about the combined facets of political influences on the destination's tourism sector, for instance, government effectiveness, social stability, safety and security, peace, regional integrity and political stability (Balli et al. 2018; Tang 2018). As a consequence, to address this demand of international visitors, cities and local governments of the Eurasia region are required to establish an effective cooperation and knowledge-sharing network. Countries in Eurasia region are in the pressing need for the development of a cooperation program, which takes into consideration the strategic goals of not only each individual country but also the Eurasia region as a whole. Based on commonality and mutual benefits, this cooperative initiative has to undergo the assessment of each local government to determine if the joint strategy is aligned with the nation's vision for economic, social and ecological developments. Also, as earlier noted, tourism providers in individual countries should be cognizant of the potential for agreements reached under the BRI to result in adverse consequences. Development of healthy institutions is an important precursor to the development of healthy tourism industries. Countries engaged with the BRI need to be sure that agreements reached under the BRI offer reciprocal benefits and do not enjoin service provision that is not consistent with national interest.

For the successful implementation of the program, voluntary participation from member countries is required, along with their active role in providing available resources, insights, ideas, as well as innovative practices and tools. This is particularly crucial for the region to successfully solve emergent complex urban issues, especially after the unexpected paradigm shift after the unprecedented global health crisis, resulted in the world's current ever-increasing adoption of digital technologies and the demand for frequent real-time information exchange. To resume the tourism economy after the COVID-19 pandemic, besides the role of local governments to provide necessary supports to BRI initiatives, the private sector and multilateral development banks are strongly encouraged to cooperate with the public sector to establish an efficient system of sharing knowledge and resources, to create the foundation for the sustainable growth of the tourism industry in Eurasia region. Future studies are suggested to investigate more comprehensively the role of infrastructure development for tourism growth, the full potentials of Eurasia as a tourist destination, as well as the most effective marketing tools and strategies to promote the tourism industry of a whole region.

References

Aharon, D.Y., Jacobi, A., Cohen, E., Tzur, J., Qadan, M.: COVID-19, government measures and hospitality industry performance. PLoS ONE **16**(8), e0255819 (2021). https://doi.org/10.1371/journal.pone.0255819

Ali, S.A., Haider, J., Ali, M., Ali, S.I., Ming, X.: Emerging tourism between Pakistan and China: tourism opportunities via China-Pakistan economic corridor. Int. Bus. Res. **10**(3), 204 (2017)

Almeida-García, F., Peláez-Fernández, M., Balbuena-Vazquez, A., Cortés-Macias, R.: Residents' perceptions of tourism development in Benalmádena (Spain). Tour. Manage. **54**, 259–274 (2016)

Arghutashvili, V.: The role of tourism in economic development of Georgia. J. Int. Manage. Stud. **18**(2) (2018)

Asia Development Bank (ADB): When Global Tourism Returns, the Silk Road Awaits. ADB Blog (2021), 24 Mar 2021. Retrieved online from: https://blogs.adb.org/blog/when-global-tourism-returns-silk-road-awaits

Atkinson, B., Wood, J., Jang, H.: Does covid-19 spark the end of globalization? In: Chaiechi, T., Wood, J. (eds.) Community Empowerment, Sustainable Cities, and Transformative Economies, pp. 535–550. Springer Publishing (2022)

Atkinson, G., Hamilton, K.: Savings, growth and the resource curse hypothesis. World Dev. **31**, 1793–1807 (2003)

Azzali, S., Kamble, Z., Wong, C., Wood, J., Thirumaran, K.: Mitigating impact from natural disasters, building resilience in tourism: the case of Kerala. In: Chaiechi, T. (ed.) Economic Effects of Natural Disasters, pp. 119–129. Elsevier Publishing (2020)
Balli F., Shahzad J., Uddin, G.: A tale of two shocks: what do we learn from the impacts of economic policy uncertainties on tourism? Tour. Manage. **68** (2018)
Bekhet, H.A., Latif, N.W.A.: The impact of technological innovation and governance institution quality on Malaysia's sustainable growth: evidence from a dynamic relationship. Technol. Soc. **54**, 27–40 (2018)
Belt and Road Initiative (BRI): Belt and Road Initiative. BRI (2022). Retrieved online at: https://www.beltroad-initiative.com/belt-and-road/
Bi, J.-W., Liu, Y., Li, H.: Daily tourism volume forecasting for tourist attractions. Ann. Tour. Res. **83**, 102923 (2020)
Chaiechi, T., Wood, J.: Community Empowerment, Sustainable Cities, and Transformative Economies. Springer Publishing (2022)
Cham, T.H., Lim, Y.M., Sia, B.C., Cheah, J.H., Ting, H.: Medical tourism destination image and its relationship with the intention to revisit: a study of Chinese medical tourists in Malaysia. J. China Tour. Res. **17**(2), 163–191 (2021)
Costa, C., Panyik, E., Buhalis, D.: Trends in European tourism planning and organisation. Channel View Publications (2013)
Crotti, R., Misrahi, T.: The Travel & Tourism Competitiveness Report 2017. Paving the Way for a More Sustainable and Inclusive Future. World Economic Forum, Geneva (2017)
Daye, M., Charman, K., Wang, Y., Suzhikova, B.: Exploring local stakeholders' views on the prospects of China's Belt & Road Initiative on tourism development in Kazakhstan. Curr. Issue Tour. **23**(15), 1948–1962 (2020)
Demir, E., Gozgor, G.: Does economic policy uncertainty affect tourism? Ann. Tour. Res. **69**, 15–17 (2018)
Du, Q., Deng, D., Wood, J.: Differences in distance and spatial effects on cross-border e-commerce and international trade: an empirical analysis of China and one-belt one-road countries. J. Glob. Inf. Manag. **30**(2), 1–24 (2021)
Erwin, E.J., Brotherson, M.J., Summers, J.A.: Understanding qualitative metasynthesis: issues and opportunities in early childhood intervention research. J. Early Interv. **33**(3), 186–200 (2011). https://doi.org/10.1177/1053815111425493
European Commission: Directorate General for Economic and Financial Affairs. European Economic Forecast: Winter 2021 (Interim); Publications Office of the European Union, Luxembourg (2021)
Financial Times: Hurdles ahead along the "New Silk Road" (2015). Retrieved from: https://www.ft.com/content/aa15bb9b-0219-320d-ab66-ae61cfb002d3
Fu, X., Ridderstaat, J., Jia, H.: Are all tourism markets equal? Linkages between market-based tourism demand, quality of life, and economic development in Hong Kong. Tourism Manag. 77, 104015 (2020). https://doi.org/10.1016/j.tourman.2019.104015
Ghalia, T., Fidrmuc, J., Samargandi, N., Sohag, K.: Institutional quality, political risk and tourism. Tourism Management Perspectives. 32, 100576 (2019)
Greiner, R.: Improving the net benefits from tourism for people living in remote Northern Australia. Sustainability 2(7), 2197–2218 (2010)
Guoqiang, C.: The belt and road: a mission to benefit two-thirds of the world. Horiz. J. Int. Relat. Sustain. Dev. **4**, 154–161 (2015)
Hajizade, E.M.: Great silk road and the "Baku-Tbilisi-Kars" project. Black Sea Sci. J. Acad. Res. **22**(4), 59–68 (2015)
Heshmati, A., Kim, J., Wood, J.: A survey of inclusive growth policy. Economies **7**(3), 65 (2019). https://doi.org/10.3390/economies7030065
Hughes, A.C.: Understanding and minimizing environmental impacts of the belt and road initiative. Conserv. Biol. J. Soc. Conserv. Biol. **33**(4), 883–894 (2019)
Hyndman, J.: The securitisation of Sri Lankan tourism in the absence of peace. Stab. Int. J. Secur. Dev. **4** (2015)
Iftikhar, H., Pinglu, C., Ullah, S., Ullah, A.: Impact of tourism on sustainable development in BRI countries: the moderating role of institutional quality. PLoS ONE **17**(4), e0263745 (2022). https://doi.org/10.1371/journal.pone.0263745
Kanwal, S., Pitafi, A., Ahmad, M., Khan, N., Ali, S., Surahio, M.: Cross-border analysis of China-Pakistan Economic Corridor development project and local residence quality of life. J. Public Aff. **20**(2), e2022 (2020)
Kaushal, V., Srivastava, S.: Hospitality and tourism industry amid COVID-19 pandemic: perspectives on challenges and learnings from India. Int. J. Hosp. Manage. **92** (2021). https://doi.org/10.1016/j.ijhm.2020.102707
Kavaliauskė, M., Kočytė, R.: Sustainable tourism development in Neringa region. Procedia Social and Behavorial Sciences, 156, 208–212 (2014)
Khalid, U., Okafor, L., Aziz, N.: Armed conflict, military expenditure and international tourism. Tourism Economics, 26, 135481661985140 (2019)
Khan, M.A., Popp, J., Talib, M.N.A., Lakner, Z., Khan, M.A., Oláh, J.: Asymmetric impact of institutional quality on tourism inflows among selected Asian Pacific countries. Sustainability **12**(3), 1223 (2020)
Kim, Y.R., Saha, S., Vertinsky, I., Park, C.: The impact of national institutional quality on international tourism inflows: a cross-country evidence. Tour. Anal. **23** (2018)
Kim, J.-S., Wood, J.: Service sector development in Asia: an important instrument of growth. Asia-Pacific Econ. Lit. **34**(1), 12–15 (2020)
Koh, S., Kwok, A.: Regional integration in central Asia: rediscovering the silk road. Tour. Manage. Persp. **22**, 64–66 (2017)
Lai, L.W.C.: Sustainable development of heritage conservation and tourism: a Hong Kong case study on colonial heritage. Sustain. Dev. **28**(5), 1181–1188 (2020)
Látková, P., Vogt, C.: Residents' attitudes toward existing and future tourism development in rural communities. J. Travel Res. **51**(1), 50–67 (2012)
Lee, H.S.: Roles of institutional quality on the relationship between tourism and economic development in Malaysia. Int. J. Psychosoc. Rehabil. **24**, 685–693 (2020)
Li, X. R.: Chinese outbound tourism 2.0, CRC Press (2016)
Li, T., Shi, H., Yang, Z., Ren, Y.: Does the belt and road initiative boost the tourism economy? Asia Pac. J.tour. Res. **25**(3), 311–322 (2020)
Liow, J.C., Liu, H., Xue, G.: Conclusion: BRI, COVID-19 and the future. In: Chinyong, J.L., Liu, H., Xue, G. (eds.) Research Handbook on the Belt and Road Initiative, pp. 439–448 (2021)
Liu, Y., Suk, S.: Influencing factors of Azerbaijan and China's sustainable tourism development strategy under the one belt one road initiative. Sustainability **14**(1), 187 (2021)
Liu, W., Dunford, M.: Inclusive globalization: unpacking China's Belt and Road Initiative. Area Development and Policy **1**(3): 323–340 (2016)
Mamirkulova, G., Mi, J., Abbas, J., Mahmood, S., Mubeen, R., Ziapour, A.: New silk road infrastructure opportunities in developing tourism environment for residents better quality of life. Glob. Ecol. Cons. **24**, e01194 (2020)
Mathew, P., Sreejesh, S.: Impact of responsible tourism on destination sustainability and quality of life of community in tourism destinations. J. Hosp. Tour. Manag. **31**, 83–89 (2017)

Meyer, D., Masehla, T.M., Kot, S.: The relationship between economic growth and economic development: a regional assessment in South Africa. J. Adv. Res. Law Econ. **8**, 377–385 (2017)

Ministry of Foreign Affairs of PRC: The Suzhou Guidelines for Cooperation between China and Eastern European Countries. Belt and Road Portal (2015). Retrieved from online https://eng.yidaiyilu.gov.cn/zchj/sbwj/1432.htm

Mushtaq, R., Thoker, A.A., Bhat, A.A.: Does institutional quality affect tourism demand? Evidence from India. J. Hosp. Tour. Insights (2020)

Orîndaru, A., Popescu, M-F., Alexoaei, A. P., Căescu, S-C., Florescu, M. S., Orzan, A. O.: Tourism in a Post-COVID-19 Era: Sustainable strategies for industry's recovery. Sustainability, **13**(12), 6781 (2021)

Pechlaner, H., Thees, H., Manske-Wang, W., Scuttari, A.: Local service industry and tourism development through the global trade and infrastructure project of the New Silk Road–the example of Georgia. Serv. Ind. J. **41**(7–8), 553–579 (2021)

Qiu, R.T.R., Park, J., Li, S., Song, H.: Social costs of tourism during the COVID-19 pandemic. Ann. Tour. Res. **84**, 102994 (2020)

Rauf, A., Liu, X., Amin, W., Rehman, O.U., Li, J., Ahmad, F.: Does sustainable growth, energy consumption and environment challenges matter for Belt and Road Initiative feat? A novel empirical investigation. J. Clean. Prod. **262**, 121344 (2020)

Sandelowski, M., Barroso, J.: Handbook for Synthesizing Qualitative Research. Springer Publishing Company, New York (2007)

Santana-Gallego, M., Ledesma-Rodríguez, F.J., Pérez-Rodríguez, J.V.: International trade and tourism flows: An extension of the gravity model. Econ. Model. **52**, 1026–1033 (2016)

Shafiee, S., Ghatari, A., Hasanzadeh, A., Jahanyan, S.: Developing a model for sustainable smart tourism destinations: a systematic review. Tour. Manage. Persp. **31**, 287–300 (2019)

Sikder, M., Wang, C., Yao, X., Xu, H., Wu, L., KwameYeboah, F., Wood, J., Zhao, Y., Dou, X.: The integrated impact of GDP growth, industrialization, energy use, and urbanization on CO_2 emissions in developing countries: evidence from the panel ARDL approach. Sci. Total Environ. (2022). https://doi.org/10.1016/j.scitotenv.2022.155795

Stoever, J.: On comprehensive wealth, institutional quality and sustainable development-quantifying the effect of institutional quality on sustainability. J. Econ. Behav. Organ. **81**(3), 794–801 (2012)

Tang, C.F.: The impacts of governance and institutions on inbound tourism demand: evidence from a dynamic panel data study. Asia Pac. J. Tour. Res. **23**(10), 1000–1007 (2018)

The Star: Belt-and-Road Spawns Healthcare Opportunities (2018). Retrieved from https://www.thestar.com.my/metro/metro-news/2018/01/25/beltandroad-spawns-healthcare-opportunitiesmedical-industry-urged-to-venture-into-crosscountry-coll/

Thirumaran, K., Zohre, M., Pourabedin, Z., Azzali, S., Sim, K.: COVID-19 in Singapore and New Zealand: newspaper portrayal, crisis management. Tour. Manage. Persp. **38**, 108849 (2021)

Thirumaran, K., Khan, G.F., Wood, J.: Tourism in the Tropics: Emerging Themes and Keyword Analysis. A Tropics Perspective. In: Wood, J., Thirumaran, K., Chaiechi, T. (eds.) Business, Industry and Trade in the Tropics, pp. 267–285. Routledge (2022)

United Nations World Tourism Organization (UNWTO): UNWTO Annual Report 2017. UNWTO (2017). Retrieved from: https://www.unwto.org/global/publication/unwto-annual-report-2017

Wang, J., Huang, X., Gong, Z., Cao, K.: Dynamic assessment of tourism carrying capacity and its impacts on tourism economic growth in urban tourism destinations in China. J. Destin. Mark. Manag. **15**, 100383 (2020a)

Wang, C., Wood, J., Geng, X., Wang, Y., Qiao, C., Long, X.: Transportation CO_2 emission decoupling: empirical evidence from countries along the belt and road. J. Clean. Proc. **263**, 121450 (2020b). https://doi.org/10.1016/j.jclepro.2020.121450

Wang, C., Wood, J., Wang, Y., Geng, X., Long, X.: CO_2 emission in transportation sector across 51 countries along the Belt and Road from 2000 to 2014. J. Clean. Prod. **263**, 122000 (2020c)

Wang, C., Zhao, Y., Wang, Y., Wood, J., Kim, C.-Y., Li, Y.: Transportation CO_2 emission decoupling: an assessment of the Eurasian logistics corridor. Transp. Res. Part D: Transp. Environ. **86**, 102486 (2020d)

Wang, P., Zhang, H., Wood, J.: Foreign direct investment, natural disasters, and economic growth of host countries. In: Chaiechi, T. (ed) Economic Effects of Natural Disasters, pp. 9?–117. Elsevier Publishing (2020e)

Wang, P., Zhang, H., Wood, J.: Achieving sustainable development goals (SDGs): a tropical region perspective. In: Wood, J., Thirumaran, K., Chaiechi, T. (eds.) Business, Industry and Trade in the Tropics, pp. 68–96. Routledge (2022)

Wood, J.: An industrial sector analysis of the factors effecting FTA negotiation outcomes. Glob. Bus. Rev. **18**(4), 1–16 (2017)

Wood, J., Khan, F.G.: International trade negotiation analysis: network and semantic knowledge infrastructure. Scientometrics **105**(1), 537–556 (2015)

Wood, J., Jang, H., Oh, K.: The U.S.—China trade war: is it more pain than gain? J. China Stud. **23**, 61–85 (2020)

Wood, J., Li, Y., Wu, J.: An analysis of the trends and determinants of intra- industry trade between china and asia pacific economic cooperation member countries. Singap. Econ. Rev. **66**(3), 743–766 (2021)

Wood, J., Jang, H., Atkinson, B.: Covid-19: its impact on trade and development in the ASEAN region. In: Wood, J., Thirumaran, K., Chaiechi, T. (eds.) Business, Industry and Trade in the Tropics, pp. 137–150. Routledge (2022a)

Wood, J., Thirumaran, K., Chaiechi, T.: Business, Industry and Trade in the Tropics. Routledge (2022b)

Wu, J., Huh, C.-G., Wood, J.: Globally Chained Economies: Unwitting Victims of the US-China Trade War. Asia-Pac. Econ. Lit. **35**(2), 60–76 (2021a). https://doi.org/10.1111/apel.12334

Wu, J., Wood, J., Huang, X.: How does GVC reconstruction affect economic growth and employment? Analysis of USA–China decoupling. Asia Pac. Econ. Lit. **35**(1), 67–81 (2021b). https://doi.org/10.1111/apel.12319

Wu, J., Wood, J., Oh, K., Jang, H.: Evaluating the cumulative impact of the US–China trade war along global value chains. World Econ. **44**(12), 3516–3533 (2021c). https://doi.org/10.1111/twec.13125

World Tourism Organization and Global Tourism Economy Research Centre: UNWTO/GTERC annual report on tourism trends. UNWTO, Madrid (2014)

Zeybek, H.: What role for railways in the eurasian supply chains? In: Railway Transportation in South Asia, pp 269–292. Springer (2021)

Zhang, H., Wang, P., Wood, J.: Does institutional quality matter for the nexus between environmental quality and economic growth? A tropics perspective. In: Wood, J., Thirumaran, K., Chaiechi, T. (eds.) Business, Industry and Trade in the Tropics, pp. 33–52. Routledge (2022)

The Eurasian Economic Union and Silk Road Tourism: Qualifying Resources, Amplifying Messages

K. Thirumaran, Redeem Faith Sabacan, Zahra Pourabedin, Hong Hanh Nguyen, Haejin Jang, Benedict Atkinson, and Jacob Wood

Abstract

The promotion of tourism can often be a dichotomy of national and intra-regional interests, especially when countries are a part of an economic regional grouping. The vast Eurasian Economic Union (EEU), which stretches from the Caspian Sea to the western borders of China, comprises five large countries all of which are situated along the ancient Silk Road. Although an ancient historical route, its modern development is still very much in its infancy, with countries in the region eager to develop their own capabilities to not only attract international visitors but also capture important a significant share of the rapidly growing tourism marketplace. This chapter uses the (Ritchie and Crouch, in The competitive destination: a sustainable tourism perspective, CABI, Wallingford, UK, 2003) destination competitive model to assess the travel and tourism opportunities that exist across the Belt and Road Initiative (BRI). The significance of this study, lies in the proposition that Eurasian Economic Union (EEU) as a regional grouping has the potential to emerge as a dynamic tourism destination. At the heart of such a believe will be the inherent need for countries to leverage their qualifying resources and amplify in unison key developmental factors such as the safety, location, interconnectedness of the different destinations within and more broadly the competitive elements of costs and exchange rates.

Keywords

Silk road • Belt and road initiative (BRI) • Eurasian economic union • Destination competitive model • Amplify • China

1 Introduction

Emerging as a major global economic power in the twenty-first century, China is intertwining its relations with countries in Asia and westward using the historical fulcrum of trade routes known as the Silk Road and the Maritime Silk Route (see Fig. 1). Fist formulated in 2013, the China led Belt and Road initiative (BRI) includes some 140 countries with intra-trade volume amounting to USD10 trillion (Hettige 2022). With the inclusion of Siberia (Russia) and Xinjiang (China) with countries that stretch all the way to the Caspian Sea, the Silk Road encompasses 448 million people and a purchasing power parity of $1702 USD per person (Devonshire-Ellis 2021). These significant figures and the prospect of countries situated along the corridor benefitting from the enormous transportation connections that exist present massive opportunities for trade, tourism, and cultural exchange.

As a regional grouping, the Eurasia Economic Union (EEU), which comprises of Armenia, Belarus, Kazakhstan, Kyrgyzstan and Russia represents a significant geoeconomic axis. Together with Uzbekistan, the region is beginning to

K. Thirumaran · H. H. Nguyen · B. Atkinson · J. Wood (✉)
JCU Singapore Business School, James Cook University Singapore, 149 Sims Drive, Singapore, Singapore
e-mail: jacob.wood@jcu.edu.au

K. Thirumaran
e-mail: k.thirumaran@jcu.edu.au

H. H. Nguyen
e-mail: honghanh.nguyen@my.jcu.edu.au

B. Atkinson
e-mail: benedict.atkinson@jcu.edu.au

R. F. Sabacan
Central Luzon, Philippines

Z. Pourabedin
College of Business, Shepherd University, Shepherdstown, WV 25443, USA
e-mail: zpourabe@shepherd.edu

H. Jang
JCU Singapore Learning Centre, James Cook University Singapore, 149 Sims Drive, Singapore, Singapore
e-mail: haejin.jang@jcu.edu.au

S. Kostopoulou et al. (eds.), *Cities' Vocabularies and the Sustainable Development of the Silkroads*, Advances in Science, Technology & Innovation, https://doi.org/10.1007/978-3-031-31027-0_23

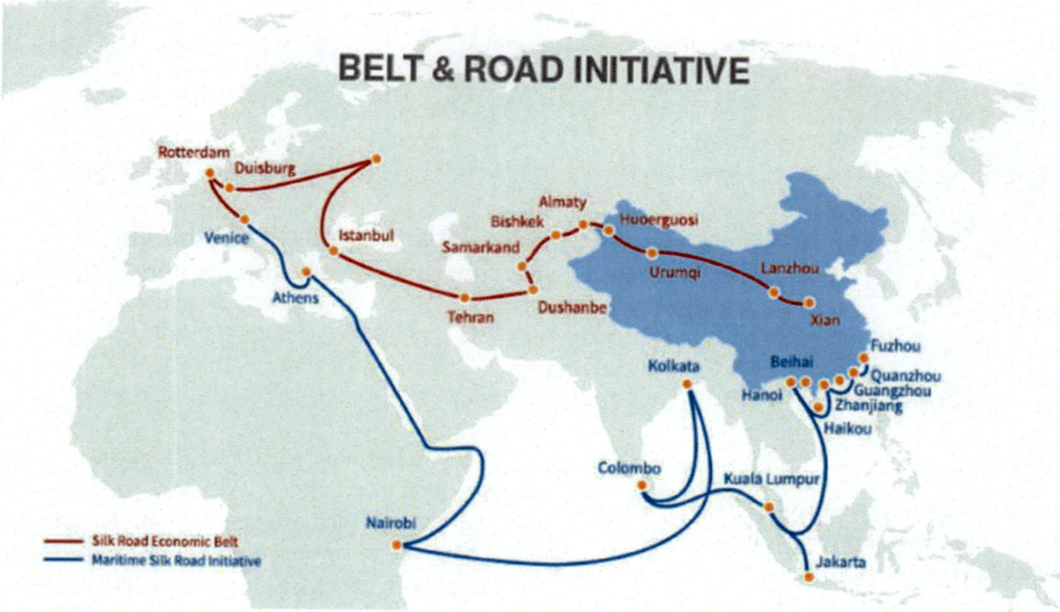

Fig. 1 Belt and road initiative

develop itself economically with a Gross Domestic Product (GDP) that is in approximately $2.7 trillion USD (Michel 2014). While the EEU already appears to moving more closely together economically, the BRI presents a significant opportunity to further enhance their unity with a single destination marketing approach for the region. The BRI contains two key facets, the Silk Road Economic Belt and a corresponding Maritime Silk Road Initiative. With the strategic development linking Europe in the west with Asia in the east occurring on both land and water there is enormous potential to grow the tourism sector and in doing so provide critical employment opportunities.

During the United Nations World Trade Organization's (UNWTO's) General Assembly in Indonesia in 1993, the concept of Silk Road Tourism was initially discussed. Motivated by a sense of renewed enthusiasm in the Silk Road, the UNWTO decided to revive the ancient routes as a tourism concept, uniting three continents once again in a project spread across many thousands of kilometres. The Silk Road represents an historic array of trade routes, which have played an integral role in driving not only cultural interaction but also important source economic cooperation connecting China in the east through to the Mediterranean Sea in the west (Gursoy and Altinay 2021). The rebirth of the Silk Road Network and its benefits requires the participation of all countries along the route (Schwab 2018).

After reviewing the literature on regional groupings and their marketing efforts, we outline the research questions and study approach. This chapter examines the key aspects of the tourism marketing initiatives that are being developed by the EEU countries. More specifically, how these nations can utilize their abundant natural and socio-cultural resources to amplify their message that the Silk Road has the potential to be a leading global tourism hub in the future. Drawing on this pretext, the chapter also provides a framework that policy makers within the EEU can embrace that BRI is developed in a manner that benefits all parties involved in the tourism sector.

The rest of the study flows as follows. Section 2 examines the issues of regionalism, regional economic cooperation, and the important roles they play in growing tourism in the Eurasian region. It also provides an assessment of the literature surrounding the Eurasian Economic Union members in regard to the Silk Road and the marketing of the Tourism sector in the region. Section 3 provides a series of further analysis and discussion regarding key policy initiatives that could be implement to further drive tourism development. Finally, Sect. 4 draws a series of relevant conclusions and details areas of research that should form key areas of focus in the future.

2 Literature Review

Since the end of the 1960s, regionalism has become one of the key theories of international relations. In the 1950s, the first wave of regionalism commenced when the dynamic momentum of European regional integration started. During the latter part of the 1980s, the world saw the proliferation of regional groupings gained momentum in different parts of the world—from Europe to North America, Latin America and South and Southeast Asia, the second wave of theorizing on regionalism began (Wood 2017; Wood and Jang 2017).

Scholars (i.e., Beeson and Stubbs 2012; Wunderlich 2007; Sridharan 2007) have provided ample definitions on the construct of regionalism (Chheang 2013). For Beeson and Stubbs (2012) and Sridharan (2007), regionalism is a state-led development initiatives in collaboration with countries of common geographic or interests. Often, in such regional organisations, we can expect a strong state actor and their institutional cooperatives to design and implement the common interests of the grouping (Wunderlich 2007). While regionalism attempts to neutralise individual state nationalism to give way to a stronger regional unity, challenges to a collective identity and collaborative initiative is where much more efforts have to be placed (Fawcett 2004).

Regionalism is an enabler of tourism (Thirumaran et al. 2022). Regional cooperation and integration was found to form a favourable condition for the development of the tourism industry when viewed as a cooperative magnet can contribute to mutual benefits generated from the industry. The collapse of the Soviet Union in 1991 resulted in the establishment of the five new countries in Central Asia. Since their independence, these countries have faced several challenges. However, the disintegration of the Soviet Union paved the way for new relations in Central Asia with other neighbouring countries such as China and Iran. Chheang (2013) noted that regional cooperation efforts in terms of tourism development and promotion is gaining more political and economic attention, specifically in the context of globalization and regionalization. For example, Kostopoulou et al. (2021), the classification of the Silk Road cultural heritage assets is considered a useful tool for strategic planning in tourism and regional planning, addressed to local authorities and tourism stakeholders.

As regards to transforming Central Asia from a "land-locked" to a land-linked" region, Gussarova et al. (2017) noted that the EEU and the Silk Road Economic Belt have been the most discussed undertakings. The same scholars compared the two major initiatives and two regional powers – Russia and China, with the EEU and the BRI, which directly influences the interests of Central Asia (Nurseiit 2020). Meanwhile, in their studies on the Silk Road and the service industries, Gursoy and Altinay (2021) concluded that the Silk Road project offers significant opportunities to develop an open-minded, all-encompassing, well-balanced, and inclusive regional economic cooperation framework that can benefit service industries and improve the quality of life of citizens of countries within the Silk Road.

Du and Zhang (2018) found that China's outward foreign direct investments (OFDIs) had a huge increase in countries participating in the BRI. Li et al. (2019) corroborate that BRI-related subsidies incentivize exporting to BRI markets, while cultural friction and ethnicity differences of owners create complications in capitalizing the policy benefits. Similarly, Quer (2021) presents the positive effects of BRI in the tourism industry, as Chinese tourists seem to prefer the countries on the BRI list. While Wilson (2016) conducted a study assessing the Russian and Chinese perspectives on the EEU. His study also aimed to ascertain the developmental mechanisms that were associated with the extended foreign policy goals of Russia and China and their impact on the strategic Russian-Chinese alliance. The author concluded that the initiatives represent a rivalry between the powerful neighbours which are largely a consequence of the growing power disparities that exist between the two countries.

Central Asia, as one of the most underserved tourist destinations in the world, faces several challenges such as security, economic development, infrastructure, and transportation to develop tourism (Koh and Kwok 2017). Barriers to accessing the China tourism market include "shortage of tourism infrastructures, unprofessional tourism workers, and low service level" in Central Asia (Li 2016, p128). However, China's motivation to develop the new Silk Road represents a broader developmental aim, however, there is no doubt that the tourism industry will be affected by the significant infrastructural developments that are proposed and regional integration that will occur as a consequence (Koh and Kwok 2017).

Most Silk Road studies have focused on trade (Du et al. 2021), infrastructure development (Sikder et al. 2022a), cleaner forms of economic development (Wang et al. 2020a, 2020b, 2020c; Sikder et al. 2022b), and China's motivations behind the project (Ye 2021). However, despite this focus, more Silk Road related research is needed within the tourism context (Liu and Suck 2021), with a recent research tourism research trends study by Thirumaran et al. (2022) revealing very little has been done with a belt and road focus.

Due to the rich history of the Silk Road countries, there is great potential to develop cultural heritage tourism in the region. In this regard, China plans to invest $500–700 million USD in Lanzu for the Silk Road Cultural Heritage Expo City (Shephard 2016) and the XinJiang Yughur Autonomous region, which will act as an important new tourism hub within the new Silk Road. The New Silk Road initiative will increase cultural engagement, and provide new opportunities for event marketing (Kuchumov and Testina 2020).

Under the New Silk Road Initiative, the tourism industry between Azerbaijan and China has developed rapidly (Liu and Suck 2021). After signing a Memorandum of Understanding between Azerbaijan and China on March 1, 2019, for the Tourism Implementation Plan, Azerbaijan has actively developed a series of marketing strategies that promote their development of electronic visas, tourism exhibitions, and direct flights with Beijing to attract Chinese tourists. Using a SWOT analysis, Liu and Suck (2021) explored the development strategy of tourism cooperation between the two countries. From this they found that the countries share a strong foundation and interests in deploying resources that nurture the development of tourism infrastructure that benefits both China and Azerbaijan. Though Azerbaijan may be lagging behind China in turns of its economic development, a strong Chinese tourist flow into the country might augur well for both partner countries and neighbouring BRI countries (Liu and Suck 2021).

Furthermore, in order to facilitate tourism development that is sustainable over the long term, it is important to investigate the attitudes of local stakeholders (Daye et al. 2020). If tourism development projects lead to economic and social benefits, residents will support the projects (Lundberg 2017; Zuo et al. 2017). Daye et al. (2020) investigated the attitudes that local stakeholders in Kazakhstan have towards the New Silk Road initiative and developments to the tourism sector in particular. When compared to other Central Asian countries along the Silk Road, Kazakhstan has relatively higher levels of economic performance (Daye et al. 2020). For other countries along the BRI, tourists in places like Kazakhstan are divided into two groups, namely European and Asian tourists. While European tourists are interested in culture and ecotourism, horse riding, and steppe safari, Asian travellers are interested in gambling, sightseeing, and shopping (Daye et al. 2020). The result of this study shows that although some stakeholders back tourism development under New Silk Road initiative, there are still some uncertainties due to the soft power diplomacy tactics often used by China.

In other empirical works Pechnaler et al. (2019) explored the tourism development work that had occurred in Georgia. In this context, the BRI initiative is proving to be a key conduit for tourism development, helping to not only grow the country into a regional tourism hub but also drive new employment opportunities and innovation among local businesses, such as small hotels. As part of this, rural tourism has continued to flourish. However, to achieve all of tourism development plans, Georgia should do more to cooperate and work with other competitors in the region, such as the Baku.

Finally, Kuchumov and Testina (2020) studied how participation in the New Silk Road initiative influences tourism flows. Their research highlighted the impact that increases in the flow of Chinese tourists will have on the development of higher quality Russian tourism offerings. However, development across the EEU does not occur without some degree of political tension, with Russia dominating the economic relationships that exist (Togt 2020). It is important to remember that all the EEU states were once part of the former Soviet Union and as such the Russian Federation has a strong desire to rekindle old ties, something that is sadly all to evident with the Russian invasion and war in Ukraine. In others countries, the Russia has played a highly influential role in Georgia's and Chechnya's domestic politics, Khazhakstan's and Belarus' civil unrest, and territorial disputes between Azerbaijan and Armenia. In all of these geo-political situations, China has erred on the Russian side of diplomacy.

3 Analysis and Discussion

In this section we discussion how the Silk Road has affected the tourism sectors of Central Asian economies. Additionally, it identifies some of the divergent resources that link the EEU and can serve as an opportunity for the region. We also document the political implications that are involved with marketing the BRI tourism sector as a single entity.

As emerging economies, much has been said about the potential for the tourism sector to drive economic growth and prosperity. Khan et al. (2020) stated that for emerging economies, tourism has recently been strongly supported by the local governments. At the heart of this support has been the sector's ability to generate employment, alleviate poverty, increase national gross domestic product, and drive development across the rural communities (e.g., Jianping et al. 2014; Deloitte 2019). In its contemporary perspective, the revival of the Silk Road heritage is expected to encompass and encourage the creation of new tourist flows, ranging from independent travelers looking forward to experience diverse cultures and alternative forms of tourism through to the corporate business traveler. However, in order to realize this potential more needs to be to develop and then launch effective strategies that cater to this diverse market opportunities.

Kazakhstan, as the EEU member that is nearest to China, its rich oil and mineral deposits in uranium, copper, lead, zinc and chromium have allowed it to develop at a faster rate than other member nations from Central Asia (Jianping et al. 2014) It is also more economically competitive, with Schwab (2018) noting that Kazakhstan placed 59th out of 140 in the 2018 Global Competitiveness Report. From a tourism perspective, the World Travel and Tourism Council (2019a) noted that the tourism sector contributed 5.7% to Kazakhstan's total GDP output. In other EEU nations, the tourism sector is also growing with Tajikistan experiencing 0.4% growth in the sector in 2018 (World Travel and Tourism Council 2019b).

From a destination management perspective, every country in the region is remarkably similar (Usmanova 2018) and there is significant growth opportunities right across the tourism sector. Currently, the area is only focusing on promoting four key tourism products, namely: medical tourism, sports tourism, ecological-based tourism and lastly, archaeological-based tourism (Xu 2020). In fact, even though political instability has undoubtedly had a negative impact on the way these destinations are perceived by international tourists, these EEU economies can do much to realign their marketing objectives and develop marketing campaigns that focus on the region's unique advantages, using culture and geo-physical landscapes (Ibragimov et al. 2021).

In terms of resources, the EEU countries possess not only unique and natural landscapes with intricate ecosystems, but they are also are well-known for some of the most diverse cultural heritage, as evidenced by the number of historical sites that have been recognized by the United Nations Educational, Scientific and Cultural Organization (UNESCO) (Xu 2020). According to Kostopoulou et al. (2021), in the broader Silk Road region, some 138 of these cultural assets are identified as being tangible, of which, 76 are part of the Built Heritage category.

As mentioned earlier, the EEU region, the tourism sector has failed to reach its full potential (Xu 2020). This is despite the fact that there is an abundance in tourism resources and a large number of distinctive tourism assets that can cater for the demands of tourists worldwide (Usmanova 2018). In the case of Kyrgyzstan, while it has also been impacted by the pandemic, there potential remains largely untapped, Jenish (2018) noting that only 20 percent of their tourism potential has been realized.

In other works, Himaz (2021) explored the challenges of driving tourism development across the EEU. As part of their research, they highlight the significant problems that have been caused by poor Chinese backed infrastructure investments increasing corporate debt, and high levels of political corruption. They also highlight the risks of contagion, and how a prolonged economic downturn in China may have wider implications for countries such as Russia and the EEU. Development of the tourism sector, like any form of growth brings with it clear winners and losers, with many studies highlighting the inequitable role that development can bring (Heshmati et al. 2019; Kim and Wood 2020; Rhee et al. 2022). As such, it is imperative that any gains achieved are appropriately redistributed across an economy. A failure to do so, can result in increased inequality, poverty, and structural unemployment. Finally, the BRI may have a detrimental impact on the environment unless effective policy regulations are implemented.

In other works, Himaz (2021) noted that external factors such as substantial changes in regulatory frameworks, institutional fragility, trade patterns, and competition may hinder the ability for BRI countries to drive development across the tourism sector. Likewise, in the absence of appropriate institutional frameworks macroeconomic change may exacerbate problems related to poverty, unemployment, and fiscal issues. Nonetheless, these shifts may also pave way for opportunities and in particular to grow foreign exchange reserves and in doing so address concerns around national debt. Finally, further research on the creation of a sustainable, responsible tourism sector across the BRI that acts with a 'moral conscience', might also be helpful (Gursoy and Altinay 2021).

4 Conclusion and Future Directions

While the BRI concept immediately conjures an economic and historical link that goes back several millennia, the modern opportunities that exist for the EEU are existing. While the Chinese economy has continued to flourish in recent years, its investments in the region have also helped many other countries to thrive. The EEU, while historically linked to Russia has opportunities through the BRI to tap into the enormous potential that China offers in driving the tourism sector.

The West, in particular the US, has cautioned many countries about the perils of engaging with China around the development of infrastructure projects that ultimately end up being so-called debt traps. Nonetheless, if the EEU countries are careful in how they manage their contractual agreements for the tourism infrastructure projects they can help to ensure that they do not fall victim to such traps and in doing so remain independent and sovereign decision makers. However, it is important that, at the same time, they do not turn their back on China and give themselves ever chance of fostering a successful and mutually beneficial relationship with China.

Despite similarities amongst them in tourism products, the EEU has an advantage of shared heritage and complementarity and in some instances, attractions of interests to Chinese tourists. In a recent 2019 document celebrating the fifth anniversary of the EEU, Tigran Sargsyan, the Chairman of the Eurasian Economic Council (EEC) highlighted the importance of people to people exchange

> This entails cooperation in education, sports, tourism, and healthcare. We understand that nowadays these key spheres not only have economic significance, but are also important for our citizens. (Eurasian Economic Union 2019, p 51)

Though the remark is targeted to citizens within the EEU grouping, it is an important statement that the leaders are conscious of leveraging on their economic union to maximize benefits for the people. This same spirit of engaging China instead of peripheral to the region but as a member of the region itself, more intricate and deeper economic ties can meaningfully spill over to higher decibels of tourism activities.

References

Beeson, M., Stubbs, R.: Introduction. In: Beeson, M., Stubbs, R. (eds.) Handbook of Asian Regionalism, pp. 1–8. Routledge, London and New York (2012)

Chheang, V.: Tourism and Regional Integration in Southeast Asia, VRF Series, 481, 1–93 (2013). https://www.ide.go.jp/library/English/Publish/Reports/Vrf/pdf/481.pdf

Daye, M., Charman, K., Wang, Y., Suzhikova, B.: Exploring the views of local stakeholders on the prospects of the China's Belt & Road Initiative on tourism development in Kazakhstan. Curr. Issue Tour. **23**(15), 1948–1962 (2020)

Deloitte: Shanghai Municipal Commission of Commerce Belt and Road Countries Investment Index Report 2018. Deloitte, Shanghai (2019)

Devonshire-Ellis, C.: The Greater Eurasian Partnership: connecting Central & South-East Asia. In: Silk Road Briefing (2021). Retrieved from https://www.silkroadbriefing.com/news/2021/04/14/the-greater-eurasian-partnership-connecting-central-south-east-asia/

Du, J., Zhang, Y.: Does the one belt one road initiative promote Chinese overseas direct investment? China Econ. Rev. **47**, 189–205 (2018)

Du, Q., Deng, D., Wood, J.: Differences in distance and spatial effects on cross-border e-commerce and international trade: an empirical analysis of China and one-belt one-road countries. J. Glob. Inf. Manag. **30**(2), 1–24 (2021)

Eurasian Economic Union: 5 Years: The EAEU Anniversary, EEU (2019). Retrieved from: https://eec.eaeunion.org

Fawcett, L.: Exploring regional domains: a comparative history of regionalism. Int. Aff. **80**(3), 429–446 (2004)

Gursoy, D., Altinay, L.: The Silk Road and the service industries. Serv. Ind. J. **41**(7–8), 441–445 (2021). https://doi.org/10.1080/02642069.2021.1928823?needAccess=true

Gussarova, A., Aminjonov, F., Khon, Y.: The Eurasian economic union and the Silk Road Economic Belt. Friedrich-Ebert-Stiftung, Almaty (2017). http://greater-europe.org/wp-content/uploads/2018/04/13620.pdf

Heshmati, A., Kim, J., Wood, J.: A survey of inclusive growth policy. Economies **7**(3), 65 (2019). https://doi.org/10.3390/economies7030065

Hettige, S.: Why should the world listen to President Xi Jinping? Adaderana Sri Lanka Business News (2022). Retrieved from http://bizenglish.adaderana.lk/why-should-the-world-listen-to-president-xi-jinping/

Himaz, R.: Challenges associated with the BRI: a review of recent economics literature. Serv. Ind. J. (2021). https://doi.org/10.1080/02642069.2019.1584193

Ibragimov, L.Z., Ubaydullayev, D.K.U., Musayev, B.M.: Opportunities and prospects for tourism development in Jizzakh region. Am. J. Interdisc. Innovations Res. **3**(02), 33–37 (2021). https://doi.org/10.37547/tajiir/Volume03Issue02-07

Jenish, N.: Tourism sector in Kyrgyzstan: Trends and challenges. University of Central Asia—Institute of Public Policy and Administration (IPPA) Working Paper No. 42 (2018). https://doi.org/10.2139/ssrn.3807841

Jianping, L., Minrong, L., Jinnan, W., Jianjian, L., Hongwen, S., Maoxing, H.: Report on Global Environmental Competitiveness (2013). Spring, Berlin; Heidelberg (2014)

Khan, A., Bibi, S., Lorenzo, A., Lyu, J., Babar, Z.U.: Tourism and development in developing economies: A policy implication perspective. Sustainability **12**(4), 1618 (2020). https://doi.org/10.3390/su12041618

Kim, J.-S., Wood, J.: Service sector development in Asia: an important instrument of growth. Asian-Pac. Econ. Lit. **34**(1), 12–15 (2020)

Koh, S.G.M., Kwok, A.O.J.: Regional integration in central Asia: rediscovering the silk road. Tour. Manage. Persp. **22**, 64–66 (2017)

Kostopoulou, S., Sofianou, P.K., Tsiokanos, K.: Silk Road heritage branding and polycentric tourism development. Sustainability **13**, 1893 (2021). https://doi.org/10.3390/su1304

Kuchumov, A., Testina, Y.: The impact of participation in the belt and road initiative on the tourist flows. In: New Silk Road: Business Cooperation and Prospective of Economic Development (NSRBCPED 2019), pp. 51–57 (2020)

Li, X.R.: Chinese Outbound Tourism 2.0. CRC Press (2016)

Li, J., Liu, B., Qian, G.: The belt and road initiative, cultural friction and ethnicity: their effects on the export performance of SMEs in China. J. World Bus. **54**(4), 350–359 (2019)

Liu, Y., Suk, S.: Influencing factors of Azerbaijan and China's sustainable tourism development strategy under the one belt one road initiative. Sustainability **14**(1), 187 (2021)

Lundberg, E.: The importance of tourism impacts for different local resident groups: a case study of a Swedish seaside destination. J. Destin. Mark. Manag. **6**(1), 46–55 (2017)

Michel, C.: How Significant Is the Eurasian Economic Union? The Diplomat (2014). Retrieved from https://thediplomat.com/2014/06/how-significant-is-the-eurasian-economic-union/

Nurseiit, A.N.: Prospects for the regional cooperation in Central Asia. Eurasian J. Bus. Econ. **13**(26), 1–22 (2020)

Pechnaler, H., Thees, H., Manske W., Scuttari, A.: Local service industry and tourism development through the global trade and infrastructure project of the New Silk Road–the example of Georgia. Serv. Ind. J. **41**(7–8), 553–579 (2019)

Quer, D.: Location decisions of Chinese firms in the global tourism industry: The role of prior international experience and diplomatic relations. J. Hospitality Tourism Manage. **46**, 62–72 (2021)

Rhee, T.-H., Wood, J., Kim, J.: Demographic transition, sustainable economic growth, and digital economy in Southeast Asian Developing Countries. Sustainability **14**(5), 2857 (2022). https://doi.org/10.3390/su14052857

Ritchie, J.R., Crouch, G.I.: The Competitive Destination: A Sustainable Tourism Perspective. CABI, Wallingford, UK (2003)

Schwab, K.: The Global Competitiveness Report 2018. World Economic Forum, Geneva (2018)

Shephard, W.: A look at 7 new cities rising along the New Silk road. Forbes (2016). Retrieved from: https://www.forbes.com/sites/wadeshepard/2016/09/20/7-new-cities-that-are-rising-along-the-new-silk-road/?sh=27eac4d41300

Sikder, M., Wang, C., Yeboah, F.K., Wood, J.: Driving factors of CO_2 emission reduction in the logistics industry: An assessment of the RCEP and SAARC economies. Environ. Dev. Sustain. (2022a). https://doi.org/10.1007/s10668-022-02840-3

Sikder, M., Wang, C., Yao, X., Xu, H., Wu, L., KwameYeboah, F., Wood, J., Zhao, Y., Dou, X.: The integrated impact of GDP growth, industrialization, energy use, and urbanization on CO_2 emissions in developing countries: evidence from the panel ARDL approach. Sci. Total Environ. (2022b). https://doi.org/10.1016/j.scitotenv.2022.155795

Sridharan, K.: Regional Cooperation in South Asia and Southeast Asia. ISEAS, Singapore (2007)

Thirumaran, K., Khan, G.F., Wood, J.: Tourism in the tropics: emerging themes and keyword analysis. A tropics perspective. In: Wood, J., Thirumaran, K., Chaiechi, T. (eds.) Business, Industry and Trade in the Tropics, pp. 267–285. Routledge (2022)

Togt, V.D.T.: EU & Eurasian Economic Union: A Common Chinese Challenge. Institut fuer Sicherheitspolitik (2020). Retrieved from https://www.clingendael.org/publication/eu-eurasian-economic-union-common-chinese-challenge

Usmanova, S.: Prospects of intensive institutional cooperation in tourism development in Central Asia: The case of Uzbekistan. J. Tourismology **4**(1), 55–66 (2018)

Wang, C., Wood, J., Geng, X., Wang, Y., Qiao, C., Long, X.: Transportation CO_2 emission decoupling: empirical evidence from countries along the belt and road. J. Clean. Prod. **263**, 121450 (2020a). https://doi.org/10.1016/j.jclepro.2020.121450

Wang, C., Wood, J., Wang, Y., Geng, X., Long, X.: CO_2 emission in transportation sector across 51 countries along the Belt and Road from 2000 to 2014. J. Clean. Prod. **263**, 122000 (2020b)

Wang, C., Zhao, Y., Wang, Y., Wood, J., Kim, C.-Y., Li, Y.: Transportation CO_2 emission decoupling: an assessment of the Eurasian logistics corridor. Transp. Res. Part d: Transp. Environ. **86**, 102486 (2020c)

Wilson, J.L.: The Eurasian economic union and China's silk road: implications for the Russian-Chinese relationship. Eur. Polit. Soc. **17**, 113–132 (2016). https://doi.org/10.1080/23745118.2016.1171288?scroll=top&needAccess=true

Wood, J.: An industrial sector analysis of the factors effecting FTA negotiation outcomes. Glob. Bus. Rev. **18**(4), 1–16 (2017)

Wood, J., Jang, H.: Brexit: The economic and political implications for Asia. Soc. Sci. **6**(41), 1–11 (2017)

World Travel & Tourism Council: Kazakhstan: 2019 Annual Research: Key Highlights. World Travel & Tourism Council, London (2019a)

World Travel & Tourism Council: Kyrgyzstan: 2019 Annual Research: Key Highlights. World Travel & Tourism Council, London (2019b)

Wunderlich, J.U.: Regionalism, Globalization and International Order. Ashgate, Hampshire (2007)

Xu, J.: Research on the construction of cultural tourism market in Shenyang based on big data. J. Phys. Conf. Ser. **1621**, 012104 (2020)

Ye, M.: Fragmented motives and policies: The belt and road initiative in China. J. East Asian Stud. **21**(2), 193–217 (2021). https://doi.org/10.1017/jea.2021.15

Zuo, B., Gursoy, D., Wall, G.: Residents' support for red tourism in China: the moderating effect of central government. Ann. Tour. Res. **64**, 51–63 (2017)

Economic Development Along the New Silk Road: A Vehicle for Driving Sustainable Tourism

Redeem Faith Sabacan, Hong Hanh Nguyen, Emiel L. Eijdenberg, Haejin Jang, K. Thirumaran, and Jacob Wood

Abstract

For over two millennia, the Silk Road has been instrumental in linking Asia in the east, with Europe in the west. It has served as an important route of integration, exchange, conflict, and economic and political dialogue. In 2013, the Chinese government made an important first step to revitalize this ancient region by establishing the "New Silk Road", also known as the Belt and Road Initiative. At the heart of this developmental plan, lies an extensive range of land and marine-based infrastructural projects that seek to ignite new trade and business activity along the road. A key aspect of this development are new opportunities to grow the region's already impressive tourism sector. As such, the aim of this study is to assess not only what role the New Silk Road Initiative will play in growing tourism activity in the region, but also to provide several practical policy suggestions that both governments and industry can adopt in the future. In particular, we suggest that to continuously boost the strategic projects that best represent New Silk Road's interests, countries need to constantly endeavor on infrastructural developments that have paved the way for the steady growth in trade between the countries, hence realizing BRI's goal of developing closer trading links between and among member countries.

Keywords

BRI • Economic development • Marine infrastructure • Silk road • Sustainability • Tourism

1 Introduction

There is enormous potential in the Central Asian region to further develop its tourism sector (Koh and Kwok 2017). Located within the heart of this area, lies the Silk Road, a route of significant and longstanding political and economic significance. While its exact origins are difficult to date, the Silk Road has been the focal point for many important events throughout human history (Kuzmina 2008), including the first-hand rise of many great Central Eurasian empires, such as the Hun and Genghis Khan-led Mongols (Beckwith 2009). However, it is the Silk Road's modern form that shapes the narrative of our study and in particular the potential tourism opportunities that lie within it. The countries along the Silk Road, can be divided into the Silk Road Economic Belt, which includes five Central Asia countries (i.e., Kazakhstan, Kyrgyzstan, Tajikistan, Uzbekistan, Turkmenistan), Central Asia (e.g., Russia, Afghanistan, India, Pakistan, Iran, Azerbaijan, Armenia, Georgia, Turkey, Saudi Arabia, Iraq and other regions), and the Asia-Europe Economic Belt, which includes Europe and North Africa (e.g., Germany, France, Britain, Italy, Ukraine, Egypt, Libya, Algeria). The BRI also includes parts of Latin America and the Caribbean region, which are now participating in the project (EUI 2022).

R. F. Sabacan
Central Luzon, Philippines

H. H. Nguyen · E. L. Eijdenberg · K. Thirumaran · J. Wood (✉)
JCU Singapore Business School, James Cook University
Singapore, 149 Sims Drive, Singapore, Singapore
e-mail: jacob.wood@jcu.edu.au

H. H. Nguyen
e-mail: honghanh.nguyen@my.jcu.edu.au

E. L. Eijdenberg
e-mail: emiel.eijdenberg@jcu.edu.au

K. Thirumaran
e-mail: k.thirumaran@jcu.edu.au

E. L. Eijdenberg
School of Business and Economics, Vrije Universiteit Amsterdam, De Boelelaan 1105, 1081 HV Amsterdam, The Netherlands

H. Jang
JCU Singapore Learning Centre, James Cook University
Singapore, 149 Sims Drive, Singapore, Singapore
e-mail: haejin.jang@jcu.edu.au

S. Kostopoulou et al. (eds.), *Cities' Vocabularies and the Sustainable Development of the Silkroads*, Advances in Science, Technology & Innovation, https://doi.org/10.1007/978-3-031-31027-0_24

Given its significance to the Eurasian region, many countries have sought to seize the economic opportunities that the Silk Road presents. As mentioned earlier, the Chinese government first significant efforts in this regard occurred in 2013 with the creation of the "New Silk Road" initiative, which is also referred to as the Belt and Road Initiative (BRI).[1] For Mamirkulova et al. (2020), the initiative represents China's attempt to encourage socio-economic development across all countries that border it, an area that includes approximately 64 percent of the global population and more than 30 percent of the world's Gross Domestic Product (GDP). From a tourism sector perspective, a key facet of the initiative is undoubtedly the land and marine infrastructure developments. Such investments in infrastructure development have been the focus of much research by tourism scholars in recent times (Kanwal et al. 2020; Wang et al. 2020a), with many arguing it is a leading antecedent for driving growth across the sector (Hestmati et al. 2019; Kim and Wood 2020) and in doing so generate new employment opportunities, growing the economy as a whole (Kanwal et al. 2019). Research on the importance of developing infrastructure initiatives in the region has also illustrated its role in nurturing cultural and historical values, and natural landscapes (Bi et al. 2020; Shafiee et al. 2019) as well as its impact on the tourist experience, the tourists themselves, tour operators and the staff involved in the construction of infrastructural facilities (Daye et al. 2019; Mamirkulova et al. 2020). Given the environmental impact that projects of this nature can cause, other studies have confirmed the need to strike the correct balance between growing infrastructural capacity and protecting key environmental resources, while also minimizing the level of CO_2 emission that emanate from these developments (Sikder et al. 2022; Wang et al. 2020b, c, d, 2022; Zhang et al. 2022). As such, any infrastructural developments that seek to grow the tourism sector must be done in an environmentally sustainable manner. The ability to develop to do so would also ensure more equitable forms of growth across the region as well (Heshmati et al. 2019).

Given this pretext, our study examines not only the important land and marine-based infrastructure projects that are being developed along the New Silk Road, but also the extent to which these developments can help to grow tourism activity in the region. To further support this analysis, we also provide a series of practical policy suggestions that not only seek to drive growth across the sector but do so in a manner that balances out important economic and environmental concerns. Our study makes contributions to existing scholarship around the importance of the newly-built BRI infrastructure to ensure sustainable tourism sector growth in the region. This article underscores that over the years, infrastructural development has always been one of the transformative processes for wealth generation (Du et al. 2021).

The next section takes a step back and puts the chapter in perspective by reflecting on the Silk Road's history. It then provides a critique of significant contemporary empirical studies on the New Silk Road initiative and tourism development. The article also traces and underscores some of the key policy initiatives that should be implemented. The chapter concludes with a series of relevant conclusions and areas of research that should form key areas of focus in the future.

2 An Historical Assessment of the Silk Road

The Silk Road has existed in many shapes for several millennia. During which time the peoples of Western Europe and East Asia have connected over land and sea for cultural, political, religious and commercial exchanges (Liu 2010; Whitfield and Sims-Williams 2004; Wood 2002). Given its long life, the Silk Road has witnessed the rise and fall of different empires (e.g., Roman Empire, 30 before the Christian era until the third century) and dynasties (e.g., Tang dynasty in seventh century); the upswing and ebbing of pandemics (e.g., Black Death in the fourteenth century); and developments of different eras and Zeitgeists (e.g., Islamic era between the eighth and thirteenth century, the European Renaissance in the fifteenth and sixteenth century). The decline of the Silk Road started in the fifteenth century and the 300 years following due to geopolitical changes (e.g., fragmentation of the Mongol Empire) around the Silk Road and altering commercial interests of societies (e.g., colonialization of Africa and the Americas) (Liu 2010; Wood 2002).

However, it was not until the Venetian merchant and explorer Marco Emilio Polo (1254–1324) took account of his Silk Road voyages (or technically "Silk Routes", as it comprised a network of different destinations in the Middle East and Asia) that reached a greater audience for which other cultures and societies were unknown at that time. Marco Polo was among the first Europeans who connected Europe in the west and Asia in the east through an autobiographic travelogue "Il Milione" (i.e., today known as "The Travels of Marco Polo"), co-written by romancer Rustichello da Pisa around 1300 (Lamer 1999; Polo 1985). The travelogue, split into different books, was a cultural masterpiece that gave a closer look inside the experiences of Marco Polo's Asia-bound travels between 1271 and 1295, and reports the lands, culture, economic activity and historical

[1] In line with numerous other scholars (e.g., Mamikulova et al. 2020; Sevilla 2017; Zhang et al. 2020), we use the terms "Belt and Road Initiative" (BRI) and the "New Silk Route" interchangeably, referring to one and the same.

events along the Silk Road (Lamer 1999; Whitfield and Sims-Williams 2004).

Marco Polo's Silk Road spanned over 6400 km across what is now known to be Turkey, Iraq, Iran, Turkmenistan, Uzbekistan, Tajikistan, Kyrgistan and China (Lamer 1999). Its name is derived from one of the most lucrative trades along the route, i.e., silk textiles. In addition to silk textiles, a broad scale of raw materials, half-fabricates and products were traded between Europe and Asia such as, and by far not limited to, spices, tea, coffee, figs, perfume, gold, glassware, rice, gems, cattle, timber and handcrafts (Christian 2000; Elisseeff 2000; Whitfield 1999). Basically everything of value was traded. Arguably the trade of paper and gunpowder have changed the course of entire societies a great deal (Christian 2000; Elisseeff 2000; Whitfield 1999). As one may expect, the Silk Road led to a growth of settlements—that developed into sizeable cities—along the route where people from different cultures interacted for business and leisure purposes (Elisseeff 2000; Liu 2010). Regarding the latter purpose, this could be assumed as the very preliminary forms of "tourist activities", although the contemporary understanding and use of the word "tourist" did not emerge until the eighteenth century, during the Industrial Revolution in West Europe (Pennant 1776).

The twentieth and twenty-first centuries marked the rise of the New Silk Road, which meant faster travel could occur over land (e.g., the railway) and sea (i.e., the Maritime Silk Road through the Suez Canal) through the use of modern technologies, such as the steam engine brought by the Industrial Revolution (Liu 2010). This development led to the rapid emergence of industrialized cities and major ports along the New Silk Road, such as Athens, Hamburg, Istanbul, Moscow, Rotterdam and Seoul, to name only a few (Chaziza 2018; Hernig 2018). The New Silk Road has undoubtedly made Shanghai and Singapore the largest and second-largest container ports of the world, respectively (World Shipping Council 2022). This industrial activity has also had economic spill-over effects into other sectors such as leisure and tourism, which have led to not only a growing middle class (Heshmati et al. 2019; Kim and Wood 2020; Rhee et al. 2022), but also the rise of luxury yachting and associated tourist activities (e.g., the arrival of high-end hotels, regattas) and greater trade flows (Wood et al. 2021; Wu et al. 2021a, b, c) in Southeast Asia. For this reason, Singapore has developed Sentosa, its leisure and tourism island partly reclaimed from the sea, which includes a Monaco-inspired marina, yacht club and luxury resorts (Eijdenberg et al. 2022).

One of the most recent developments involving the New Silk Road in 2013 is the Belt and Road Initiative (BRI), which is China's leading international trade strategy in the region, with its government working hard to strengthen its commercial, infrastructural and cultural ties with an increasing number of countries along the route (Cheng 2016; Huang 2016; Liu and Dunford 2016). BRI's land network branches out into different "corridors", which include: (1) the New Eurasia Land Bridge corridor (i.e., connecting China with Europe through Russia), (2) the China – Mongolia – Russian Federation corridor, (3) the China – Central Asia – West Asia corridor, (4) the China – Indochina Peninsula corridor, (5) the China – Pakistan corridor, and (6) the Bangladesh – China – Myanmar corridor. Conversely, BRI's "21st Century Maritime Silk Road" connects China with Europe via ports in Southeast Asia, India, Africa and the Mediterranean (World Tourism Organization 2019). Hence, the BRI promotes economic development and inter-regional connectivity by using existing trade connections from the historical Silk Road, but also expanding by the inclusion of new countries and regions (e.g., West Africa) (Beeson 2018; Han and Webber 2020). In conjunction with increasing globalization, the BRI has been at the forefront of recent socio-economic developments (Qiu et al. 2021) across a wide range of industries including tourism, although a few scholars take a nuanced standpoint by stating that significant positive effects in certain industries are yet to be seen (Li et al. 2020; Niu and Li 2019). The next section documents the ways in which new infrastructural projects are driving tourism activities along the New Silk Road.

3 The New Silk Road Initiative and Sustainable Development in Tourism

Growth in trade through the BRI goes hand-in-hand with tourism development (Çalışkan et al. 2019). In the context of the China–Pakistan Economic Corridor (CPEC), which is part of the BRI, research by Nazneen et al. (2019) on the attitudes that Pakistani residents' have towards tourism, showed that they have a favourable attitude towards the sector even if its costs can sometimes outweigh the advantages associated with its development. In a similar study, Mamirkulova et al. (2020) found that the BRI's tourism development projects had a positive effect on Kazakh residents' quality of life. Their research showed that communities that are used to interacting with other cultures are often the ones most welcoming and willing to incorporate new cultural practices into their lives and in doing so embrace new tourism opportunities (Mamirkulova et al. 2020). From the perspective of a destination's policy makers, it is crucial that they do not underestimate the indispensable role that local residents play when considering the implementation of new tourism policy or the launch of new sustainable forms of economic development. A fundamental aim of all these initiatives is to protect both the residents' quality of life and cultural values, whilst also yielding highly desirable economic outcomes from the tourism infrastructure development. As a result, destinations are strongly encouraged to focus on stimulating the positive impact that all three key pillars can bring to sustainable development, namely the important social, environmental and economic factors (Mamirkulova et al. 2020).

In order to facilitate new tourism developments along the New Silk Road initiative, investments in new land and maritime infrastructure are needed (World Tourism Organization 2019). China's desire to drive new maritime investments stems from the Chinese Government's 18th Party congress, in which it highlighted the country's aspiration to become a strong maritime power for the first time. As part of this, they proposed three so-called" blue corridors", these include (1) connecting China-Pakistan, and the Bangladesh-China-India-Myanmar routes through the South China Sea and the Indian Ocean, and then passing through the Mediterranean Sea and into the Atlantic and Africa; (2) linking China, the South Pacific and then Australia; (3) a passage beginning in China through to the north-west part of Russia's northern coast, and then on to the Nordic region and parts of Europe through the Arctic Ocean (Song and Fabinyi 2022). As a consequence of these aims, China seeks to not only achieve ocean-based prosperity and security, but also secure greener forms of development, and new types of collaborative governance. Included within these aims, are attempts to generate a stronger tourism sector across the region. China has indicated that it will work with countries along the BRI to create new marine tourism routes and high-quality tourism products, which in turn can help to establish more effective platforms for information sharing (Song and Fabinyi 2022). To help achieve these aims, the Chinese Government has been signing tourism promotion and visa exemption agreements with the BRI countries, with more than 95 exemption treaties signed between 2013 and 2020 (MFA 2022). These visa exemption agreements have made it easier for Chinese tourists to travel to other BRI countries, which in turn generated new opportunities, such as the Tourism Cooperation Plan (2017–2022), involving the Chinese and Cambodian Governments (MFA 2017). While these are important new developments, it is important to note that tourism has always played an important role in increasing Gross Domestic Product (GDP) and labour demand creation in the region (Chen and Yang 2010) with coastal communities also benefiting as new jobs and business opportunities have become available (Lee and Chang 2008). In more recent times, other countries within the BRI, like Cambodia, have benefited from new tourism activities, through either the higher wages that Chinese Casinos pay local workers in the country or the higher incomes that local land owners can enjoy by renting out their properties to Chinese investors (Po and Heng 2019).

In addition to the tourism activities that have stemmed from the maritime-focused developments, have been the launching of new flights between China and other associated countries (Zhou 2016). Additionally, the 12th China Cruise Shipping Conference and International Cruise Expo, which was hosted in Shanghai in 2017, saw a number of bilateral marine transportation network agreements signed involving China and 36 BRI countries (World Tourism Organization 2019). The main aim of these agreements is to encourage the sustainable development of port services, taxation and maritime cooperation by regulating preferential treatments and establishing new initiatives to facilitate the growth of this area. For instance, cruise tourism is considered a rising tourism product in Central Asia. However, in terms of economic benefit, the extent to which each port can profit from cruise tourism is a different question. The successful management of each port greatly depends on a variety of factors, including tourism management strategies, port fees, retail quality, as well as the quality of other facilities provided. Due to these above-mentioned factors, measuring the impact of cruise tourism on a consistent basis is a crucial requirement for the sustainable development of this sector in the area (World Tourism Organization 2019).

Thanks to the continuous efforts of Chinese government to strengthen transportation links and cooperative initiatives with New Silk Road countries, the Central Asian region is expected to be surrounded by a wide range of promising tourism opportunities. As part of China's tourism development plans for the BRI, the country hopes to attract up to 85 million visitors from the region, while also sending some 150 million Chinese tourists to various destinations along the New Silk Road by 2022 (Yang 2017). Despite significant investment in in the region, the BRI is still at an early developmental stage. In order to achieve all of its developmental aims, member nations need to be flexible and adaptable, in response to the possible evolvement of governmental policies in China (Liu and Dunford 2016).

Over time, infrastructural development, will play a crucial role in growing the scale and overall capabilities of the tourism industry (World Tourism Organization 2019). For instance, highly effective infrastructure creation has helped to grow new trade routes and the emergence of new commercial hubs in places like Delhi, Jaipur and Lahore, which have in turn acted as important gateways into South to Central Asia. As Engdahl (2012) noted, this New Silk Road initiative will undoubtedly transform the Eurasia and Africa continents through the establishment of more interconnected economic zones and stronger communication and transportation linkages.

Another key success story of how the BRI initiative and other local infrastructure projects has led to the development of new tourism opportunities is Singapore. Despite the country's smaller geographic presence and lack of natural assets (Wong et al. 2020, 2021), the country has done incredibly well to attract a growing number of international visitors for over three decades, and in doing so establish itself as a global tourism leader. Much of this has come by way of strong support from local government, effective long-term planning and active role of Singapore Tourism Board (Henderson 2010). By adopting a modern and sustainable developmental approach, Singapore has built a series of highly acclaimed resorts and attractions on its small, nearby island of Sentosa. In order to connect tourists to the island, road and light rail infrastructure projects have been completed. This infrastructure means tourists can easily enjoy water parks, a maritime museum, shops, spas, hotels, dining outlets, a casino and an internationally acclaimed Universal Studios Park (Henderson 2010). As a standalone tourist attraction, in 2017, Sentosa attracted one-third of the total number of international visitors to Singapore, equivalent to around 7 million visitors (Tan 2018).

4 Key Policy Initiatives on Infrastructure Development Along the New Silk Road

Policy coordination is pivotal to expanding common interests, developing mutual trust, and establishing agreements about where further cooperation opportunities may lie (Wood and Khan 2015; Atkinson et al. 2022; Wood et al. 2022). Like any regional development initiative, it is critical that countries along the Belt and Road identify common ground while setting aside political and economic differences (Wood and Jang 2017). Moreover, improving inter-governmental communication around developing more aligned macroeconomic policy and developmental strategies should be an important priority, so as to not only capacity build but to do so in a consistent, coordinated, and inclusive manner (Hong 2017; Wang et al. 2020e).

In order to further develop the tourism capabilities of the region, BRI member countries are encouraged to develop long-term strategic plans. While the infrastructural investments provide socio-economic and political benefits, it is imperative that the environmental implications of such infrastructural investments are not ignored (Sikder et al. 2022). As such, member countries are strongly recommended to follow relevant United Nations World Tourism Organizations (UNWTO) guidelines around benchmarking, measuring and managing the impact that tourist attractions have on the local environment. It is important to understand there are significant environmental risks associated with the construction of roads, railways, pipelines and other power generating/transmitting infrastructure. Moreover, the expansion of ports in the area also raise concerns around the potential damage that could occur in coral areas, which are already severely threatened by global warming and climate change (World Tourism Organization 2019). As a consequence, to ensure the protection of natural ecosystems in the region, it is important that environmental impact assessments are conducted from an early stage. Given these important environmental considerations, there has been a conscious effort since 2017 to promote the construction of ecofriendly infrastructure, green trade, green investment, green financing, sustainable modes of production and consumption along the BRI (Cheshmehzangi et al. 2021). These domains are all necessary in order to execute and realize the full potential of green developments. As Cheshmehzangi et al. (2021) noted, with a number of already developed partnerships, and some at the beginning stages, there is a tendency that the BRI could actually become an influencing mechanism for the promotion of green development, green policy coordination, and green transition. As part of examining the infrastructural developments that are driving tourism along the BRI it is also important to also assess how individual countries in the region have also stepped in to spur e-commerce trade, economic growth and further develop their own tourism sectors (Du et al. 2021). One such example is Azerbaijan. Located in the crossroads of the Silk Road, the country has in the last decade invested heavily on ports, roads and railways connecting the main trading artery in the east from China to the west, Georgia, Turkey and the EU including Russia in the North. First built in 1878, today the rail network has been extensively developed over the years, with the new Baku-Tbilisi-Kars railway, connecting Georgia, Azerbaijan and Turkey and in doing so saves approximately 7500 km of travel for goods being sent between China and Europe,). In addition to this, with the completion of 50 major road projects and the construction and repair of some 300 bridges, Azerbaijan has positioned itself as a significant economic hub in the region. By 2020, its trade volume had increased significantly through the development of such projects, with 84.1% of the product flowing along the various completed road projects (Explore Azerbaijan 2022). In other developments along the Silk Road, China has invested $20 billion invested in Israel, with half of this money going to targeted infrastructure projects (Mohnblatt 2022). Similar investments have been made in Turkey, United Arab Emirates, Iran, and Saudi Arabia to enhance both the maritime Silk Route and the land-based Silk Road. The Middle East's geographic location is of great strategic significance moving forward. Situated at the intersections of Asia, Africa and Europe, the region serves as a junction for the five seas as well as important maritime routes, including the Strait of Hormuz

5 Conclusions and Future Research Directions

To facilitate the construction of key international arteries, and to build an infrastructure network that not only links all the sub regions in Asia, but also extends out to Europe and

Africa, a closer alignment of infrastructure development planning and technical standards is needed across the BRI countries. The foregoing initiatives necessitate that individual nation's sovereignty and security concerns remain paramount. This was underscored in the 2030 Agenda, the Addis Ababa Action Agenda (AAAA) of the Third International Conference on Financing for Development, which consists of seven action areas to support, complement and help to contextualize the means of implementation targets in the 2030 Agenda. For instance, the AAAA emphasizes that investing in sustainable and resilient infrastructure (e.g., transport, energy, water and sanitation for all) is necessary in realizing many SDGs (United Nations 2015a, b). Better understanding the importance of key infrastructure developments is vital for developing tourism practices along the Silk Road that are economically and environmentally sustainable (Mamirkulova et al. 2020).

Our study also highlighted the importance of environmental sustainable infrastructure development. Infrastructural development in the New Silk Road boosts the strategic projects that best represent their interests. In particular, these infrastructure developments have paved the way for achieving steady growth in trade between the countries, hence realizing BRI's goal of establishing closer trading links between and among member countries. As such, more research on the "Green BRI", including research on renewable energy use (Zhang and Zhang 2018); cleaner forms of production (Qi et al. 2019), policy arrangements (Alkon et al. 2019; Ascensao et al. 2018), and cooperation and coordination by countries on environmental policies and regulations (Cheshmehzangi et al. 2021) is also needed.

References

Alkon, M., He, X.G., Paris, A.R., Liao, W., Hodson, T., Wanders, N., Wang, Y.: Water security implications of coal-fired power plants financed through China's mar Belt and Road initiative. Energy Policy **132**, 1101–1109 (2019)

Ascensao, F., Fahrig, L., Clevenger, A.P., Corlett, R.T., Jaeger, J.A.G., Laurence, W.F., Pereira, H.M.: Environmental challenges for the belt and road initiative. Nat. Sustain. **1**(5), 206–209 (2018)

Atkinson, B., Wood, J., Jang, H.: Does covid-19 spark the end of globalization? In: Chaiechi, T., Wood, J. (eds.) Community Empowerment, Sustainable Cities, and Transformative Economies, pp. 535–550. Springer Publishing (2022)

Beckwith, C.I.: Empires of the Silk Road: A history of Central Eurasia from the Bronze Age to the present. Princeton University Press (2009)

Beeson, M.: Geoeconomics with Chinese characteristics: the BRI and China's evolving grand strategy. Econ. Polit. Stud. **6**(3), 240–256 (2018)

Bi, J.-W., Liu, Y., Li, H.: Daily tourism volume forecasting for tourist attractions. Ann. Tour. Res. **83**, 102923 (2020)

Çalışkan, U., Saltik, I.A., Ceylan, R., Bahar, O.: Panel cointegration analysis of relationship between international trade and tourism: case of Turkey and Silk Road countries. Tour. Manage. Persp. **31**, 361–369 (2019)

Chaziza, M.: The Chinese Maritime Silk Road initiative: the role of the Mediterranean. Mediterr. q. **29**(2), 54–69 (2018)

Chen, K.H., Yang, H.Y.: Appraising the economic impact of the "opening up to mainland Chinese tourist arrivals" policy on Taiwan with a tourism-CGE model. Asia Pac. J. Tour. Res. **15**(2), 155–175 (2010). https://doi.org/10.1080/10941661003629961

Cheng, L.K.: Three questions on China's "Belt and road initiative." China Econ. Rev. **40**, 309–313 (2016)

Cheshmehzangi, A., Xie, L., Tan-Mullins, M.: Pioneering a green belt and road initiative (BRI) alignment between China and other members: mapping BRI's sustainability plan. Blue-Green Syst. **3** (1), 52 (2021)

Christian, D.: Silk roads or steppe roads? The silk roads in world history. J. World Hist. **11**(1), 1–26 (2000)

Daye, M., Charman, K., Wang, Y., Suzhikova, B.: Exploring local stakeholders' views on the prospects of China's Belt & Road Initiative on tourism development in Kazakhstan. Curr. Issue Tour. **15**(1), 1948–1962 (2019)

Du, Q., Deng, D., Wood, J.: Differences in distance and spatial effects on cross-border e-commerce and international trade: an empirical analysis of China and one-belt one-road Countries. J. Glob. Inf. Manag. **30**(2), 1–24 (2021)

Eijdenberg, E.L., Thirumaran, K., Wong, C.: Emerging yachting tourism and luxury policy mobility: Singapore and the Southeast Asian Extension. In: Gladkikh, T., Séraphin, H., Gladkikh, V., Vo-Thanh, T. (eds.) Luxury Yachting, pp. 173–194. Palgrave Macmillan (2022)

Elisseeff, V.: The Silk Roads: Highways of Culture and Commerce. Berghahn Books (2000)

Engdahl, F.W.: Eurasian economic boom and geopolitics: China's land bridge to Europe: The China–Turkey highspeed railway. Global Research (2012). www.globalresearch.ca

European University Institute (EUI): The Belt and Road Initiative in Latin America: How China Makes Friends and What This Means for the Region. EUI (2022). Retrieved online from: https://blogs.eui.eu/latin-american-working-group/the-belt-and-road-initiative-in-latin-america-how-china-makes-friends-and-what-this-means-for-the-region/#:~:text=Initially%2C%20the%20BRI%20was%20an,increasing%20attention%20towards%20the%20region

Explore Azerbaijan: The Silk Road: Why advanced transport infrastructure is key to the future of Azerbaijan. BBC (2022). Retrieved online from https://www.bbc.com/storyworks/azerbaijan-the-opportunity-awaits/transport-and-infrastructure

Han, X., Webber, M.: From Chinese dam building in Africa to the Belt and Road Initiative: assembling infrastructure projects and their linkages. Polit. Geogr. **77**, 102102 (2020)

Henderson, J.C.: New visitor attractions in Singapore and sustainable destination development. Worldwide Hosp. Tour. Themes **2**(3), 251–261 (2010). https://doi.org/10.1108/17554211011052195

Hernig, M.: Die Renaissance der Seidenstraße: Der Weg des chinesischen Drachens ins Herz Europas. FinanzBuch Verlag (2018)

Heshmati, A., Kim, J., Wood, J.: A survey of inclusive growth policy. Economies **7**(3), 65 (2019)

Hong, P.: Strengthening National Policy Capacity for Jointly Building the Belt and Road Towards the Sustainable Development Goals. Belt and Road Forum for International Cooperation, Beijing China (2017)

Huang, Y.: Understanding China's Belt & Road Initiative: motivation, framework and assessment. China Econ. Rev. **40**, 314–321 (2016)

Kanwal, S., Pitafi, A.H., Ahmad, M., Khan, N.A., Ali, S.M., Surahio, M.K.: Cross-border analysis of China Pakistan Economic Corridor development project and local residence quality of life. J. Public Aff. **20**(2), e2022 (2019)

Kanwal, S., Rasheed, M.I., Pitafi, A.H., Pitafi, A., Ren, M.: Road and transport infrastructure development and community support for tourism: the role of perceived benefits, and community satisfaction. Tour. Manage. **77**, 104014 (2020)

Kim, J.-S., Wood, J.: Service sector development in Asia: an important instrument of growth. Asia-Pac. Econ. Lit. **34**(1), 12–15 (2020)

Koh, S.G.M., Kwok, A.O.J.: Regional integration in Central Asia: rediscovering the Silk Road. Tour. Manage. Persp. **22**, 64–66 (2017)

Kuzmina, E.E.: The Prehistory of the Silk Road. University of Pennsylvania Press (2008)

Lamer, J.: Marco Polo and the Discovery of the World. Yale University Press (1999)

Lee, C.C., Chang, C.P.: Tourism development and economic growth: a closer look at panels. Tour. Manage. **29**(1), 180–192 (2008)

Li, T., Shi, H., Yang, Z., Ren, Y.: Does the belt and road initiative boost tourism economy? Asia Pac. J. Tour. Res. **25**(3), 311–322 (2020)

Liu, X.: The Silk Road in World History. Oxford University Press (2010)

Liu, W., Dunford, M.: Inclusive globalization: unpacking China's belt and road initiative. Area Dev. Policy **1**(3), 323–340 (2016)

Mamirkulova, G., Mi, J., Abbas, J., Mahmood, S., Mubeen, R., Ziapour, A.: New Silk Road infrastructure opportunities in developing tourism environment for residents' better quality of life. Glob. Ecol. Conserv. **24**, e01194 (2020)

Ministry of Foreign Affairs (MFA): Joint Press communiqu´e of the People's Republic of China and the Kingdom of Cambodia-Ministry of Foreign Affairs of the People's Republic of China (2017). Retrieved online: https://www.mfa.gov.cn/web/zyxw/t1462712.shtml

Ministry of Foreign Affairs (MFA): List of Sino-Foreign Mutual Visa Exemption Agreements (English Version) (2022). Retrieved online: http://cs.mfa.gov.cn/wgrlh/bgzl/t902373.shtml

Mohnblatt, D.: China seeks Silk Road Revival through heavy investment in Middle East. The Jerusalem Post (2022). Retrieved from https://www.jpost.com/middle-east/article-705616

Nazneen, S., Xu, H., Ud din, N.: Cross-border infrastructural development and residents' perceived tourism impacts: A case of China-Pakistan Economic Corridor (CPEC). Inter. J. Tourism. Res. **21**(3), 334–343 (2019)

Niu, S., Li, N.: Sino-Turkish tourism cooperation under the framework of the Belt and Road Initiative. Asian J. Middle Eastern Islamic Stud. **13**(3), 445–461 (2019)

Pennant, T.: A tour in Scotland 1769. B. White (1776)

Po, S., Heng, K.: Assessing the impacts of Chinese Investments in Cambodia: the case of preah Sihanoukville Province. In: Pacific Forum, 19 (2019). Retrieved online from: https://cicp.org.kh/publications/assessing-the-impacts-of-chinese-investments-in-cambodiathecase-of-preahsihanoukville-province/

Polo, M.: The Travels of Marco Polo. In: Murray, H. (trans.). Harper & Brothers (1985)

Qi, S., Peng, H., Zhang, Y.: Energy intensity convergence in Belt and Road Initiative (BRI) countries: what role does China-BRI trade play? J. Clean. Prod. **239**, 118022 (2019)

Qiu, H., Chen, D., Li, Q., Peng, K., Jenkins, C.L.: Assessing the belt and road initiatives impact on tourism development in China: insights from MOOCs learning platform. J. Hosp. Tour. Technol. **12** (4), 641–657 (2021)

Rhee, T.-H., Wood, J., Kim, J.: Demographic transition, sustainable economic growth, and digital economy in Southeast Asian Developing Countries. Sustainability **14**(5), 2857 (2022). https://doi.org/10.3390/su14052857

Sevilla, H.A., Jr.: China's New Silk Route Initiative: Political and economic implications for the Middle East and Southeast Asia. Asian J. Middle Eastern Islamic Stud. **11**(1), 83–106 (2017)

Shafiee, S., Rajabzadeh Ghatari, A., Hasanzadeh, A., Jahanyan, S.: Developing a model for sustainable smart tourism destinations: a systematic review. Tour. Manage. Persp. **31**, 287–300 (2019)

Sikder, M., Wang, C., Yao, X., Xu, H., Wu, L., Kwame Yeboah, F., Wood, J., Zhao, Y., Dou, X.: The integrated impact of GDP growth, industrialization, energy use, and urbanization on CO_2 emissions in developing countries: evidence from the panel ARDL approach. Sci. Total Environ. Early Cite (2022). https://doi.org/10.1016/j.scitotenv.2022.155795

Song, A.Y., Fabinyi, M.: China's 21st century maritime silk road: challenges and opportunities to coastal livelihoods in ASEAN countries. Mar. Policy **136**(104923), 1–11 (2022)

Tan, H.T.: Resorts World Sentosa: Asia's Lifestyle Destination [Interview]. The World Folio (2018), 15 Oct 2018. Found online at: https://www.theworldfolio.com/interviews/resorts-world-sentosa-asias-lifestyle-destination/4566/

United Nations: Transforming Our World: The 2030 Agenda for Sustainable Development (2015a). http://www.un.org/ga/search/view_doc.asp?symbol=A/RES/70/1&Lang=E

United Nations: Addis Ababa action agenda of the Third International Conference on Financing for Development. United Nations (2015b). Retrieved online: http://www.un.org/esa/ffd/wp-content/uploads/2015b/08/AAAA_Outcome.pdf

Wang, J., Huang, X., Gong, Z., Cao, K.: Dynamic assessment of tourism carrying capacity and its impacts on tourism economic growth in urban tourism destinations in China. J. Destination Mark. Manage. **15**, 100383 (2020a)

Wang, C., Wood, J., Geng, X., Wang, Y., Qiao, C., Long, X.: Transportation CO_2 emission decoupling: Empirical evidence from countries along the belt and road. J. Clean. Prod. **263**, 121450 (2020b)

Wang, C., Wood, J., Wang, Y., Geng, X., Long, X.: CO_2 emission in transportation sector across 51 countries along the Belt and Road from 2000 to 2014. J. Clean. Prod. **263**, 122000 (2020c)

Wang, C., Zhao, Y., Wang, Y., Wood, J., Kim, C.-Y., Li, Y.: Transportation CO_2 emission decoupling: an assessment of the Eurasian logistics corridor. Transp. Res. Part d: Transp. Environ. **86**, 102486 (2020d)

Wang, P., Zhang, H., Wood, J.: Foreign direct investment, natural disasters, and economic growth of host countries. In: Chaiechi, T. (eds.) Economic Effects of Natural Disasters, pp. 97–117. Elsevier Publishing (2020e)

Wang, P., Zhang, H., Wood, J.: Achieving sustainable development goals (SDGs): a tropical region perspective. In: Wood, J., Thirumaran, K., Chaiechi, T. (eds.) Business, Industry and Trade in the Tropics, pp. 68–96. Routledge (2022b)

Whitfield, S.: Life along the Silk Road. John Murray (1999)

Whitfield, S., Sims-Williams, U.: The Silk Road: Trade, Travel, War and Faith. Serindia Publications (2004)

Wong, C., Wood, J., Paturi, S.: Vertical farming: an assessment of Singapore City. ETropic Electron. J. Stud. Trop. **19**(2), 228–248 (2020)

Wong, C., Wood, J., Paturi, S.: Managing Waste in the Smart City of Singapore. In: Azzali, S., Thirumaran, K. (eds.) Tropical Constrained Environments and Sustainable Adaptations: Managing the Asian Century, pp 225–241. Springer Publishing (2021)

Wood, J., Jang, H.: Brexit: the economic and political implications for Asia. Soc. Sci. **6**(41), 1–11 (2017)

Wood, J., Khan, F.G.: International trade negotiation analysis: network and semantic knowledge infrastructure. Scientometrics **105**(1), 537–556 (2015)

Wood, J., Li, Y., Wu, J.: An analysis of the trends and determinants of intra- industry trade between China and Asia Pacific economic cooperation member countries. Singap. Econ. Rev. **66**(3), 743–766 (2021)

Wood, J., Jang, H., Atkinson, B.: Covid-19: its impact on trade and development in the ASEAN region. In: Wood, J., Thirumaran, K., Chaiechi, T. (eds.) Business, Industry and Trade in the Tropics, pp. 137–150. Routledge (2022)

Wood, F.: The Silk Road: Two Thousand Years in the Heart of Asia. University of California Press (2002)

World Shipping Council: The Top 50 Container Ports (2022). Retrieved from: https://www.worldshipping.org/top-50-ports

World Tourism Organization: The 21st Century Maritime Silk Road—Tourism Opportunities and Impacts. United Nations World Tourism Organization (2019)

Wu, J., Huh, C.-G., Wood, J.: Globally chained economies: unwitting victims of the US-China trade war. Asian-Pac. Econ. Lit. **35**(2), 60–76 (2021a)

Wu, J., Wood, J., Huang, X.: How does GVC reconstruction affect economic growth and employment? Analysis of USA-China decoupling. Asia Pac. Econ. Lit. **35**(1), 67–81 (2021b)

Wu, J., Wood, J., Oh, K., Jang, H.: Evaluating the cumulative impact of the US–China trade war along global value chains. World Econ. **44** (12), 3516–3533 (2021c)

Yang, F.: New pact to boost tourism under Belt and Road Initiative. China Daily (2017). www.wap.chinadaily.com.cn, 25 Apr 2022

Zhang, Y.J., Zhang, K.B.: The linkage of CO_2 emissions for China, EU and USA: evidence from the regional and sectoral analyses. Environ. Sci. Pollut. Res. **25**(20), 20179–20192 (2018)

Zhang, Y., Tsang, D., Fuschi, D.L.: Chinese multinationals on the New Silk Route: managing political risk by branding the nation. Thunderbird Int. Bus. Rev. **62**(3), 291–303 (2020)

Zhang, H., Wang, P., Wood, J.: Does institutional quality matter for the nexus between environmental quality and economic growth? A tropics perspective. In: Wood, J., Thirumaran, K., ChaiechiT. (eds.) Business, Industry and Trade in the Tropics, pp. 33–52. Routledge (2022)

Zhou, S.: Belt and Road Initiative takes tourism to the skies. China Daily USA (2016). www.english.gov.cn

A Series of Case Studies: Silk Road Tourism

Thessaloniki on the Silk Road: Exhibition Tourism at Thessaloniki International Fair

Stella Kostopoulou, Styliani Lefaki, Dimitrios Kyriakou, Evina Sofianou, and Paraskevi Papadopoulou

Abstract

Historically the longest network of trade routes in the world, the Silk Road served as a bridge between East and West, facilitating trade and fostering religious, cultural and technological exchange. Nowadays, the revival of the Silk Road offers a wide and rich array of destinations and emerging tourism products, providing visitors the opportunity to experience a unique network of destinations linked by a common history and encompassing many countries with a diverse range of economies, policies, societies and cultures. Greece is historically linked to the Silk Road countries due to its strategic location at the east–west crossroads. This article focuses on Thessaloniki, the metropolitan center of northern Greece, a major transportation node and commercial hub on the Silk Routes, with diachronic links with the Silk Road countries. These relationships are also reflected in the city's exhibition events, which attract participants from Silk Road countries, who also act as leisure tourists. In this perspective, the methodological approach of the research focuses on the Thessaloniki International Fair (TIF), the largest exhibition event in Greece, to examine the visitor flows of exhibitions from the Silk Road countries since the 1920s. The aim of the study is to examine the exhibition participation of Silk Road countries in TIF thematic exhibitions, to highlight alternative visitation patterns based on the common Silk Road heritage and contemporary exhibition trade fairs. The research adopts a twofold perspective, addressing the chronological and the country visitor analysis. The examination of Silk Road countries participating at TIF allowed to identify various visiting patters proposed such as tourism flows of high, medium and low frequency. Therefore, the patterns proposed can be used as a tool to systematically organise and further develop a wide range of exhibition tourism programs, policy and planning goals through continuous networking and collaboration among Silk Road countries to further enhance Silk Road tourism development.

Keywords

Silk road • Thessaloniki • Exhibition tourism

1 Introduction

Widely recognized as a vast historical trade route, the longest and oldest worldwide, the Silk Road consisted of many land and sea trade routes connecting the great ancient civilizations of the east and west. It is generally believed that the first activities along the Silk Road began in the Han dynasty of China (206 BC–220 AD) and were inspired by the mission of Ambassador Zhang Qian (138–139 BC). However, archeological studies have shown that the route existed thousands of years earlier, when Alexander the Great of the ancient Greek kingdom of Macedon crossed part of it in his eastern conquests (Green 1991; Hammond 1989; Kostopoulou et al. 2016). For many centuries, Chinese silk, spices and other valuable goods were transported to Europe, while China received gold and other precious productsfrom the West (Stanojevic 2016). The Silk Road also promoted

S. Kostopoulou (✉) · D. Kyriakou · P. Papadopoulou
School of Economics, Aristotle University of Thessaloniki, Thessaloniki, Greece
e-mail: kostos@econ.auth.gr

D. Kyriakou
e-mail: dkyriakou@econ.auth.gr

P. Papadopoulou
e-mail: ppparask@econ.auth.gr

S. Lefaki
School of Architecture, Aristotle University of Thessaloniki, Thessaloniki, Greece
e-mail: slefaki@arch.auth.gr

E. Sofianou
School of Civil Engineering, Aristotle University of Thessaloniki, Thessaloniki, Greece
e-mail: sofianou@civil.auth.gr

S. Kostopoulou et al. (eds.), *Cities' Vocabularies and the Sustainable Development of the Silkroads*, Advances in Science, Technology & Innovation, https://doi.org/10.1007/978-3-031-31027-0_25

cultural, religious, linguistic and informationalexchanges between the thriving economies of China, Persia, India, Arabia, Greece and Rome, serving as a bridge between people and civilizations. Many cities were built along the safe long-distance travel (Wojciech 2006) Silk Routes, initially as small transport gateways that grew over time into significant commercial and trade hubs (Vasiliev and Shmigelskaia 2016).

The term Silk Road "Die Seidenstraße" was introduced in 1877 by the German geologist and geographer Ferdinand von Richthofen (1833–1905). Ferdinand von Richthofen (1833–1905) coined the term "Die Seidenstraße" in 1877 to refer to the Silk Road (Chin 2013), also used in the plural (Waugh 2007). The Silk Road got its name from the first product to be exported from China—and a very valuable one —silk. Richthofen established the groundwork for an entire path of historical research, that later became interdisciplinary (Kylasov 2019; Kostopoulou et al. 2021).Today, the Silk Road covers a vast geographical area and a great number of countries, which are diverse in terms of their economy, politics, society and culture, with a multitude of natural and man-made obstacles including corresponding interactions and interrelationships.

Thessaloniki is a city with a history of 2350 years that has developed over the centuries in the coexistence of different cultures. This multicultural character has been developed by its historical context and circumstances and its geographical and strategic location at an important maritime and road crossroads linking Europe to the East and thus constituting a hub of the Silk Road network.

Hellenistic and Roman period: Thessaloniki was founded in 315 BC as a new urban center by Kassandros, king of Macedon, who named the city in honor of his wife, sister of Alexander the Great. The new city developed rapidly, both in demographic and economic terms, acquiring great charm for its natural advantages, its port and its fortifications (Papagianopoulos 1982), and soon became an important trading center and an important military-naval base. Already from the first centuries of its existence, the city solidified a vivid image. Ships from all over the Aegean Sea, from the coasts of Syria, Phenicia, Egypt and other places unloaded products from abroad in the port. Sailors or other travelers became carriers of elements of new cultures and new religious beliefs (Vakalopoulos 1983).

After its conquest by the Romans, Thessaloniki retained a sort of administrative autonomy. During his tenure as Caesar, the tetrarch Gaius Galerius (250–311 AD) established his seat in Thessaloniki and built, among other things, a palace, a hippodrome, a triumphal arch (camara) and the rotunda. It is important to note that with the rise of Roman rule, the political and economic center shifted to the West as all countries worked to bring Rome to power. Macedonia was the center of commerce during this period with the construction of the ancient Via Egnatia, which served as a bridge over which troops and goods passed from Italy to the east and vice versa (Christodoulou 1933). The metropolitan character of Thessaloniki was reinforced by the construction of the world known military road and it gradually became an important commercial, cultural and military center, an important stop on this Roman road. The branches of the Via Egnatia and the connections to the north, south and east had a direct connection with Thessaloniki, in whose port the products were transported by land by caravans (Vasdravellis 1959).

Furthermore, with the Roman conquests tothe north, Thessaloniki's hinterland expanded to greater depth, while at the same time the safe harbor became more and more important, as the city was located at the end of a natural road leading from the Danube and the areas at Thermaikos bay. Thessaloniki was therefore the center of east–west, but also north–south transport, a natural exit for central Europe to the Mediterranean and the right transit and transport point for fast connection and transport from the Danube countries to the Near and East Center (Vakalopoulos 1983). Israelis, Egyptians, Syrians, Rhodes and Romans settled in the city and devoted themselves to navigation, trade and crafts, giving the city a cosmopolitan character. Fabrics were the subject of lively trade at the time, as silk yarns and fabrics were imported from China (Christodoulou 1933).

Byzantine and Ottoman period: The historical character of Thessaloniki is undoubtedly linked to the Byzantine period, when Constantinople and Thessaloniki were the main commercial and industrial centers of the empire (Christodoulou 1933). Constantine the Great built a new port outside the city walls on the south western side of the city during his stay in Thessaloniki in 322–323. Thessaloniki served as a center for import and export trade in a wide range of the entire Balkan area, with the city market frequented by merchants from the Balkan hinterland and the east (Svoronos 1983). From the seventh to the tenth century, the Arabs, the Slavs, the Bulgarians and the Byzantine iconoclastic period, Thessaloniki continued to develop and often had a leading role, giving grandly demonstrates its importance and position within the Byzantine state. During this period, many important churches and other monuments were built.

A notable factor in the commercial development of the city during the Byzantine period were the trade fairs (bazaars) dating from the eleventh century AD. The first mentions concern the fair of Agios Dimitrios, the patron saint of the city, which was taking place every October for about a month as part of the celebrations in honor of the saint, therefore called "Dimitria", attracting many people from the Christian world (Vasdravellis 1959). Trade fairs, local (bazaars) and hyperlocal, also took place during the

Ottoman period, empires continued to develop along the land trade network and always played a very important role for the Balkan region as they gave impetus to trade and transport of a large part of the goods by caravan (Kourkouridis et al. 2021).

During the 14th and early fifteenth centuries, the transit trade was conducted mainly by the Venetians, who transported large quantities of goods, mainly grain, from Serbia and Bulgaria west through Thessaloniki. With the conquest of Constantinople by the Turks, this situation changed and for some time the intense commercial traffic in the port was interrupted. Furthermore, the economic importance of the city decreased and the population decreased significantly, so that Thessaloniki slowly and steadily regained its old rhythm (Vasdravellis 1959).

Under Ottoman rule (1430–1912) the urban development of Thessaloniki was encouraged and the population increased significantly, with the city progressively taking on a multicultural and multireligious character with Muslim, Christian and Jewish populations, who brought new impulses for hyperlocal trade. The Ottomans reorganized the city's main feature as a market and trading center. The market and the city center were enriched with commercial, religious and public buildings such as bezestens, khans, imaret, mosques, covered markets and public baths, which became focal points of daily life in the city (Kaya 2014).

In the eighteenth century, the city was characterized by economic prosperity. Many Western countries established consular posts there and their diplomatic missions added to the cosmopolitan character of the city, that remained the main port of Central Macedonia. At the end of the nineteenth century, Thessaloniki had the appearance of a modern European city. In less than 200 years, the city was transformed from a Byzantine Orthodox city with a dominant Greek language into a multi-ethnic, multi-religious and multilingual port city. Wool and silk of the highest quality were manufactured at factories and were in high demand in the markets of Constantinople and Turkey. Sericulture flourished in the Thessaloniki region, and local silk products competed with those from Bursa (Vakalopoulos 1973). It is worth noting that the presence of Jews in Thessaloniki had a significant impact on the history of the city. Thessaloniki Jews were skilled merchants and also had a reputation as carpet makers, producing rugs with a unique character, made from local felt and decorated with colorful patterns. They also produced blue and green serge for the 40,000 Janissaries of the Turkish Empire and blue silk peştemals (towels to carry bathroom utensils) (Kostopoulou et al. 2016).

Twentieth–twenty-first century: Thessaloniki was freed from the Ottoman Empire and incorporated into the Greek state in 1912, following the Balkan Wars. Nearly the whole city's historic district was destroyed in the disastrous fire of 1917, which also rendered 73,000 people homeless and destroyed numerous priceless remnants of the city's rich legacy. Thousands of refugees from Asia Minor, Eastern Thrace, and Pontus moved in the city after the Treaty of Lausanne and the population exchange between Greece and Turkey in 1922, while the city's Muslim population vanished. Numerous districts and suburbs of Thessaloniki were created as a result of the influx of thousands of refugees from Asia Minor. During the Second World War the city's Jewish community was almost completely wiped off, since thousands of Jews were deported from Thessaloniki to Nazi concentration camps in Germany in 1943 (TIF 2020).

After the war, the rapid growth of the city was favored by the intense industrialization that took place after 1960. At the same time, Thessaloniki, also due to its extensive port, became one of the commercial centers and most important businesses in the Balkans and invests mainly in construction and housing. Rapid urbanization has been aided by the accumulation of rural populations, although the growth of the city has not taken into account its historic character and most of its ecological and urban characteristics, leading to problems and an ecological devaluation of the urban area of Thessaloniki. Despite this, Thessaloniki managed to preserve some of its cultural testimonies and parts of its architectural heritage.

Thessaloniki, a historic city, has always been an important economic and commercial center as well as a hub for caravans and maritime trade between East and West, bringing traders from many countries to the region. Thessaloniki, now the second largest city in Greece, has thus become a booming place for the exchange of culture, goods, science and technology between East and West. As a major hub of the Silk Roads, the city has rich tangible and intangible cultural heritage related to the Silk Road heritage, some of them being revived, such as Silk Road historical sites like caravanserai, bezestens, khans, events, fairs, etc. (Kostopoulou et al. 2016). Today, with a population of over one million people, the city is considered not only a major commercial center in the Balkans, but also one of the largest centers of education in South Eastern Europe with nearly 150,000 students in various universities and research institutes, including Aristotle University, the largest university in the Balkans, and over 5000 researchers, more than the EU regional average. Thessaloniki today attracts a growing number of tourists, especially from the Balkans, and has a thriving cultural activity (TIF 2020).

Thessaloniki attracts a large number of tourists due to meetings, incentives, conferences and exhibitions. Thessaloniki International Fair S.A. plays a leading role in this tourism, as a contemporary product of the city's long commercial history. The significance of this story can certainly be documented from the Byzantine era, when important fairs and bazaars were held, directly connected to the Silk Road network. These fairs and bazaars attracted the attention of visitors from many different areas which formed the core of wider local and hyper-local business networks and over time transformed Thessaloniki into a travel destination (TIF 2020).

Thessaloniki is the headquarters of TIF-HELEXPO SA, the main Greek organizer of exhibitions and conferences, and one of the most important in the Eastern Mediterranean and South Eastern Europe, with over 90 years of continuous activity. TIF-HELEXPO organizes a significant number of national and international fairs which are major industrial events, the most famous of which is the Thessaloniki International Fair (TIF), an annual international exhibition held in the first week of September since 1926. Since its inception, TIF has been an important event for the country and the entire neighbouring region, introducing innovations often presented for the first time in Greece and the Balkans (TIF 2019). With a focus on entrepreneurship, innovation and culture, Thessaloniki International Fair (TIF) attracts the participation of representative companies and government agencies from Greece and the global market, highlighting its dynamism on the international fair tourism market. With thousands of visitors each year (210,000 in 2016), the event has become a benchmark for Thessaloniki's business tourism and synonymous with the economic and cultural history of the city. In addition to its commercial and tourism significance, the event also has political significance, with the Prime Minister traditionally announcing a series of program statements at its opening ceremony (TIF 2019).

With the aim of strengthening its role as a leading regional trade fair and conference center in the wider Balkan region, TIF focuses on building strong partnerships and active involvement in the geographic area of the Balkans and beyond (TIF 2019). Apart from the wide participation of exhibitors, the most notable element of this fair is its internationalization as also important foreign markets show their support through their national participation, including many Silk Road countries, such as Russia, China, India, Iran, Armenia, Indonesia, Egypt and Turkey. Although TIF plays a significant role in the city's economy, there is limited research on the potential of Thessaloniki as an exhibition tourism destination based on the participation of Silk Road countries, designating a rather unexplored field of tourism study.

The aim of this paper is to explore exhibition tourism created by the participation of Silk Road countries in TIF exhibitions, to highlight alternative visitation patterns based on the common Silk Road legacy and contemporary exhibition fairs. Silk Road is a complex tourism system, on various spatial scales (international, national, interregional, regional, local) and requires to be systematically organized. In order to develop a methodological approach enabling country-specific, yet comparable Silk Road exhibition tourism analysis, the concept of networking and matrix construction is implied, as a tool to explore international cooperation in the tourism sector. In this perspective, the methodological approach in this article focuses on the Thessaloniki International Fair (TIF), to examine the flows of exhibition visitors from the Silk Road countries since the 1920s. This approach includes the secondary study of research on TIF HELEXPO archives, to examine the multifaceted role of the TIF from its establishment to date, as well as the "Honored Country" strategy at the TIF and its role.

In order to meet the stated goals, the paper sets the following methodological steps: (i) delineation of the study periods; (ii) identification of the countries related to the Silk Road heritage that have participated at TIF; (iii) collection of information and data on their population and tourism characteristics; (iv) classification of countries that have participated at TIF and creation of a Silk Road tourism inventory; and, (v) visualization of the potential exhibition tourism networks. The proposed patterns can be used as a tool to organize and systematically advance a wide range of exhibition tourism programs, policy objectives and planning through continuous networking and cooperation with the countries of the Silk Road to further enhance Silk Road tourism development.

2 Thessaloniki as an Exhibition Tourism Destination

2.1 Exhibitions and Tourism Development

Exhibitions are professionally organised large-scale events where products and services are exhibited. There are data and findings that illustrate that since ancient times people have gathered for meetings, conferences and fairs, mainly for social, sporting, political or religious reasons. The range and frequency of these events expanded as cities emerged as regional hubs, and several associations and organisations routinely sponsored exhibitions (Walker and Walker 2011). Germany started hosting modern exhibitions in the 1850s, and the Crystal Palace hosted the first international exhibition in 1851 (Tafesse and Skallerud 2017).

Exhibitions are part of the MICE industry, which is used in short for Meetings (all kinds of professional meeting packages), Incentives (organization of Incentive tours/trips for companies), Conferences/Convention (organization of large scale gatherings such as conferences and all packages) and Exhibitions/Events (organization of national and global trade fairs in various trade fair districts) (Dwyer et al. 2001). MICE includes corporate events and activities, also known as 'Meeting or Event Industry', a subsegment of business travel, which may also include a leisure component, a set of meetings, conferences, exhibitions, corporate activities, hotel industry, various transportation, travel associations, professional conference organizers, entertainment industries, travel agencies and professionals (Jamgade 2018). The main purpose of MICE events is to bring together interest groups for a specific purpose and create a networking base for business, industry, government and academia to engage in meaningful discussions. MICE specialist travel providers are usually affiliated with major corporate agencies as this needs to be planned and organized in advance. Travellers who attend MICE activities have a purpose beyond leisure tourism and are actually business travellers.

Exhibitions are regular market events over a period of time, where companies present their main assortment and sell it primarily to professional buyers at random (Kirchgeorg 2005). Most exhibitions are held at the same place each year and are usually held by trade and industry associations, government agencies and specialized exhibition organizations. Exhibitors, companies that announce noveltiesand/or innovations, usually rent an area from the organizer and set up a stand to meet potential buyers. Buyers or the general public need an invitation or a ticket to attend the exhibition. Expos attract significant numbers of visitors to the destination, potentially making substantial economic contributions to the host city or region, as well as enhanceing its image and reputation as a leading corporate event destination (Deery and Jago 2010; Getz 2008) building future visits to the destination (Arnegger and Herz 2016). Given that exhibitions have a direct impact on hospitality destinations in terms of economic benefits, it is possible to make significant investments in purpose-built infrastructure and government support to develop the potential of a destination as a series of corporate events (Weber and Ladkin 2011).

Hosting an exhibition is considered an important strategy not only to increase the number and income from tourists in the short term, but also to improve the image of the destination and build a brand for sustainable development (Jago et al. 2010; Yu et al. 2012). Exhibitors and participants, both short or long-distance travelers, will request hotel rooms and food and drink to accommodate themselves. Exhibition activities increase the economic production of related industries such as transportation, hotels, restaurants, tourism promotion, various manufacturers, exporters, importers, traders, industry groups, governments, etc. (Tafesse and Skallerud 2017). Exhibition tourism contributes significantly to the country's economy as its representatives spend more than the average leisure tourist (Wan 2011). Trade fairs also extend the average length of stay for visitors, while also creating additional business opportunities for related industries thus increasing employment (Mistilis and Dwyer 1999; Oppermann 1996).

The number of international exhibition activities is rapidly increasing, and so is the participation rate (Chun 2016). Nearly 32,000 exhibitions are organized annually all over the world, with the worldwide exhibition business contributing USD 325 billion to the global economy. In 2018, 4.5 million exhibitors participated in exhibitions, and 300 million people attended to source their business needs (UFI 2019). Many large-scale exhibitions are frequently visited by tourists. In order to obtain information more cheaply and to maintain relationships with other attendees, exhibitors and attendees frequent these events on a regular basis (Godar and O'Connor 2001).

Exhibitions have gained recognition as a sizable market segment and crucial driver to regional economies in the past few decades, boosting local economies in the form of income generation, opening up of employment opportunities in industries that provide related hospitality services, such as lodging, food and beverage, convention services, transportation, tourism and entertainment.

On the part of the organizing agencies and the beneficiaries (local, regional, national), the advantages of exhibitions can be summed up, especially their economic significance for the host towns. Particularly, the exhibition organizers process orders, pay taxes and make investments in the infrastructure of the event. The local, regional industry, the government, and the private sector, on the other hand, stand to gain. Additionally, human resources are needed for planning and sustaining trade shows, which leads to the creation of jobs. These actions lead to revenue generation for the city, since private spending multiplies the positive impact (UFI 2010).

Additional benefits for destinations of a growing MICE sector include:

(i) Other than the economic and employment benefits, Exhibition tourism supports local economies in numerous additional ways, including business opportunities, knowledge sharing, training opportunities, skill improvement, and more, in addition to the economic and employment advantages.
(ii) Exhibitions, which are a part of the MICE industry, also encourage local government and private sector investments, which improve the overall hospitality environment of the destination country (Kumar et al. 2014).

According to Kowalik (2012), exhibitions and events in general are a way for a city to promote and "sell" itself, in an effort seeking to attract visitors, businesses, and investments. A city gains the attributes of a commercial, technological, and business hub through the organization of an exhibition. More jobs are produced as a result, local businesses are strengthened, and the tourism industry is growing. According to several studies, exhibitions have a substantial multiplier effect, where seven additional euros are flowing back into the local economy (UFI 2010). More than just a commercial endeavor, organizing exhibitions is more than simply an economic activity, it contributes to the growth of the city and the surrounding region (Kourkouridis et al. 2016). The economic effects of TIF-HELEXPO are significant for Thessaloniki, according to research by Kourkouridis et al. (2016), as exhibition activity contributes around 2.5% of the GDP of the Regional Unit of Thessaloniki. The population of the city increases dramatically during TIF, sizable fraction of them are tourists (Sfakianakis 2000).

2.2 Thessaloniki International Fair

2.2.1 Historical Background of TIF-HELEXPO

Thessaloniki International Fair S.A (hereinafter referred to as "TIF-HELEXPO") is the national organizer ofexhibitions and conferences in Greece and the owner of two major exhibitions and conference venues, in Thessaloniki and in Athens. Over the years, TIF-HELEXPO has given businesses, manufacturers, traders, dealers, vendors, exhibitors, and visitors with the place to communicate, network, collaborate and promote antagonism with each other. It has also helped to fortify their relationships both domestically and internationally. Thessaloniki International Fair (TIF), which takes place every September and is a significant yearly event with political significance, is the most well-known event, hosted by TIF-HELEXPO, with the Prime Minister traditionally announcing at the opening ceremony a series of programming statements (TIF 2020).

The idea of the first Thessaloniki Trade Fair was conceived in 1925 by Nikolaos Germanos (1864–1935), a member of the Greek Parliament, and a group of Thessaloniki business people and journalists who were driven to build and carry out the idea an inaugural International Fair in Thessaloniki (TIF). On April 28th, 1925, Nikolaos Germanos submitted a petition to the Ministry of National Economy, requesting authorization to organize the country's first international trade fair (TIF 2019, p. 6). On October 3rd, 1926, the inaugural Thessaloniki International Fair was launched, planned to be an annual event, with the goal of enhancing commercial links with foreign countries and businesses, marking a new era for the regional economy (TIF 2020, p. 3). However, Thessaloniki Trade Fair was not fully recognized as such until 1929, when it was formally acknowledged as 'International' by the Union of International Fairs (Foires Internationals Union) (Kousidonis 2006). Since then, it has developed into the most significant trade event of Greece, with implications for the country's economy and politics, as well as for alarger region of southeastern Europe, acting as a forum for cooperation and synergy (TIF 2019). The Fair includes a political component that is still present today ever since Prime Minister EleftheriosVenizelos visited the event in 1928.

Thessaloniki International Fair (TIF) moved to its present buildings in 1937. During its initial period of events, from 1926 to 1939, TIF played two roles: on the one hand, it had an economic character, as indicated by the need to strengthen trade ties with the Balkan region and Eastern Europe, and on the other, it had a political character, evidenced by the official participations of states presenting their advances and new products (Kourkouridis et al. 2016).

Its operations were temporarily suspended because of World War II and the ensuing Greek Civil War. Due to the Second World War and the destruction of infrastructure caused by the war, the next TIF took place 11 years after the previous one in 1939. The first post-war TIF took place in 1951 in new facilities near the current Thessaloniki fairgrounds and was funded by the Marshall Plan to restore the looted exhibition building (TIF 2019). This event—and everything that followed until 1970—took place in the context of Greece's post-war productive restructuring, as well as the Cold War transition period, when exhibition activities were characterized by the national participation of countries with great financial power, such as the United States and the Soviet Union (Kourkouridis et al. 2021).

During the 1950s and 1960s, TIF attracted an average of 4000 Greek exhibitors per year, while international exhibitors attracted an average of 1500 per event. In 1967 the number of visitors realized a new record of 1.7 million. During this time, TIF has evolved from a popular festival to a city landmark (Kostopoulou et al. 2016).

The largest TIF stands were built in the 1970s and are still operational today, while the first sector fairs, the "International Fur Fair" and the "International Marble Fair", were held in 1973 (TIF 2019). Since 1970 the exhibition activity has gradually taken on a new form, with fairs taking center stage. These fairs focused on a specific sector and attracted the attention of entrepreneurs in that sector. As a result, the role of the general fair has decreased, while the role of trade fairs specialized in specific sectors has grown (Kourkouridis et al. 2021).

In addition to the International Fair, TIF's installations have hosted several field exhibitions which started developing in Thessaloniki since 1973, simultaneously with the "Thessaloniki International Fair". These exhibitions followed the national trends and gradual development of the exhibition industry and specialised in events that attracted professionals from diverse sectors (TIF 2020). In addition, as part of the Thessaloniki International Fair, other events such as the Internationally renowned Film Festival and Song Festival have begun. In 1974 the first industrial fair was born, the most important of which was and still remains the textile industry. Sector exhibitions have been held since 1974, allowing the exhibition space to be used practically all year round.

In 1977, Thessaloniki International Fair changed from a legal entity under public law to a public limited company with the distinctive title "Helexpo-TIF SA" (Pozrikidis 2013). The continuous development of the TIF from 1950 to 1990, as well as the general development in the field (after 1970), increased the need for a larger area of exhibition halls. From 1950 to the end of the 1990s, this resulted in an expansion of the exhibition center and the construction of new pavilions.

"TIF HELEXPO SA" continued its activities until recently (2010) in the same specific form of exhibition, however, after the financial crisis and the increased demand for extroversion in the Greek economy, attempts were made to reinforce the international character of the exhibitions in Thessaloniki. As a result, more organized and systematic efforts have been made to attract foreign business visitors (Kourkouridis et al. 2021, p. 188). Apart from this, during the economic crisis in Greece, TIF-HELEXPO expanded its activities to new areas of interest such as thematic events (Christmas theme park "Astero Cosmos", "Factory and Chocolate Museum", etc.) (TIF 2020, p. 4), "Beer Festival", "Burger Festival", "Flower Festival" and other festivals and open space events that attract many visitors and have a great impact on the city's traditional exhibition business, creating a joyful atmosphere and transforming the easily accessible exhibition area where the festival takes place into a bustling center of the city (TIF 2020, p. 5).

TIF is located in a central urban area of Thessaloniki, outside its historical walls, and covers an area of almost 180,000 m^2 (Figs. 1, 2), of which 62,000 m^2 are covered with exhibition spaces, congresses and others. The premises are located in the city center, at the crossroads of the most central avenues and streets of the city. On-site facilities include 44,000 square meters of exhibition space (Fig. 2), two convention centers, one of which, Ioannis Vellidis, which is housed in its own building with a capacity of 2400 people and is one of the largest major congress venues in the Balkans with 16 shops, a large gym, a small open-air cinema, a small to medium-sized museum of modern art and two organized parking lots: an open space (210 spaces) and an underground one (420 or 430 seats) in the basement of the Ioannis Vellidis Palais des Congrès (Fig. 1) (Kousidonis 2006). There are a number of additional company-managed properties on the TIF-HELEXPO site. Specifically, the Company owns 11 properties and one outdoor cinema under

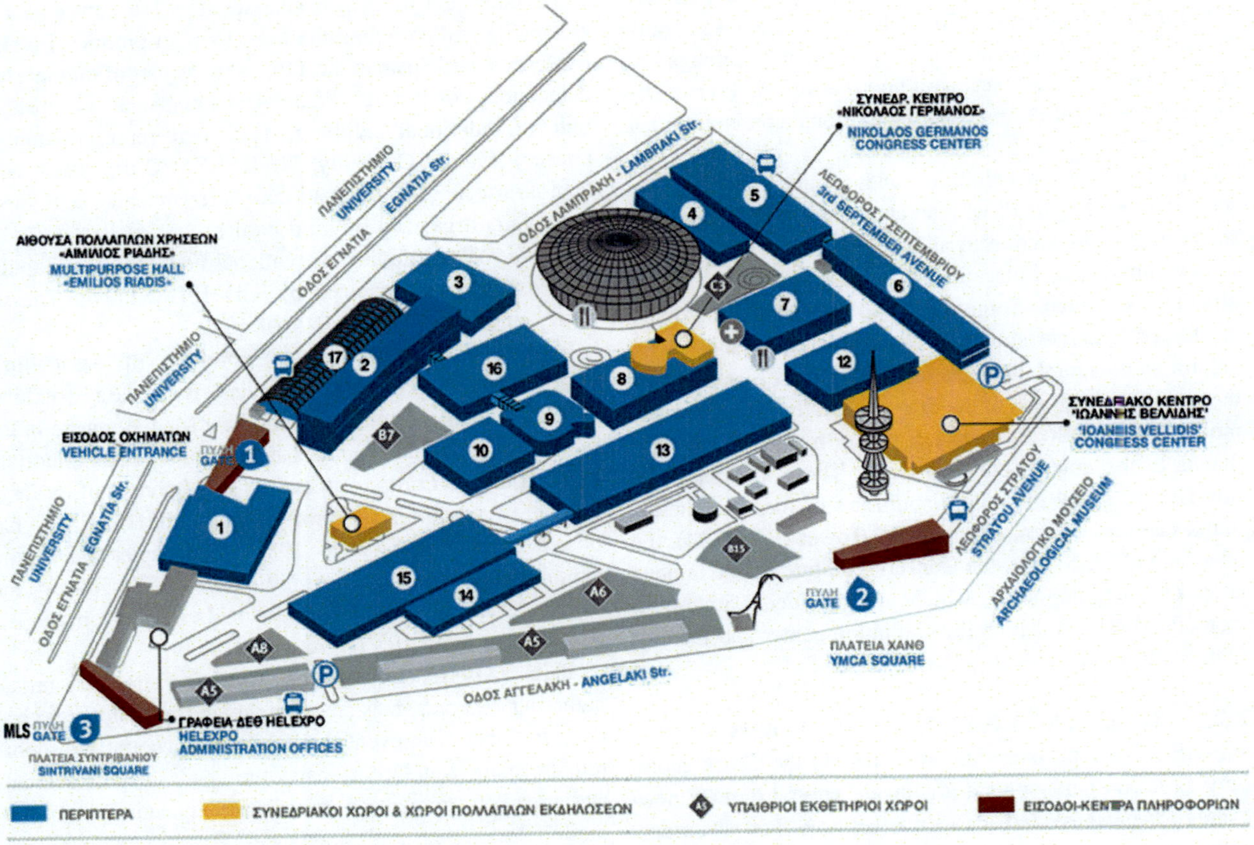

Fig. 1 TIF topographic scheme. *Source* TIF-HELEXPO, https://www.helexpo.gr/en/thessaloniki-exhibitional-area

Fig. 2 Aerial view of TIF installations. *Source* TIF-HELEXPO, https://www.helexpo.gr/en/thessaloniki_home

commercial lease. Additionally, TIF-HELEXPO has granted the Macedonian Museum of Contemporary Art the use of a 4000 m^2 building until 2020. In this way, the Thessaloniki International Exhibition Center, which belongs to TIF-HELEXPO S.A., acts as a hub that offers services that successfully meet the needs of exhibitions, conferences and cultural events.

TIF has been an exhibition and commercial center, as well as a city landmark with an abundance of contemporary monuments and architectural uses (pavilions designed by famous architects, Museum of Contemporary Art, etc.), as already mentioned. Some of the oldest exhibition halls, reinforced concrete structures from the 1950s and 1960s, exhibit architectural elements of great interest that represent the architectural trends of the time, especially the modernism movement (TIF 2020, p. 11). It is worth pointing out that thanks to the gradual construction of Thessaloniki Exhibition Center, many important landmarks of modern Thessaloniki have been born, e.g. the OTE Tower (Figs. 3, 4), the gates at YMCA Square, the CorTen sculpture by the famous artist G. Zoggolopoulos, which are considered important iconic landmarks for both the exhibition center and the city (TIF 2020, p. 38).

2.2.2 TIF-HELEXPO Exhibitions and Events

TIF-HELEXPO plays an important role in the Greek economy as it constantly puts a significant effort to create new events for business and innovative services that encourage extroversion and help companies create and develop business networks. In addition to promoting and strengthening hotel and tourism businesses, the TIF contributes to enhancing and enriching the local tourism product. Foreign exhibitors participate in the TIF, either as part of official state delegations, or through bilateral chambers, or as direct or indirect individual exhibitors (via their representation in Greece) (Fig. 5). Currently, TIF-HELEXPO organizes more than fourteen (14) international fairs annually and hosts many fairs organized by third parties.

Besides Thessaloniki International Fair, the most popular field exhibitions organized by TIF-HELEXPO in Thessaloniki since 1985 are: (TIF 2020, p. 4).

AGROTICA (International Exhibition of Agricultural Machinery, Equipment and Supplies) is the largest exhibition in the Balkans and the South–East Mediterranean in the field of agricultural economics and one of the most important exhibitions in Europe, attracting the interest of a large number of exhibitors and professionals. TIF HELEXPO has hosted AGROTICA since 1985. It occupies the entire exhibition center with an exhibition area of 34,500 m^2 and welcomes around 120,000 professional visitors from the EU and 30 other countries (Agrotica 2018).

PHILOXENIA (International Tourism Fair) is an annual gathering of companies, institutions and tourism organizations in Greece, an event that takes into account the global tourism market and coordinates the needs of professionals and travellers to make a contribution meaningful in promoting destinations and experiences. Tourism entrepreneurs, decision-makers, accommodation buyers from all over the

Fig. 3 View of TIF exhibition. *Source* https://www.typosthes.gr/thessaloniki/249124_thessaloniki-epistrefei-fetos-i-deth-me-timomeni-hora-tin-ellada

Fig. 4 TIF south entrance, OTE tower and part of exhibition infrastructure. *Source* https://www.helexpo.gr/

world, but also simple visitors looking for information for their next vacation form the profile of a show associated with the tourism industry (Philoxenia n.d.).

HOTELIA (International Exhibition of Hotel Equipment) focuses on the latest advances and trends in the hotel, catering and coffee sectors and is aimed at hoteliers, architects, decorators, technical companies, catering and tourism professionals. Side events are an important element of HOTELIA and give the exhibition an informative and instructive dimension. HOTELIA organizes a series of innovative events, follows visitors from the Balkans and encourages Greek companies to be more open-minded by organizing B2B meetings with invited international professional visitors. The shared location with the PHILOXENIA tourism fair provides a strategic advantage as the percentage of visitors from the country's tourism industry increases significantly (Philoxenia n.d.).

FRESKON (International Trade Fair for Fresh Fruits and Vegetables) has been the premier event in the fresh production industry in recent years and the largest gathering of producers, international retail groups and commercial traders in South Eastern Europe. Freskon is committed to displaying

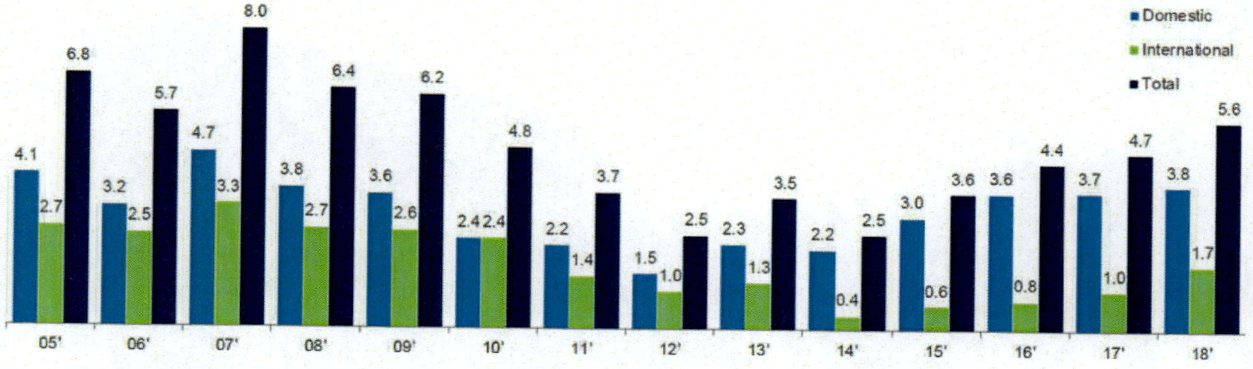

Fig. 5 TIF—HELEXPO: total number of exhibitors (2005–2018)—(in '000'). *Source* TIF (2019), p. 18

excellent fruit and vegetables, cultivating trade and export, and building partnerships through a quality exhibition with a targeted B2B program of international meetings, conferences and lectures, parallel events and innovative product presentations.

ZOOTECHNIA (International Livestock and Poultry Fair) is a specialized event for livestock in Greece and the Balkans. Poultry farms participate and the latest developments in the sector are presented, on a wide range of topics, such as the main types of production like animals, machinery, equipment, services and research results (Zootechnia n.d.).

ARTOZYMA (International Exhibition of Bakery, Pastry, Machinery, Equipment, Ready Products) takes place every 2 years.

THESSALONIKI INTERNATIONAL BOOK FAIR began in 2004 with the National Book Center as organizer, before being taken over by the Hellenic Foundation for Culture in 2014. It was a cultural institution of great importance and originality, serving as the only annual gathering of all book actors from Greece, the Balkans, Western Europe, the southeastern Mediterranean and the world. entire (Thessaloniki Book Fair n.d.).

The organization of ART THESSALONIKI INTERNATIONAL CONTEMPORARY ART FAIR is a catalytic and multifaceted event that promotes dialogue, artistic mobility and exchange, innovation challenges, neighborhoods and artistic exchange, above all the appreciation of diversity cultural phenomena (ART_THESSALONIKI INTERNATIONAL FAIR n.d.).

The SPORTEXPO exhibition main objectives are to present sport items to the general public, to inform the general public about less common sports and developments in the field, to create conditions for business contacts and cooperation, to promote and to organize Greek regions as sports tourism centers and holding parallel events as well as to the activation of school units. Only Greece and Cyprus have registered to participate.

INFACOMA Expo was founded with the aim of supporting innovations in the construction sector. At Infacoma, professional visitors and the general public can discover trends in the fields of restoration, construction, "green" construction, innovative products and solutions and proposals for modern architecture. Infacoma is held in conjunction with the international Aquatherm exhibition, covering professional, industrial and residential heating, air conditioning, ventilation, water management, water supply, use of alternative energy sources, equipment for spa facilities, saunas and swimming pools for professional spaces, hotels and residences.

The annual KOSMIMA exhibition is the meeting place for jewelery and watchmaking professionals and brings togerher manufacturers, workshops, designers, importers, agents and jewelery wholesalers. KOSMIMA goal is to open new markets and promote Greek jewelry in international markets. A total of 2000 experts from Greece and the Balkans participated in KOSMIMA 2019.

In Fig. 6 the top 4 exhibitions and events of TIF are presented by number of visitors for the period 2005–2018 (TIF 2019, p. 18).

In Fig. 7 the main TIF facilities hosting the exhibitions, congresses and secondary uses as well as the main tourist flows are depicted.

The international character of the exhibitions in Thessaloniki in the 50, 60 and 70 s was reflected in the national participation of leading and financially strong countries in special national pavilions, such as the American pavilion or the USSR pavilion. After a period of decline in national participation and rise in trade exhibitions, the institution of the honored country was created to allow the honored country to present its main achievements (economic, cultural, etc.), as well as its main development activities, business and the strengthening of economic, trade, investment and cultural relations between foreign countries and Greece (Kourkouridis et al. 2021).

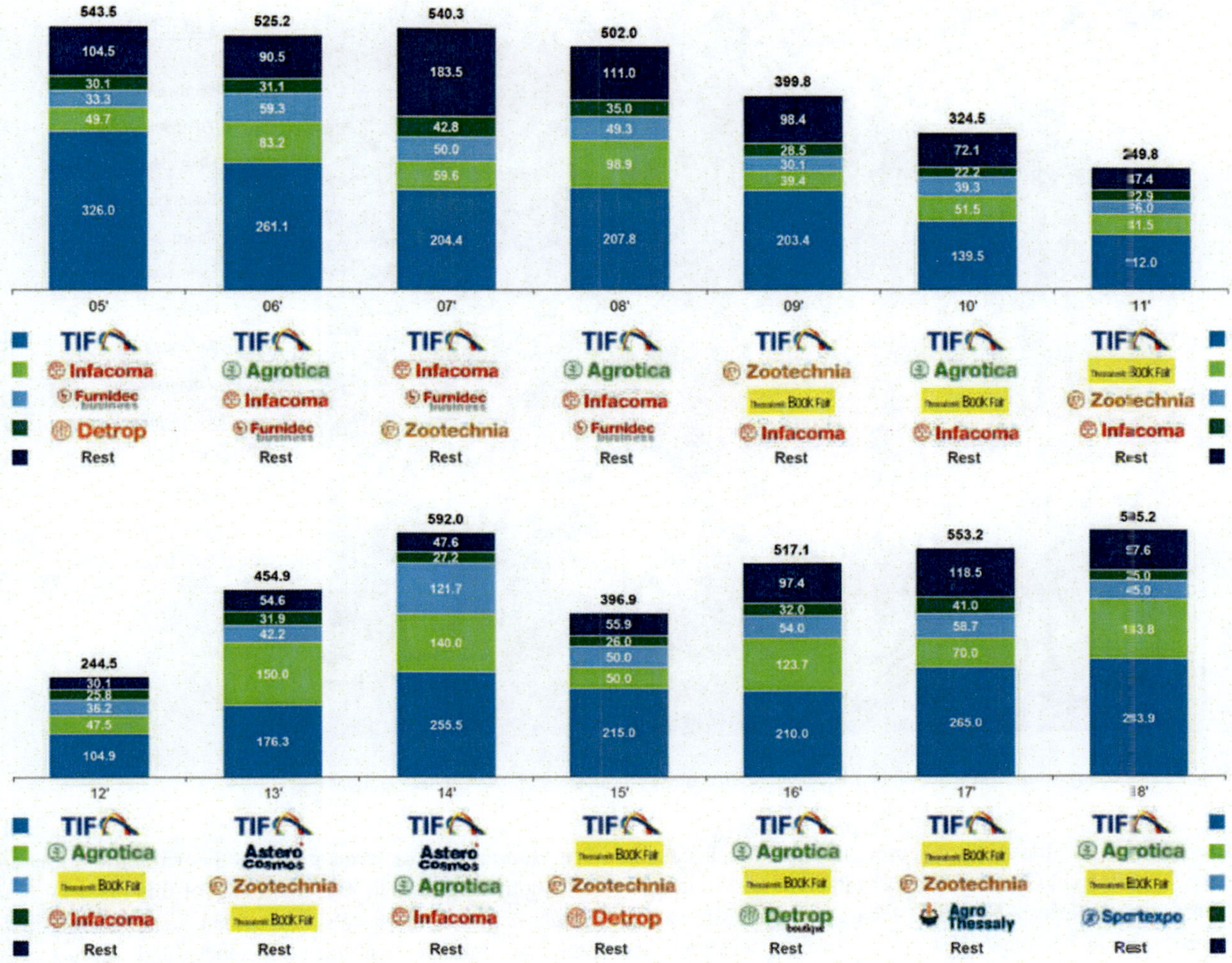

Fig. 6 TIF—HELEXPO: top 4 exhibitions and events by number of visitors (2005–2018)—(in '000'). *Source* TIF (2019), p. 18. Infosystem was an International Exhibition of Information Systems that was last held in 2010. Star World ("Asterocosmos") was last organized by the TIF-Helexpo in 2014. Since then (2015 onwards), although the event is still hosted in the Company's premises, it is held by a 3rd-party organizer, that rents Helexpo's facilities

The institution of the Honored Country was established in 1998 as an evolution of official government participation (Kourkouridis et al. 2021), with China being the first Honored Country. It has been used on several occasions as an indicative means to designate with precision Thessaloniki and TIF as an international commercial crossroads between East and West and has always shown its global character. Although this concept began to decline in previous decades, attempts have been made to revive it in recent years. This was attempted by selecting three geopolitically powerful countries as honored countries as well as potential economic partners for Greece. The reopening of the Honored Country institute began with Russia in 2016, China in 2017, the United States in 2018 and India in 2019 (TIF 2019).

Furthermore, TIF-HELEXPO actively supports the expansion and internationalization of its network and its activities through strategic membership of national and international associations and organizations, such as: Union des Foires Internationales (UFI), International Association of Exhibitions and Events (IAEE), International Congress and Congress and Convention Association (ICCA), Central European Fair Alliance (CEFA), International Association of Convention Centers (AIPC), International Tourism Trade Fair Association (ITTFA), Hellenic Association of Professional Congress Organizers (HAPCO). Under its mission and vision for opening new markets and connecting business entities, TIF-HELEXPO, is also a member of various commercial and industrial chambers located in Thessaloniki and Athens, domestic (Thessaloniki Chamber of Commerce and Industry, Union of Hellenic Chambers, Thessaloniki's Chamber of Small and Medium Enterprises, Thessaloniki's Chamber of Tradesmen) and international (German–Greek

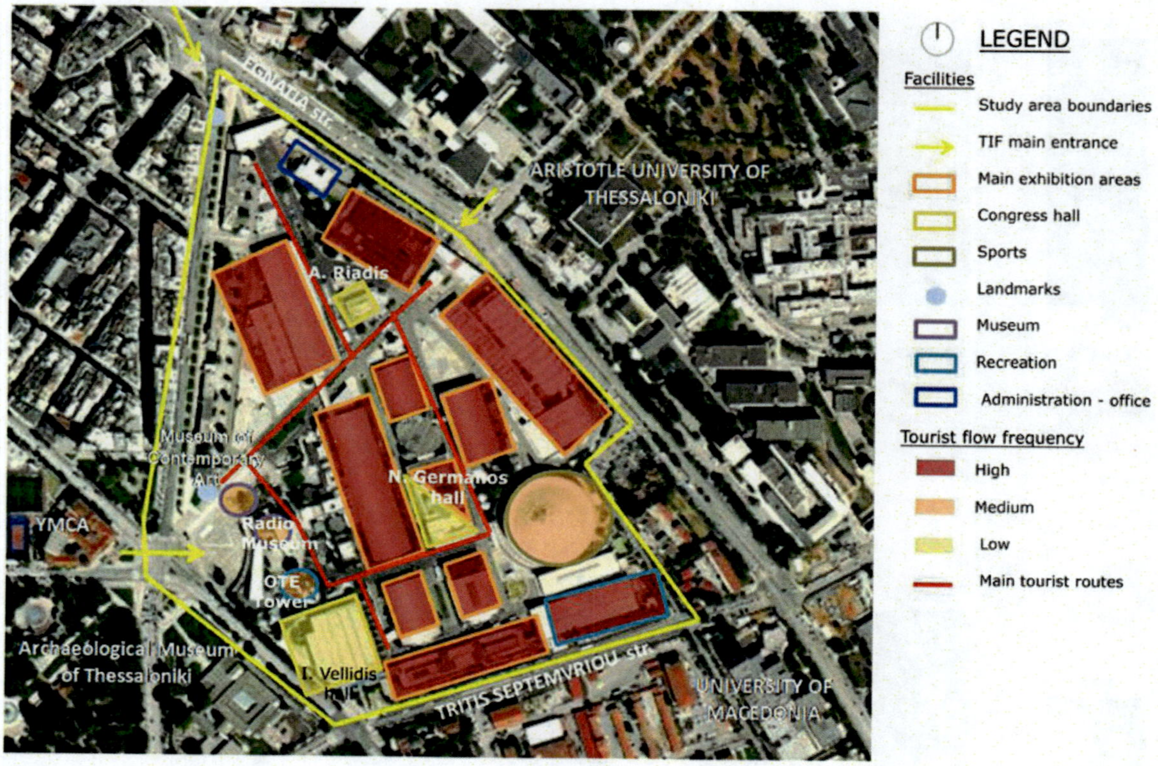

Fig. 7 TIF infrastructure and tourist flow frequency. *Source* Idem processus. Map background: Bing Maps

Commercial and Industrial Chamber, Italian–Greek Chamber of Commerce, Greek–French Commercial and Industrial Chamber, American–Hellenic Chamber of Commerce, Arab–Hellenic Chamber of Commerce and Development) and particularly with Silk Road countries (Chinese–Greek Commercial and Industrial Chamber, Russian–Greek Chamber of Commerce (TIF 2019).

With the constant aim of becoming a leader in the trade fair sector, mainly in the Balkans, the company has signed a series of Memoranda of Understanding (MoU) for cooperation with foreign organizations and companies operating in the Balkans and in the rest of the world, especially with Silk Road countries such as: Euro Expo (Moscow, Russian Federation), Roscongress Foundation (Russian Federation), Iran International Exhibitions Company (Tehran, Iran), Shenzhen Convention and Exhibition (PR China) According to these MoUs, the line joint agreement of the actions provide for the exchange of exhibition spaces, exhibitors and visitors (TIF 2019).

3 Methodology

As tourism generates opportunities for synergies especially in the cultural sector, the proposed methodological approach is used for the visualization of Silk Road countries' flows to TIF, serving as an evaluation of the exhibition tourism development potential. The main aim of this research is to further explore the city's Silk Road exhibition tourism potential through customized products and services. To partially obtain this goal is to analyse through time the flows from each country to TIF and relative observations led to notable results.

The following tasks are outlined in the paper in order to meet the stated goals: (i) delineating the study periods; (ii) identifying the countries related to the Silk Road heritage that have participated at TIF; (iii) gathering information and data on their population and tourism characteristics; (iv) classifying participating countries at TIF and developing a Silk Road tourism inventory; and, (v) visualizing potential exhibition tourism networks. Through ongoing networking and cooperation with the countries of the Silk Road, the proposed patterns can be used as a tool to organize and systematically promote a wide range of exhibition tourism programs, policy objectives and planning thereby enhancing Silk Road tourism development.

In order to explore and address the research hypotheses, secondary research initially involved a review of the literature in scientific publications, articles, books, studies, historical sources. Then a thorough and in-depth analysis of the Silk Road countries' participation in the general TIF Fair, the sectoral TIF HELEXPO Exhibitions and the Trade Visitor

Programme was carried out. The information was obtained from TIF secondary sources, internet research and contact with the managers of sectoral exhibitions. The result of this process was the compilation of A database containing the information acquired for analysis was created as a result of this approach.

The first methodological phase was to construct a matrix for selected time periods based on the initial database information enriched with more specified criteria. Silk Road countries were approached according to a number of attributes in order to attempt Silk Road tourism networking. The networking criteria were chosen in line with the aims of the proposed methodology and organized into three categories:

1. Attributes connected to the Silk Road heritage
   a. member countries of the UNESCO Silk Road network
   b. member countries of the UNWTO Silk Road network
   c. member countries of both UNESCO and UNWTO Silk Road networks.
2. Attributes referring to visits of Silk Road countries to TIF
   a. Visits to exhibitions, total (1929–2019)
   b. Visits to main TIF exhibition (1929–2019)
   c. Visits to main TIF exhibition (2010–2019)
   d. TIF honored countries.
3. Attributes referring to population data of Silk Road countries
   a. number of inhabitants (population census 2018 where available)
4. Attributes referring to tourism development data of Silk Road countries
   a. International tourism (number of departures, year of reference 2018)
   b. International tourism expenditure (bn. $, year of reference 2018).

The first category of attributes includes data on the relationship of selected countries to the Silk Road identity, and in particular whether they are Silk Road Member States as defined by UNWTO (https://www.unwto.org/silk-road), if they are included in the list of UNESCO Silk Road countries (https://en.unesco.org/silkroad/countries), or if they are included in both lists.

Motivated by the growing interest in the Silk Road for cultural exchange, trade and tourism, the United Nations World Tourism Organization (UNWTO) decided to revitalize the historic Silk Road as a tourism concept and to unite three continents in a project that extends over 12,000 km. With an initiative dating back to 1994, UNWTO today has a network of 34 Silk Road member countries working together to promote the development of Silk Road tourism (https://www.unwto.org/silk-road). The UNESCO Silk Road Program aims to revitalize the historical networks of the Silk Road through a digital platform to enhance the common heritage of societies and encourage joint actions of interconnected cultures emerging along the Silk Roads. Today, more than 40 countries along the Silk Roads are included in the platform (https://en.unesco.org/silkroad/unesco-silk-roads-programme-0). In this context, the countries included in the UNWTO or UNESCO lists are considered in this study to be closely related to the Silk Road, while the countries participating in both lists are considered to have a greater impact and significance. Based on the secondary research, all countries that participated in the TIF at least once from the first TIF in 1926 until 2019, were recorded. Visits were recorded, as well as for the period 2010–2019. In total, 42 Silk Road countries that participated in the TIF were included in the final matrix.

Following the above analysis, a unique five (5) digit ID number is assigned to each Silk Road country. The ID code is regenerated from the combination of data divided into three main categories: "Silk Road Country Profile Data", "TIF Participation Data" according to the respective classification and "Population data" always according to the respective classification. Starting with the category: "Silk Road Country Profile Data", the first digit refers to the network in which each country participates. More specifically, member countries of the UNWTO Silk Road network are assigned digit 1, the UNESCO Silk Road Program network member countries are assigned digit 2, and member countries of both networks are assigned digit 3. The following digit refers to the category: "TIF participation data", and more specifically whether if each Silk Road country has visited TIF at least once from its inception in 1926–2019 (Table 1). Based on the number of visits recorded, the countries were ranked according to the proposed methodology. The same procedure applies to visits for the period 2010–2019 (Table 2). All visits to the TIF for the periods 1926–2019 and 2010–2019 have been classified according to Tables 1, 2 in order to facilitate the elaboration of the coding process in the matrix. The next digit depends on whether the Silk Road country was a TIF honored country (1) or not (0). Regarding the category "Population data", the population of each Silk Road country was normalized according to the latest national census available for the majority of them (base year 2018) and organized into three (3) classes (categories) as shown in Table 3.

Accordingly, population data was elaborated also based on the above methodological step in order to normalize all data. More specifically, the population classes are set as follows (Table 3):

Table 1 Classification of visits at TIF (1926–2019)—class total

Number of visits at TIF 1926–2019	Class (*t*)
0	0
[1, 25)	1
[25, 50)	2
[50, 75)	3
[75, 100)	4

Source Idem processus

Table 2 Classification of visits at TIF (2010–2019)—class last decade

Number of TIF exhibition visits 2010–2019	Class (*d*)
0	0
[1, 5)	1
[5, 10)	2
[10, 15)	3
[15, 20)	4

Source Idem processus

Table 3 Silk road countries' population classification

Class 1:	[0–150.000)
Class 2:	[150.000–1.000.000)
Class 3:	≥ 1.000.000

Source Idem processus

The ID code is completed with the serial number of each Silk Road country in respect to the group it belongs based on the previous coding procedure.

Finally, the proposed methodology can be summarized in the following scheme (Fig. 8).

4 Thessaloniki International Fair on the Silk Road

4.1 Silk Road Countries at Thessaloniki Internationa Fair (TIF)

The results of the secondary research record the participation in Thessaloniki International Fair (TIF) events of the following non-EU Silk Road countries, as listed by the international organizations UNWTO and UNESCO (Fig. 9) such as: Azerbaijan, Egypt, Armenia, Afghanistan, Vietnam, Georgia, United Arab Emirates, Japan, India, Indonesia, Jordan, Iraq, Iran, Israel, Kazakhstan, Qatar, China, Korea, Kuwait, Lebanon, Malaysia, Bangladesh, Nepal, Uzbekistan, Ukraine, Pakistan, Russia, Saudi Arabia, Sri Lanka, Syria, Taiwan, Thailand, Turkey and the Philippines.

4.2 Silk Road Countries as Honored Countries at Thessaloniki International Fair (TIF)

In Table 4 the 'Honored Country' scheme per event is presented PR China and the Russian Federation, two of the world's largest economies, have participated in the TIF, four and five times respectively during the last ten years. There have been honored countries (along with India) that demonstrates how the TIF prioritizes improving economic and trade relations with these countries (HELEXPO S.A. 2016).

4.2.1 Russian Federation

Russia, that has been participating in TIF since its first event, was the honored country in 2016 (HELEXPO S.A. 2016). More than 100 Russian exhibitors participated representing agricultural production and processing, food, energy and industry, Russian pavilion occupied over 3000 sq.m. of exhibition space. Furthermore, a Greek-Russian Business Forum titled "New impulse to trade and economic cooperation between Russia and Greece" was held, with the aim to explore the prospects for cooperation between the two countries and strengthen financial ties.

Trade relations between Greece and the Russian Federation

As a result of Russia's countermeasures, exports of commodities (including petroleum products) from Greece to Russia since 2013, fell by 60.30% and could not be replaced by other commodities. Greece's exports fell by more than half in 2020 compared to 2013, while Greece's imports from Russia fell by 55.70%. Overall, bilateral trade in 2020 reached € 3.09 billion, a significant decrease (−27.90%) compared to 2019. The bilateral trade deficit balance improved slightly (deficit decreased by 28.00%). (Financial and Commercial Office of the Hellenic Embassy in Moscow 2021) (Table 5) (Ministry of Foreign Affairs 2021).

Tourism

According to records from the Bank of Greece (cited in Financial and Commercial Office of the Hellenic Embassy in Moscow 2021), the number of Russian tourists who visited Greece in 2020 amounted to 26,000, corresponding to 0.35% of total arrivals, ~95.5% less compared to 2019. However, in comparison to 2013, arrivals have decreased by more than 70% (Table 6).

4.2.2 People's Republic of China

PR China started taking part in the TIF General Exhibition in 1957, and has since participated a total of 17 times. It was

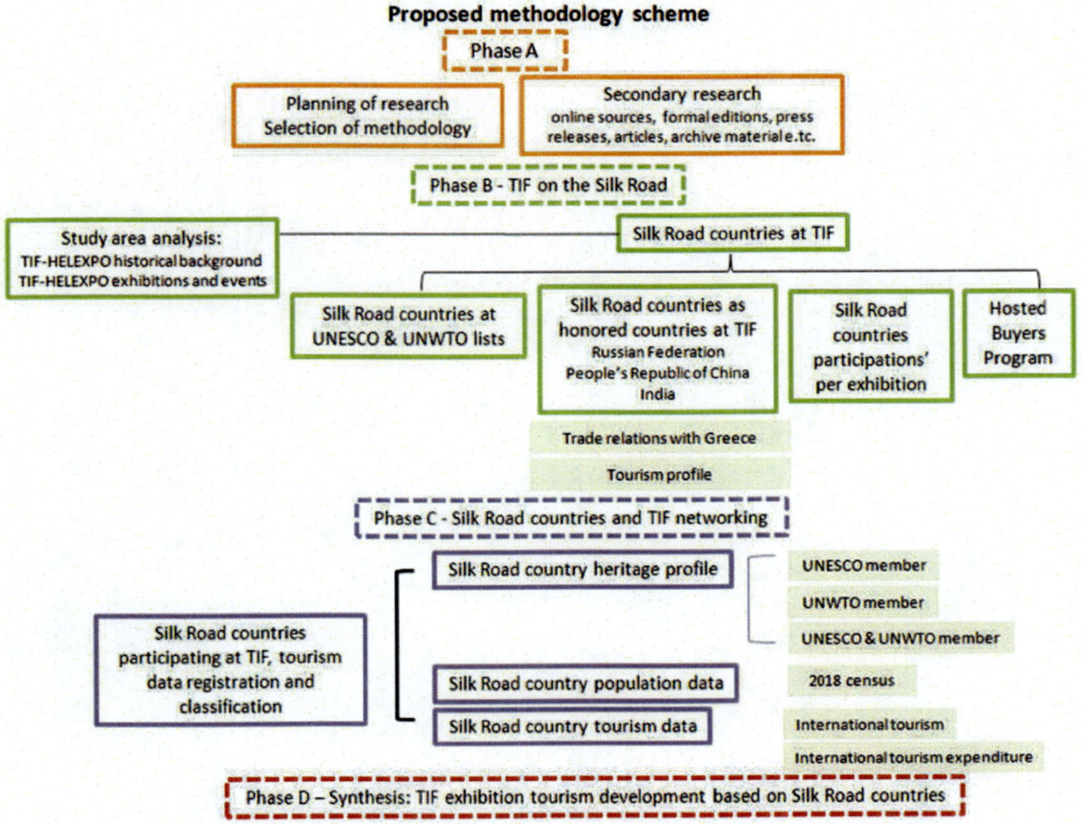

Fig. 8 Proposed methodology scheme. *Source* Idem processus

the first TIF honored country in 1998, showing the results of the initial stage of China's reform and opening-up process. As a result, TIF acted as a crucial gateway for Chinese commodities into Europe. China was the honored country again at the 82nd TIF, when PR China the second-largest economy in the world, participated with 167 Chinese companies in the fields of energy, food, tourism, scientific innovation, culture, real estate and health care to show its expectations of Sino-Greece cooperation. The signing of the Memorandum of Understanding (MoU) between China Development Bank (CDB) and Public Power Corporation (PPC) was one of the major events of the 82nd TIF.

Trade relations between Greece and PR China

Greece–PR China key pillars of bilateral economic relations with China is to strengthen bilateral trade, and promote foreign investment and cooperation in the shipping sector. As an example of cooperation consider COSCO's investment in the port of Piraeus, with the potential to become the most important port in the Mediterranean and Europe, and a major hub for "Belt and Road" cooperation. The railway running from Piraeus to Central and Eastern Europe will connect the land-based with the maritime Silk Road. As an important link of China–Europe Land-Sea Express Route, Thessaloniki will surely embrace wider development prospects.

In 2018, China emerged as the fourth most important partner of Greece in terms of imports, occupying 6.5% of the country's total imports (Financial and Commercial Office of the Hellenic Embassy in Beijing 2019). Table 7 presents the Greece–PR China trade balance for the past six years which appears strongly deficient. This fact indicates demonstrates the need to intensify the efforts of Greek companies to increase the country's exports to China.

During the years 2014–2019, there has been a recovery and dynamic growth of the Sino-Greek trade, with a total value of almost € 5 billion in 2019, about 80% more than the corresponding value in 2014. Starting from 2016 and after three consecutive years of negative course (2013–2015), the value of Greek exports to PR China began to follow impressive growth rates. In particular, the value of Greek exports to China in 2018, almost doubled, compared to 2017 (91% increase), marking a record level. In 2019, the value of Greek exports showed a marginal decline of −1.1% and amounted to € 892.5 million (Financial and Commercial Office of the Hellenic Embassy in Beijing 2019). This year was the year of China-Greece Cultural Exchanges and

Fig. 9 Silk road (non-EU) countries, as listed by UNWTO and UNESCO, that have participated in Thessaloniki International Fair (TIF). *Source* Idem processus from UNWTOb (n.d.) and UNESCO (n.d.) programmes

Table 4 'Honored country' scheme at TIF (per event)

Year	Trade fair	Date	Honored country
1998	63rd	5–14 September 1998	PR China
2000	65th	2–11 September 2000	USA
2002	67th	7–15 September 2002	France
2008	73rd	6–14 September 2008	Italy
2009	74th	5–13 September 2009	India
2010	75th	11–19 September 2010	Hungary
2011	76th	10–18 September 2011	Serbia
2016	81st	10–18 September 2016	Russian Federation
2017	82nd	9–17 September 2017	PR China
2018	83rd	8–16 September 2018	USA
2019	84th	7–15 September 2019	India
2020	Thessaloniki Helexpo Forum	5–13 September 2020	Germany
2021	85th	11–19 September 2021	Greece

Source TIF-HELEXPO (idem processus)

Cultural Industry Cooperation. In this context, China's as a TIF honored country has acquired strategic and far-reaching significance.

Tourism

Greece is one of the most attractive European tourist destinations for Chinese travelers. In 2019, the number of Chinese tourists arriving at Greek airports amounted to about 164,000, showing an increase of 9.3%, compared to 2018 (Financial and Commercial Office of the Hellenic Embassy in Beijing 2019), while in 2018, Chinese tourist visits were increased by 30% compared to 2017. As of 30.9.2017, an Athens-Beijing direct connection has been launched by Air China, contributing significantly to the increase in visitors (Table 8).

Table 5 Bilateral trade Greece–Russian Federation (with petroleum products)

(In millions Euro)	2013*	2016	2017	2018	2019	2020	Change 2020/2013	Change 2020/2019
Greek exports	40,611	21,518	23,406	23,184	20,855	1614	− 60.30%	− 22.60%
Greek imports	6.62157	2.82188	3.4261	4.14512	4.08251	2.93346	− 55.70%	− 28.10%
Trade volume	7.02768	3.03706	3.66016	4.37696	4.29106	3.09486	− 56.00%	− 27.90%
Trade balance	− 6.21546	− 2.6067	− 3.19204	− 3.91328	− 3.87396	− 2.77206	− 55.40%	− 28.40%

* Year before the Western sanctions and Russian countermeasures
Source Hellenic Statistical Authority (in Ministry of Foreign Affairs 2021)

Table 6 Russian tourist arrivals

(In thousands)	2013*	2014	2015	2016	2017	2018	2019	2020
Russian Federation arrivals to Greece	1.353	1.250	513	596	589	520	583	26
Change %		− 7.59	− 59.98	16.12	− 1.14	− 11.63	12.05	− 95.5
Percentage on total arrivals to Greece %	6.72	5.15	1.96	2.12	1.95	1.57	1.71	0.35

* Year before the Western sanctions and Russian countermeasures
Source Bank of Greece (in Ministry of Foreign Affairs 2021)

Table 7 Bilateral trade Greece–PR China

(In millions Euro)	2014	2015	2016	2017	2018	2019
Greek exports	273.5	228.2	328.1	472.6	902.3	892.5
Greek imports	2.4943	2.5516	2.8869	2.7134	3.5988	4.0613
Trade volume	2.7678	2.7798	3.2150	3.1861	4.5012	4.9537
Trade balance	− 2.2208	− 2.3234	− 2.559	− 2.241	− 2.6965	− 3.1688

Source Hellenic statistical authority (in Financial and Commercial Office, Hellenic Embassy in Beijing 2021)

Table 8 Chinese tourist arrivals

	2015	2016	2017	2018	2019
Chinese arrivals to Greece	55.097	80.000	110.000	150.000	164.000
Change %	16	45.2	27.3	36.4	9.3
Percentage on total arrivals to Greece %	0.23	0.28	0.36	0.45	0.49

Source Hellenic Statistical Authority (in Financial and Commercial Office, Hellenic Embassy in Beijing 2021)

4.2.3 India

Since the beginning of the fair, India has participated in the TIF General Exhibition a total of twenty times, with its first year being in 1977. India traditionally has a large participation in the exhibition every year through the Indian Trade Promotion Organization (ITPO) and in cooperation with the Embassy of India in Greece. The number of Indian exhibitors is usually around 60 companies. It is considered one of the most successful international participations due to the arrival of Indian companies interested in the Greek market and respectively attracting the interest of the Greek public visiting the TIF exhibition (TIF HELEXPO SA, Department of International Holdings). The 84th TIF, when India was an honored country, was attended by 1600 Hindi exhibitors (direct and indirect).

Trade relations between Greece and India

In recent years, after the coordinated efforts of both the public and private sector, there is a steady annual increase in Greek exports, both in goods and services (transport, tourist flows, etc.). The bilateral trade is finally showing steadily increasing trends and in 2018 Greek exports exceeded in value 1.25 billion euros. The annual increase in exports of Greek goods over the last three years has marked remarkable growth rates, such as 55.8 (2017) or 52.9% (2018) (Financial

Table 9 Trade balance Greece–India (goods)

(In millions USD)	2015	2016	2017	2018	2019
Exports (value)	60.7	62.4	97.4	144.02	123.71
Imports (value)	285.2	297.6	361.27	503.37	445.53
Trade balance (value)	− 224.5	− 235.2	− 263.8	− 359.35	− 321.83

Source Hindi Ministry of Commerce (in Financial and Commercial Office of Hellenic Embassy in N. Delhi 2019)

and Commercial Office of the Hellenic Embassy in N. Delhi 2019) (Tables 9, 10).

Tourism

Regarding the Indian tourist flows to Greece, their steady increase on an annual basis is underlined. The applications of Indians for tourist visas to enter Greece amounted to 33,000 in the year 2017, to 38,000 in 2018 and 40,500 in 2019 (Financial and Commercial Office of the Hellenic Embassy in N. Delhi 2019). Additionally, according to Hellenic Statistical Authority (2019), the arrivals of Indian tourists in Greek accommodation in 2018 amounted to 82,776 thousand which in turn corresponded to 188,470 thousand overnight stays.

The Silk Road countries are represented at the TIF by a minor percentage (10%) of the total number of exhibitors. PR China (50%) is the country with the most exhibitors (50%), followed by India (21%) and the Russian Federation (14%) (Table 11). These countries were honored countries, resulting in more exhibitors at the TIF in the respective years. India, PR China and the Russian Federation were then singled out as having the highest participations, as measured by the number of exhibitors/visitors.

4.3 Participations of Silk Road Countries Per Sectoral Exhibition (2010–2020)

In Table 12 a summary of the participating Silk Road countries per exhibition during the decade 2010–2020 is provided.

4.4 Hosted Byers Program

The Trade Fair Organizers, establish programs to welcome foreign professional visitors to their events, in order to differentiate themselves and offer maximum level of satisfaction to their exhibitors. The organizers welcome and host selected Commercial Visitors who are potential buyers of respective industry products/services represented at each industry exhibition, who make pre-arranged appointments (B2B meetings) with the exhibitors. The objective of these meetings is to conclude business agreements that will help companies (Exhibitors) to increase their export activities (https://www.iee.org.gr/el/survey-on-the-touristic-dimension/). According to Kourkouridis et al. (2018), in addition to the economic and commercial impact on companies participating in trade fairs, the Hosted Buyers program also has a significant tourism impact. For the period 2014–2016, trade visitors came from 54 different countries, 35 of them were Silk Road countries: Egypt, Albania, Algeria, Australia, Austria, Belgium, Bosnia and Herzegovina, Bulgaria, France, Germany, Denmark, Switzerland, Estonia, UAE, USA, Ireland, India, Jordan, Iraq, Israel, Italy, Canada, Qatar, PR China, Kuwait, Croatia, Cyprus, Costa Rica, Latvia, Belarus, Lebanon, Libya, Lithuania, Luxembourg, Great Britain, Moldova, Norway, South Africa, The Netherlands, Hungary, Ukraine, Northern Macedonia, Poland, Russian Federation, Saudi Arabia, Serbia, Slovenia, Sweden, Turkey, Czech Republic, Tunisia, Hong Kong.

In Fig. 10 the countries that participated in the Commercial Visitors Programme classified:

Table 10 Trade balance Greece–India (services)

(In millions USD)	2016	2017	2018	2019
Exports (value)	367.3	498.0	652.1	717.4
Imports (value)	29.3	38.0	43.4	65.5
Trade balance (value)	338.0	460.0	608.7	651.9

Source Bank of Greece (in Financial and Commercial Office of Hellenic Embassy in N. Delhi 2019)

Table 11 Number of participations/exhibitors

Country	81st TIF	82nd TIF	84th TIF	Total 81st, 82nd, 84th TIF
Bangladesh			1	1
Egypt	3	6		9
India	4	5	48	57
Indonesia	1		1	2
Iran	2	1	1	4
Korea			1	1
Pakistan	1	1		2
PR China		102	38	140
Russian Federation	39		1	40
Turkey	8	13		21

Source Idem processus after authors' communication with TIF

5 Results Analysis

In this section the relevant results analysis is presented. Specifically, in Table 13 the coding methodology for Silk Road countries' TIF participation is summarized, based on the data presented in the methodology section.

According to Table 13, Turkey is the country that has participated the most times at TIF (47 times), followed by Israel (39 times), India (23 times), Egypt and Russian Federation (23 times), PR China (21 times) and Syria (20 times). Japan and Indonesia (17 times), Pakistan (14 times) and Iran (11 times), indicated as countries with High frequency of participation. The next group of countries, Iraq, Ukraine, Armenia, Bangladesh, Korea, Saudi Arabia, Uzbekistan, Kuwait, Azerbaijan, Thailand, Malaysia, Lebanon, United Arab Emirates) were among the countries that participated less than ten times and are indicated as countries with Medium frequency of participation. Finally, Taiwan, Sri Lanka, Nepal, Kazakhstan, Georgia and Vietnam had the fewest entries, while Afghanistan and Qatar only participated in the Hosted Buyers Program, indicated as countries with Low frequency of participation.

It should be noted that due to the heterogeneity among the selected Silk Road country profiles and the limited availability of national data (especially tourism data) in many cases, population and tourism data were not included in the coding methodology, however they could be taken into consideration in cases of smaller country sample examined. Furthermore, the classification of TIF participant countries, normalization and coding are introduced in this paper as a methodological approach for elaboration of additional variables and further research (i.e. visits to branch exhibitions, chronological periods).

TIF visits were summarized and reproduced in the following graphs. Visits to TIF in Thessaloniki were visualized for the period 1926–2019 (Fig. 11) and for the period 2010–2019 (Fig. 12). As a result, weighted graphs were produced, where links represent the flows from the Silk Road countries to TIF, depicted as the nodes of each graph, and the weights represent the number of visits and more specifically, the visit classification formed based on the proposed methodology.

Figure 13, in relation to Table 14, depicts the Silk Road countries that participated in TIF HELEXPO events (both in the general TIF exhibition and the sectoral ones) from 1926 to 2019 as well as the total number of countries participations made by each country. The countries are ranked from the highest to the lowest in terms of the number of their participations and are then grouped according to the frequency of their participation in three categories: high, medium and low frequency.

The examination of Silk Road countries participating at TIF allowed to identify participations (1) of high frequency; (2) of medium frequency and (3) of low frequency.

High frequency of participation countries

Turkey, has been a regular participant in the TIF General Exhibition since 1933 and has participated forty-seven times in the TIF HELEXPO events on a regular basis in recent years five times during 2010–2019 and is the country with the highest participation frequency. In terms of sectoral-industry fairs Turkey participated every year in PHILOXENIA all ten years from 2010 to 2019, in AGROTICA every year from 2016 to 2020, in ZOOTECHNIA in 2017, 2018 and 2019, in ARTOZYMA in 2020, at the Thessaloniki International Book Fair in 2014 2016, 2017 and 2018 and at ART THESSALONIKI (2016–2019). Also, trade visitors from Turkey participated every year in the industry fairs from 2014 to 2019.

Israel, has participated a total of thirty-nine times in the TIF HELEXPO events, thirty-six times at the General Exhibition of TIF and three times at the sectoral exhibitions. However, Israel's participation in the TIF General Exhibition has been on the decline as it has never participated at the TIF General Exhibition between 2010 and 2019. It

Table 12 Number of participations/exhibitors

Sectoral exhibition	Year	Afghanistan	Azerbaijan	Egypt	Armenia	Vietnam	Bulgaria	Georgia	UAE	Japan	India
Hotelia	2016	0	0	0	0	0	0	0	0	0	0
	2018	0	0	0	0	0	1	0	0	0	0
	2020	0	0	0	0	0	0	0	0	0	0
Philoxenia	2010	0	0	0	0	0	1	0	0	0	0
	2011	0	0	0	0	0	1	0	0	0	0
	2012	0	0	0	0	0	1	0	0	0	0
	2013	0	0	0	0	0	1	0	0	0	0
	2014	0	0	1	0	0	1	0	0	0	0
	2015	0	1	0	0	0	1	0	0	0	0
	2016	0	0	0	0	0	1	0	0	0	1
	2017	0	0	0	0	0	1	0	0	0	0
	2018	0	0	0	0	0	1	0	0	0	0
Agrotica	2019	0	0	0	0	0	1	0	0	0	0
	2016	0	0	0	0	0	0	0	0	0	0
	2018	0	0	0	0	0	1	0	0	0	1
Zootechnia	2020	0	0	1	0	0	1	0	0	0	0
	2017	0	0	0	0	0	1	0	0	0	0
Artozyma	2019	0	0	1	0	0	1	0	0	0	0
	2016	0	0	0	0	0	0	0	0	0	0
	2018	0	0	0	0	0	1	0	0	0	0
Frescon	2020	0	0	0	0	0	0	0	0	0	0
	2016	0	0	0	0	0	1	0	1	0	0
	2017	0	0	0	0	0	1	0	0	0	0
	2018	0	0	0	0	0	1	0	0	0	0
Art	2019	0	0	0	0	0	0	0	0	0	1
TBF	2016–2019	0	0	0	0	0	0	0	0	1	0
	2014	0	0	0	0	0	0	0	0	0	0
	2015	No data	No data	No data	No data	No data	No data	No data	No data	No data	No data
	2016	0	0	1	0	0	0	0	0	0	0
	2017	0	0	1	0	0	0	0	0	0	0
	2018	0	1	1	0	0	0	0	0	0	0

Sectoral exhibition	Year	Indonesia	Jordan	Irak	Iran	Israel	Kazakhstan	Gaboji	Quatar	China	Kirghizstan
Hotelia	2016	0	0	0	0	0	0	0	0	0	0
	2018	0	0	0	0	0	0	0	0	0	0
	2020	0	0	0	0	0	0	0	0	0	0
Philoxenia	2010	0	0	0	0	0	0	0	0	0	0
	2011	0	0	0	0	0	0	0	0	0	0
	2012	0	1	0	0	0	0	0	0	0	0
	2013	0	0	0	0	0	0	0	0	0	0
	2014	0	0	0	0	0	0	0	0	0	0
	2015	1	0	1	0	0	0	0	0	0	0
	2016	0	0	0	0	0	0	0	0	0	0
	2017	1	0	0	0	0	0	0	0	0	0
	2018	0	0	0	0	0	1	0	0	0	0

(continued)

Table 12 (continued)

Sectoral exhibition	Year	Indonesia	Jordan	Irak	Iran	Israel	Kazakhstan	Gaboji	Quatar	China	Kirghizstan
Agrotica	2019	0	0	0	0	0	0	0	0	0	0
	2016	0	0	0	0	0	0	0	0	0	0
	2018	0	0	0	0	0	0	0	0	0	0
Zootechnia	2020	0	0	0	0	0	0	0	0	0	0
	2017	0	0	0	0	0	0	0	0	0	0
Artozyma	2019	0	0	0	0	0	0	0	0	1	0
	2016	0	0	0	0	0	0	0	0	0	0
	2018	0	0	0	0	0	0	0	0	0	0
Frescon	2020	0	0	0	0	0	0	0	0	0	0
	2016	0	0	0	0	0	0	0	0	0	0
	2017	0	0	1	0	0	0	0	0	0	0
	2018	0	0	0	0	1	0	0	0	1	0
Art	2019	0	0	0	0	0	0	0	0	0	0
TBF	2016–2019	0	0	0	0	1	0	0	0	1	0
	2014	1	0	0	1	1	0	0	0	0	0
	2015	No data	No data	No data	No data	No data	No data	No data	No data	No data	No data
	2016	0	0	0	1	1	0	0	0	1	0
	2017	0	0	0	1	1	0	0	0	1	0
	2018	0	0	0	1	1	0	0	0	1	0

Source Idem processus

participated at the sectoral Exhibition FRESCON of 2018, in ART THESSALONIKI (2016–2019), at the Thessaloniki International Book Fair the years 2014 and 2016–2018. However, in recent years, from 2014 to 2020, trade visitors from Israel have frequently participated.

India has participated a total of twenty-three times at the TIF HELEXPO events, twenty times at the General Exhibition of TIF and is on the rise. For the last ten years India participates almost every year at the General Exhibition of TIF and has a large participation (TIF HELEXPO A.E, Department of International Holdings). In 2019, India was the Honored Country of the 84th TIF and participated in it with 1600 exhibitors. In recent years, Exhibitors from India have taken part in the industry exhibitions such as PHILOXENIA in 2016, AGROTICA in 2018 and FRESCON in 2019.

Egypt has participated twenty-three times at the TIF HELEXPO events, for the first time in 1972 during the organization of the 37th TIF General Exhibition. Egypt's participation has been on the rise since 1983. Egypt has participated at the most recent TIF events (2010–2019) four times. Regarding the Sectoral Exhibitions and according to available data, Egypt has participated in Sectoral Exhibitions, in 2014 at PHILOXENIA, in 2019 at ZOOTECHNIA and from 2016 to 2018 at the Thessaloniki International Book Fair. Egypt participated every year in the Hosted Buyers Program since 2014 when the Program systematically started, until 2020 (Kourkouridis et al. 2018).

The Russian Federation has participated a total of twenty-three times in the TIF HELEXPO events, fifteen times at the general exhibition of TIF till 2015 and since then it participates almost every year. In 2016 at the 81st TIF the Russian Federation was an Honored Country, a fact that shows the special emphasis on strengthening cooperation. The Russian Federation participated at sectoral fairs such as PHILOXENIA from 2010 to 2013, ART THESSALONIKI (2016–2019) and the Thessaloniki International Book Fair in 2016, while trade visitors participate every year in the sectoral fairs from 2014 to 2019.

PR China has participated a total of twenty-one times at the TIF HELEXPO events, seventeen times at the General Exhibition of TIF, every year for the period 2016–2019. In 2017 at the 82nd TIF PR China was an Honored Country, a fact that shows the special emphasis on strengthening economic and trade relations with China. At the sectoral exhibitions there was participation from PR China in 2019 at ZOOTECHNIA, in 2018 at FRESCON and at ART THESSALONIKI (2016–2019). In the years 2014–2016 and 2017–2019 Chinese trade visitors participate in the Hosted Buyers Program.

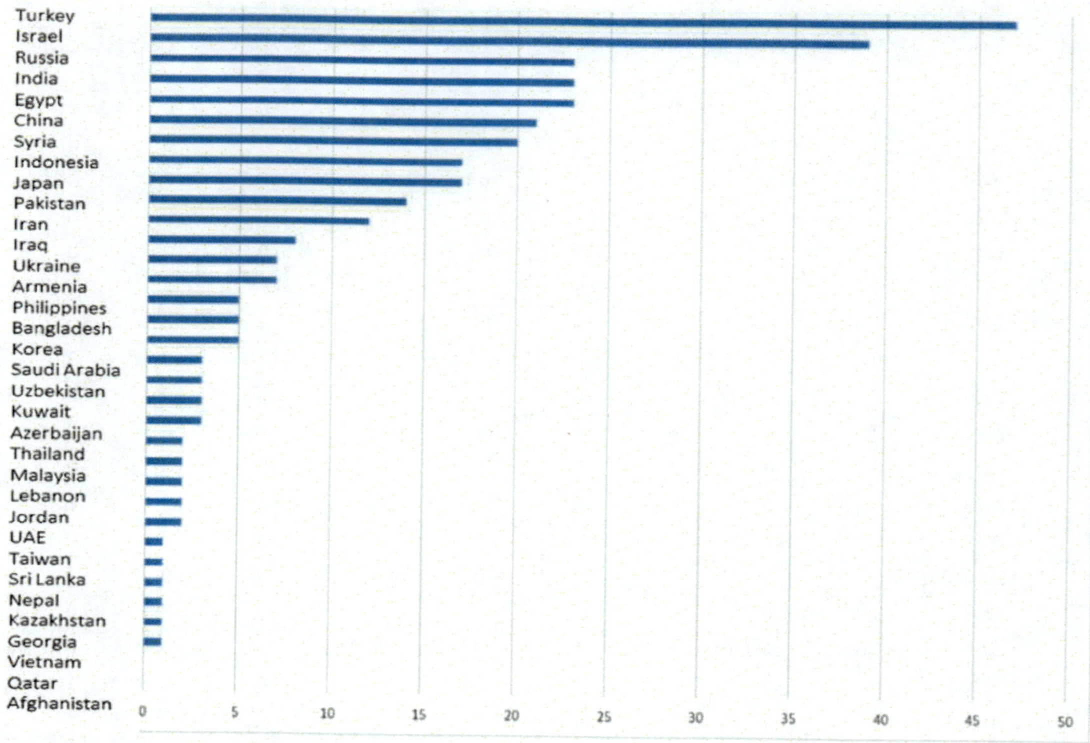

*TIF exhibitions 2020: not realized due to Covid-19

Fig. 10 Number of participations per Silk Road country at TIF—all exhibitions (1926–2019*). *Source* Idem processus. *TIF exhibitions 2020: not realized due to Covid-19

Syria has participated a total of twenty times at the TIF HELEXPO events, twenty times at the General Exhibition of TIF the first time in 1965. Syria had a continuous presence at the TIF until 2011 which was the last year it participated. From the available data, there are no Syrian entries at the sectoral fairs and the Hosted Buyers Program recorded.

Medium frequency of participation countries

Indonesia has participated a total of seventeen times at the TIF HELEXPO events, fifteen times at the TIF General Exhibition since 1996. In recent years Indonesia participates at the TIF General Exhibition almost every year but with a relatively small number of exhibitors. Indonesia participates in the TIF through its Embassy in Greece, either as an Embassy that with its participation represents its country, or by organizing group participation with Indonesian companies. The last time Indonesia participated with about seven companies (small and medium enterprises) (TIF HELEXPO A.E, Department of International Holdings). Exhibitors from Indonesia have also participated at the PHILOXENIA industry-sectoral exhibition in 2015 and 2017.

Japan which has participated a total of seventeen times at the General Exhibition of TIF. While its participation was relatively frequent until 2007, Japan did not participate at the TIF General Exhibition ever since. Exhibitors from Japan participated at the industry exhibition ART THESSALONIKI (2016–2019) and Hosted Buyers Program for 2017.

Pakistan has taken part at the TIF General Exhibition fourteen times, the first being in 1995, however it has not participated since 2013. In 2019 there was a Pakistani presence in the Hosted Buyers Program, while it did not participate at the TIF HELEPXO industry-sectoral fairs.

Iran's participation in the TIF is also noteworthy, as it has participated every year since 2016 (TIF Helexpo Press Release 2017). According to TIF-HELEXPO, both the public and commercial sectors have a significant interest in boosting bilateral trade and economic relations, as well as a wide range of joint actions and exchange of know-how about exhibitions. In this context, TIF-HELEXPO went on to sign a Memorandum of Cooperation with Iran's Organization of International Exhibitions. In the first phase of this collaboration, each of the two exhibition organizations is to participate each year in at least one of the other's events. At the same time, they exchange commercial and economic information, as well as know-how for each country's exhibition industry and inform each other about exhibitions or other events where the two countries' products can be displayed

Table 13 Silk road countries participating at TIF—coding methodology

Country	Silk road country profile data			TIF participation data				Population data		Tourism development data		Coding
	UNWTO (1)*	UNESCO (2)**	UNWTO + UNESCO (3)	Total TIF visits 1926–2019	Class (t)	TIF visits 2010–2019	Class (d)	Population (thousands) ***	Class (p)	International tourism (number of departures)	International tourism expenditure (bn. $)***	Code
Afghanistan	0	2	0	1	1	0	0	38,928.34	1	4,908,000	2,3	2.0.0.1.1
Armenia	1	2	3	23	1	6	2	2963.23	1	6,180,000 (2014)	2,7	3.1.0.1.3
Azerbaijan	1	2	3	8	1	1	1	10,110.12	1	1,623,000	1,3	3.1.1.0.1.1
Bangladesh	1	0	0	16	1	5	2	164,689.38	2	n/a	0,2	1.1.0.2.1
Cambodia	0	2	0	0	0	0	0	16,718.97	1	n/a	6,1	2.0.0.1.8
China	1	2	3	52	3	13	3	1,402,112.00	3	3,698,000	0,5	3.1.1.3.1
Egypt	1	2	3	77	4	15	4	102,334.40	1	18,904,000	18	3.1.0.1.2
Georgia	1	2	3	6	1	1	1	3714.00	1	18,954,000	20,2	3.1.0.1.4
India	0	2	0	72	3	19	4	1,380,004.39	3	26,296,000	21,3	2.1.1.3.1
Indonesia	1	2	3	56	3	15	4	273,523.62	2	9,468,000	10,3	3.1.0.2.1
Iran	1	2	3	40	2	11	3	83,992.95	1	n/a	1,4	3.1.0.1.7
Iraq	1	2	3	16	1	4	2	40,222.50	1	n/a	7,9	3.1.0.1.6
Israel	1	0	0	64	3	10	3	9216.90	1	7,243,000	8,7	1.1.0.1.1
Japan	1	2	3	23	1	3	1	125,836.02	1	8,473,000	7,9	3.1.0.1.5
Jordan	0	2	0	6	1	0	0	10,203.14	1	10,646,000	2,7	2.1.0.1.3
Kazakhstan	1	2	3	3	1	0	0	18,754.44	1	n/a	n/a	3.0.0.1.1
Korea	1	2	3	12	1	2	1	51,780.58	1	149,720,000	277,3	3.1.0.1.8
Kuwait	0	2	0	5	1	1	1	4270.56	1	28,696,000	35,1	2.1.0.1.4
Kyrgyzstan	1	2	3	0	0	0	0	6591.60	1	n/a	14,3	3.0.0.1.1
Lebanon	0	2	0	11	1	1	1	6825.44	1	n/a	6,3	2.1.0.1.5
Malaysia	1	2	3	9	1	2	1	32,366.00	1	n/a	12,1	3.1.0.1.9
Mongolia	1	2	3	0	0	0	0	3278.29	1	n/a	0,8	3.0.1.1.12
Myanmar	0	2	0	0	0	0	0	54,409.79	1	1,197,000 (2017)	n/a	2.0.0.1.9
Nepal	0	2	0	1	1	1	1	29,136.81	1	8,594,000	2,2	2.0.0.1.3
Oman	0	2	0	0	0	0	0	5106.62	1	27,977,000	7,9	2.0.0.1.10
Pakistan	1	2	3	46	2	14	3	220,892.33	2	n/a	1,9	3.1.0.2.2
Philippines	0	2	0	4	1	1	0	109,581.09	1	41,964,000	34,3	2.1.0.1.7
Qatar	0	2	0	0	0	0	0	2881.06	1	27,425,000	16,7	2.0.0.1.11
Russia	1	2	3	80	4	16	1	144,104.08	1	6,259,000 (2011)	n/a	3.1.1.1.1
Saudi Arabia	1	2	3	10	1	2	0	34,813.87	1	n/a	19,4	3.1.0.1.11
Singapore	0	2	0	0	0	0	0	5685.81	1	9,966,000	12,1	2.0.0.1.13
Sri Lanka	0	2	0	1	1	0	0	21,919.00	1	8,383,000	4,6	2.0.0.1.4
Syria	1	2	3	38	2	12	0	17,500.66	1	3,078,000	11,9	3.1.0.1.12
Tajikistan	1	2	3	0	0	0	0	9537.64	1	n/a	n/a	3.0.0.1.2
Thailand	0	2	0	7	1	2	0	69,799.98	1	n/a	n/a	2.1.0.1.6
Turkey	1	2	3	96	4	18	0	84,339.07	1	n/a	n/a	3.1.0.1.13
Turkmenistan	1	2	3	0	0	0	0	6031.19	1	n/a	n/a	3.4.4.1.1
UAE	0	2	0	13	1	1	0	9890.40	1	n/a	n/a	2.1.0.1.2
Ukraine	1	0	0	28	2	6	0	44,134.69	1	n/a	n/a	1.1.0.1.2
Uzbekistan	1	2	3	7	1	1	0	34,232.05	1	n/a	n/a	3.1.0.1.10
Vietnam	0	2	0	4	1	1	0	97,338.58	1	n/a	n/a	2.1.0.1.1
Yemen	0	2	0	0	0	0	0	29,825.97	1	n/a	n/a	2.0.0.1.14

Source Idem processus

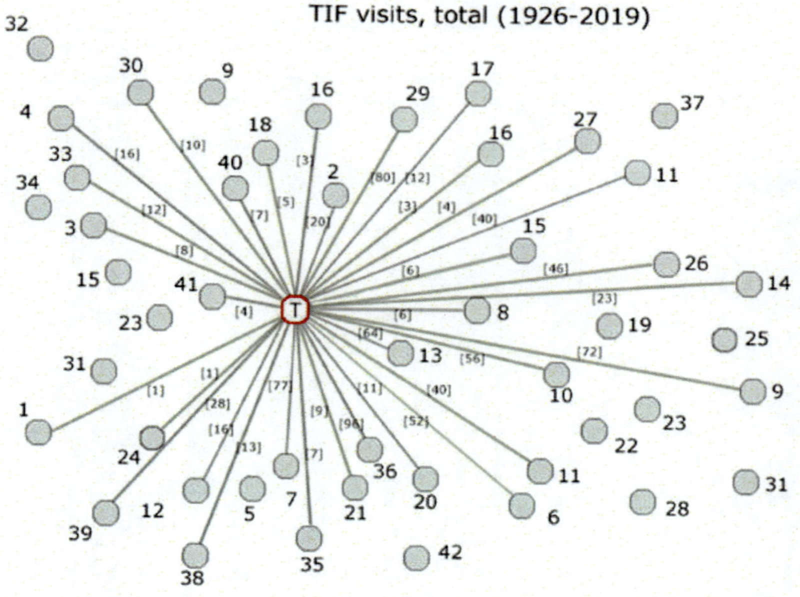

A/A	Country	A/A	Country
1	Afghanistan	21	Malaysia
2	Armenia	22	Mongolia
3	Azerbaijan	23	Myanmar
4	Bangladesh	24	Nepal
5	Cambodia	25	Oman
6	China	26	Pakistan
7	Egypt	27	Philippines
8	Georgia	28	Qatar
9	India	29	Russia
10	Indonesia	30	Saudi Arabia
11	Iran	31	Singapore
12	Iraq	32	Sri Lanka
13	Israel	33	Syria
14	Japan	34	Tajikistan
15	Jordan	35	Thailand
16	Kazakhstan	36	Turkey
17	Korea	37	Turkmenistan
18	Kuwait	38	UAE
19	Kyrgistan	39	Ukraine
20	Lebanon	40	Uzbekistan
		41	Vietnam
		42	Yemen

Fig. 11 Total visits at TIF exhibitions, 1926–2019. *Source* Idem processus

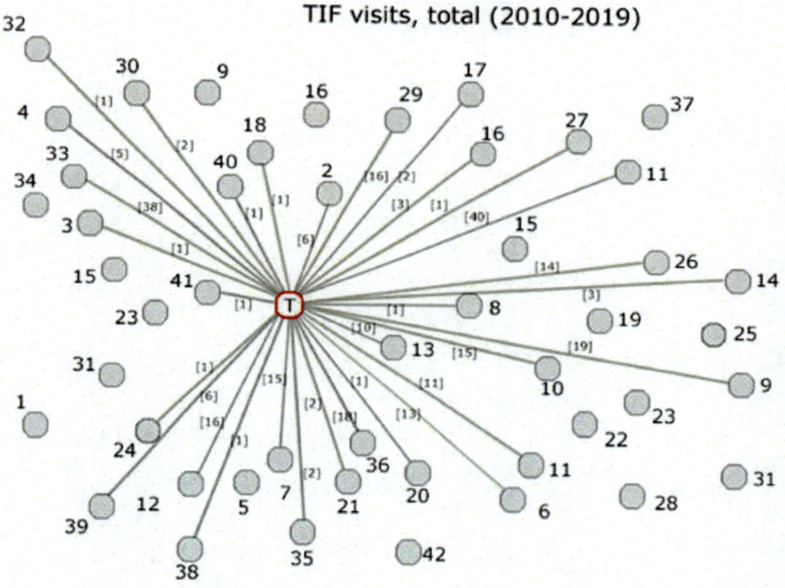

A/A	Country	A/A	Country
1	Afghanistan	21	Malaysia
2	Armenia	22	Mongolia
3	Azerbaijan	23	Myanmar
4	Bangladesh	24	Nepal
5	Cambodia	25	Oman
6	China	26	Pakistan
7	Egypt	27	Philippines
8	Georgia	28	Qatar
9	India	29	Russia
10	Indonesia	30	Saudi Arabia
11	Iran	31	Singapore
12	Iraq	32	Sri Lanka
13	Israel	33	Syria
14	Japan	34	Tajikistan
15	Jordan	35	Thailand
16	Kazakhstan	36	Turkey
17	Korea	37	Turkmenistan
18	Kuwait	38	UAE
19	Kyrgistan	39	Ukraine
20	Lebanon	40	Uzbekistan
		41	Vietnam
		42	Yemen

Fig. 12 Total visits at TIF exhibitions, 2010–2019. *Source* Idem processus

and promoted, offering mutual support and technical assistance.

Iran participated twelve times at the TIF-HELEXPO events. At the general exhibition of TIF According to TIF-HELEXPO, both the public and commercial sectors have a significant interest in boosting bilateral trade and economic relations, as well as a wide range of joint actions and exchange of know-how at the exhibition level. In this context, TIF-HELEXPO went on to the sign a memorandum of cooperation with the Organization of International Exhibitions of Iran. In the first phase of this collaboration, each of the two exhibition organizations will participate in at least one of the other's event each year. At the same time, they will exchange commercial and economic information, as well as know-how for the exhibition industry of each country, while they will inform each other about exhibitions or other events where the products of the two countries can be displayed and promoted, offering mutual support and

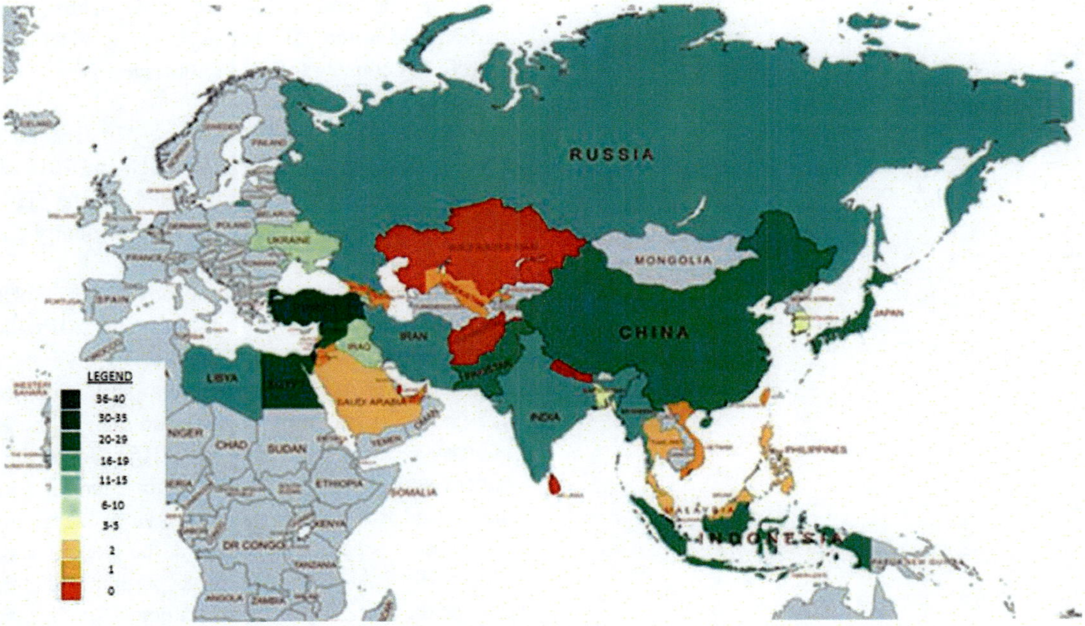

Fig. 13 Silk road countries participation at TIF general exhibition 1926–2019. *Source* Idem processus

technical help (TIF Press Release 2016). Iran also participated at the Thessaloniki International Book Fair in 2014 and 2016–2018 and in the Hosted Buyers Program in 2016.

Pakistan and Syria took part four times until 2011 (Pozrikidis 2012).

Low frequency of participation countries

Iraq, Azerbaijan, Ukraine and Uzbekistan each participated only once in the last ten years.

These countries due to their low frequency of participation so far do not seem likely to be major business tourism markets for Thessaloniki and therefore should be the subject of research on the possibilities of further strengthening trade-tourism relations with Thessaloniki and Greece. The countries with the lowest participation at TIF events (less than ten times) are the following:

Iraq, which has participated a total of eight times at the TIF General Exhibition. The first time Iraq participated was in 1974. Iraq had a declining presence since 1983 and in 2013 was the last time it participated at the TIF General Exhibition. Exhibitors from Iraq have participated at PHILOXENIA, in 2015 and at FRESCON, in 2017. Trade visitors from Iraq for the years 2014–2016 participated in the industry-sectoral exhibitions of TIF-HELEXPO.

Armenia participated in the TIF General Exhibition three times between 2015 and 2017 (TIF Helexpo Press Release 2017).

Armenia has participated seven times at the TIF-HELEXPO events. It participated at the TIF General Exhibition as follows: first participation in 1994 and then in 1998, 1999 and 2005. For the next nine years Armenia did not participate at the TIF General Exhibition, but reappears and participates in the recent years 2016 and 2017 as well as the years 2018 and 2019 where trade visitors from Armenia participated in the sectoral Exhibitions of TIF-HELEXPO.

Ukraine has participated at the TIF General Exhibition seven times. The last time it participated was in 2017. From 2007 onwards its participation followed a declining course. Ukraine also participated at the Thessaloniki International Book Fair in 2016 and 2017.

Bangladesh has taken part twice in the TIF, in 2018 and 2019, has little participation at the General Exhibition of TIF, since it has participated five times, with the first time being in 1980 while in the last ten years it has participated twice, last time in 2018. It also participated at the Thessaloniki International Book Fair in 2016.

The Philippines have participated once at the TIF General Exhibition, in 2002. However, in 2016, 2017 and 2018 the Philippines participate at the PHILOXENIA tourism industry exhibition.

Korea has little participation at the General Exhibition of TIF, since it has participated four times and all of them before 2008. It participated at the Thessaloniki International Book Fair in the years 2016–2018.

Thailand has participated twice at the TIF general exhibition, in 1992 and 2015. From the available data, there are no Thai entries in the industry exhibitions and the Hosted Buyers Program.

Table 14 TIF 1926–2019—frequency of participation

High	
Israel	36
Turkey	30
Egypt	20
India	20
Syria	20
PR China	17
Japan	16
Indonesia	15
Russian Federation	15
Pakistan	14
Iran	11
Medium	
Armenia	7
Ukraine	7
Iraq	6
Bangladesh	5
Korea	4
Kuwait	3
Lebanon	2
Malaysia	2
Saudi Arabia	2
Thailand	2
Philippines	2
Low	
Azerbaijan	1
Vietnam	1
Georgia	1
UAE	1
Jordan	1
Uzbekistan	1
Taiwan	1
Afghanistan	0
Kazakhstan	0
Qatar	0
Nepal	0
Sri Lanka	0

Source Idem processus

Saudi Arabia has participated twice at the TIF General Exhibition, in 1999 and 2002. Trade visitors from Saudi Arabia participate at the trade fairs of 2015, 2019 and 2020. It also participated at the Thessaloniki International Book Fair in 2018.

Kuwait has little participation at the TIF General Exhibition, as it has participated three times in 1976, 1992 and 2000. In the years 2014–2016, trade visitors from Kuwait participated at the TIF industry-sectoral exhibitions. This is therefore a country that does not have frequent participation at the TIF.

Lebanon has participated at the TIF General Exhibition once in 2004. In the years 2014–2016 and 2019–2020, trade visitors from Lebanon participated at the TIF industry-sectoral exhibitions. This is therefore a country that does not have frequent participation in the TIF.

Malaysia has participated at the TIF General Exhibition twice in 1990 and 1991. In 2018 and 2019 there was a presence of Malaysia in the Hosted Buyers Program.

The United Arab Emirates have participated once at the TIF General Exhibition, in 2001 and once at the FRESCON industry-sectoral exhibition in 2016. However, business-trade visitors from the United Arab Emirates participate in the Hosted Buyers Program every year, except in 2016.

Jordan has participated once at the General Exhibition of TIF, in 1995. Also, in 2012, it has participated at the sectoral Exhibition PHILOXENIA. Commercial-trade visitors from Jordan for the years 2014–2016 and 2020 participated at the industry-sectoral exhibitions of TIF-HELEXPO.

Georgia has participated once at the TIF General Exhibition, in 1996. According to available data, there were no entries from Georgia at the TIF industry-sectoral exhibitions. However, in 2019 and 2020, trade visitors from Georgia participated at the TIF industry-sectoral exhibitions of TIF-HELEXPO.

Taiwan (Chinese Taipei) has once participated AT the TIF general exhibition. From the available data, there are no Taiwanese participations at the sectoral fairs and the Hosted Buyers Program.

The following countries are countries that despite their small participation have recently started to participate so it is likely that this picture will change in the future:

Azerbaijan is a country which, according to available data, has started participating at TIF General Exhibition and Sectoral Exhibitions during the last five years. Specifically, it has participated only once at the General Exhibition of TIF, in 2015 and in the same year also participated at the sectoral Exhibition PHILOXENIA. In 2018, publishers from Azerbaijan visited the Thessaloniki International Book Fair. Also commercial-trade visitors participated in the years 2018–2020 at the sectoral Exhibitions of TIF-HELEXPO.

Uzbekistan is a country which, according to available data, started participating at the TIF General Exhibition in 2018. Exhibitors from Uzbekistan also participated at the PHILOXENIA Industry Exhibition in 2016 and 2019.

Vietnam has participated once at the TIF General Exhibition, in 2018. According to available data, there were no entries from Vietnam at the TIF industry-sectoral exhibitions or in the Hosted Buyers Program.

The following are the countries that participate only through the industry-sectoral fairs or the Hosted Buyers Program:

Kazakhstan, Nepal and Sri Lanka have participated at the PHILOXENIA Industry Exhibition once, for the years 2018, 2019 and 2018 respectively. Also from the available data of the Hosted Buyers Program, Afghanistan has participated in 2018, Jordan in 2014, 2018, 2019 and 2020 and Kazakhstan in 2016 and 2018.

6 Discussion

Based on the proposed methodology, the results of the secondary research provide with intriguing findings. According to the aforementioned categorization, Silk Road country visitor analysis at TIF is introduced in this research as a methodological approach for the investigation of flow patterns and creation of alternative tourism routes in order to highlight Thessaloniki as a Silk Road tourism destination. In this aspect, the coding methodology and matrix construction is introduced as an appropriate methodological tool for the analysis of various chronological periods and in this case, numerous exhibitions and events. The connection of each country to the Silk Road heritage, as described by UNESCO and UNWTO, is one of its key characteristics.

Countries that have received both UNESCO and UNWTO recognition as Silk Road nations are seen as having a stronger relation to the Silk Road legacy than those participating in only one list. Thus, visitors from countries with a code starting by '3' and higher participation at major TIF or sectoral exhibitions (such as China, Egypt, India, Indonesia), can be considered to be the main Silk Road tourist 'players', and it is necessary to plan adequate tourism products and services for them.

On the other hand, visitors from countries with a lower tie to the Silk Road heritage—codes starting by '1' or '2'—and lower frequency of visits to TIF (such as Lebanon or the United Arab Emirates), could be considered as potential Silk Road tourists, and more emphasis should be given to the creation of tourism packages attractive to them as well. Additionally, countries who have participated at TIF since its inception, but had no participation in the past ten years (i.e. Jordan) should receive special consideration. The design of tailored tourism products in this case could boost flows from such countries too, along with countries included in UNESCO or UNWTO lists but who have never visited TIF (i.e. Qatar, Nepal). The population and tourism profiles of each country serve as additional but important guiding principles for the design of alternative Silk Road tourism packages. Finally, the tourism flow graphs are used for the depiction of the results and the interpretation of the findings.

7 Concluding Remarks

This study aimed to provide a methodological framework analysis of exhibition tourism flows by using the Thessaloniki International Fair (TIF) as a case study. The research adopts a twofold perspective, addressing a chronological (1926–2019 and 2010–2019) and a country visitor analysis. The main aim of the research is to support building Silk Road exhibition tourism networks based on the participation at TIF-HELEXPO exhibitions and events. Based on the case of Thessaloniki International Fair, it was able to designate Thessaloniki as an exhibition tourism destination on the Silk Road. This study could help to understand this better while providing the opportunity to establish Thessaloniki as a tourism center on the Silk Road, merging the legacy of the Silk Road legacy and the modern Silk Road times ahead. In order to further build a unified "Silk Road" exhibition tourism brand and enhance the effectiveness of global exhibition tourism collaboration, the ultimate goal is to propose various tourism products and services with potential for functional specialization Moreover, to further establish the relevant Silk Road exhibition tourism brand into the market, relevant stakeholders should try and focus additionally into designing, mapping and implementing tourism activities before, and during the exhibition, and thus generate tourism after the exhibition. This could be achieved by utilizing for example follow-up tourism products, such as visits in neighboring destinations in order to further distinguish and establish Silk Road exhibition tourism brand.

Turkey, the People's Republic of China, India, the Russian Federation, Indonesia, Egypt, Iran and Israel are among the countries that have developed significant links with TIF in both historical and recent years. In particular, they are the countries with the highest participation in TIF events. The Russian Federation, the People's Republic of China and India were also honored countries, indicating that particular emphasis is placed on strengthening relations with these countries. Emphasis is also placed on cooperation with Iran, as evidenced by the signing of a memorandum on upgrading bilateral economic and trade relations. Indonesia, also participates in TIF events on a regular and organized basis.

As for the tourism dynamics of these countries, China's tourism expenditure in 2018 was US $277.3 billion (UNWTOa, n.d.) with 1497.2 million (The World Bank, n.d) outbound trips. These figures are the most significant of the Silk Road countries considered in this analysis making PR China arguably the most important business tourism market for Thessaloniki. The Russian Federation is also a very important market for business tourism in Thessaloniki with the international tourism expenditure of Russian travelers reaching 34.3 billion US dollars and 41.964.000 outbound trips in 2018 (The World Bank, n.d). India is also a country with significant tourism potential, 26,296,000 outbound trips and $ 26.3 billion in international tourism spending. Egypt regularly participates in the TIF, but has a moderate interest in terms of tourist arrivals (6,180,000, outbound trips). Indonesia, Iran and Turkey have moderate tourism potential. Indonesia recorded 9,468,000 outbound trips and $9.5 billion in tourism spending in 2018. Iran recorded 7,243,000 outbound trips and $8.7 billion in tourism spending, while Turkey recorded 8,383,000 outbound trips and $4.6 billion in tourism expenditure.

Ending, in these modern times whereas the Silk Road is being revived as a tourist route that enables visitors to experience a network of destinations united by a common history, it can be an important means of enriching the travel experience and, in doing so, can attract a large part of the international tourism market (both professional and common), making a significant contribution to the economic and sustainable tourism development of destinations.

It should be emphasized that the study has limitations due to the heterogeneity among the selected Silk Road countries' profiles and limited availability of national data (especially tourism data). Thus, in many cases these attributes were not taken into consideration for the coding methodology. Moreover, relevant data discovered could not be accessed due to portal restrictions or the use of other native languages, difficult to interpret. Regarding sectoral exhibitions data, it was not possible to locate them i.e. past data with lists of exhibitors by country or by year, since they were not available in list form.

On the other side, this study might be used as a tool to systematically organise and further develop a wide range of exhibition tourism programs, policies and planning goals through continuous networking and collaboration among Silk Road countries. Specifically, it might help in the creation of specialized tourism products that could increase flows from countries along the Silk Road, creating mutual benefits for each country. Additionally, in can be underlined that the exhibition industry, both local, regional, national, cross border and international, can promote the Silk Road tourism industry efficiently, such as cluster and synergies effect, spillover effect to represent the different and diverse effects with any kind of impact. Secondly, it could promote research and analysis of a possible application of Thessaloniki's local practice based on the Silk Road tourism supply and demand system of exhibition industry and its relevant value chain. Relatively, specific research with the collaboration of local academic institutions could develop strategies, both sustainable and viable in tourism and economic terms, during the development stage, aside with the adoption of a tourism marketing strategy with branding and regional cooperation which is in line with the actual situation of Thessaloniki positioning itself as a Silk Road point of reference destination. Finally, by fusing the past with the future of the Silk Road, this study could help in understanding and providing the opportunity to establish Thessaloniki as a tourism hub on the Silk Road.

Acknowledgements This research is cofunded by Joint Operational Programme Black Sea Basin 2014–2020 Subsidy Contract No. 99372/31-08-2018 in the context of the Project BSB570 SILC—Silk Road Local Culture. Joint Operational Programme Black Sea Basin 2014–2020 is co-financed by the European Union through the European Neighbourhood Instrument and by the participating countries: Armenia, Bulgaria, Georgia, Greece, Republic of Moldova, Romania, Turkey and Ukraine. This publication has been produced with the financial assistance of the European Union. The contents of this publication are the sole responsibility of the authors and can in no way be taken to reflect the views of the European Union.

References

Agrotica: Agrotica International Fair on Agricultural Machinery, Equipment and Supplies. HELEXPO S.A. https://agrotica.helexpo.gr/el/2018_Introduction (2018)

Arnegger, J., Herz, M.: Economic and destination image impacts of mega-events in emerging tourist destinations. J. Destin. Market. Manag. **5**, 7 (2016). https://doi.org/10.1016/j.jdmm.2015.11.007

ART_THESSALONIKI INTERNATIONAL FAIR: Art_Thessaloniki International Contemporary Fair (TIF)—Introduction. https://art-thessaloniki.helexpo.gr/el/introduction (n.d.)

Chin, T.: The invention of the silk road, 1877. Crit. Inquiry **40**(1), 194–219 (2013)

Christodoulou, G.: Thessaloniki is a city of trade: history of trade from its inception to modern times. Typ. N. Nikolaidou publications (In Greek), Thessaloniki (1933)

Chun, C.S.: The influence of trade show selection criteria on re-participation intention of exhibitors. J. Int. Trade Commer. **12** (1), 65–75 (2016)

Dalkrani, V., Kourkouridis, D., Pozrikidis, K.: Research on the Tourism Dimension of the HELEXPO Hosted Buyers Program (in Greek). https://www.iee.org.gr/el/survey-on-the-touristic-dimension/ (n.d.)

Deery, M., Jago, L.: Social impacts of events and the role of anti-social behaviour. Int. J. Event Fest. Manag. **1**(1), 8–28 (2010). https://doi.org/10.1108/17852951011029289

Dwyer, L., Mistilis, N., Forsyth, P., Rao, P.: International pricecompetitiveness of Australia's MICE industry. Int. J. Tourism Res. **3**, 123–139 (2001)

Financial and Commercial Office of Hellenic Embassy in Russia: Business Guide. https://www.agora.mfa.gr/ta-grafeia-oikonomikon-emporikon-upotheseon/grafeia-ana-xora/infofile/72159 (2021)

Financial and Commercial Office, Hellenic Embassy of Beijing: 2019 Report. https://www.agora.mfa.gr/infofiles/%CE%95%CF%84%CE%AE%CF%83%CE%B9%CE%B1%20%CE%88%CE%BA%CE%B8%CE%B5%CF%83%CE%B7%202019%20cn.pdf (2020)

Financial and Commercial Office, Hellenic Embassy of N. Delhi: 2019 Report. https://www.agora.mfa.gr/ta-grafeia-oikonomikon-emporikon-upotheseon/grafeia-ana-xora/infofile/72159 (2020)

Getz, D.: Event tourism: definition, evolution, and research. Tour. Manag. **29**(3), 403–428 (2008). https://doi.org/10.1016/j.tourman.2007.07.017

Godar, S.H., O'Connor, P.J.: Same time next year: buyer trade show motive. Ind. Market. Manag. **4**, 77–86 (2001). https://doi.org/10.1016/S0019-8501(99)00100-5

Green, P.: Alexander of Macedon, 356–323 B.C. A Historical Biography. University of California Press, Los Angeles and London (1991)

Hammond, N.G.L.: The Macedonian State: The Origins, Institutions, and History, Oxford (1989)

HELEXPO S.A.: HELEXPO Profile. https://www.helexpo.gr/ (n.d.)

HELEXPO S.A.: Russia as Honoured country at 81st TIF (in Greek). https://www.helexpo.gr/el/content/%CE%B7-%CF%81%CF%89%CF%83%CE%AF%CE%B1-%CF%84%CE%B9%CE%BC%CF%8E%CE%BC%CE%B5%CE%BD%CE%B7-%CF%87%CF%8E%CF%81%CE%B1-%CF%83%CF%84%CE%B7%CE%BD-81%CE%B7-%CE%B4%CE%B5%CE%B8 (2016)

Hellenic Statistical Authority: "Greece–India through numbers" (in Greek and Hindi languages). ELSTAT. https://www.statistics.gr/documents/20181/16786229/INDIA_india.pdf/c04751c0-5375-efc0-c815948a884366f4?t=1567515404675 (2019). Accessed 28 Jan 2022

Jago, L., Dwyer, L., Lipman, G., Van Lill, D., Voster, S.: Optimising the potential of mega-events: an overview. Int. J. Event Fest. Manag. **1**(3), 220–237 (2010)

Jamgade, S.: Scope of MICE tourism as an emerging hospitality industry. UAS JMC **04**(02), 26–29 (2018)

Kaya, D.A.: Living off others' aid: the socioeconomic structure of salonica's jews in the mid-nineteenth century. Jewish Hist. **28**(3/4), 313–336 (2014)

Kirchgeorg, M.: Characteristics and forms of trade shows (2005). https://doi.org/10.1007/978-3-663-05658-4_3

Kostopoulou, S., Toufengopoulou, A., Kyriakou, D., Malisiova, S., Sofianou, E., Xanthopoulou, V.: Western Silk Road in Greece. https://webunwto.s3.eu-west-1.amazonaws.com/2019-09/aristotleuniversity-nationalswotanalysis-westernsilkroadgreece.pdf (2016)

Kostopoulou, S., Sofianou, E., Kyriakou, D.: Silk road regionalism and polycentric tourism development. In: Katsoni, V., van Zyl, C. (eds.) Culture and Tourism in a Smart, Globalized, and Sustainable World. Springer Proceedings in Business and Economics. Springer, Cham (2021). https://doi.org/10.1007/978-3-030-72469-6_49

Kourkouridis, D., Frangopoulos, I., Karanikolas, N., Seretis, S.: Exhibition and city: a relationship in negotiation. The case of the Thessaloniki international fair. In: Conference: 14th International Conference of European Regional Studies Association (ERSA), Greek Brunch. Panteion University of Social Sciences, Athens (2016)

Kourkouridis, D., Dalkrani, V., Pozrikidis, K., Frangopoulos, I.: Hosted buyers program (HBP)—tourism development and the city TIF-HELEXPO HBP for the period 2014–2016. In: Katsoni, V., Veander, K. (eds.) Innovative Approaches to Tourism and Leisure (2018). Springer, New York, pp. 537–551

Kourkouridis, D., Dalkrani, V., Pozrikidis, K., Frangopoulos, I.: Trade fairs, tourism and city: Thessaloniki international fair and the concept of honoured countries. Urban Reg. Plan. **6**(4), 185–194 (2021). https://doi.org/10.11648/j.urp.20210604.19

Kousidonis, C.T.: Thessaloniki and the Thessaloniki international trade fair. A case of conflict? In: Proceedings of the 46th Congress of the European Regional Science Association: "Enlargement, Southern Europe and the Mediterranean", August 30th–September 3rd, 2006, Volos, Greece, European Regional Science Association (ERSA), Louvain-la-Neuve (2006)

Kowalik, I.: Influence of trade fairs on a Host City Brand. Public Policy Admin. **11**(4), 629–640 (2012)

Kumar, J., Hussain, K., Ali, F.: A review of cogent reflection on the economic impact assessment of conferences: MICE tourism. SHS Web Confer. **12**, 01006 (2014). https://doi.org/10.1051/shsconf/20141201006

Kylasov, A.V.: Traditional sports and games along the silk roads. Int. J. Ethnosport Tradit. Games **1**, 1–10 (2019)

Ministry of Foreign Affairs: Factsheet on Russia. https://www.agora.mfa.gr/ta-grafeia-oikonomikon-emporikon-upotheseon/grafeia-ana-xora/infofile/77516 (2021)

Mistilis, N., Dwyer, L.: Information technology and service standards in mice tourism. J. Convent. Exhibit. Manag. **2**(1), 55–64 (1999)

Oppermann, M.: Information sources used revisited. Hosp. Res. J. **20** (1), 119–122 (1996). https://doi.org/10.1177/109634809602000110

Papagianopoulos, A.: History of Thessaloniki, Rekos Editions (In Greek) (1982)

Philoxenia: Philoxenia Tourism Exhibition. HELEXPO S.A. https://philoxenia.helexpo.gr/el (n.d.)

Pozrikidis, K.: 75 Years Multiplied by 15 Days Thessaloniki International Fair. TIF-Helexpo Publications SA, Thessaloniki (2012)

Pozrikidis, K.: Stories of the Exhibition. TIF HELEXPO SA Publications, Thessaloniki. https://www.helexpo.gr/el/Fair_Stories (2013)

Sfakianakis, M.: Alternative forms of tourism. Ellin Publications, Athens (2000)

Stanojevic, N.: The new silk road and Russian interests in Central Asia. Rev. Int. Affairs RIA **LXVII**(1161), 142–161 (2016)

Svoronos, N.: Diagram of economic history of Thessaloniki (4th–19th century). Archeology **7**, 66–71 (1983)

Tafesse, W., Skallerud, K.: A systematic review of the trade show marketing literature: 1980–2014. Ind. Market. Manag. **63**, 18–30 (2017)

The World Bank: International Tourism, Number of Departures. https://data.worldbank.org/indicator/ST.INT.DPRT (n.d.)

Thessaloniki Book Fair: Thessaloniki Book Fair profile. https://www.thessalonikibookfair.gr/about/ (n.d.)

TIF Helexpo Press Release: 82nd TIF: The largest of the Decade. https://www.helexpo.gr/el/press20170905 (2017)

TIF: Corporate Profile. https://www.helexpo.gr/el/2019_TIF_HELEXPO_SA_Corporate_profile_March (2019)

TIF-HELEXPO: International architectural design competition for the Thessaloniki ConfexPark. In: Process of Obtaining Endorsemenr from the International Union of Architects (UIA), Provisional programme, September 2020, TIF-HELEXPO S.A. https://espazium.s3.eu-central-1.amazonaws.com/files/2020-09/ConfEx_Park_provisional_Programme.pdf (2020)

UFI, The Global Association of Exhibition Industry: The Role of Exhibitions in the Marketing Mix. https://www.ufi.org/archive/ufi-online-course/UFI_education.pdf (2010)

UFI, The Global Association of Exhibition Industry: Global Economic Impact of Exhibitions. https://www.aeo.org.uk/__media/PDFs/Oxford-Economics.pdf (2019)

UNESCO: Silk Roads Programme: Countries. https://en.unesco.org/silkroad/countries (n.d.)

UNWTOa: Outbound Tourism by Country. https://www.unwto.org/country-profile-outbound-tourism (n.d.)

UNWTOb: SILK ROAD. https://www.unwto.org/silk-road (n.d.)

Vakalopoulos, A.: History of Thessaloniki 315 BC-1983. Kyriakidis Bros Publications SA, Thessaloniki (1983)

Vakalopoulos, A.: History of Macedonia 1354–1833. http://promacedonia.org/en/av/av_14_2_c.htm (1973)

Vasdravellis, I.: The Port of Thessaloniki. Publications of the Free Zone Organization and Port of Thessaloniki, Thessaloniki (1959)

Vasiliev, I.A., Shmigelskaia, N.A.: The revival of the silk road: brief overview of the 4th China–Eurasia legal forum, Vestnik of Saint Petersburg University. Russ. Law Surv. **753**, 94–101 (2016). https://doi.org/10.21638/11701/spbu14.2016.209

Walker, J.R., Walker, J.T.: Tourism Concept and Practices. Dorling Kindersley Pvt, Ltd, Noida (2011), pp. 294–295

Wan, Y.: Assessing the strengths and weaknesses of macao as an attractive meeting and convention destination: perspectives of key informants. J. Convent. Event Tour. **12**, 129–151 (2011)

Weber, K., Ladkin, A.: Career identity and its relation to career anchors and career satisfaction: the case of convention and exhibition industry professionals in Asia. Asia Pacif. J. Tour. Res. **16**, 167–182 (2011). https://doi.org/10.1080/10941665.2011.556339

Wojciech, H.: Why is the silk road so important? Silk road initiative newsletter. www.silkroad.undp.org.cn (2006). Accessed 16 June 2019

Yu, L., Wang, C., Seo, J.: Mega event and destination brand: 2010 Shanghai Expo. Int. J. Event Fest. Manag. **3**(1), 46–65 (2012). https://doi.org/10.1108/17582951211210933

Zootechnia: Zootechnia Exhibition on stock-farming and aviculture. HELEXPO S.A. https://zootechnia.helexpo.gr/el/Sales-Kit (n.d.)

Silk Road Tourism Multiplier Effect at Interregional Level: The Case of Two Greek Regions

Paris Tsartas, Evangelia Stergiou, Efthymia Sarantakou, Stella Kostopoulou, Alexios-Patapios Kontis, and Konstantinos Giannopoulos

Abstract

This paper aims to assess the tourism growth potential of Silk Road cultural assets in the Regions of Central Macedonia and Eastern Macedonia and Thrace in Greece, by introducing a conceptual framework supporting interregional strategic planning towards Silk Road tourism development in the two regions. Based on the review of the literature and according to the results of the multicriteria analysis conducted, the optimal strategy to be implemented at the regional and interregional level towards Silk Road tourism growth is assessed. The Analytical Hierarchy Process (AHP), widely used as a multiple-criteria decision-making tool for sustainability indicators, is applied in the research for computation of weights for a number of various criteria selected. The analysis points out that Silk Road tourism clusters activated at both Regions are the optimal strategy to promote and establish Silk Road tourism in the regions involved. The clustering of Silk Road tourism activities in less developed areas, intensifies cooperation and partnerships between business communities in neighboring regions and serves as a vehicle for stimulating economic growth.

Keywords

Tourism multiplier • Interregional • Silk road • Greece

1 Introduction

Over the last decades, tourism has been widely recognized as an engine of economic growth and prosperity (Lee and Brahmasrene 2013) representing an important revenue stream both at local and national level, as well as a significant source of employment and entrepreneurship. At global level, tourism industry accounts for 5% of total GDP and more than 25% of world exports of services (UNWTO 2019). Moreover, tourism growth has been established as a common strategic tool for economic development worldwide (Andereck et al. 2005; Matarrita-Cascante 2010), thus extensive research has been produced upon the relationship between tourism development and economic growth. At regional level, tourism has been generally acknowledged as a key driver for the socio-economic development (Tsartas 2010), and therefore largely used to reduce regional disparities and to improve the prosperity of local communities.

The outbreak of the Covid-19 pandemic in 2020 that acutely affected and significantly altered growth around the world, imposed an unprecedented challenge to the tourism industry, that has proven to be unprepared, especially at mass tourism destinations. Among the new tourism priorities created as a result of the Covid-19 pandemic, diverse tourism products as a means of sustainable tourism development are expected to play an important role in the recovery of the tourism industry. According to a UNWTO (2019) report the tourism industry must implement sustainable strategies and tools to lessen its detrimental impact on the environment,

P. Tsartas · K. Giannopoulos
Harokopio University, Athens, Greece
e-mail: tsartas@hua.gr

K. Giannopoulos
e-mail: giannopoulos.k@hotmail.com

E. Stergiou (✉) · S. Kostopoulou
School of Economics, Aristotle University of Thessaloniki, Thessaloniki, Greece
e-mail: evangevs@econ.auth.gr

S. Kostopoulou
e-mail: kostos@econ.auth.gr

E. Sarantakou
University of West Attica, Athens, Greece
e-mail: esarantakou@uniwa.gr

A.-P. Kontis
University of the Aegean, Mitilini, Greece
e-mail: apkontis@aegean.gr

S. Kostopoulou et al. (eds.), *Cities' Vocabularies and the Sustainable Development of the Silkroads*, Advances in Science, Technology & Innovation, https://doi.org/10.1007/978-3-031-31027-0_26

community life standards and social integration. As a result, the tourism policy paradigm needs to shift from a growth-oriented model to one that emphasizes the compatibility of its quality of growth with the quality of life of the residents (World Economic Forum 2019).

The promotion of cultural heritage is a foundational component of the tourism development strategies of an increasing number of locations worldwide, and the number of cultural attractions is expanding quickly. Cultural heritage is a crucial component of functional networking, as areas with a shared legacy may interact with one another as a content-related planning concept, and tourism is regarded as key indicator of integration into a networked spatial system. Both nodes (attractions) and links (tourism flows) that connect the place of origin with the destination can be used to depict the tourist process (Zeng 2018). In this context, cultural tourism enables cross-cultural exchange and encourages local development to prosper as a main driver of employment and investment. The possible synergies that can result from combining tourism with culture may be an essential factor in boosting the attractiveness and competitiveness of specific destinations (Mizzau and Montanari 2008).

Tourism-driven cultural heritage, creates jobs, drives innovation, fosters the development of public services and local enterprises, creates wealth, and encourages civic engagement (Vasile and Banica 2019). Several studies have shown that heritage routes can serve as key drivers for economic development and poverty alleviation in rural areas (Snowball and Courtney 2010). Heritage routes also help expand the geographic extent of tourism, and promote local heritage that benefits local communities. This approach allows us to combine secondary attractions of limited pulling power, with primary attractions that act as key waypoints (Ivanovic 2008). Therefore, the diversification of cultural heritage tourism products should be a two-fold strategy. First, it needs to be developed as an innovation tool which will increase competitiveness and attraction quality (Booysen 2012). Second, it should serve as a means for job creation and poverty reduction, extending the economic impact of tourism to a wider range of destinations (Rogerson and Visser 2011). Cultural heritage is acknowledged as both a driving force and a catalyst for socioeconomic development, and heritage-based development projects ought to be planned and implemented in developing and transitional countries (Loulanski 2006).

In this paper Silk Road cultural heritage, identified over a large variety of regions and countries, is examined. The historic Silk Road, actually a vast network of routes, was the first global trade route in human history to connect East and West, crossing numerous settlements with rich history and heritage. With a reach and significance considerably broader than the simple exchange of goods, Silk Road was also serving as a vehicle for the exchange of arts, religion, cultures, ideas and technology (Kostopoulou et al. 2016; Kostopoulou 2019). The Silk Road is currently flourishing as a tapestry of diverse cultural tourism destinations, spanning numerous countries and regions, with abundant natural and cultural resources. The revival of the Silk Road legacy from a new perspective can create enormous prospects for cultural heritage tourism development inspiring growing tourism flows of independent travelers, who wish to experience alternative forms of tourism, that depart from the mass-tourism model (Kostopoulou et al. 2021). Silk Road cultural heritage assets can greatly advance local culture, tourism, and economies. As a result, they consist part of the local cultural identity, and integral part of each region's branding and originality as well as being major points of interest for tourists. The possibilities for tourism potential along the Silk Road destinations are rich and diverse, but the development of tourism, and especially alternative forms such as cultural tourism, is being a priority for many regions and communities struggling to recover from the effects of the Covid-19 pandemic.

The paper highlights the assessment of tourism growth potential of the Silk Road cultural assets by introducing a novel conceptual framework. Silk Road tourism potential multiplier effect is explored at the interregional level using various criteria and parameters introduced through an input–output matrix analysis. A comparative analysis of the included regional development theories depicts the optimal interregional strategic planning towards Silk Road tourism development. The Analytical Hierarchy Process (AHP), a popular multiple-criteria decision-making tool, is applied in the research to compute weights for the many criteria selected. AHP, developed by Saaty (1980) as a multiple-criteria decision-making technique for dealing with complex problems with several subjective criteria, is widely used in many fields to plan, choose the best alternatives, for resource allocation, resolving conflicts and optimization (Vaidya and Kumar 2006). Pairwise comparisons are carried out by the AHP to identify the relative importance of some specified items and assess alternatives to make the best decision. The AHP has been applied in a variety of contexts, including the domains of energy renewal, education, urban areas and mobility, health technology assessment and quality improvement (for an annotated bibliography see Ponsiglione et al. 2022; Madzík and Falát 2022) and more recently in tourism topics (Božić et al. 2018; Navale et al. 2021; Bagchi et al. 2021; Ferreira and Sánchez-Martín 2022; Göral 2020).

In the paper the Region of Central Macedonia and the Region of Eastern Macedonia and Thrace in Greece are used as the case study to highlight the conceptual framework introduced towards tourism growth potential of the Silk Road cultural assets. The perspectives developed during the implementation of the SILC- Silk Road Local Culture[1] project (2017–2021), which involves the two regions, serve as the basis for the analysis. The SILC project was funded by the Joint Operational Programme Black Sea Basin 2014–2020, and was coordinated by Aristotle University of Thessaloniki. By involving partners from five countries (Greece, Armenia, Georgia, Romania and Bulgaria), SILC project sought to strengthen and diversify the tourism offer of Silk Road destinations within the eligible areas, as well as improve interregional cooperation and partnerships. In parallel, SILC activities are expected to contribute in 'bridging the prosperity gap' through job retention and creation at local level, by supporting existing traditional enterprises and encouraging small business creation (micro-entrepreneurship). The untapped tourism growth potential is anticipated to improve both the cooperation process for the goals of the overall harmonious cultural development over the SILC project study areas, as well as the openness of these areas to European and international tourism. The analysis shows that the best way to develop Silk Road tourism at both the Regions of Central Macedonia and Eastern Macedonia and Thrace, is to activate Silk Road tourism clusters. It is intended that Silk Road tourism clusters will establish and sustain cooperative relationships and synergies, share the diverse Silk Road brand, and exchange skills and experience, while promoting cultural tourism aspects. The paper is organized as follows: the introductory section is followed by a brief review of the theoretical background upon tourism and regional development, the review of the literature regarding the relationship between tourism and growth concept as well as a brief presentation of tourism growth evaluation methods. In the third section, an overview of the study areas' profile includes among others the tangible and intangible Silk Road cultural resources in the regions involved. In the fourth section the methodology of the Analytical Hierarchy Process (AHP) applied is fully described, leading to the final sections including the main results and conclusions of this research.

2 Tourism and Regional Development

There is a general consensus that tourism benefits need to be spatially distributed (Crozier 2011). Tourism has proven to be a powerful sector leading to economic growth allowing for the diversification of local productive systems, attracting foreign exchange and capital resulting in the improvement of regional development as well as the stimulation of employment, income, and domestic household consumption (Meyer and Meyer 2015). In addition, tourism attracts people even to remote destinations with unique cultural and natural assets (Tsartas et al. 2013), is labour intensive, motivates small-scale business activities, and encourages infrastructure and communication network creation.

There is a wealth of empirical research related to the relationship between tourism and regional development most of which focus on the role of tourism in reducing regional disparities that leads to balanced economic growth. Although tourism is recognized as an important driver of local development based on the literature review conducted, there are few concrete theoretical studies that attempt to interpret the relationship between regional development and tourism (Martin 2015). The difficulty in developing a coherent theoretical framework in tourism studies stems from the complex nature of tourism, which intersects with many industries and other socioeconomic elements of supply and demand, making it very difficult to establish an integrated tourism development theory (Ioannides et al. 2014).

On empirical grounds, there is an extensive literature based on sophisticated statistical and econometric tools, investigating the impact of tourism on regional development, with the majority of researchers supporting the positive economic impact of tourism, while research on regional economic development and tourism is rather limited. Pascariu and Ţigănaşu (2014) examined the prospects for sustainable tourism in France and Romania based on the new economic geography. The authors found that, in order to create more sustainable types of tourism, it is essential to emphasize on remote and peripheral areas, strongly involve local authorities, professional agents and organizations as well as local people, aiming to identify the untapped assets of the destination and change them into potential attractive points. Nilsson et al. (2005) explored core–periphery relationship in Sweden by analyzing tourism management and enterprises' networking. The authors supported that core-periphery effects in Swedish regions follow the same pattern as regions worldwide. In addition, they argued that within the European Union, less developed regions are mostly situated on the periphery, whereas more developed regions with increased market accessibility tend to be centrally located. The authors conclude to the overall European experience of uneven regional development in the periphery, compared to the central regions, pointing out the importance of this issue for European institutions and policymakers. Andraz et al. (2015) applied calculated by VAR methodology in order to assess the contribution of tourism to regional disparities in Portugal. The authors found that tourism has significantly contributed to the concentration of economic activity in the largest regions while reducing the divergence between smaller regions.

[1] Grant Contract from ENI No 99372/31.08.2018 eMS BSB570.

2.1 The Relationship Between Tourism and Growth Concept: Literature Review

Tourism development, as an economic activity, has been defined by its importance in creating new jobs and supporting the development of local communities. Many countries and regions have succeeded in expanding their economies by increasing economic participation through the development of tourism.

Rostow's (1960) structuralist model identifies the path of economic growth that societies follow through their economic transition. This pathway includes five stages beginning with the traditional society, followed by the stages of the conditions for take-off, the take-off, the drive to maturity, resulting in the final stage of high mass consumption. Although Rostow's model accounts for economic growth, in general and not specifically tourism growth, it is an essential tool for explaining tourism growth. Tourism may provide a pathway to economic growth in destination, from non-tourism areas through the prerequisite stage of take off where little tourism activity takes place, to maturity stage when a significant number of travelers dominate the destination, resulting in the final stage of high consumption in mass tourism destinations. This shows that Rostow theoretically laid the foundation for most of the models that approached tourism development.

Building on previous research and the product life cycle concept, Butler (1980) introduced a general model of the destinations' evolution. According to Butler's model, the arrivals at tourist destinations follow the S-shaped curve of the product life cycle through six stages: Exploration, Involvement, Development, Consolidation, Stagnation, Post-stagnation. Butler's work is groundbreaking, but its approach is largely conceptual and lacks an essential empirical foundation applicable to all destinations. Agarwal (2002) supported that tourism destinations may experience a reorientation stage before evolving on their life cycle path. Ma and Hassink (2013) used evolutionary economic geography to distinguish between lifecycle and evolutionary process. Kristjánsdóttir (2016) explored the tourist arrivals at Iceland and concluded to an exponential growth correlated with economic growth and public infrastructure development. Garcia-Ayllon (2016) conducted geo location analysis depicting the evolution of La Manga landscape in Spain. The author pointed out that destinations' life cycle stage is strongly associated with governmental policies as well as market competitiveness. Singh (2020) illustrated how various destinations may deviate from the typical life cycle curve due to environmental factors such as climate crisis.

The concept of growth poles, introduced by Perroux (1950) and further developed by Boudeville (1966), describes growth poles as urban centres polarizing wider areas and manifesting a growth process resulting in regional concentration of entities and population. The level of regional centralization is defined by the market share of the leading economic sector and may be balanced out through the promotion of new growth poles, resulting in the reshaping of the regional spatial pattern. Despite the strong debate upon the effect of growth pole policies (Richardson 1975) there is an extensive literature regarding their role on tourism growth. Mustatea (2013) explored the growth policy of Iaşi Metropolitan Area as the dominating regional growth pole in Romania, aiming to polarize the peripheral centers. The author found that the leading growth pole should generate effects on the surrounding peripheral areas, however the strategy producing this impact was not recorded. However, little research examines the potential impact of growth pole strategies before their implementation. Parr (1999) highlighted that growth-pole policy has not yet been assessed in terms of concrete conceptual grounds.

Diffusion is described as a path through which economic growth spreads out to various locations (Rostow 1960). A substantial precondition for generating diffusion is innovation, which may can consists of tangible aspects, or activities such as tourism (Sarre 1977). Diffusion in tourism is approached by both demand and supply perspective. Regarding demand, tourism behavior and decision upon visit specific destinations is investigated. On the other hand, supply side regards to the structure of the tourism industry at destinations. Diffusion process mainly depends on the supply side where a critical mass of local stakeholders and entities invest on tourism facilities and products. Consequently, once a limited share of suppliers create products as well as facilities attractive to tourists, other individuals follow the innovation, creating a neighborhood effect on innovation (Hägerstrand 1967). The aforementioned diffusion of innovation is followed by structural changes of the social systems, and the adoption of different cultural aspects (Sethna et al. 1980; Rogers 1995). Miossec (1977) designed a novel tourism diffusion model, depicting the long term destination evolution and pointed out the tourism facilities differentiation as well as the consequent reshaping of behavior among tourism, local stakeholders and the host society.

According to the Economic Base model, all economic activities can be sorted into two mutually exclusive sectors —one basic and one non-basic sector (McCann 2001). The Economic Base theory suggests that a basic sector- which can be tourism as empirical data has shown- can be the one which exports goods to the international market, while the non-basic sector is made up of the subsectors which play role mainly in the internal/local market of the region. Activities belonging to the basic sector are usually involved in manufacturing, mining, as well as tourism, while sectors mainly comprising non basic activities include retail and other services. Despite its wide application, the economic base model is rather limited in the case of the tourism industry (Lagos 2009), since the particular model and its multipliers estimate the aggregate effects on the entire economy, without allowing further analysis at sectoral level or of different types of effects (direct, indirect, and induced).

Porter (1998) is a major contributor to the literature, providing a comprehensive understanding of national and regional competitiveness and extending the concept of business clusters. Business cluster theory suggests that sustainable regional development is closely linked to the concentration of successful and innovative business clusters that can enhance regional competitiveness. Business cluster theory addresses challenging issues that have characterized tourism policy and academic literature over the last few decades: Sustainability and Locality. According to Lagos and Courtis (2008), the networking among universities, research institutions and tourism stakeholders are fundamental precondition for the creation of tourism clusters focusing on capacity building, know-how exchange innovation.

Exploring the tourism demand side, tourists prefer specific destinations based on their judgment regarding internal and external factors (Tsartas et al. 2016). In particular, tourists' decision-making process depends on individual internal characteristics (e.g. motivations) as well as external influences (e.g. distance, destination's authenticity) (Belk 1975). The conceptual framework of travel trends and individual decision making includes three basic theories related to motivations: (a) the psychographic profile described by Plog (1974); (b) types of tourism experiences evolved by Cohen (1974) typologies; and (c) sociopsychological preferences based on past experiences, as illustrated by Pearce (1989). However, given the complex nature of modern tourism structure, focusing only on the socio-psychological aspects cannot deeply approach the tourists' decision-making process. The above theoretical approaches have two common characteristics: (a) They relate through tourists' motivations—demand and supply characteristics of all regions (b) They question the overall tourism development model which was traditionally adopted till the 80's (the 3 s model: sun, sand and sea). In this regard, they have played a crucial role in the discussion for new models of sustainable tourism development.

2.2 Tourism Growth Evaluation Methods: A Brief Review

The tourism growth evaluation models and methods annexed below, differ in terms of geographical scalability, need for empirical data, levels of development, patterns of development, and scientific approach (economics, development, planning etc.).

1. Tourism Gross Domestic Product. In the Gross Domestic Product of a nation, we calculate the net percentage generated by the tourism sector (income, expenditure and value added method). This traditional evaluation system is mainly used to assess the importance of the tourism sector in different economies and to make the necessary comparisons between them. It is also calculated at the regional level.
2. The Input–Output Matrix specialized for tourism. In the matrix used at the nation level we can estimate the tourism product and the income deriving from the sector itself and also from the interrelationships that other sectors of the economy (agriculture, trade, industry, transport) have with the tourism sector. In many cases, Input–Output Matrix method is closely related to Tourism GDP method. In some cases, it is also used as a method to assess the economic and developmental importance of tourism at the regional level. Both Fletcher (1989) and Richardson (1972) highly recommended the input–output (I–O) model for tourism impact analysis, as a popular tool for determining the size of the multiplier and the impact of visitor expenditure on a nation's economy (Kottke 1988; Randall and Warf 1996; Saayman et al. 2000; Zhou et al. 1997). However, the I–O models relies on the economy being sizable enough for inter-industry linkages to exist. Since regional and municipal areas typically lack the comprehensive economic data necessary to build a formal input–output model, I–O models are less useful as the research area coverage becomes more limited (Kostopoulou and Kalogirou 2011).
3. Evaluation of tourism resources, public expenditure, investments and sustainability policies in tourism: multi-criteria analysis. This type of evaluation uses cost–benefit and multi-criteria analysis to measure the positive or negative impact of the development of: tourism resources, public expenditure in tourism, tourism investments and sustainability policies. In all cases, the criteria selected are scientifically interdisciplinary, combining in that aspect development, environment, planning, economic, cultural and sustainability parameters. This type of approach is suitable for local regional development projects or investments, taking into account sustainability parameters in the planning and evaluation processes, for example, Involvement of local stakeholders, bottom-up approach and evaluation processes in planning, sustainability parameters.
4. Satellite Account (TSA) method. This type of assessment is based on a multidimensional framework that includes all nationwide economic transactions related to the tourism sector (and all sectors statistically related to the tourism sector). Research began in the 60's in France, and after three decades it has been accepted that this type

of measuring and implementing tourism statistics is the most valid way of registering and evaluating the economic flows that make up the overall tourism sector (at the national level, per year). In 2011, the EU adopted a special directive on the issue of implementation of the TSA by all member countries. Although not easy to adapt, TSA is implemented in a large number of countries all over the world when counting and estimating tourism statistics. Although TSA is a very useful tool, it is mainly used at the national level, with a number of case studies at the regional level.

5. Computable General Equilibrium (CGE) methods. This type of evaluation is more dynamic and is increasingly used worldwide to simulate the economic impact of various changes, across multiple sectors (Dwyer et al. 2005). However, it is also more complex in its implementation described as a mathematical model that compares and calculates growth resulting from an investment or an event (e.g. Olympic Games) related with a sector like tourism. Interestingly, the method can be useful in calculations and estimates of regional or local development plans and investments or events. On the other hand, the complexity of implementing this model and the need for data can be a problem. The commodities flow, labor market and national accounts data used in CGE models all contain a high degree of empirical information. Although the CGE approach to economic impact estimation is considered conceptually superior to the I–O approach, Jago and Dwyer (2006) point out that CGE models are occasionally criticized as being time-consuming and expensive, and as a result, are not typically available for regions, in sub-state level (Kostopoulou and Kalogirou 2011).
6. Tourism Multiplier method. This widely used method calculates the impact of tourism on income (Tourism Income Multiplier) and employment (Tourism Employment Multiplier). The multiplier effect was primarily used to determine the economic impact of visitor spending, by multiplying the number of tourists by the average spending per visitor (Stynes 1997). In recent decades, a growing body of literature examines tourism multiplier effect from a theoretical or empirical perspective (Ntibanyurwa et al. 2006; Van Leeuwen et al. 2009; Pascariu and Tiganasu 2014). Multipliers are used at the national, regional and local levels as an effective tool to enable development decisions on the potential impact or growth. They are also used to evaluate the contribution of specific projects related to tourism products. Tourism is considered a development factor if the value of the Tourism Income Multiplier is ≥ 1. There are also estimates for positive effects on employment, the decline in unemployment and income of local population when the Income Multiplier is >1.

In Greece, the tourism income multiplier evaluation is estimated at 2.2–2.65 (IOBE 2012; SETE 2017), which means that for every euro spent there is an increase in GDP of 2.2 to 2.65 euros. The impact on employment (direct and indirect) according to this estimate (Multiplier 2.2–2.65) is also positive at the national level (especially during summer period) and at the regional and local level. It is also estimated (Kouzelis 1997; Kasimis 2008) that investments in tourism have a positive effect on local income and reduce unemployment The tourism multiplier index varies—but is generally positive—depending on the specific development characteristics of the country, while at the regional level is lower than at the national level (about 0.5). The positive effects of the development of special interest tourism products (golf, scuba diving, windsurfing, horse riding, etc.) have also been evaluated using the tourism income multiplier method for Greece and elsewhere.

3 Study Areas' Profile

In this paper the Greek Regions of Central Macedonia and the Region of Eastern Macedonia and Thrace are used as the study areas for applying the proposed methodology. Data recorded have shown that throughout both regions there are numerous tangible and intangible Silk Road cultural resources identified in urban centres, as well as spread out in rural areas. Taking into consideration the wide range of cultural assets connected to the Silk Road, most of them still untapped, there are considerable opportunities of promoting diversified Silk Road tourism products in the two regions. In this aspect, there are various economic activities that can be largely benefited by their integration to the tourism industry and supply chain resulting in the diversification of regional economies.

3.1 The Region of Central Macedonia

The geographical location of the Region of Central Macedonia, Thessaloniki's port, new transport networks (such as Egnatia) and the fact that the Region's northern border is a trade gateway for the Balkans markets, have created many competitive advantages for the Region as a services hub for Eastern European countries and a connecting route between Europe and Asia.

The Region of Central Macedonia includes a rich variety of tourist destinations, accounting for 10% of total national tourism receipts in 2020 (SETE 2020). The coastal zone of the Region is a major tourist destination, accounting for 85.8% of total overnight stays over the Region (SETE 2020). Due to its strategic geographical position at the crossroads of Europe and Asia, the Region of Central Macedonia has historically been a transport and trade centre and today it is a commercial gateway and services hub for the Balkan and Eastern European countries.

The main Silk Road cultural heritage assets identified in the Region are:

1. Chryssalis silk factory (not in operation), an industrial heritage asset in Goumenissa, Kilkis Regional Unit
2. Spinning-weaving mill YFANET industrial historical monument, in Thessaloniki city
3. Old Silk mill Benozilio, in Pilea-Chortiatis nearby Thessaloniki city
4. Palaiochori village with rich tradition in the craft of weaving in Chalkidiki Regional Unit
5. “Vermion” Spinnery, a spinning mill (not in operation) in Veroia, capital city of Imathia Regional Unit
6. Textile factory ‘Longou-Tourpali’ in Naoussa city, Imathia Regional Unit
7. Yarn factory “G. Tsitsis”, the first water driven textile factory in Edessa city, Pella Regional Unit
8. Ottoman monuments, bezestens in the cities of Thessaloniki and Serres, khans, hammams, such as the Bey Hammam, mosques and imarets, throughout the region
9. Events and festivals related to the Silk Road, e.g., the 8th UNWTO international Meeting Silk Road Tourism in Thessaloniki in 2018, the Thessaloniki International Fair 2017 themed on China and the Silk Road
10. Traditional markets, such as Kapani and Modiano markets, in Thessaloniki

3.2 The Region of Eastern Macedonia and Thrace

The Region of Eastern Macedonia and Thrace, situated on the northeastern border of Greece, a national and EU border, the Region of Eastern Macedonia and Thrace is an area of geopolitical importance between Greece, Bulgaria and Turkey. Since the European enlargement in 2007, the Region s has served as a conduit for social and commercial activity connecting Greece with the Balkan and European territories. The Region of Eastern Macedonia and Thrace is home to 608,182 people, or 5.6% of the Greek population, and occupies 10.7% of the country's total surface area (ELSTAT 2011).

Agriculture predominates in the Region of Eastern Macedonia and Thrace. Between 2008 and 2019, the Region's GDP decreased by 29.0% between 2008 and 2019 (€9.5 ml in 2008 and €6.97 ml in 2019) as a result of the economic crisis (ELSTAT 2020). In 2019, the region contributed 3.8% of the national GDP. The region was rated lowest in terms of GDP per capita with 11,639, much below the national average (17,092) (ELSTAT 2020). However, the European integration, infrastructure investments, the gradual exploitation of tourism potential and geopolitical advantages, might significantly contribute to its economic growth. In 2020, the Region of Eastern Macedonia and Thrace contributed 2% of Greece's national tourism revenue. Considerable opportunities for tourism development are identified due to significant natural and cultural resources, e.g., to the islands of Thassos and Samothrace, Nestos and Evros rivers, etc.

The main Silk Road cultural heritage assets located in the Region are:

1. Art Silk Museum of Soufli, as a working thematical museum dedicated to silk.manufacturing and processing in Evros Regional Unit
2. Tzivre Silk mill, in Soufli town, one of the most visible symbols of the industrial silk production in Thrace
3. Modern silk worm farm, in Soufli town, Evros Regional Unit
4. Cocoon houses (silkworm hatching sites), can be discovered throughout Evros Regional Unit
5. Abdera, a significant historical economic, commercial and cultural center in Xanthi Regional Unit
6. Tobacco Museum of the city of Xanthi, in Kougioumtzoglou Mansion (early tobacco tradesmen) Xanthi Regional Unit
7. Tobacco Museums and historical warehouses, in the city of Kavala, Kavala Regional Unit, which show how oriental tobacco is processed and emphasizing the social history of Kavala
8. Ottoman monuments, bezestens, khans, hammams, mosques and imarets can be found throughout the region
9. Events and festivals associated with the Silk Road, such as the 1st International Western Silk Road Workshop in Alexandroupolis, Soufli Silk Festival
10. Traditional markets, such as the Komotini Old Market and Xanthi bazaar, as well as manufacturing sites including past and active manufacturing sites such as Silk House Tsiakiris, Bouroulitis Silk Soufli and Silk Kalfas.

4 Methodology

The proposed methodology has been designed by creating a nexus among the main criteria for Silk Road cultural heritage tourism sustainable development in the study areas, namely the Regions of Central Macedonia and Eastern Macedonia and Thrace, and the theoretical approaches to tourism growth potential. The main findings of this study, reached by applying a multi-criteria analysis, indicate the strategy that should be followed in order to achieve positive impacts on the growth of interregional cultural tourism development and to produce tangible changes in the local economies.

Analytic Hierarchy Process

In this study, the Analytical Hierarchy Process (AHP) was used to calculate weights for a selected set of criteria. The AHP method is one of the most commonly used decision support tools today, with widespread usage in research and practical applications (Madzík and Falát 2022). The initial logic of the method was introduced by the American mathematician Saaty (1980) and was soon generalized as a universal decision support tool, a weighting and scoring model widely used as a multiple-criteria decision-making tool (Arranz-López et al. 2017) mainly for sustainability indicators. AHP follows a hierarchical structure, that is consistent with the framework of most sustainability initiatives. It is a decision-making technique, applied worldwide to calculate relative weights for various indicators (Vaidya and Kumar 2006; Qureshi and Harrison 2003; Yu et al. 2010; Jain et al. 2014; Panda et al. 2016; Ameen and Mourshed 2019). The AHP includes pair-wise comparisons and rating of factors that contribute to effective and focused assessment (Saaty 1990), without considering other elements.

The primary AHP stage is the Pair-wise Comparison (PC) method, which compares each index to each other (Saaty 1994). In this study, the authors used Saaty's 9-point scale, Saaty to rate the importance of each dimension. In order to obtain reliable weight for the included criteria, it is important to measure the reliability and validity of the model results. A Consistency Ratio (CR), which calculates the degree of relative judgments inconsistency, is used in AHP, to quantify the overall consistency of judgment. According to Saaty (1990), if CR > 0.1 is insufficient and the procedure requires rejection or repeating all judgments, for which CR exceeded the accepted level and reconstruction of the pair-wise comparison matrix.

The AHP stages can be described as follows: Building a decision-making hierarchical model, preparing a research questionnaire based on a 9-point comparison scale, data collecting (expressing judgments by pairwise comparisons), calculating weights (priorities), determining the consistency of judgments using CR and combining the results from various respondents. Experts express their judgments as pairwise comparisons of the elements included the hierarchical model (Prusak et al. 2016). Pairwise comparisons are performed using the special bipolar 9-point comparison scale proposed by Saaty (1977, 1980, 1990, 1994, 2008, 2013) (Table 1), known as the fundamental scale method (Nilsson et al. 2016; Evelyn and Edmondyeboah 2016; Leal 2020).

The applied measurement scale (Table 1) aimed to assist the authors upon the required decision-making resulting in the weights for each criterion. It depicts the relative importance of each criterion compared to other criteria considered in the process of weighting setting. The measurement scale includes nine points. According to Anselin et al. (1989) nine-point comparison scale is the optimum scale to be used since nine criteria/indicators are the most that can simultaneously be compared and effectively ranked by individuals. The scale points range from 1, indicating equal importance to 9, referring to absolute or extreme importance of one criterion over another. The score values of 2, 4, 6, and 8 are intermediate values between two adjacent judgments, when compromise is necessary. Relative criteria weights have been computed by firstly applying AHP separately for the authors and then averaged to set the final weights for the determinant factors of Silk Road tourism growth.

The next step in the proposed methodology includes the score of the selected criteria separately for each of the two Regions of Central Macedonia and Eastern Macedonia and Thrace. The criteria assessment was based on (a) primary data deriving from the research field conducted throughout the SILC project duration, (b) available secondary data

Table 1 Saaty's nine-point ratio scale (Saaty 1977, 1980, 1990, 1994, 2008, 2013)

Degree of importance	Definition	Explanation
1	Equal importance	Two criteria contribute equally to the objective or judgment favors both criteria equally
3	Weak or moderate importance of one over another	Experience and judgment slightly favor one criterion over the other
5	Essential or strong importance	Experience and judgment strongly favor one criterion over the other
7	Demonstrated or very strong importance	One criterion is strongly favored over another and its dominance is demonstrated in practice
9	Absolutely or extreme importance	The evidence favoring one criterion over another is of the highest possible order of affirmation
2, 4, 6, 8	Intermediate values between the two adjacent judgments	Absolute judgment cannot be given and a compromise is necessary

sources and (c) authors' evaluation, being members of the scientific team of SILC project.

Finally, after having matched each of the selected development theories with the interrelated criteria, a final score is computed for each of the theories by multiplying the relative weight by the scores assigned to the corresponding criteria. The ranking of the theories' scores indicates the SILC tourism growth strategy to be followed in each region.

Research flow

Based on the literature review conducted, specific development theories were selected, taking into account the following factors:

1. The scalability of the proposed scientific theory regarding its capability to be used as a tool for development in various regions and tourism destinations.
2. The relationship between the region's resources and the fundamental elements of the theory suggested for implementation.
3. The implementation procedure and in particular the required stakeholder's involvement both from the tourism sector (international Level) and the regional and local tourism entrepreneurs.
4. The various assessment and evaluation methods for the potential regional tourism growth (i.e. multipliers, Satellite accounts etc.)
5. The suitability of the proposed theory for both regions and for particular tourism products or clusters of them.
6. The application of the suggested theory to various tourism development patterns (massive, sustainable etc.)

Aligned with the concept and the data produced from the activities of SILC project, specific remarks are included in our analysis that meet the needs and the limitations to be taken into consideration:

1. SILC project has an interregional strategy since it involves several countries and distinct regions along the Black Sea macroregion under different development modes and characteristics.
2. The eligible regions involved in the project may also have different characteristics when it comes to tourism development.
3. SILC Project should be viewed as a planned intervention that emphasizes the improvement of the sustainable development criteria for the participating regions.
4. The project adopts a strategic approach, with the goal of fostering a direct link between two fundamental pillars: Sustainable Tourism and Local Culture.
5. The project cannot be viewed as a conventional (international, national, or regional) investment and thus the evaluation cannot be limited to economic projections and scenarios.
6. An interdisciplinary, multisectoral and multicriteria evaluation is essential to provide accurate estimations for tourism growth potential in the project regions.

The following development theories are discussed in the analysis:

- Modernization theory
- Economic base theory
- Growth pole theory
- Innovation diffusion theories
- Business clusters theory
- Motivation theories

Based on the relevant academic literate review and the aforementioned comments regarding the specifications of the SILC project and Silk Road cultural heritage tourism development in the Regions of Central Macedonia and Eastern Macedonia and Thrace, the criteria below are integrated in the analysis (Fig. 1).

The next step in this analysis includes an input–output matrix aimed at identifying key development factors corresponding to each of the selected theories related to tourism growth (Table 2).

5 Results Analysis

The pair-wise comparisons were made using a guide table (Table 1). The relative importance between the two criteria is estimated by AHP tool on a scale ranging from 1 to 9. Table 3 depicts the scale on which various criteria are assessed according to numerical categories, as illustrated by the fundamental assessment model of Saaty (1994), introduced in the previous Sect. 4. Methodology. The assumption made is that if element A is more important than element B and is rated at 3, then B is less important than A and should be rated at 1/3 (reciprocal value). For example, authors used the measurement scale introduced and assessed criterion 11. Specialized Marketing Policies for Silk Road tourism over criterion 1. Tourism Destination Characteristics. The respective score recorded is 5, indicating that criterion 11 is essentially or strongly more important than criterion 1. The evaluation of criterion 3. Tourism Infrastructure over 11. Specialized Marketing Policies for Silk Road tourism, received the reciprocal value of 3 (1/3) since tourism infrastructure was considered moderately or slightly less important compared to criterion 11.

Fig. 1 Criteria towards sustainable Silk Road tourism development

Table 2 Matching the selected theories of development with the selected criteria

	Modernization theory	Economic Base Theory	Growth Pole Theory	Innovation Diffusion Theories	Business Clusters Theory	Motivations Theories
Tourism Destination Characteristics		■	■			■
Cultural Tourism Resources	■	■	■			■
Tourism Infrastructure		■		■		
Human Capital		■			■	
Stakeholders involved in cultural tourism promotion	■		■	■	■	
Synergies with other regions	■			■	■	
Local Tourism Planning and Management of the Destination- DMO involvement	■	■	■	■	■	
Tourism Sustainability policies and measures applied in Destination	■	■	■	■	■	■
Specialized Marketing policies for SIT markets and target groups	■			■	■	■
Specialized Marketing Policies for cultural tourism market				■	■	■
Specialized Marketing Policies for Silk Road tourism					■	■
Tourism as key factor in the Regions pursuit for sustainability	■	■				

Table 3 Pair-wise comparisons for the analysis criteria

Silk road tourism growth	1	2	3	4	5	6	7	8	9	10	11	12
Tourism destination characteristics (1)	1.00	0.50	0.50	2.00	1.00	2.00	1.00	0.33	0.33	0.25	0.20	2.00
Cultural tourism resources (2)	2.00	1.00	2.00	4.00	5.00	3.00	2.00	2.00	2.00	2.00	1.00	3.00
Tourism infrastructure (3)	2.00	0.50	1.00	1.00	2.00	2.00	1.00	1.00	1.00	0.50	0.33	2.00
Human capital (4)	0.50	0.25	1.00	1.00	1.00	0.50	0.50	0.50	0.50	0.33	0.25	0.50
Stakeholders involved in cultural tourism promotion (5)	1.00	0.20	0.50	1.00	1.00	1.00	2.00	0.50	0.50	0.50	0.50	0.33
Synergies with other regions (6)	0.50	0.50	0.50	2.00	1.00	1.00	2.00	1.00	1.00	1.00	0.50	1.00
Local tourism planning and management of the destination—DMO involvement (7)	1.00	0.50	1.00	2.00	0.50	0.50	1.00	1.00	1.00	0.50	0.33	1.00
Tourism sustainability policies and measures applied in destination (8)	3.00	1.00	2.00	2.00	0.50	1.00	1.00	1.00	1.00	1.00	1.00	2.00
Specialized marketing policies for SIT markets and target groups (9)	3.00	0.50	1.00	2.00	2.00	1.00	1.00	1.00	1.00	0.50	0.33	2.00
Specialized marketing policies for cultural tourism market (10)	4.00	0.50	2.00	3.00	2.00	1.00	1.00	2.00	2.00	1.00	0.33	2.00
Specialized marketing policies for silk road tourism (11)	5.00	1.00	3.00	4.00	2.00	2.00	1.00	3.00	3.00	3.00	1.00	1.00
Tourism as key factor in the regions pursuit for sustainability (12)	0.50	0.33	0.50	2.00	3.00	1.00	0.50	0.50	0.50	0.50	1.00	1.00

The overall assessment was made under the hypothesis that the comparative analysis takes into account strictly Silk Road tourism growth in the involved regions.

The results of the pair-wise comparisons revealed reliable judgments as supported by a Consistency Ratio (CR) of less than 0.1. Consistency ratio is **0.08**, thus this research is considered valid (Saaty 1994). CR is calculated as follows:

$$\mathrm{CR} = \mathrm{CI}/\mathrm{RCI}$$

where RCI is a random consistency index and CI is a consistency index, computed by:

$$\mathrm{CI} = (\lambda \max - n)/(n - 1)$$

where max is the main eigenvalue and n is number of elements to be compared.

The synthesis of the pair-wise comparisons in Fig. 2 depicts the criteria weighting for Silk Road tourism growth potential.

'Cultural tourism resources' has the highest priority with 14.8% of the total weighting, followed by 'Specialized Marketing Policies for Silk Road tourism' with 13.7% of the total, 'Specialized Marketing Policies for cultural tourism market' with 10.4% and 'Tourism Sustainability policies and measures applied in Destination' with 9.2%.

In the final step of this analysis, the selected criteria were scored for the Regions of Central Macedonia and Eastern Macedonia and Thrace, according to the indicators depicted in Table 4 *The criteria were rated on a scale from 1-Non existent to 10-Excellent.*

Based on the matrix in Table 2 illustrating the key development factors corresponding to each of the selected theories related to tourism growth and the above criteria score, the final table (Table 5) highlights the ranking of the development theories approaching the Silk Road tourism growth in the two regions.

According to the final calculated output depicted in Table 5, the top ranked strategy to follow in order to have positive impacts on the interregional cultural tourism development and tangible results on the local economies of the two regions, is the Theory of Business Clusters followed by the Theory of Motivations. It should be emphasized that the aforementioned preferred strategies apply to both the Regions of Central Macedonia and Eastern Macedonia and Thrace. The bottom ranked strategy for the Region of Central Macedonia is the Economic Base Theory, whereas for the Eastern Macedonia and Thrace is the Growth Pole Theory.

6 Conclusions and Recommendations

Cultural tourism development based on the SILC Silk Road Local Culture project activities is considered to have significant potential to enrich and differentiate the tourism product of

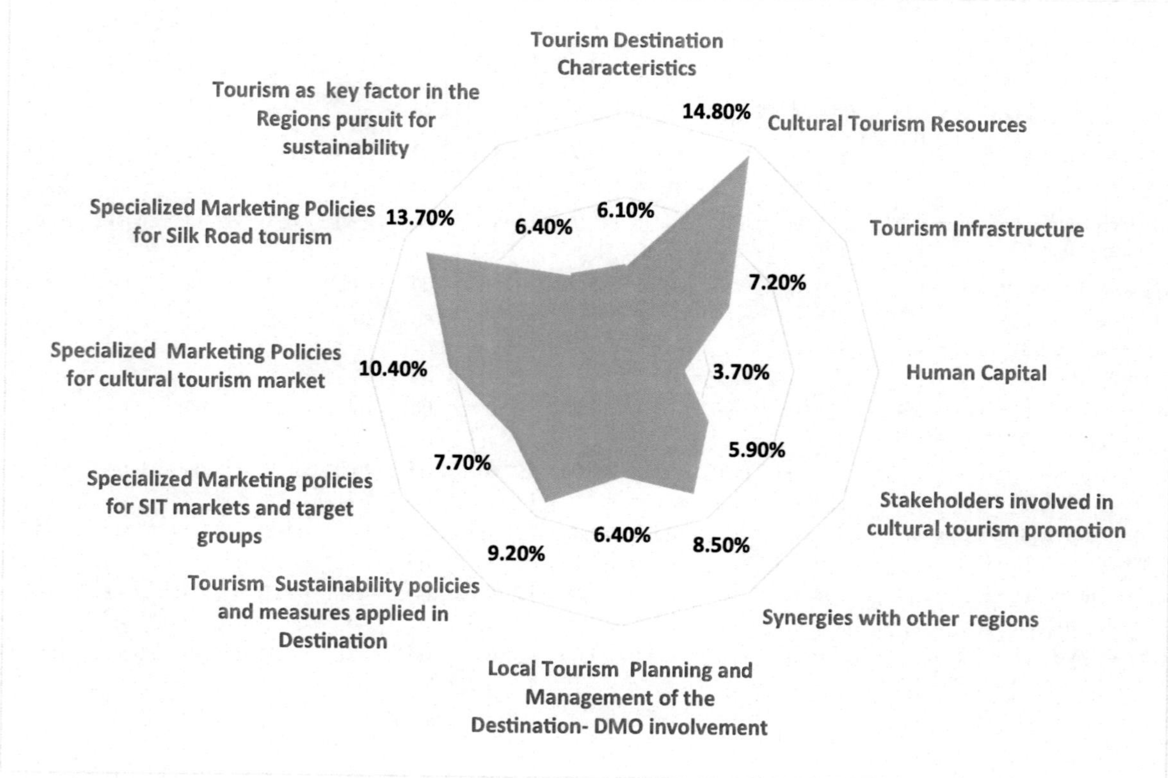

Fig. 2 Criteria weights computed by pair-wise comparisons of AHP

the participating regions and to provide sustainable development perspectives for less developed areas. Furthermore, the establishment of the Silk Road brand is expected to add significant promotional value to the regions involved, favoring the spatial dispersal of visitors, the diversification of tourism offer, and the opening to new niche markets.

This paper aims to assess the tourism growth potential of Silk Road cultural assets in the Regions of Central Macedonia and Eastern Macedonia and Thrace, by introducing a conceptual framework that supports interregional strategic planning towards Silk Road tourism development in both regions. The concluding remarks, based on the documented review of the literature and the extensive conceptual framework analysis conducted, emphasize the need to better integrate the Silk Road local cultural assets in interregional cultural routes, based on interregional synergies, common themes, and complementarities, in order to achieve positive impacts on cultural tourism development and tangible results on local economies. The main output of this research encapsulates the optimal strategy to be implemented at the regional and interregional level towards Silk Road tourism growth.

The analysis conducted argues that Silk Road tourism clusters activated at both the Regions of Central Macedonia and Eastern Macedonia and Thrace, are the optimal strategy for promoting and establishing Silk Road tourism in the regions involved. The clustering of activities and attractions in less developed areas, fosters cooperation and partnerships between communities in neighboring regions and serves as a vehicle for stimulating economic development through tourism.

Tourist industry is changing the methods of production and consumption, leading to greater competitiveness and introducing new marketing methods, for the development of quality development and global market standards. Silk Road tourism clusters are expected to build and maintain mutually beneficial partnerships and synergies, share the Silk Road diversified brand aspect, and exchange skills and experience, while promoting cultural tourism aspects. The development of successful and long-lasting partnerships will provide a vibrant environment for entrepreneurship that will inevitably lead to the creation of core, supportive and complementary industries. Silk Road tourism clusters can collaborate with

Table 4 Indicators for regional criteria assessment

Criteria	Scoring indicators
Tourism destination characteristics (1)	Destination's tourism life cycle (Bulter model), prevailing types of tourism. Supported by INSETE data (https://insete.gr/perifereies/)
Cultural tourism resources (2)	Primary data collected within SILC project. In-depth analysis of Silk Road cultural heritage assets
Tourism infrastructure (3)	Accommodation facilities registered at the Hotel Association of Greece, food and beverage facilities, communication facilities, railways, roads, waterways, SIT facilities, number of interventions (state or private investments tourism)
Human capital (4)	Percentage of population with higher education, NEETS, unemployment rate, number of universities and research institutions in the destination. The majority of data are provided by the National Statistical Authority of Greece
Stakeholders involved in cultural tourism promotion (5)	Primary data collected within SILC project
Synergies with other regions (6)	Due to lack of available data on regional linkages and synergies, an indicator about the participation of regions in interregional projects is included. The data library of https://keep.eu/programmes/ provides information upon the partners location out of all cross-border programs since 2000
Local tourism planning and management of the destination—DMO involvement (7)	Number of tourism strategies and actions included and funded by Regional Operational Programs 2021–2027. Existence or not of DMOs or similar organizations
Tourism sustainability policies and measures applied at destination (8)	Number of policies towards regional sustainable development included and funded by Regional Operational Programs 2021–2027 aligned with Agenda 2030
Specialized marketing policies for SIT markets and target groups (9)	Number of strategies implemented by local key stakeholders for various types of tourism (Cultural, Sport, Cycling, Hiking Tourism, Ecotourism, Agrotourism, Mountain, Sailing, Bird Watching, Cultural Heritage, Nature, Wellness, Health Tourism). Supported data by INSETE
Specialized marketing policies for cultural tourism market (10)	Number of strategies implemented by local key stakeholders for cultural tourism development. Supported data by INSETE
Specialized marketing policies for silk road tourism (11)	Number of local stakeholders promoting interregionally and internationally Silk Road products
Tourism as key factor in the regions pursuit for sustainability (12)	Tourism contribution to regional GDP and employment

Table 5 Total score and ranking of the development theories in every region

	Central macedonia	Eastern macedonia and thrace	Central macedonia	Eastern macedonia and thrace
	Total score		Ranking	
Modernization theory	3904	3398	4	3
Economic base theory	3542	2861	6	5
Growth pole theory	2736	2414	5	6
Innovation diffusion theories	3980	2986	3	4
Business clusters theory	**4759**	**3496**	**1**	**1**
Motivation theories	**4147**	**3407**	**2**	**2**

other local and interregional, governmental and non-governmental organizations in order to broadly spread the Silk Road entrepreneurial spirit.

References

Agarwal, S.: Restructuring seaside tourism: the resort lifecycle. Ann. Tour. Res. **29**(1), 25–55 (2002)

Ameen, R., Mourshed, M.: Urban sustainability assessment framework development: the ranking and weighting of sustainability indicators using analytic hierarchy process. Sustain. Cit. Soc. **44**, 356–366 (2019). https://doi.org/10.1016/j.scs.2018.10.020

Andereck, K.L., Valentine, K.M., Knopf, R.C., Vogt, C.A.: Residents' perceptions of community tourism impacts. Ann. Tour. Res. **32**(4), 1056–1076 (2005)

Andraz, J.M., Norte, N.M., Gonçalves, H.S.: Effects of tourism on regional asymmetries: empirical evidence for Portugal. Tour. Manag. **50**, 257–267 (2015)

Anselin, A., Meire, P.M., Anselin, L.: Multicriteria techniques in ecological evaluation: an example using the analytical hierarchy process. Biol. Conser. **49**(3), 215–229 (1989)

Arranz-López, A., Soria-Lara, J.A., López-Escolano, C., Campos, Á.P.: Retail mobility environments a methodological framework for integrating retail activity and non-motorised accessibility in Zaragoza. Spain. J. Transp. Geogr. **58**, 92–103 (2017)

Association of Greek Tourism Enterprises (SETE): The contribution of tourism in the Greek economy. URL: https://insete.gr/wp-content/uploads/2020/02/Tourism-and-Greek-Economy_2017-2018.pdf (2017). Accessed 26 Feb 2022

Association of Greek Tourism Enterprises (SETE): The contribution of tourism in the Greek economy. https://insete.gr/wp-content/uploads/2021/11/21_11_Tourism_and_Greek_Economy_2019-2020.pdf (2020). Accessed 26 Feb 2022

Bagchi, S., Uddin, K.M.S.: How to rank tourist destinations: a literature review. Int. J. Res. GRANTHAALAYAH **9**(8), 193–209 (2021). https://doi.org/10.29121/granthaalayah.v9.i8.2021.4178

Belk, R.W.: Situational variables and consumer behavior. J. Consum. Res. **2**(3), 157–164 (1975)

Booysen, I.: Innovation in tourism: a new focus for research and policy development in South Africa. Afr. Insight **42**(2), 112–126 (2012)

Boudeville, J.-R.: Problems of Regional Economic Planning. Edinburgh University Press, Scotland (1966)

Božić, S., Vujičić, M.D., Kennell, J., Besermenji, S., Solarević, M.: Sun, sea and shrines: application of analytic hierarchy process (AHP) to assess the attractiveness of six cultural heritage sites in Phuket: Thailand. Geogr. Pannon. **22**, 121–138 (2018)

Butler, R.W.: The concept of a tourist area cycle of evolution: implications for management of resources. Can. Geogr. **24**, 5–12 (1980). https://doi.org/10.1111/j.1541-0064.1980.tb00970.x

Cohen, E.: Who is a tourist? A conceptual clarification. Sociol. Rev. **22**, 527–555 (1974). https://doi.org/10.1111/j.1467-954X.1974.tb00507.x

Crozier, M.N.: Cross-border tourism planning and development: the case of the Lake!Gariep Initiative. Master Thesis Dissertation for the Degree of Masters of Arts: Development Studies Submitted to The School of Development Studies, Nelson Mandela Metropolitan University (2011)

Dwyer, L., Forsyth, P., Spurr, R., Ho, T.: The economic impacts and benefits of tourism in Australia: a general equilibrium approach. Sustainable Tourism CRC, Gold Coast, Australia (2005)

Evelyn, E., Edmondyeboah, N.: Ranking agricultural supply chain risk in Ghana: an AHP approach. Int. J. Econ. Comm. Manag. **3**(2), 1–12 (2016)

Ferreira, D.I.R., Sánchez-Martín, J.-M.: The assessment of the tourism potential of the tagus international nature reserve landscapes using methods based on the opinion of the demand. Land **11**(1), 68 (2022). https://doi.org/10.3390/land11010068

Fletcher, J.: Input–output analysis and tourism impact studies. Ann. Tour. Res. **16**, 514–529 (1989)

Garcia-Ayllon, S.: Geographic information system (GIS) analysis of impacts in the tourism area life cycle (TALC) of a Mediterranean Resort. Int. J. Tour. Res. **18**, 186–196 (2016). https://doi.org/10.1002/jtr.2046

Göral, R.: Prioritizing the factors which affect the selection of hotels by consumers traveling for vacation with analytic hierarchy process (AHP) method. J. Tour. Manag. Res. **7**, 11–31 (2020)

Hägerstrand, T.: Innovation Diffusion as a Spatial Process. University of Chicago Press, Chicago (1967)

Hellenic Statistical Authority (ELSTAT). https://www.statistics.gr/en/2011-census-pop-hous (2011). Accessed 1 March 2022

Hellenic Statistical Authority (ELSTAT). https://www.statistics.gr/en/statistics/eco (2020). Accessed 1 March 2022

Institute of Economic and Industrial Research (IOBE): The Effect of Tourism in the Geek Economy, Athens (2012)

Ioannides, D., Halkier, H., Lew, A.A.: Special issue introduction: evolutionary economic geography and the economies of tourism destinations. Tour. Geogr. **16**(4), 535–539 (2014)

Ivanovic, M.: Cultural Tourism. Juta, Cape Town (2008)

Jago, L., Dwyer, L.: Economic evaluation of special events. A practitioner's guide, Common Ground Publishing Pty. Ltd., Cooperative Research Centre for Sustainable Tourism, Australia (2006)

Jain, S., Aggarwal, P., Kumar, P., Singhal, S., Sharma, P.: Identifying public preferences using multi-criteria decision making for assessing the shift of urban commuters from private to public transport: a case study of Delhi. Transport. Res. F Traffic Psychol. Behav. **24**, 60–70 (2014)

Kasimis, C.: Survival and expansion: migrants in Greek rural regions. Populat. Space Place **14**(6), 511–524 (2008). https://doi.org/10.1002/PSP.513

Kostopoulou, S.: Silk road cultural heritage tourism network. In: Proceedings of the 5th International Conference of International Association for Silk Road Studies IASS SUN "Silk Road: Connecting Cultures, Languages, and Ideas", Moscow, Russia, pp. 232–260 (2019)

Kostopoulou, S., Kalogirou, S.: The spatial-economic impact of cultural events. Int. J. Sustain. Dev. **14**(3–4), 309–331 (2011)

Kostopoulou, S., Sofianou, P.-K., Tsiokanos, K.: Silk road heritage branding and polycentric tourism development. Sustainability **13**, 1893 (2021). https://doi.org/10.3390/su13041893

Kostopoulou, S., Toufengopoulou, A., Kyriakou, D., Malisiova, S., Sofianou, E., Xanthopoulou-Tsitsoni, V.: The western silk road in Greece. Aristotle University of Thessaloniki, Silk Road Programme, Western Silk Road Tourism Initiative, A UNWTO-EU Initiative. https://webunwto.s3.eu-west-1.amazonaws.com/2019-09/aristotleuniversitynationalswotanalysis-westernsilkroadgreece.pdf (2016). Accessed 8 March 2022

Kottke, M.: Estimating economic impacts of tourism. Ann. Tour. Res. **15**, 122–133 (1988)

Kouzelis, A.: Impacts of currency policies in the recent performances of the tourism sector. ITEP, Essays 1, Athens (1997)

Kristjánsdóttir, H.: Can the Butler's tourist area cycle of evolution be applied to find the maximum tourism level? A comparison of Norway and Iceland to other OECD countries. Scand. J. Hosp. Tour. **16**(1), 61–75 (2016)

Lagos, D.: Multiplier analysis and tourism development. In: Proceeding of the 2nd International Conference on Quantitative and Qualitative Methodologies in the Economic and Administrative Sciences. TEI of Athens, 25–27/05/2009, pp. 287–293 (2009)

Lagos, D., Courtis, P.: Clusters and competitiveness in the Greek tourism sector. Eur. Res. Stud. **XI**, 1–2 (2008)
Lee, J.W., Brahmasrene, T.: Investigating the influence of tourism on economic growth and carbon emissions: evidence from panel analysis of the European Union. Tourism. Manag. **38**, 69–76 (2013)
Leal, J.E.: AHP-express: a simplified version of the analytical hierarchy process method. MethodsX **7**, 100748 (2020)
Loulanski, T.: Cultural heritage in socio-economic development local and global perspectives. Environ. J. **34**(2), 51–69 (2006)
Ma, J., Hassink, R.: An evolutionary perspective on tourism area development. Ann. Tour. Res. **41**(suppl C), 89–109 (2013)
Madzík, P., Falát, L.: State-of-the-art on analytic hierarchy process in the last 40 years: literature review based on latent dirichlet allocation topic modelling. PLoS ONE **17**(5), e0268777 (2022). https://doi.org/10.1371/journal.pone.0268777
Martin, R.: Rebalancing the spatial economy: the challenge for regional theory. Territ. Polit. Gover. **3**(3), 235–272 (2015)
Matarrita-Cascante, D.: Beyond growth: reaching tourism-led development. Ann. Tour. Res. **37**(4), 1141–1163 (2010)
McCann, P.: Urban and Regional Economics. Oxford University Press, Oxford (2001)
Meyer, D.F., Meyer, N.: The role and impact of tourism on local economic development: a comparative study. Afr. J. Phys. Health Educ. Recreat. Dance **21**(1:1), 197–214 (2015)
Miossec, J.M.: Un model de l'espace touristique. Espace Géogr. **6**, 41–48 (1977). https://doi.org/10.3406/spgeo.1977.1690
Mizzau, L., Montanari, F.: Cultural districts and the challenge of authenticity: the case of Piedmont, Italy. J. Econ. Geogr. **8**, 651–673 (2008). https://doi.org/10.1093/jeg/lbn027
Mustatea, N.M.: Growth poles: an alternative to reducing regional disparities. Case study: Iaşi growth pole. Roman. Rev. Reg. Stud. **IX**(1), 51–60 (2013)
Navale, S.B., Suryawanshi, R.S., Bhagat, V.S.: A morphological study for identified the potential of ganpatipule coastal tourism in Western Maharashtra (India): analytical hierarchy process. Appl. Ecol. Environ. Sci. **9**(3):322–335 (2021). https://doi.org/10.12691/aees-9-3-2
Nilsson, P.Å., Petersen, T., Wanhill, S.: Public support for tourism SMEs in peripheral areas: the Arjeplog project, northern Sweden. Serv. Ind. J. **25**(4), 579–599 (2005)
Nilsson, H., Nordström, E.-M., Öhman, K.: Decision support for participatory forest planning using AHP and TOPSIS. Forests **7**(5), 100 (2016). https://doi.org/10.3390/f7050100
Ntibanyurwa, A.: Tourism as a factor of development. In: Brebbia, C. A., Pineda, F.D. (eds.) Sustainable Tourism II, pp. 73–85. Wit Press, Southampton (2006)
Panda, S., Chakraborty, M., Misra, S.K.: Assessment of social sustainable development in urban India by a composite index. Int. J. Sustain. Built Environ. **5**, 435–450 (2016)
Parr, J.B.: Growth-pole strategies in regional economic planning: a retrospective view. Part 2. Implementation and outcome. Urban Stud. **36**(8), 1247–1268 (1999)
Pascariu, G.C., Ţigănaşu, R.: Tourism and sustainable regional development in Romania and France: an approach from the perspective of new economic geography. Amfit. Econ. **16**(8), 1089–1109 (2014)
Pearce, D.G.: International and domestic tourism: interfaces and issues. GeoJournal **19**(3), 257–262 (1989)
Perroux, F.: Economic space: theory and applications. Q. J. Econ. **64**, 89–104 (1950)
Plog, S.: Why destination areas rise and fall in popularity. Cornell Hotel Restaur. Admin. Quart. **14**(4), 55–58 (1974)
Ponsiglione, A.M., Amato, F., Cozzolino, S., Russo, G., Romano, M., Improta, G.A.: Hybrid analytic hierarchy process and likert scale approach for the quality assessment of medical education programs. Mathematics **10**, 1426 (2022). https://doi.org/10.3390/math10091426
Porter, M.E.: Clusters and the new economics of competition. Harvard Bus. Rev. **76**(6), 77–90 (1998)
Prusak, A., Stefanów, P., Strojny, J., Garcia-Melon, M.: The influence of the form of the 9-point scale in the AHP method on the consistency of judgments. Mod. Manag. Rev. **23**(3), 97–114 (2016). https://doi.org/10.7862/rz.2016.mmr.30
Qureshi, M.E., Harrison, S.R.: Application of the analytic hierarchy process to riparian revegetation policy options. Small Scale for. **2**, 441 (2003). https://doi.org/10.1007/s11842-003-0030-6
Randall, J.E., Warf, B.: Economic impacts of AAG conferences. Profess. Geogr. **48**(3), 272–284 (1996)
Richardson, H.W.: Input–Output and Regional Economics. Weidenfeld and Nicolson, London (1972)
Richardson, H.W.: Growth pole spillovers: the dynamics of backwash and spread. Reg. Stud. **41**(suppl1), S27–S35 (1975)
Rogers, E.M.: Diffusion of Innovations, 4th edn. Free Press, New York (1995)
Rogerson, C.M., Visser, G.: African tourism geographies: existing paths and new directions. Tijdschr. Econ. Soc. Geogr. **102**, 251–259 (2011). https://doi.org/10.1111/j.1467-9663.2011.00651.x
Rostow, W.W.: The stages of economic growth: a non-communist manifesto. University Press, Cambridge (1960)
Saaty, T.L.: A scaling method for priorities in hierarchical structures. J. Math. Psychol. **15**(3), 234–281 (1977)
Saaty, T.L.: The Analytic Hierarchy Process. Mc-Graw-Hill, New York (1980)
Saaty, T.L.: How to make a decision: the analytic hierarchy process. Eur. J. Oper. Res. **48**(1), 9–26 (1990)
Saaty, T.L.: How to make a decision: the analytic hierarchy process. Interfaces **24**, 19–43 (1994). https://doi.org/10.1287/inte.24.6.19
Saaty, T.L.: Decision making with the analytic hierarchy process. Int. J. Serv. Sci. **1**(1), 83–98 (2008)
Saaty, T.L.: Theory and applications of the analytic network process: decision making with benefits, opportunities, costs and risks. RWS Publications, Pittsburg (2013)
Saayman, A., Saayman, M., Naudé, W.A.: The impact of tourist spending in South Africa: spatial implications. South Afr. J. Econ. Manag. Sci. **3**(3), 369–386 (2000)
Sarre, P.: Innovation diffusion. In: The Open University (ed) Spatial Analysis Movement Patterns Units, pp. 18–21. Open University, Milton Keynes, UK (1977)
Sethna, R.J.: Social impact of tourism in selected Caribbean countries. In: Hawkins, D.E., Shafer, E.L., Rovelstad, J.M. (eds.) Tourism Planning and Development Issues, pp. 239–249. George Washington University, Washington, DC (1980)
Singh, S.: Time, tourism area 'life-cycle', evolution and heritage. J. Herit. Tour. **16**, 1–12 (2020). https://doi.org/10.1080/1743873X.2020.1766475
Snowball, J., Courtney, S.: Cultural heritage routes in South Africa: effective tools for heritage conservation and local economic development. Dev. South. Afr. **27**, 563–576 (2010). https://doi.org/10.1080/0376835X.2010.508589
Stynes, D.J.: Economic impacts of tourism. http://citeseerx.ist.psu.edu/viewdoc/download?doi=10.1.1.541.2793andrep=rep1&type=pdf (1997). Accessed 22 Mar 2022
Tsartas, P.: Greek Tourism Development: Characteristics, Investigations, Proposals. Kritiki, Athens (2010)
Tsartas, P., Coccossis, H., Vasileiou, M.: Regional perspectives on tourism geographies: the case of Greece (2013)
Tsartas, P., Sarantakou, E., Kontis, I.A.-P.: Tourism development models in Greece: trends and challenges in an effort to change the paradigm. In: Rontos, K., José António, F., Tsartas, P. (eds.)

Modeling and New Trends in Tourism: A Contribution to Social and Economic Development. Nova Publishers, New York, (2016)
Vaidya, O.S., Kumar, S.: Analytic hierarchy process: an overview of applications. Eur. J. Oper. Res. **169**, 1–29 (2006). https://doi.org/10.1016/j.ejor.2004.04.028
Van Leeuwen, E.S., Nijkamp, P., Rietveld, P.: A meta-analytic comparison of regional output multipliers at different spatial levels: economic impacts of tourism. In: Matias, Á., Nijkamp, P., Sarmento, M. (eds.) Advances in Tourism Economics; New Developments, pp. 13–33. Physica-Verlag HD, Heidelberg (2009)
Vasile, V., Bănică, E.: Cultural heritage tourism export and local development. Performance indicators and policy challenges for Romania. In: Valentina, V. (ed.) Springer Proceedings in Business and Economics, Caring and Sharing: The Cultural Heritage Environment as an Agent for Change, pp. 271–290. Springer, New York (2019)
World Economic Forum: The Travel and Tourism Competitiveness Report. http://reports.weforum.org/travel-and-tourism-competitiveness-report-2019 (2019)
World Tourism Organization (UNWTO): International Tourism Highlights, 2019 Edition, Madrid (2019). https://doi.org/10.18111/9789284421152
Yu, L., Hou, X., Gao, M., Shi, P.: Assessment of coastal zone sustainable development: a case study of Yantai, China. Ecol. Indicat. **10**(6), 1218–1225 (2010). https://doi.org/10.1016/j.ecolind.2010.04.003
Zeng, B.: Pattern of Chinese tourist flows in Japan: a social network analysis perspective. Tour. Geogr. **20**(5), 810–832 (2018). https://doi.org/10.1080/14616688.2018.1496470
Zhou, D., Yanagida, J.F., Chakravorty, U., Lueng, P.: Estimating economic impacts from tourism. Ann. Tour. Res. **24**(1), 76–89 (1997)

Entrepreneurial Networks as Accelerators for Cooperation in Silk Road Tourism: The SILC Project

Evangelia Stergiou, Stella Kostopoulou, and Nikos C. Varsakelis

Abstract

The tourism literature gradually adopts network analysis to encounter the complexity of managing tourism destinations. The establishment of networks is now widely acknowledged to help expand tourism benefits, strengthen investment opportunities, and promote interregional and international collaboration. Silk Road, the trade network that historically bridged East and West, currently involving a wide range of countries with a diverse array of tourism destinations and products, provides a contemporary tourism field in this regard. Silk Road tourism networks can bring together a variety of locations, activities, and cultural assets through an extended collaboration among distinct destinations and the involvement of local stakeholders. These networks could increase the interest in the tourism industry and the growing market of independent travellers who are looking for new experiences. This paper presents the case study of Silk Road cultural heritage tourism entrepreneurship networks, proposed in the SILC-Silk Road Local Culture project funded by the Joint Operational Program 'Black Sea Basin 2014–2020' with participating countries: Greece, Armenia, Bulgaria, Georgia, Romania. The Black Sea Basin, at the crossroads of ancient civilizations, exhibits a rich and diversified Silk Road cultural heritage, which could support an emerging and diverse tourism market. We argue that the establishment of SILCNET, SILC project entrepreneurial network within the SILC partners' eligible areas, will help to spread tourism benefits over the broader Black Sea region, offering additional opportunities for local firms to expand their markets, while also launching new collaborations, thus serving as an incubator for new entrepreneurial networking.

Keywords

Entrepreneurial networks • Cultural tourism • Interregional cooperation • Silk Road

E. Stergiou (✉) · S. Kostopoulou · N. C. Varsakelis
School of Economics, Aristotle University of Thessaloniki, Thessaloniki, Greece
e-mail: evangevs@econ.auth.gr

S. Kostopoulou
e-mail: kostos@econ.auth.gr

N. C. Varsakelis
e-mail: barsak@econ.auth.gr

1 Introduction

Tourism has grown to become one of the world's largest and fastest-growing industries in recent decades, and it is likely to develop even more in the coming years, with constant spatial expansion and product diversification once the COVID19 pandemic effect on travel will be hopefully over. New tourism trends have gradually resulted, in a transition away from the mass tourism model of the post-war period, toward more individualistic creative tourism patterns, marked by greater flexibility and more qualitative experiences in cultural heritage tourism products. The cultural identity of a region is clearly derived from the regions' tangible and intangible cultural heritage, "the creative expressions of people's existence in the past, near past and present that have been passed on to the present generation by past generations" (UNESCAP 2008). Cultural identity and heritage enhance the destination attractiveness and are essential for the increase of new tourist flows and the creation of niche tourism markets. Contemporary travelers are increasingly seeking authentic experiences, such as meeting and learning about local people and culture. Hence, tourism destinations should adapt to the new trends by offering high-quality products that promote local customs and traditions and investing in sustainability aspects such as the preservation of heritage, landscape, and local culture.

To ensure sustainable tourism development, policymakers and stakeholders should design and implement policies

S. Kostopoulou et al. (eds.), *Cities' Vocabularies and the Sustainable Development of the Silkroads*, Advances in Science, Technology & Innovation, https://doi.org/10.1007/978-3-031-31027-0_27

to build a proper tourism environment that adheres to world standards and new trends. Tourism, as a temporary interaction between visitors and host destinations, brings together a variety of stakeholders, including tourism companies, travelers, national tourism offices, infrastructure suppliers and various types of supporting organizations that are directly or indirectly involved. Because contemporary tourism activity involves the interaction of numerous economic agents (firms, consumers, sectors, regions, countries), the tourism system complexity has risen at an unprecedented rate. Thus, tourism could be considered as a network of interconnected organizations that interact to produce and deliver a service (Omerzel and Kregar 2016; Van der Zee and Vanneste 2015; Scott et al. 2008).

Over the past years, social network analysis has emerged as an attractive research methodology in social sciences, due to its flexibility and originality, and its ability to systematize relationships, and visually represent social structures, facilitating interactions' analysis and examining structural patterns (Holgado 2016). The application of Complex Social Network Analysis (CSNA) in economic analysis provides additional inputs besides the existing parametric and non-parametric techniques. The value of CSNA lies in its ability to map landscapes and highlight links, interactions, and behavioral patterns. During the last decades, an increasing bulk of the literature used social network analysis to explain various economic phenomena such as strategic alliances, knowledge diffusion, and research collaborations. Hagedoorn et al. (2000), Orsenigo et al. (2001), Grabher and Powell (2005), Malerba and Vonortas (2009), Vonortas and Zirulia (2011), and Angelidis et al. (2020) provide broad coverage of theoretical concepts and empirical analysis. The concept of networks was introduced and garnered prominence in the tourism literature to encounter the complexity of managing tourism destinations (van der Zee and Vanneste 2015). Networking and knowledge sharing can significantly improve the performance and competitiveness of tourism SMEs by enabling them to access resources and information systems, as well as by combining talents and assets traditionally exploited by large firms with economies of scale (Tayler and McRae-Williams 2005; Braun 2007; Fuglsang and Eide 2013). The development of local tourism products and markets requires a long-term integrated approach and commitment to the collaboration of private, public, and non-profit sector organizations. The strategic mix of organizations, agents, and networks to be involved varies according to the territory covered by the plan.

The emergence of the network approach is demonstrated in this paper with the example of a proposed entrepreneurial Silk Road network over the Black Sea Basin. The Silk Road territory, an array of rising destinations along the ancient trade routes that historically served as a bridge between East and West, a vast trade network connecting the ancient empires of China, India, Persia, and Rome, may provide a contemporary approach in this regard. The historic Silk Road is currently reviving as a burgeoning cultural tourism field, involving a wide range of countries with a diverse tapestry of tourism destinations and products (Chin 2013; Kostopoulou 2019, 2020). These destinations are endowed with the natural and cultural heritage of various landscapes, peoples, and cultures, including unique monuments, archeological sites and architectural complexes, arts, music, customs, traditions, clothing, lifestyles, agricultural techniques, city organization, etc. most of them listed in the UNESCO World Heritage List.

Due to its natural environment and rich historical and cultural heritage, the Silk Road plays an important role in the creation and development of new tourism routes. Therefore, it could be leveraged as a unique branding, marketing, and development instrument to advance a wide spectrum of tourism policy objectives. For example, international conferences, festivals, and other events are organized in different parts of the destination countries, to attract and involve more visitors. Silk Road tourism development is significant because it directly promotes economic development and facilitates political dialogue between different countries and the integration of their cultural heritage into prospective international tour routes.

The Black Sea Basin, at the crossroads of ancient civilizations, exhibits a rich and diversified Silk Road cultural heritage that could support a diverse and emerging tourism market. Several tangible and intangible cultural assets over the area have been designated as part of humanity's legacy, and the plethora of cultural expressions provides significant prospects for tourism development and collaboration. Thus, despite the lack of comprehensive regional data, the number of international visitors to the Black Sea Basin appears to have gradually increased. The development of tourism networks along the Black Sea Basin is based on the idea of bringing together a variety of locations, activities, and tourism assets under a Silk Road integrated theme, hence stimulating economic growth through the expansion of auxiliary products and related services (Greffe 1994). Network members should strive together to establish a mutually beneficial platform for all stakeholders involved in culture and tourism.

In this paper, we present and discuss the development of Silk Road cultural heritage tourism networks over Black Sea Basin proposed in the SILC-Silk Road Local Culture project.[1] The SILC project aims at establishing cross-border

[1] The SILC—Silk Road Local Culture (2017–2021) project, an Interreg ENI-CBC "Black Sea Basin" Programme 2014–2020 project (eMS Code BSB570/Contract No. 99372/31.08.2018), was financed by the Joint Operational Program Black Sea Basin 2014–2020 and coordinated by the Aristotle University of Thessaloniki, School of Economics.

networks of local culture and tourism businesses to facilitate interregional cooperation and partnership development for Silk Road cultural tourism. As a result, we expect the project to mobilize individuals, businesses and communities, across the study areas to build cross-border collaborative projects. Member countries (hereafter referred to as partner countries) of the SILC project have their distinctive place and traditions in the global tourism market. In the paper, we also stress the importance of connecting those emerging Silk Road destinations, often not widely recognized, to increase interest in the tourism industry and in the growing market of independent travelers looking for new places to visit. We argue that the establishment of networks could help to expand tourism benefits over the broader Black Sea Basin, strengthen investment opportunities and promote interregional and international collaboration, in the framework of sustainable destination management. The research methodology of this conceptual study is based on the descriptive-analytical approach, conducted upon secondary data used in the SILC-Silk Road Local Culture project. The paper is structured as follows: the second section briefly discusses the tourism stakeholders theory; the third section offers a review of the literature on tourism networks; the fourth section presents an extended examination of the case study on a Black Sea Silk Road network. Finally, section five offers some concluding remarks and policy recommendations.

2 Tourism Stakeholders

2.1 Conceptual Framework

In this section, we provide an overview of the definitions of "stakeholders", their identification, and their role in the organization of cultural heritage sites. Identification of stakeholders is critical for tourism destination management, particularly in terms of synergistic activities, which have a significant impact on the sustainability agenda (Maiden 2008). However, there is no consensus in the literature about the methodology to identify stakeholders nor a universally accepted classification of stakeholders (Krce Miočić et al. 2016).

Freeman (1984: 46) defined stakeholders as 'any group or individuals who can affect or be affected by the achievement of the organization's objectives. This definition is also applicable to tourism. According to Johnson et al. (2006), stakeholders refer to individuals or groups who rely on the organization to achieve their own goals and on whom, in turn, the organization is reliant. Alike the heterogeneous and complex tourism product (Middleton and Clarke 2001; Ferri and Aiello 2017), the tourism ecosystem also displays the same features consisting of a large number of firms, the majority of which are small and medium-sized enterprises (SMEs) (Scott, Parfitt and Laws 2000). According to BPP Learning Media (2010), there are three broad types of stakeholders in an organization, categorized as follows (Rahman 2017): (a) the internal stakeholders (e.g. employees and management), (b) the connected stakeholders (e.g. customers, suppliers, shareholders, fanciers), and (c) the external stakeholders (e.g. community, government, pressure groups).

Many authors attempted to classify stakeholders in the tourist destination at various levels. Swarbrooke (2001) classified stakeholders into five main categories: governments, tourists, host communities, tourism business, and other sectors to pursue sustainable tourism development. Cooper et al. (2005) proposed locals, tourists, the tourism industry, the public sector, and others as the key stakeholders in tourism (Slivar 2018). The World Tourism Organization (UNWTO 2005) identified tourism professionals, public authorities, residents, and media as significant stakeholders in tourism destinations who may affect directly or indirectly tourism planning and development, having a wide range of responsibilities and sometimes competing aims and objectives. Yet, interest groups and individuals, particularly local inhabitants and indigenous communities, require proper recognition as stakeholders in their own right (Macbeth et al. 2002).

Rahman (2017) distinguishes the following groups of stakeholders in tourism development. **National governments** design tourism activities and strategies at the national level and formulate and implement the appropriate policies. National governments develop and maintain infrastructures such as roads, railways, and ports; promote destinations to local and international tourists; improve accessibility between destinations; and ensure the safety and security of tourists. **Local governments** promote local tourist attractions, provide infrastructure and services to support tourism, and manage the negative impacts of tourism. They comprehend the variety of unique characteristics of a destination and create authentic experiences for visitors. **Tourism establishments and enterprises** contribute to tourism development since traveling abroad would be nearly impossible without tourism organizations and private sector entrepreneurship. Visitors have to travel from one location to another by planes, trains, and buses, visit restaurants that provide food and beverages, hotels for accommodation, tour operators, travel agencies, and tourist information centers. **Travelers**, including business travelers, and visitors to tourism destinations, sites, and attractions, are perhaps the most crucial stakeholders. The traveler visits attractions, spends money, and may communicate the experience at the destination through word of mouth and social media, affecting the destination's appeal depending on the positive or negative experience. **Local people** are also important for tourism development. They might have legitimate concerns on tourism development that is detrimental to their interest, especially if tourism affects their standard of living. Locals could add value to any tourist destination with their unique values and traditions. Tourists may also benefit from local people in various ways, such as interaction with them. Thus, they must take part in the decision-making process when appropriate. **Tourism employees, professionals, and consultants** also play a significant role in tourism development. Restaurants, hotels, transportation, and other related businesses will undoubtedly struggle if they employ an inadequate number of employees, or if there are shortages of skilled people in the market.

There are some other stakeholders that tourism planners should also consider. The national and international financial organizations, the trade unions of tourism employees and professionals, the tourism educational centers, and tourism research institutions contribute to tourism development. Boronyak et al. (2010) expanded the range of potential stakeholders who may be involved in sustainable tourism in the framework of the Community-Based Tourism model by adding NGO's (social and environment), education (public and training), and communities. Even though tourism planners should consider a variety of stakeholders in tourism planning and development, not all stakeholders would have the same authority and interest in a particular tourism project. Therefore, planners should identify the most significant stakeholders in local tourism development and devise an effective communication strategy. Different stakeholders might have diverse goals and interests, while their activities could affect the sustainability of tourism (Scott et al. 2000; Krce et al. 2016; Slivar 2018).

Stakeholders' knowledge and experience in tourism management, participation in tourism planning and development processes, and community involvement, are essential in tourism destination management (Hardy and Beeton 2001; Leiper 1995). However, while each stakeholder group plays an important role in the development of tourism, some stakeholders are more significant in determining the success of operations than others (Vincent 1990). Governments are concerned about infrastructures, such as transportation and accommodation, to achieve sustainable tourism development goals (Hardy and Beeton 2001). Alternatively, host communities are often increasingly concerned about local issues, such as the effects of tourism on their community, their quality of life, and sustainability (Getz and Timur 2005). Tourism businesses are typically more concerned about issues that directly affect their operation, such as product, marketing, and customer satisfaction or dissatisfaction, whereas tourists are usually interested in the quality of the destination's tourism product (Hardy and Beeton 2001). Each group of stakeholders is a vital component of the tourism destination because the initiatives and opinions of stakeholders are external to the strategic planning and management processes (Dill 1975). Robson and Robson (1996: 543) asserted that the involvement of stakeholders in tourism has 'the potential to provide a framework within which sustainable tourism development can be delivered' (Dabphet 2013).

The local community's interests should be prioritized among other stakeholders, yet, "only to a certain point whereas their aspirations do not negatively affect other sustainability elements" (Galant et al. 2016). Nature has no direct stakeholders. Thus it is difficult to be adequately represented, which leads us back to stress the priority of its preservation. Sustainable or unsustainable tourism is the consequence of social relationships and proper coordination among all key players (Gergen 2009; Garcia-Rosell 2014). Hence it is significant to deeply explore the many possible interactions to get the most out of collaborations (Slivar 2018).

2.2 Identification of Stakeholders Involved in the Organization of Cultural Heritage Sites

Local governments at cultural heritage sites are responsible for determining policies and plans, setting and enforcing standards related to facilities and services, managing infrastructure construction, protecting cultural and natural tourism resources according to UNESCO criteria, managing tourism resources in a sustainable manner, auditing the restoration of historical buildings, encouraging the participation of all stakeholders in the planning process, and providing community integration for the sustainability of the tourism industry (Anuar et al. 2012). Similarly, the private sector is responsible for developing and maintaining high-quality tourism products and services such as accommodation, food and beverages, local tours, transportation facilities, and recreational activities. The private sector should be involved in the maintenance process of buildings that are used as hotels when these buildings are historical structures (Turker et al. 2016).

Stakeholders in traditions-based tourism could be viewed from two perspectives: that of the emitting destination (i.e. the place of origin of tourists) and that of the receptive destination (i.e. the destination which offers cultural or rural tourism products and is visited by tourists). The character of the emitting destination differentiates the institutional interrelations. State regulatory authorities, such as the national administration of tourism, ministries, and public agencies, fall in the scope of both aspects. In the emitting destination, these state institutions are linked to the network members such as non-governmental organizations, travel agencies associations, tour operators, etc. In turn, in a receptive destination, state regulatory bodies (national administration of tourism, ministries, public agencies) develop mediating relationships with local authorities: mayoralties, municipalities, regional entities, and non-governmental organizations, (hotels and restaurants associations, industry associations, etc.). The cooperation between network members, organizations, and institutions, evidenced by direct and indirect links, contributes to raise public awareness and the development of optimal solutions. The involvement of the state, local governments, civic associations, and non-governmental organizations (NGOs) could lead to better synchronization and interaction among all stakeholders, resulting in an optimal environment for cultural tourism development.

Potential users (tourists) are the driving force for the development of cultural tourism, since their interest and, above all, their solvent demand contribute to its development. While the local population's interest in tourism development is mainly economic, user motivation is related to the satisfaction of moral, educational, aesthetic, and cognitive needs. The fulfillment of moral interests and needs is linked to the formation of spiritual values, rationalizing leisure time, and gaining knowledge of a distinct social environment seen through the prism of family ties and customs. Consequently, the establishment and maintenance of an active network of stakeholders, contributes to tourism development only if it is:

- Inclusive, by reflecting a diverse range of stakeholders and their respective needs and aims,
- Sustainable, by guaranteeing a balanced approach of the interests of the stakeholders involved,
- Impactful, by securing that actions are coordinated and targeted to the needs of destinations and the satisfaction of the tourist visiting,
- Beneficial, by spreading the positive impacts of tourism proportionally and as evenly as feasible,
- Resilient, by ensuring synergetic and collaborative planning that can adapt to changes, of both internal and external environment of a destination's tourism sector, whether expected or unexpected,
- Innovative, by constantly aspiring for improvement of the tourism offer and synchronization with the contemporary trends of the targeted markets and international tourism.

3 Silk Road Tourism Networks

3.1 The Network as Organizational Framework

In simple and general terms, a network is a set of elements, called vertices or nodes, with connections between them, called edges (Newman 2003: 167). Social network analysis examines the structure of relationships between entities. These entities could be individuals, groups, organizations, regions, countries, etc. A network consists of entities and relationships between them. The nodes represent the entities, and the number of nodes is the order of the network. Two nodes connected by a link are adjacent or neighboring. In a directed graph, the order of the two nodes is important: L_{ij} stands for a link from i to j, and $L_{ij} = L_{ji}$.

To visualize a network, we draw a dot for each node and connect two dots with a line if the two corresponding nodes are connected by a link. If the network represents connections between firms, the linkages in the network could represent communication channel or business transaction. The shortest paths play an important role in communication within a network. Suppose one needs to send a data packet from one computer to another through the Internet: the geodesic distance provides an optimal pathway since it allows quick transfer and save resources. Thus, shortest paths play a significant role in a network's internal structure. Finally, a network is dense when the number of links of each node is close to the maximal number of nodes. On the other hand, a network is sparse when is connected with a small number of links only. Positive network effects are expected to be lower in sparse networks than in dense networks.

In the business context, a network is defined as 'an arrangement of inter-organization cooperation and collaboration' Hall (2005: p. 179). Nordin (2003) discusses the benefits of cooperation networks in an industry, such as the effective exploitation of collective efficiencies, like shared infrastructure and service specialization. The establishment of a business network accommodates the information and knowledge transfer across the network, resulting in business cooperation opportunities. Communication between firms—nodes requires: (a) the establishment of a communication channel—this is the link—the edge between two nodes-firms, (b) a certain level of absorptive capacity for the nodes —firms, in other words, a common semantic understanding, and (c) a mechanism to clear the noise that may distort the message in the communication. Hence, business networks enhance interactive learning, reduce the cost of knowledge and information transfer because communication channels between agents produce limited errors or noise. Besides, tacit knowledge is transferred without semantic difficulties since the agents operate in the same social and economic environment.

In the tourism academic literature, tourism networks are seen as systems that organize and connect tourism destinations, generating benefits for the involved tourism businesses, improving destination performance and quality, and encouraging the provision of unforgettable experiences for tourists (Zach and Racherla 2011: 98). Benefits are attributed to tourism networks for their role in the development of tourism destinations within an ever-changing, complex, and competitive global market (Albrecht 2013; Cawley et al. 2007), as well as the integration, performance, and quality enhancement of tourism destinations (Zach and Racherla 2011). In addition, based on Porter (1990), tourism destinations and organizations can develop competitive advantages by combining the knowledge, skills, capital, and other resources of several stakeholders (Nordin 2003; Bramwell and Lane 2000; Manhas et al. 2014). Tourism destinations seem to have a better chance of being competitive on a national and global scale when their businesses compete and collaborate at the same time' (Novelli et al. 2006: 1142), even though empirical evidence supports that possible network benefits in tourist destinations are limited (van der Zee and Vanneste 2015; Żemła 2016: 10). Cross-border, inter-regional, and/or transnational tourism networks, like the destinations along the Silk Road, may be advantageous because transaction costs will be decreased, and the productivity and efficiency of tourism business activities within the networks will be improved (Zeng 2008). Silk Road Networks, due to their uniqueness, can draw external interest and resources, as well as mobilize local actors and potential members around a common vision and strategy.

3.2 Silk Road Tourism Networks

Tourism networks are a source of concern for regional and local economic development because they include tourism planning into a collaborative policy approach to facilitate the areas to help localities benefit from combined marketing efforts (Rogerson 2007). This collaboration adds value by creating and establishing cross-border business opportunities, a common framework and vision for tourism development within territories with common Silk Road cultural heritage characteristics and challenges, transfer cross-border knowledge in the tourism and cultural sectors in the project areas. However, the adoption of a homogeneous management system in Silk Road networks is not efficient due to political concerns, economic disparities, and social challenges, as well as diversity in local conditions and cultural values norms (Go and Pine 1995; Manhas et al. 2014).

The development of tourism networks along the Silk Road areas would be better understood at a regional level. The Western Silk Road (WSR) Tourism Development Initiative, launched in 2016 by the United Nations World Tourism Organization (UNWTO) and the European Union (EU), constitutes a joint tourism project, a key segment of the Silk Road Programme focusing on the impact of the Silk Road on European destinations (UNWTO 2017). The WSR Initiative aims at revitalizing the Silk Road heritage throughout Europe, from the Caspian Sea, around the Black Sea, and along the Mediterranean basin (Kostopoulou et al. 2016).

Regional stakeholders and local firms involved in Silk Road tourism could expand their business opportunities and potential if they collaborate with firms from other regions related to Silk Road tourism. In the following section, we will examine how the establishment of a network between stakeholders and business firms in selected regions throughout the Black Sea Basin could upgrade these regions in the global tourism market accelerating productivity and growth potential. Eligible regions included in the SILC- Silk Road Local Culture project are used to highlight the tourism development potential of an entrepreneurial network SILCNET.

4 SILCNET: A Black Sea Silk Road Entrepreneurial Network

4.1 SILC Project Entrepreneurial Network: Goals and Objectives

This section introduces the SILCNET cooperation network among stakeholders in Black Sea regions, aiming to promote Silk Road cultural tourism. The analysis is based on the activities carried out within the SILC—Silk Road Local Culture (2017–2021) project. The partners involved were five countries (Greece, as lead partner, Armenia, Bulgaria, Georgia, and Romania), represented by three universities (Greece, Armenia, Bulgaria), one NGO (Romania), and one institute (Georgia). The SILC project aimed to strengthen and diversify the tourism offer of Silk Road destinations over the eligible areas, as well as to promote interregional cooperation and cross-border partnerships by reinforcing tourism development and cooperation of tourism entrepreneurship related to the Silk Road cultural heritage, contributing to the economic and social development throughout the Black Sea Zone. Furthermore, SILC project also aimed to strengthen the Western Silk Road objectives promoted by UNWTO and the EU, and develop new instruments for entrepreneurship in Silk Road tourism, such as company visibility and networking.

One of the main goals of SILC project is to create networks and encourage local entrepreneurship so that positive economic impacts and other externalities can be exploited. Furthermore, the project aims to share experience and jointly develop local tourism products and services based on the Silk Road cultural footprint, as well as to address the development of cross-border tourism products and transnational thematic routes through the diversification of the tourism product that Silk Road destinations offer within the project study areas.

The proposed Black Sea interregional network 'SILCNET' aims to create cross-border business opportunities in the tourism and cultural sectors, by structuring the network in a way to ensure a clear cross-border impact on Silk Road tourism entrepreneurial initiatives. The fundamental goal of SILCNET network is to create a consistent strategy to develop and implement a joint marketing plan promoting cultural and tourism entrepreneurial initiatives and perspective development schemes, to increase Silk Road brand awareness. This collaborative network facilitates the exchange of experiences and expertise, as well as professional cross-border cooperation, with the aim of adopting a coherent joint marketing strategy for promoting Silk Road entrepreneurial initiatives. SILCNET is expected to boost the economic impact of the Silk Road brand in the project eligible areas. The objective is to create and maintain mutually beneficial partnerships and synergies, while sharing the Silk Road diversified brand element. Furthermore, cross-regional integration could lead to the emergence of sustainable clusters in the Silk Road tourism and culture sectors. Finally, the SILCNET network will interact with other local and interregional networks in Black Sea areas, regional governmental and non-governmental organizations to widely promote the Silk Road entrepreneurial spirit.

The SILCNET network aims to stimulate:

- The promotion of diversified Silk Road tourism products, which SILCNET members would pool and promote in niche markets of special interests.
- The development of Silk Road tourism products and common service standards across borders.
- The exchange and dissemination of knowledge and expertise.
- The intensification of cooperation and management proficiency of SILCNET members in cultural tourism issues.
- The entrepreneurial reinforcement over the Black Sea Silk Road.

The SILCNET could be visualized as a network with two interconnected subnetworks. The first layer is the business tourism layer which includes hotels and other accommodation establishments, tourism agencies, transportation companies, etc. The second layer consists of industrial firms that produce added value for the tourism sector such as food, beverages, textiles, etc. Figure 1 depicts the Silk Road Tourism network for the selected regions. To build the linkages, we use the data from Tables 1 and 2. Unfortunately, data in Table 1 on the global value chains come from OECD data base where data for Armenia and Georgia are not available. The tourism industry subnetwork is at the top and the industry subnetwork is at the bottom. The figure clearly shows that the tourism and industry subnetworks, as well as the whole network, are highly sparse.

For comparison purposes, we present the intermediate value-added consumption in Italy, one of the leading European countries in tourism. The total intermediate value-added consumption was 47.1%, the domestic share of intermediate value-added consumption was 43.1%, while the primary and secondary sectors contributed 21.35%, and the food industry 13%. As a result, a sparse network like the one shown in Fig. 1 does not offer growth potential to the specific regions in Silk Road tourism.

The significantly low level of connectivity and networking among SILC stakeholders is recorded in the qualitative research conducted among key stakeholders from the Region of Central Macedonia and the Region of Eastern Macedonia and Thrace to investigate the Silk Road cultural assets' assessment and the overall perspectives upon the growth potential of Silk Road cultural footprint. The sample consisted of 135 stakeholders, 90 of them located in the Region of Central Macedonia and 45 in the Region of Eastern Macedonia and Thrace.

Figure 2 illustrates the indicators used to capture the knowledge of the respondents on Silk Road History, Geography, Tourism Products, Tourism Destinations, National Silk Road Programs, and International Silk Road Programs. Most of the stakeholders responded that they have

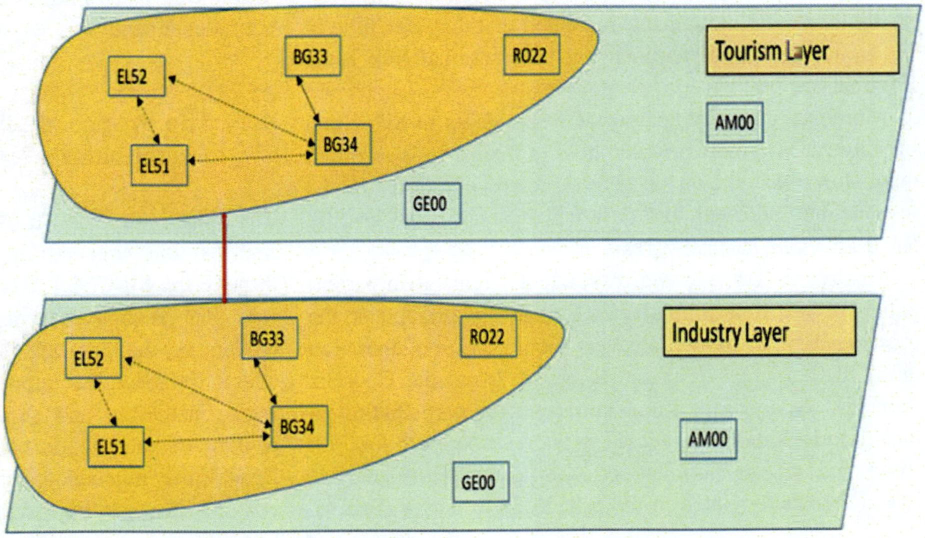

Fig. 1 Current network in the examined areas

Table 1 Tourism industry in the global value chain (2014)

	Output (in millions US$)	% of intermediate value-added consumption	% of domestic intermediate value-added consumption	% of intermediate value added consumption from Bulgaria	% of intermediate value added consumption from Greece	% of intermediate value added consumption from Romania
Bulgaria	2384	48.9	41.6	N/a	0.19	0.038
Greece	26,200	42.6	35.7	0.52	N/a	0.013
Romania	6762	49.27	43.7	0.02	0.003	N/a

Table 2 Intermediate consumption of specific sectors

	% of intermediate value-added consumption from the primary sector and from the secondary sector	% of intermediate value added consumption from food industry
Bulgaria	15.2	2.9
Greece	22.1	6
Romania	20.4	8

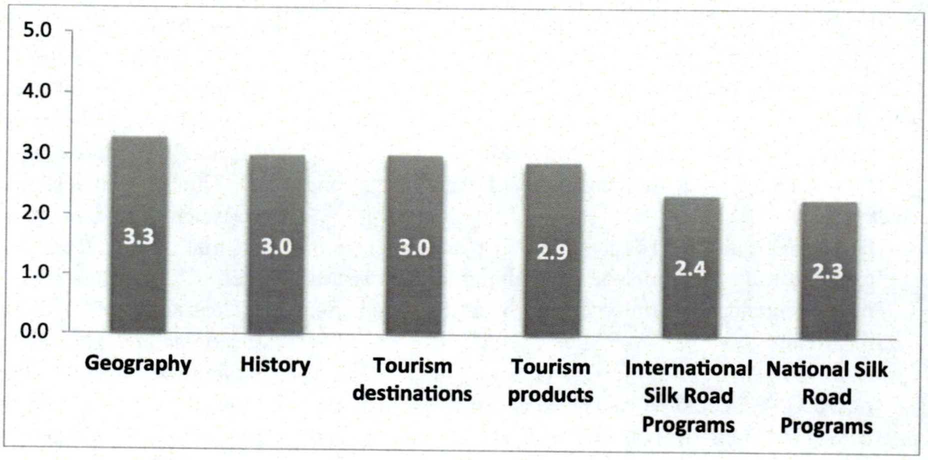

Fig. 2 Knowledge about Silk Road. *Note* History, Geography, Tourism Products, Tourism Destinations, National Silk Road Programs, and International Silk Road Programs-On a scale from 1-Nonexistent to 5-Excellent)

fair knowledge regarding Silk Road History, Geography, Tourism Products, and Tourism Destinations. However, nearly 2 out of 3 participants indicated that they had little or no knowledge about national and international Silk Road Programs.

Moreover, the participation of respondents in Silk Road initiative/activity/network/plan held in their area is recorded quite low. As shown in Fig. 3, only one-fifth of the respondents indicated that they had been involved in Silk Road activities and networks.

Finally, in terms of the respondents' interest in participating in Silk Road initiatives/activities/networks/plans, the most positively assessed answers include: (a) collaboration with SILC partners (Greece, Bulgaria, Romania, Georgia, Armenia) on the Silk Road initiative, (b) participation in local/interregional/international networks related to the Silk Road and (c) development of joint cross-border tourism actions/products related to the Silk Road initiative (Fig. 4). On the other hand, participants had no intention of investing to join Silk Road initiatives, highlighting their reluctance to act independently.

4.2 The Proposed Silk Road Tourism Stakeholder's Network

The objective of the SILC-Silk Road Tourism project is to improve the connectivity of the current sparse network through the SILCNET-Silk Road Tourism Network. Therefore, the creation of synergies among the tourism stakeholders of the five SILC partners would create network effects, increase the intermediate consumption of the network businesses and stimulate the regional income and welfare of the inhabitants. The proposed SILCNET-Silk Road Entrepreneurship Network is expected to serve this goal, as shown in Figs. 5 and 6.

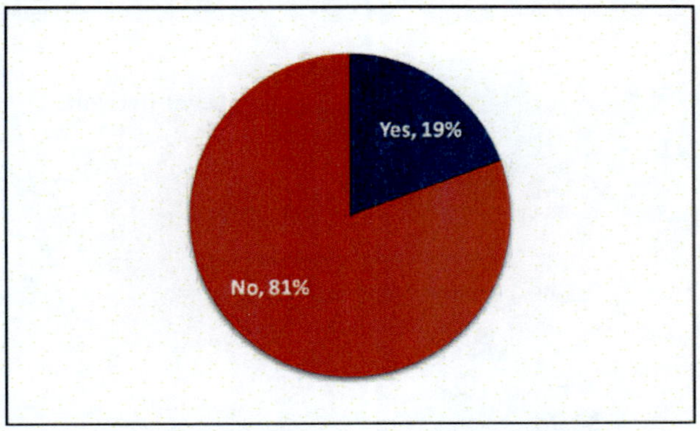

Fig. 3 Involvement in Silk Road initiative/activity/network/plan

The overview of the stakeholder's definition provided a list of stakeholders who could be involved in Silk Road tourism. However, each type of stakeholder will be involved in a distinct layer of the SILCNET—Silk Road Entrepreneurship Network. The proposed inter-regional SILCNET network is structured in two core-sub networks to ensure a clear cross-border impact on Silk Road tourism entrepreneurial initiatives.

The first core—sub-network is composed of the partners' representatives and strategic key actors involved in Silk Road tourism and cultural sectors. The first core—subnetwork will operate in two layers, at the partner and project level. The network at the project level consists of the partners' representatives and the strategic key actors, whereas the stakeholders' network at the partner level coordinates the activities and procedures regarding local entrepreneurial involvement and participation. The second core—subnetwork consists of the Silk Road local firms in tourism and culture, as well as cultural organizations over the project areas, which will operate in the five local entrepreneurial networks also including industries which are complementary to the tourism sector. The SILCNET objectives are the following:

- Dissemination of the project's results and objectives to all potential stakeholders to strengthen their involvement in the development of Silk Road tourism;
- Support the development of local entrepreneurship and partnerships among collective organizations such as Chambers of Commerce and tourism agents associations, and individual firms to achieve the goals of the Silk Road tourism initiative;
- Strengthen the involvement of regional and international stakeholders;
- Promotion of important Silk Road-related cultural, historical, and geographical characteristics of the areas involved;
- Development of joint strategies and policies to increase Silk Road brand awareness;

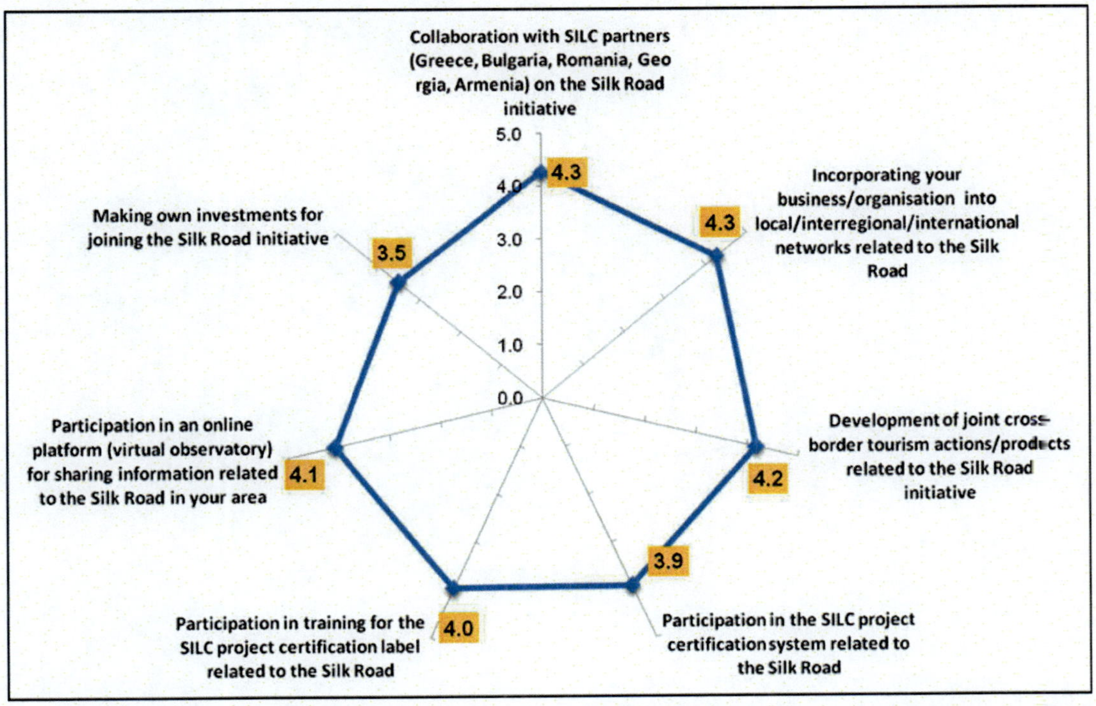

Fig. 4 Interest to participate in initiatives/activities/networks/plans in relation to the Silk Road. *Note* on a scale from 1-totally disagree to 5-totally agree)

Fig. 5 The stakeholder's network at partner level

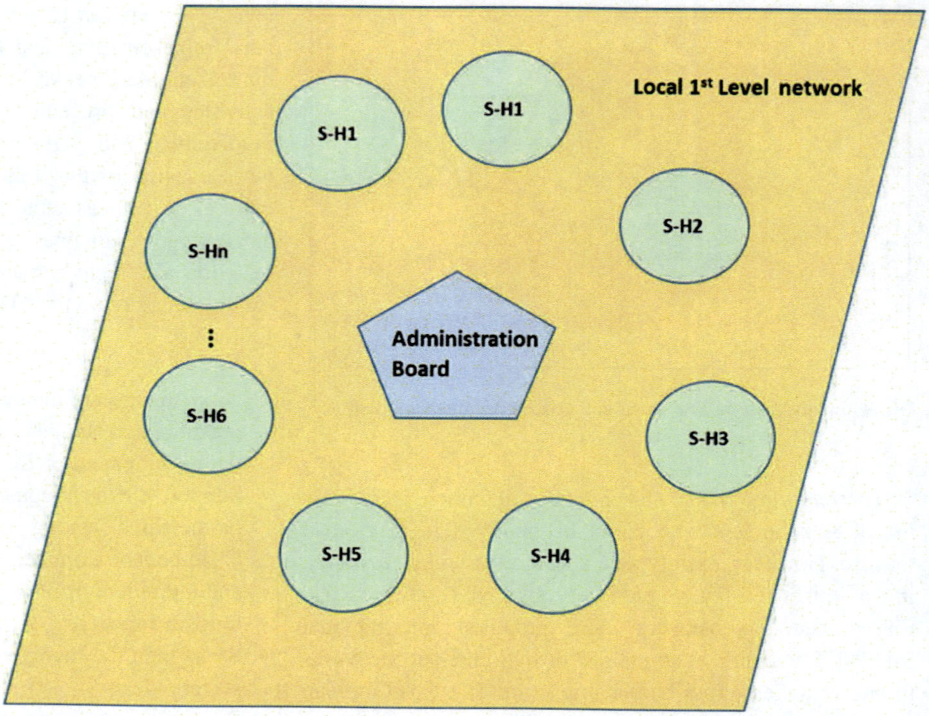

Fig. 6 The stakeholder's network at project level

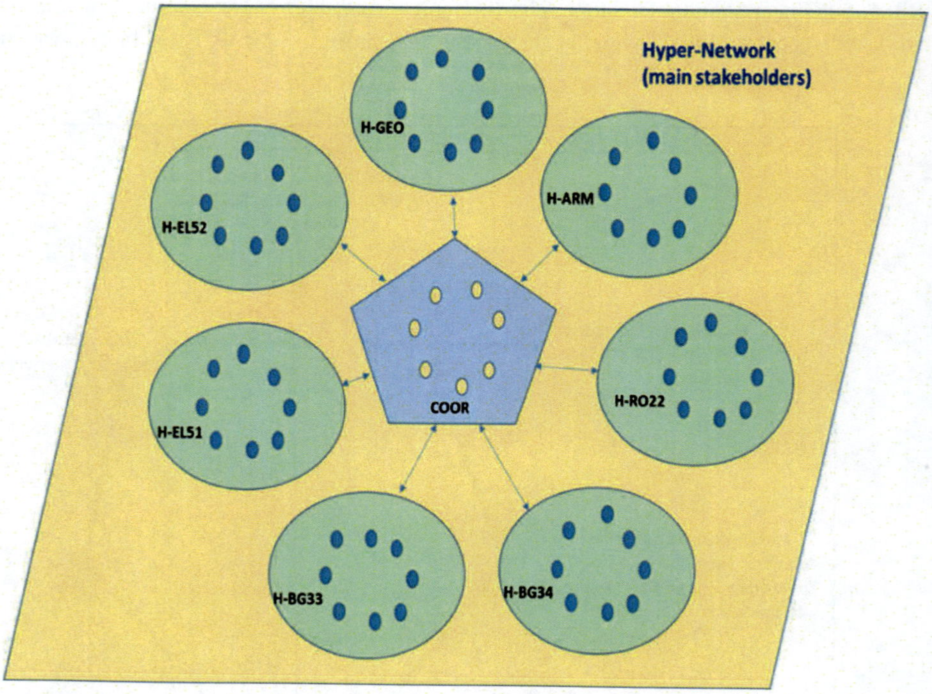

- Encouragement and advice from national authorities and stakeholders in initiating, implementing, and promoting activities related to the Silk Road tourism;
- Promotion of the intercultural dialogue, reconciliation, and cross-border cooperation among entrepreneurs who share the Silk Road spirit.

Each partner is responsible to manage and coordinate their local network and collaborate to provide feedback and establish synergies within the transnational network. In addition, the network operation may provide interesting inputs for a wider audience of managers, experts, and researchers.

The target groups of the Silk Road Entrepreneurship Network include entrepreneurs in the tourism and culture sectors, strategic business representatives and support organizations, local key agents, policymakers and stakeholders involved in the decision-making process, local authorities, sectoral agencies, NGOs, and education/training centers. Digital tools will be developed for the two types of networks, enabling network members to communicate via platforms. SILCNET members will participate in public consultations to facilitate the exchange of knowledge and expertise on issues that arise throughout the project implementation. Each network subsystem could publish documents that share good practices and feedback receiving.

Based on network theory, the factors that are expected to contribute to the success of SILCNET are:

- **Organization Structure**: A well-defined organizational structure facilitates communication clarity and ensures the efficient operation and coexistence of the two sub-networks. The network of organizations usually suffers from low efficiency because the organizational structure is not well defined, which leads to problems in communication among network members, resulting in poor network effects and in some cases, collapse of the network.
- **Communication Channels**: A significant aspect of entrepreneurial networks is the existence of effective communication channels. As entrepreneurial networks expand, the complexity of the network increases and the communication among the agents becomes more critical to network efficiency. The proposed SILCNET Entrepreneurial Network, both hyper and local ones, will meet the expected communication needs based on the Virtual Observatory platform[2] created under the project. Therefore, firms, official agents, and associations could communicate with their counterparts from other project areas either directly through the Observatory or the Coordination Committee. A co-formulated communication protocol defines the types of information to be communicated over the noise-free network and identifies the person(s) responsible for communication topics.
- **Network Operation**: The operational mode of a network is to ensure flexibility, particularly for the entrepreneurial network. Local entrepreneurial networks will grow as more entrepreneurs will be motivated to join the Silk Road tourism development. Therefore, the network's organizational structure and operation should have the characteristics of a scale-free network.
- **Coordination Committee**: The Coordination Committee should maintain the central position in the network and ensure the success of SILCNET. The committee will serve as a central hub for the exchange of expertise, good practices, and information among local networks. Moreover, the committee will be responsible in pursuing joint planned network actions.

The sustainability of the project will be ensured by the main Stakeholders Network, which is envisioned to be institutionalized as a non-governmental non-profit entity upon project completion. Figure 5 presents the structure of the Stakeholders Network at partner level. The S-H1, S-H2, …, S-Hn denote the official local, regional and national stakeholders of the network.

A cooperation/formation agreement determines the relationships between the network members. This agreement will be realized through the establishment of a legal entity between the formal stakeholders. The role of the administration board is to promote the implementation of the project results for the development of Silk Road tourism products. The main actions to be initiated by the local Stakeholders Network over the Greek eligible regions, namely the Region of Central Macedonia (RCM) and the Region of Eastern Macedonia and Thrace (REMTh), include:

1. Integration of stakeholders in Silk Road tourism development;
2. Promotion of the recording of buildings and intangible assets related to the Silk Road;
3. Recognition by local stakeholders of thematic cultural and natural heritage routes;
4. Experience improvement in cultural heritage and tourism development exploitation;
5. Planning of alternative tourism products and experiences development, as well as coordination with Silk Road tourism and vice versa;
6. RCM and REMTh positioning in the Silk Road tourism map;
7. SICNET Entrepreneurial Network governance;

[2] https://silc-tourismobservatory.econ.auth.gr.

8. Development of a Silk Road tourism cluster to include agri-food, creative and tourism firms;
9. Use of ICT in Silk Road tourism;
10. Share of information and drafting for EU structural funds funding proposals about Silk Road assets;
11. Promotion of higher education and research institutes network;
12. Promotion for a potential Silk Road tourism railway route connecting Silk Road sites throughout RCM and REMTh;
13. Promotion for intercontinental direct flights to Thessaloniki following the completion of a newly operated runway;
14. Promotion of Thessaloniki as a homeport for cruise ships in North Aegean and Black Sea areas;
15. Development of a plan to raise awareness of new entrepreneurial opportunities in Silk Road tourism related activities among private sector firms and information on existing firms about the possibility to expand tourism season and destinations to meet the demand for alternative experiences and facilitate spatial sprawl efforts.

The Stakeholders Network at the project level is concerted by a Coordination Committee consisting of the partners' representatives. The role of the Coordination Committee is twofold: First, to serve as a mediator between the local networks. Second, to promote planned synergies and joint actions. The planned initiatives of the Stakeholders Network at the project level for the sustainability of the SILCNET network include:

1. Elaboration of a plan to raise Regional Stakeholders' awareness of network benefits;
2. Enhancement of business collaboration between firms across the project area;
3. Elaboration of research linking the Silk Road heritage of the project partners;
4. Elaboration of a study and a report on the establishment of an intra- and inter-regional network of Silk Road-related museums and cultural assets;
5. Creation of new knowledge and development of Silk Road Tourism Intelligence on both tourism demand and supply for destinations involved to complement UNWTO efforts in this area;
6. Organization of a network between educational and research institutes in the project area;
7. Coordination of Technology Transfer Offices to prepare programs for new Silk Road Tourism entrepreneurial opportunities;
8. Preparation of a proposal plan in collaboration with Thessaloniki Port Authorities and the local tourism industry to promote Thessaloniki as a homeport for cruises related to the Silk Road in the North Aegean and Black Seas area;
9. Establishment of a network between pre-incubators in the project area;
10. Preparation of joint proposals for funding for the restoration of Silk Road-related cultural and historic landmarks.

4.3 The SILCNET Entrepreneurial Network

The SILCNET Entrepreneurial Network aims to provide long-run directions for building and maintaining synergies for mutual economic benefit, sharing Silk Road diversified brand aspects, exchanging expertise and experience, and promoting cultural tourism elements. The strategic partnerships developed will create a vibrant environment for entrepreneurship that will lead to growing core and complementary industries. The Silk Road entrepreneurial network in the project regions may collaborate with other local and interregional stakeholders to spread the Silk Road entrepreneurial vision. The establishment of partnerships among tourism stakeholders of the regions involved will increase the intra-regional value-added exchange contributing to regional income and well-being improvement.

Figure 7 illustrates the structure of the Interregional Entrepreneurial SILCNET Network, which consists of the five Local Entrepreneurial SILCNET Networks. The network consists of enterprises and organizations involved in the cultural tourism sector willing to participate in the project's Silk Road initiative. The yellow bubbles represent the local Entrepreneurial Network, and the green bubbles represent the individual firms and agents. The following agents shall participate in the Entrepreneurial SILCNET Networks:

i. Companies directly related to tourism: hotels, tourism agencies, transport companies, recreation companies, restaurants and bars.
ii. Companies in the agrifood sector: primary sector, food manufacturing firms.
iii. Cultural agencies: creative industries, festival organizers.
iv. Cultural organziations: museums, galleries.

The cooperation/formation agreement specifies the relationships between the network members. The network members will have the opportunity to receive the SILCNET Label, prepared within the scope of the project, if they meet the requirements. Participation in the local Entrepreneurial Network will have no charges. Each Any company could join the network after attending a webinar about the Silk Road and the scope of the entrepreneurial network.

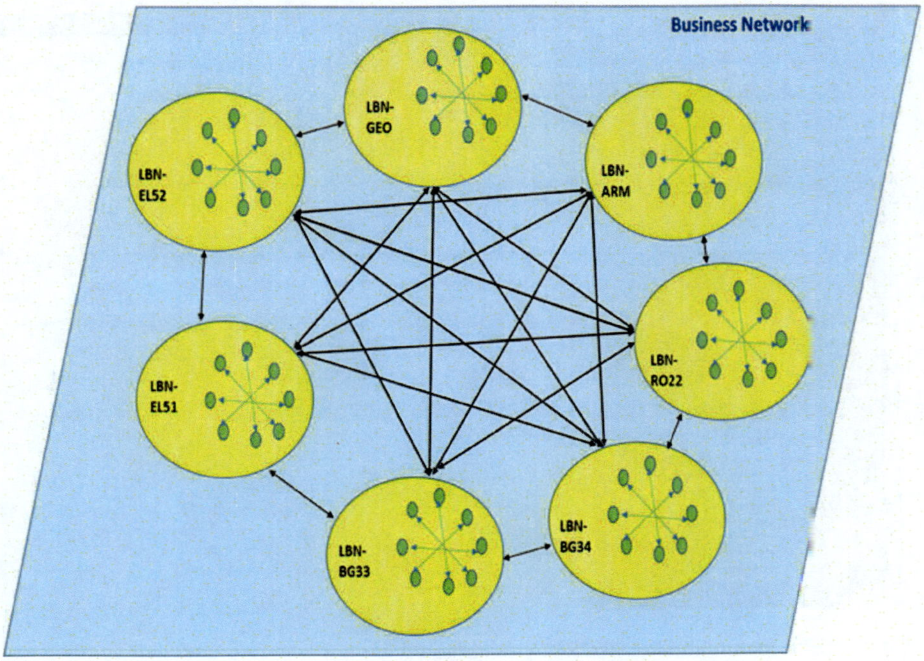

Fig. 7 SILCNET entrepreneurial network

The SILCNET entrepreneurs will have access to the following digital tools[3] included in the SILC project outputs (Fig. 8):

Profile Setting: Companies, Upon invitation and after acceptance of their application for network membership by the local SILC coordinator, companies should set up the entrepreneurial/institutional profile by outlining their activities and interests.

Search for other tourism or culture stakeholders: Companies could browse the online stakeholders' network using advanced search functions to find partners (academics, companies) to pursue joint activities. This function will allow to contact these partners via the platform for further cooperation.

Open calls for cooperation: Enterprises/institutions wishing to carry out joint activities will be able to launch a call for cooperation specifying: (a) the topic, (b) the research challenges, (c) existing business partners and staff, (d) the expertise required, (e) the annotated business fields, and (f) the funding opportunities and the period of implementation.

Consulting calls for cooperation: Companies could browse for cooperation calls if they overlap with their own needs, and may also be experts in specific fields.

The transnational entrepreneurial network will create new opportunities in the destinations of the regions examined and offer solutions to common challenges while promoting cross-border mutual understanding. These opportunities could contribute to sustainable and profitable tourism. Besides, they will strengthen and diversify the region's tourism products through the joint promotion of business and entrepreneurship in the tourism and culture sectors. The expected benefits for members are:

i. **Turnover increments** because of
   - increase in local sales;
   - access to new markets;
   - exclusive collaboration opportunities.
ii. **Cost reduction** because of the implementation of collective promotion programs;
iii. **Promotion** of innovation progams aiming at improving business competitiveness.
   - joint promotion of the Silk Road tourism product in the distribution channels;
   - joint negotiation of participation in tourism shows, festivals and fairs;
   - joint presentations at tourism fairs;
   - participation in joint missions;
   - design, creation and distribution of common promotional material;
   - application of common quality control;
   - increase in employment;
   - application of a common e-commerce platform;
   - training in quality and compliance with quality standards matters;
   - certification and use of the SILC label.

[3] https://silc-tourismobservatory.econ.auth.gr.

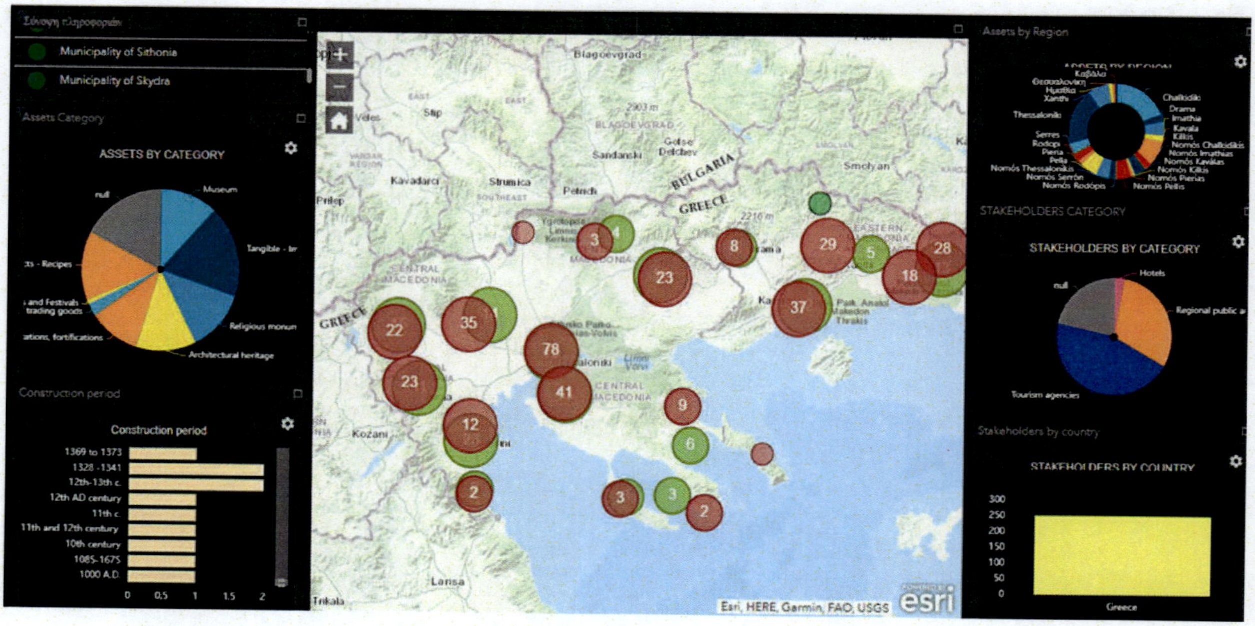

Fig. 8 Digital tools facilitating SILCNET communication

As already mentioned, participation in the network will be flexible. The participating firms in relevant sectors may collaborate with other regional firms either vertically or horizontally. In addition, the interregional network allows local firms to expand their market and reach partner markets. The organization of B2B meetings set by the Coordinating Committee could facilitate this process. Finally, synergies and activities can be vertical when firms participate in clusters.

5 Conclusions and Suggestions

This paper concludes that in a highly competitive global tourism market, defining the concept of Silk Road tourism network is challenging and requires a strategic marketing approach. The historic Silk Road, a network of ancient trans-continental trade routes that have been a major factor of economic, geopolitical, and historical-cultural progress over East and West for several centuries, has marked the history of travel and trade. Nowadays, the Silk Road offers visitors the opportunity to experience a unique network of emerging destinations with diverse cultural heritage and several tourist attractions spread across the ancient routes. This requires the effective collaboration of a chain of tourism resources, that form a network of destinations across several countries. The transnational network of destinations could open up new tourism markets and economic opportunities within Silk Road areas. Incorporating tourism activities and resources into an integrated route can further stimulate cooperation and partnership between destinations (Meyer 2004; Manhas et al. 2014; Kostopoulou 2019).

The Silk Road exhibits positive prospects for joint marketing and promotional initiatives as a unique network of places united by a similar cultural heritage. However, the success of Silk Road tourism requires cooperation, coordination, expertise, and management of diverse destinations, as well as the participation and support from stakeholders. We argue that sustainable tourism development involves both personal and contextual factors, and therefore, the perception of sustainable practices in one region may differ from another. Recognizing the key stakeholders and understanding how each group of stakeholders pursues its interests in sustainable tourism development is unavoidable (Dabphet 2013). The Silk Road has the potential to become one of the most dynamic global tourism brands, but only if local tourism and culture entrepreneurs work together. Currently, some regions are using the Silk Road to market their destinations, but their efforts will not reach their full potential without coordinated communication strategies. Alliances formed through collaborative procedures could offer additional competitive benefits alongside individual national and regional initiatives.

SILC-Silk Road Local Culture project provides a platform for Silk Road tourism networking throughout the Black Sea eligible areas. Several Silk Road sites in the area offer a diverse range of opportunities for the development of various forms of cultural heritage tourism. Yet, the area has not been adequately branded and marketed as a Silk Road cultural heritage tourism destination. SILC project explores new tourism opportunities for the project's area destinations and local tourism stakeholders interested in cross-border collaborations and developing novel products based on the Silk Road legacy. SILC project implementation is expected to stimulate the Silk Road economic impact in the Black Sea Basin. The SILC project offers the opening of new tourism market segments for the project partners, to promote a diversified tourism brand that corresponds to the local Silk Road cultural footprint. SILCNET members could develop and implement sustainable and consistent marketing strategies required to enhance the brand image of the Silk Road, improve the efficiency of marketing and promotion efforts, and spread the Silk Road entrepreneurial spirit in the Black Sea regions.

SILC project encourages the creation of a collaborative entrepreneurial dialogue and common methodologies to raise awareness of Silk Road tourism products. Interregional businesses will join forces to establish a strong Silk Road brand by identifying and promoting the unique Silk Road entrepreneurial aspect. The jointly planned collaborative marketing approach is expected to promote a diversified tourism offering suited to the local Silk Road cultural footprint, building on the growing awareness of the Silk Road as a driver for interregional development. The networks of partners, supporters, and key stakeholders will facilitate the dissemination of project outputs outside the collaboration region and scope by bringing together research institutions, cultural organizations, private/public authorities, and SMEs. Local communities should support the design and implementation of tourism development projects, and national and regional authorities should facilitate the process. Future projects should be carried out in other countries to expand the project's goals, principles, and efforts to achieve desired results.

The successful design and implementation of Silk Road Cultural Heritage Tourism Networks requires broad collaboration between academia, the tourism industry, local authorities, and communities, extending beyond conferences and workshops to include transnational tourism development planning initiatives. Silk Road local cultures over the Black Sea region are parts of the *genius loci* of the region and can be used as anchor points for the development of local cultural and entrepreneurial tourism activities by establishing culture and tourism networks. SILCNET, thus, provides to small and medium enterprises and young entrepreneurs the opportunity to innovate, grow, and expand their existing activities, while also launching new collaborations, thus serving as an incubator for new entrepreneurial opportunities.

Acknowledgements This research is cofunded by Joint Operational Programme Black Sea Basin 2014–2020 Subsidy Contract No. 99372/31-08-2018 in the context of the Project BSB570 SILC—Silk Road Local Culture. Joint Operational Programme Black Sea Basin 2014–2020 is co-financed by the European Union through the European Neighborhood Instrument and by the participating countries: Armenia, Bulgaria, Georgia, Greece, Republic of Moldova, Romania, Turkey and Ukraine. This publication has been produced with the financial assistance of the European Union. The contents of this publication are the sole responsibility of the authors and can in no way be taken to reflect the views of the European Union.

References

Albrecht, J.N.: Networking for sustainable tourism—towards a research agenda. J. Sustain. Tour. **21**(5), 639–657 (2013). https://doi.org/10.1080/09669582.2012.721788

Angelidis, G., Ioannidis, E., Makris, G., Antoniou, I., Varsakelis, N.: Competitive conditions in global value chain networks: An assessment using entropy and network analysis. Entropy **22**(10), 1068 (2020). https://doi.org/10.3390/e22101068

Anuar, A.N.A., Ahmad, H., Jusoh, H., Hussain, M.Y.: Understanding the role of stakeholder in the formation of tourist friendly destination concept. J. Manag. Sustain. **2**(2), 69–74 (2012)

Boronyak, L., Sally Asker, S., Naomi Carrard, K., Michael Paddon, M.: Effective community based tourism: A best practice manual for Peru. APEC Tourism Working Group August 2010 Sustainable Tourism Cooperative Research Centre 2010, Gold Coast Campus, Griffith University QLD 4222 Australia APEC Secretariat 2010, APEC Publication number: APEC#210-TC-03.4 ISBNs: 9781921 785139 (2010)

BPP Learning Media: Business Essentials: Business Strategy, 2nd edn. BPP Learning Media Ltd., London (2010)

Bramwell, B., Lane, B. (Eds.): Tourism Collaboration and Partnerships: Politics, Practice and Sustainability (Aspects of Tourism No. 2). Channel View Publications (2000)

Braun, P.: Creating value to tourism products through tourism networks and clusters: Uncovering destination value chains. In: Ghose, A. (ed.) In Business Networking — Trends and Cases, pp. 75–96. ICFAI University Press, Hyderabad, India (2007).

Cawley, M., Marsat, J.-B., Gillmor, D.A.: Promoting integrated rural tourism: Comparative perspectives on institutional networking in France and Ireland. Tourism Geogr. an Int. J. Tourism Space, Place Environ. **9**(4), 405–420 (2007). https://doi.org/10.1080/14616680701647626

Chin, T.: The invention of the Silk Road, 1877. Critical Inquiry, vol. 40, No. 1 (Autumn 2013), pp. 194–219. The University of Chicago Press (2013). https://doi.org/10.1086/673232 https://www.jstor.org/stable/10.1086/673232

Cooper, C., Fletcher, J., Gilbert, D., Fyall, A., Wanhill, S.: Tourism Principles and Practice, 3rd edn. Pearson Education Limited, Harlow (2005)

Dabphet, S.: The key stakeholders in the implementation of sustainable tourism development in two rural towns of Thailand. Int. J. Bus. Tourism Appl. Sci. (2013). Assessed 20 Aug 2021 http://www.ijbts-journal.com/images/main_1366796758/0029-Siripen.pdf

Dill, W.R.: Public participation in corporate planning—Strategic management in a Kibitzer's World. Long Range Plan. **8**(1), 57–63 (1975)

Ferri, M.A., Aiello, L.: Tourism destination management in sustainability development perspective, the role of entrepreneurship and networking ability: Tourist kit. World Review of Entrepreneurship, Management and Sustainable Development, vol. 13, No. 5–6, pp. 647–664 (2017). Assessed 28 Aug 2021 at https://www.inderscienceonline.com/doi/abs/10.1504/WREMSD.2017.086334

Freeman, R.E.: Strategic management: A stakeholder approach. Pitman, Boston (1984)

Fuglsang, L., Eide, D.: The experience turn as 'Bandwagon': Understanding network formation and innovation as practice. Eur. Urban Reg. Stud. **20**(4), 417–434 (2013)

Galant, A., Periša, A., Slivar, I.: Aligning wishes and environment sustainability: A focus on higher tourism consumption. In: International Conference on Hospitality, Leisure, Sports, and Tourism (HLST 2016). Thailand Bangkok, pp. 141–159 (19–21 Jan) (2016).

Garcia-Rosell, J.C.: A multi-stakeholder perspective on sustainable tourism management research and education. Polit. Policy Gov. Sustain. Tourism **14**, 122–137 (2014)

Gergen, K.J.: An invitation to social construction, 2nd edn. Sage, London (2009)

Getz, D., Timur, S.: Stakeholder involvement in sustainable tourism: Balancing the voices. In: Theobal, W.F. (ed.) Global Tourism, 3rd edn, pp. 230–247. Butterworth-Heineman, Maryland Heights (2005)

Go, M.F., Pine, R.: Globalization Strategy in the Hotel Industry. Routledge, London (1995)

Grabher, G., Powell, W.W.: Networks I & II. Critical Studies in Economic Institutions Series. Edward Elgar, Cheltenham. ISBN: 978 1 84376 035 1 E (2005)

Greffe, X.: Is rural tourism a lever for economic and social development? J. Sustain. Tour. **2**, 23–40 (1994)

Hagedoorn, J., Link, A., Vonortas, N.: Research partnerships. Res. Policy **29**, 567–586 (2000)

Hall, C.M.: Tourism: Rethinking the social science of mobility. Prentice-Hall, Harlow (2005)

Hardy, A.L., Beeton, R.J.S.: Sustainable tourism or maintainable tourism: Managing resources for more than average outcomes. J. Sustain. Tour. **9**(3), 168–192 (2001)

Holgado, D.: Analyzing social networks. Revista Hispana Para El Análisis De Redes Sociales **27**(2), 141–145 (2016)

Johnson, G., Scholes, K., Whittington, R.: Exploring Corporate Strategy: Text and Cases, 7th edn. Prentice Hall, UK (2006)

Kostopoulou, S.: "Silk Road Cultural Heritage Tourism Network". In: Proceedings 5th International Conference of International Association for Silk Road Studies IASS SUN 'Silk Road: Connecting Cultures, Languages, and Ideas', pp. 232–260 (26–28 Sept 2019), Moscow, Russia (2019)

Kostopoulou, S.: Past, present and future of Silk Road tourism. SILKRODIA, the Silk-Road Universities Network's Web Magazine **2**(2), 140–156 (2020)

Kostopoulou, S., Toufengopoulou, A., Kyriakou, D., Malisiova, S., Sofianou, E., Xanthopoulou–Tsitsoni, V.: The Western Silk Road in Greece. Aristotle University of Thessaloniki, Silk Road Programme 2016, Western Silk Road Tourism Initiative, A UNWTO-EU Initiative (2016). Available at https://webunwto.s3.eu-west-1.amazonaws.com/2019-09/aristotleuniversity-nationalswotanalysis-westernsilkroadgreece.pdf

Krce Miočić, B., Razović, M., Klarin, T.: Management of sustainable tourism destination through stakeholder cooperation. Manag. J. Contemp. Manag. Issues **21**(2), 99–120 (2016)

Leiper, N.: Tourism management. RMIT Press, Melbourne (1995)

Macbeth, J., Burns, G.L., Chandler, L., Revitt, M., Veitch, S.: Community as tourism object: Associated disciplinary understandings. In: 12th International Research Conference of the Council for Australian University Tourism and Hospitality Education (CAUTHE), (6–9 Feb) Fremantle, Western Australia (2002)

Maiden, J.A.: Participation in sustainable tourism development: Stakeholders and partnership working. Ph.D. Thesis, Cardiff University (2008). Assessed 30 May 2020 at https://orca.cf.ac.uk/55742/1/U584263.pdf

Malerba, F., Vonortas, N.S.: Innovation networks in industries. Edward Elgar, Cheltenham, UK (2009)

Manhas, P.S., Kour, P., Bhagata, A.: Silk route in the light of circuit tourism: An avenue of tourism internationalization. Procedia—Soc. Behav. Sci. **144**, 143–150 (2014). https://doi.org/10.1016/j.sbspro.2014.07.283

Meyer, D.: Tourism Routes and Getaways: Key Issues for the Development of Tourism Routes and Gateways and Their Potential for Pro-Poor Tourism. Overseas Development Institute, London. Background paper, output of the ODI project: Pro-poor Tourism Pilots in Southern Africa (2004). Retrieved 20 June 2019 at https://www.odi.org/sites/odi.org.uk/files/odi-assets/publications-opinion-files/4040.pdf

Middleton, V.T.C., Clarke, J.: Marketing in travel and tourism, 3rd edn., p. 2001. Butterworth-Heinemann, Oxford (2001)

Newman, M.E.J.: The structure and function of complex networks. SIAM Rev. **45**(2), 167–256 (2003)

Nordin, S.: Tourism Clustering and Innovation. European Tourism Research Institute, MidSweden University, Sweden (2003)

Novelli, M., Schmitz, B., Spencer, T.: Networks, clusters and innovation in tourism: A UK experience. Tour. Manage. **27**(6), 1141–1152 (2006). https://doi.org/10.1016/j.tourman.2005.11.011

Omerzel, D.G., Kregar, T.B.: Networking in the tourism industry: The case of three neighboring Mediterranean tourism destinations. OUR SEA: Int. J. Maritime Sci. Technol. **63**(2), 66–72 (2016). https://hrcak.srce.hr/158193

Orsenigo, L., Pammolli, F., Riccaboni, M.: Technological change and network dynamics: Lessons from the pharmaceutical industry. Res. Policy **30**(3), 485–508 (2001)

Porter, M.E.: Competitive advantage of nations. Free Press, New York (1990)

Rahman, M.: Stakeholders in Tourism Development (2017). Assessed 14 Aug 2018 at https://www.howandwhat.net/stakeholders-tourism-development/

Robson, J., Robson, I.: From shareholders to stakeholders: Critical issues for tourism marketers. Tour. Manage. **17**(7), 533–540 (1996)

Rogerson, C.M.: Tourism routes as vehicles for local economic development in South Africa: The example of the Magaliesberg Meander. Urban Forum **18**(2), 49–68 (Springer Netherlands) (2007)

Scott, N., Parfitt, N., Laws, E.: Destination management: Co-operative marketing, a case study of the Port Douglas brand. In: Faulkner, B., Moscardo, G., Laws, E. (eds.) Tourism in the twenty first century: Lessons from experience, pp. 198–221. Continuum, London (2000)

Scott, N., Cooper, C., Baggio, R.: Network Analysis and Tourism: From Theory to Practice. Channel View, Clevedon, UK (2008)

Slivar, I.: Stakeholders in a tourist destination—Matrix of possible relationships towards sustainability. Open J. Res. Econ. **1**(1), 1–10 (2018)

Swarbrooke, J.: Sustainable Tourism Management, 2nd edn. CAB International, London (2001)

Tayler, P., McRae-Williams, P.: Internal versus External Economies of Scale. Working Paper, Centre for Regional Innovation & Competitiveness (2005)

Turker, N., Alaeddinoglu, F., Can, A.S.: The Role of Stakeholders in Sustainable Tourism Development in Safranbolu, Turkey. HLST-Summer 12–14 July, 2016, Kyoto, Japan, ISSN 2309-3757 (2016)

UNESCAP. (2008). Cultural Tourism Sites Management, a Training Manual for Trainers in the Greater Mekong Subregion. https://www.unescap.org/publications. Accessed 05 Sept 2018

UNWTO: Report on the WTO Survey on the Implementation of the Global Code of Ethics for Tourism, Note by the Secretary-General. World Tourism Organization General Assembly, Dakar, Senegal, 25 Nov–2 Dec 2005, Sixteenth session, Provisional agenda item 20, Activities of the World Committee on Tourism Ethics, Addendum 1, A/16/20 Add.1, Madrid, July (2005). https://www.e-unwto.org/doi/pdf/10.18111/unwtogad.2005.1.m32874412w500336

Van der Zee, E., Vanneste, D.: Tourism networks unravelled; a review of the literature on networks in tourism management studies. Tourism Manag. Perspect. **15**, 46–56 (2015)

Vincent, D.R.: The Information-Based Corporation-Stakeholders Economic and the Technology Investment. Dow Jones-Irwin, Ilinois (1990)

Vonortas, N., Zirulia, L.: Business network literature review and building of conceptual models of networks and Knowledge Intensive Entrepreneurship. Working Paper for the project 'Advancing Knowledge-Intensive Entrepreneurship and Innovation for Economic Growth and Social Well-being in Europe', European Commission, DG Research and Innovation (2011)

World Tourism Organization (UNWTO): Research on the Potential of the Western Silk Road Highlights. WP3. Western Silk Road Tourism Development. Grant Agreement, Enhancing the Understanding of European Tourism. Co-funded by the COSME programme of the European Union, June 2017 (2017)

Zach, F., Racherla, P.: Assessing the value of collaborations in tourism networks: A case study of Elkhart County, Indiana. J. Travel Tour. Mark. **28**(1), 97–110 (2011). https://doi.org/10.1080/10548408.2011.535446

Żemła, M.: Tourism destination: The networking approach. Moravian Geogr. Rep. **24**(4), 2–14 (2016)

Zeng, D.Z. (ed.): Knowledge, Technology, and Cluster-Based Growth. World Bank Institute Learning Resources Series, Washington (2008)

The City of Ohrid as a UNESCO World Cultural Heritage Site: Challenges and Solutions for Cultural Tourism Development

Sasho Korunovski and Michael Risteski

Abstract

The City of Ohrid represents an important site with UNESCO status of mixed heritage located on the historic route of Via Egnatia. It is the most important tourist center in the Republic of North Macedonia and therefore faces numerous development challenges. Namely, urban and tourism development has caused several risks and dangers to its protection status in the past which is also prevalent to the present day. Since its proclamation as a World Cultural Heritage Site the city of Ohrid and its surroundings had constantly been trying to find a balance between the increase in tourist visits on one side, and the protection of its cultural and natural values, on the other. This paper will explore the developmental path of this area with its different socio-economic and environmental challenges and impacts in its pursuit to achieve long-term sustainability. In that sense, several strategic documents will be analyzed in which the most recent protection measures and activities have been envisioned to be implemented in the near future. This will also include sustainable solutions and recommendations for improvements mainly in a tourist-developmental and preservation sense. This is important for the city to transform itself into a sustainable, yet highly competitive global tourist destination.

Keywords

Ohrid • Protection • Cultural heritage • Tourism development

S. Korunovski (✉) · M. Risteski
University St. Kliment Ohridski, Bitola, Republic of North Macedonia
e-mail: saso.korunovski@uklo.edu.mk

M. Risteski
e-mail: michael.risteski@uklo.edu.mk

1 Introduction

World Heritage Sites (WHS) represent an important potential for tourism development of many regions/countries in the world. The designation "UNESCO World Heritage Site" is an important promotional tool for the development of tourism. Many destinations have built their tourist brand around these designations. Such a site is the city of Ohrid in the Republic of North Macedonia. The city with its surroundings has been proclaimed a World Cultural and Natural Site since 1979 and 1980 (Mixed Heritage Status). The Convention concerning the Protection of the World's Cultural and Natural Heritage was adopted by the United Nations Educational, Scientific and Cultural Organization (UNESCO) in 1976 with the purpose of "identification, protection, conservation, presentation and transmission to future generations of cultural and natural heritage of outstanding universal value" (Leask and Fyall 2006). Sites are being designated through a set of determined criteria which must be followed. The designation process is being conducted through the system of inscription of the site in the World Heritage List (WHL). This includes adopting a set of operational guidelines and periodic reporting processes. The state of conservation of World Heritage properties is being conducted by the World Heritage Committee through monitoring processes which can be either periodic or reactive (UNESCO World Heritage Centre 2012). The Ohrid region has been inscribed in the World Heritage List since 1979 (Decision No. 03COM XII.46), for its natural values according to the selection criterion (viii): "to be outstanding examples representing major stages of earth's history, including the record of life, significant on-going geological processes in the development of landforms, or significant geomorphic or physiographic features" (UNESCO World Heritage Committee 1979). This inscription has been widened in the following year on its cultural values (Decision No. 04COM V.13), according to the following selection criteria: (i): "to represent a masterpiece of human creative

S. Kostopoulou et al. (eds.), *Cities' Vocabularies and the Sustainable Development of the Silkroads*, Advances in Science, Technology & Innovation, https://doi.org/10.1007/978-3-031-31027-0_28

genius"; (ii): "to bear a unique or at least exceptional testimony to a cultural tradition or to a civilization which is living or which has disappeared"; (iii): "to be an outstanding example of a type of building, architectural or technological ensemble or landscape which illustrates (a) significant stage (s) in human history" (UNESCO World Heritage Committee 1980). Hence, the Ohrid region has been inscribed as a WHS with mixed heritage status (cultural and natural heritage). A total of 39 sites globally have been proclaimed as properties with mixed heritage status during the time of writing this paper (UNESCO World Heritage Convention 2021).

In the past two decades the municipality of Ohrid showed the highest increase in tourism turnover on the national level. Almost all World Heritage Sites face the problem of balancing tourism and conservation/protection of its values. This partially stems from the large number of stakeholders involved, often with opposing and conflicting interests and agendas (McKercher et al. 2005). In this sense, the aim of the paper is to determine appropriate development directions of tourism in the municipality which will incorporate sustainable development of its cultural values. The situation in historic urban centers such as the Ohrid Old city core is especially problematic. Tourist activity in such spaces can be considered as sustainable if it does not involve a deterioration of the physical, socioeconomic and symbolic characteristics as a whole (García-Hernández et al. 2017). In this sense, a detailed situation analysis has been conducted of local, regional and national development strategies and management plans in the Ohrid Region in order to determine the past and current interactions between tourism development and cultural heritage protection. Furthermore, the current state of protection of the Ohrid region must also be included and examined. From such an analysis different gaps and weaknesses in tourism development and protection of cultural heritage have come to light that might endanger the cultural values of the property listed as a WHS. In this direction, the challenges that the municipality of Ohrid currently faces in terms of tourism development and cultural heritage protection are identified in separate proposed tourism zones. This analysis is complemented by a tourism valorization process of the most important cultural complexes and attractions located in the municipality of Ohrid that represent an important development pillar for tourism in the region. The valorization process includes identifying value parameters, analysis of their uniqueness as well as development of benefits/assumptions that have the function of solutions to the identified challenges. The valorization process used in the paper derives from previous research activities conducted as part of strategies and studies for development in the Ohrid region where the authors were directly involved. A detailed inventory of cultural heritage resources is provided within the *Study on tourist potentials in the Southwest Planning Region* (Marinoski et al. 2013). These resources were systematized in groups and sub-groups along with information provided about their characteristics based on cultural value elements (Throsby 2006). Furthermore, the findings from a conducted study for creating a *Register of potentials for the development of rural tourism in the Southwest Planning Region* were also taken into account, having in mind that the identified and valorized cultural values could be used as complementary categories within rural tourism activities (Jankuloski et al. 2016). Findings were also used from the *Sub-strategy for development of cultural tourism in the Republic of N. Macedonia* for the purposes of valorization of cultural values, with special emphasis on the Ohrid region as a World Heritage Site (Korunovski et al. 2019). An important outcome of tourism valorization is the prioritizing of values, which is essential for creating tourism development directions for policy makers (Risteski 2020). In this sense, the valorization of cultural values within this paper aims to the government's efforts for maintaining the Outstanding Universal Value (OUV) of the WHS status of the Ohrid region. This will allow appropriate improvement measures and activities to be undertaken in the domains of cultural heritage protection and sustainable tourism development.

2 Theoretical Concept

Tourism has been developing globally on an unprecedented scale in the pre-COVID-19 era. This stems from the fact that tourist demand is constantly expanding and is directed towards new experiences. Culture and tourism have always formed an interconnected and interdependent unity. Nowadays, they coexist in a distinctive type of tourism activity. Cultural tourism can be considered as being a growing type of tourism which accounts for approximately 40% of the total number of tourist trips worldwide (Brida et al. 2013). It can be envisaged that cultural tourism will continue to grow and develop in the future. Cultural tourism has a specific place in most countries' tourism policy as well as in their marketing and promotional planning. The most important priorities in policy agendas for most countries that are developing cultural tourism are: understanding tourist behavior; understanding cultural identity and diversity; diversification of activities; and balancing protection and promotion (UNWTO 2018). Policy makers in destinations must constantly develop new ways of creating attractive tourism activities and products in order to be competitive on the tourism market. For this to be achieved, it is necessary to combine cultural and natural values into innovative tourism products. It is not enough to offer solely cultural values. In this sense, the Ohrid region has excellent preconditions for creating competitive tourism products and activities (Risteski et al. 2020). World Heritage Sites usually combine a larger number of cultural and natural values on a relatively small area (Korunovski 2011). For areas, or properties to be listed as World Heritage Sites, they must fulfill certain conditions, such as "integrity", "authenticity" and "outstanding universal value". These conditions at the time of inscription must be maintained and enhanced in the future (UNESCO World Heritage Centre 2008). With the listing of a WHS, the state is obligated to take measures for safeguarding the site (Alberts and Hazen 2010). This can often be a challenge for the state bodies when wanting to enhance tourism on the site itself, given the fact that many stakeholders are present in the protected area or are indirectly connected to it. The involved parties belong to a wide range of domains and often have contradictory interests and agendas (Risteski et al. 2012). It is required to implement strict sustainable planning and management steps, provided in the UNESCO World Heritage Sustainable Tourism Toolkits (UNESCO World Heritage Committee 2012). Designating a property as a WHS represents a type of tourism brand which has positive implications for tourism growth (Fyall and Rakic 2006). Therefore, the benefits are evident by themselves. Tourism plays an important role in the social, cultural and economic development in the Ohrid region. In this sense, tourism generates a large portion of the revenue in the municipality of Ohrid. In 2019 this revenue amounted to 33.8% nationally—the highest in the country. Furthermore, tourism leads to numerous social benefits for the locals. For example, a study on the locals' perception about tourism impact on the quality of life shows mainly positive social benefits. 78.4% of the locals regard tourism to play an important role in widening the possibilities for shopping, eating in restaurants and amusement, 85.6% think that they have some use of tourism development, while 46.6% answered that tourism has improved the quality of public services (Andreeski et al. 2020). However, during the peak of the tourist season (June, July and August) there are several negative social impacts, such as traffic congestion, crowding, noise pollution, increase in generated solid waste and increased water use. The increase in tourism visits shows the most negative impacts on the cultural and natural environment—changed physiognomy of the Old city core with inadequate billboards; installation of solar panels and satellite dishes; renovation of old houses with inappropriate materials; inappropriate conversion of old buildings for commercial use; crowded cultural sites; increased pollution; illegal buildings on the shores of Lake Ohrid; overfishing and illegal fishing due to a higher demand for fish specialties; overflowing of wastewater from the wastewater collection system around the Ohrid Lake due to the system's insufficient capacity. Hence, this region must take into account tourism development on one side, and the protection and sustainable development of cultural (and natural) values on the other. Tourism destinations can be very attractive due to their culture, which can lead to transforming conventional tourism into creative tourism, providing more authentic experiences. On the other hand, tourism (if poorly managed) can cause negative impacts on culture and cultural heritage and possibly threaten the long term sustainability of both sectors: tourism and culture (UNWTO 2018). The concept of sustainability in both cultural heritage and tourism can be identified as a synthesis of multiple dimensions: economic, physical, social, cultural and environmental (Loulanski and Loulanski 2011). Protection of vulnerable cultural heritage sites is accompanied by different visitor management strategies. Visitor management in cultural heritage sites should be based on the principles of sustainability and commercial viability. Conservation and protection activities should be primary goals but commercial services should not be ceased/decreased because they offer economic benefits and employment in the area (World Tourism Organization 2007). The revenues created by the development of cultural tourism can contribute in subsidizing the processes of conservation and protection of cultural heritage in the World Heritage Sites (Smith 2003). Within the municipality of Ohrid the generated revenues by foreign tourists represented 60.4% of the total amount (domestic and foreign tourists) in 2017 (Trajkov et al. 2019). However, the created revenues in the municipality of Ohrid amount to only about 1.4% of the country's GDP. Tourism is still very low on the government's list of priorities, which partially stems from the fact that a separate tourism ministry currently does not exist. The newly adopted Law for managing the world heritage of the Ohrid region, the Management Plan for the World Natural and Cultural Heritage in the Ohrid Region and the Strategy for tourism development of the municipality of Ohrid, shows that the government (local, regional and national) is clearly shifting its priorities towards protection of the Ohrid region WHS and sustainable tourism development.

Many properties listed in the World Heritage List face multiple pressures of different nature. Most of the sites encounter problems that are related to the implementation of management systems and plans (Ringbeck 2018). According to a research study by the World Heritage Committee in 2016, the most reported factors that pressured the WHL properties in 2016 (as well as the listed properties in danger) are the inappropriately implemented management systems/plans (UNESCO World Heritage Committee 2016). The percentage of the reported properties affected by this factor accounted for 72.4% of all the listed properties (natural and cultural alike) worldwide. More precisely, the top 3 factors that had a negative impact on the listed cultural properties were management systems/plans (78.1% of the reported properties affected), housing (43.7%) and management activities which accounted for 24% of the affected properties. The logical conclusion is that the World Cultural Heritage of the Ohrid region faces similar pressures on its listed property. Namely, a Reactive Monitoring Mission was conducted in 2017 in order to evaluate the state of protection and the pressures/threads on the Outstanding Universal Value of the listed property of the Ohrid region. There were some significant shortcomings identified (Decision No. 43COM 7B.36), related to some large infrastructural projects (constructing a ski-center and A3 road, building a railroad) that have since been abandoned (UNESCO World Heritage Committee 2019). However, there were also findings related to different management plans and activities that have not yet been completely implemented in the efforts to alleviate the complex situation in the Ohrid region. This could result in the Ohrid region being listed in the World Heritage List in Danger. Therefore, it is needed to examine the past and current strategic documents and management plans and activities that have been implemented in the Ohrid region in order to identify the factors that have negative impacts on its WHS status. Important strategic documents in this sense are the newly adopted Management Plan for the World Natural and Cultural Heritage in the Ohrid Region 2020–2029 and consequently the Strategy for Tourism Development in the Municipality of Ohrid 2020–2025 (Andreeski et al. 2020).

3 Situation Analysis

From a tourism developmental point of view, the Ohrid area is located in the Southwest Planning Region (SWPR) of the country. This region is the most developed tourist region in the Republic of North Macedonia which comprises a total of 9 municipalities (Sekulovska et al. 2018). The state of tourism development in the SWPR consists of both the tourism infrastructure and the realized tourism turnover (Marinoski and Risteski 2019). The municipality of Ohrid accounts for 72.7% of the realized tourist visits in the SWPR and 68.3% of the overnight stays in 2019 (Andreeski et al. 2020). Furthermore, accommodation facilities in the municipality represent 70.2% of the region's total available capacities.

The situation analysis explores the main planning and strategic documents that have been implemented in the municipality of Ohrid, regarding tourism development and tourism's impact on the cultural values. This is being done in order to determine existing gaps and insufficiencies in the implementation of the created strategies/plans. The focus of this analysis is primarily put on those activities and goals that have partially been realized or have not been realized at all.

Firstly, the national strategies for tourism development (2009–2013 and 2016–2021) must be examined. This analysis includes activities that are related to the advancement of cultural values and their inclusion in tourism products/activities.

In the *National Strategy for Tourism Development 2009–2013* the following **partially implemented** activities are important to be mentioned: activities for developing products and their marketing related to accommodation in monasteries, developing centers for learning traditional crafts and selling craft-related souvenirs (Flaneri 2009). In addition, the following **unrealized activities** could be observed: no signage for landscape routes (including cultural landscapes), no detailed calendar of events has been published, while programs for training of employees in the domain of cultural heritage have not been implemented.

Within the *National Strategy for Tourism 2016–2021* the following activities have **partially been implemented**: creating a Destination Management Organization (DMO) in the region where the municipality of Ohrid would be included, creating national tourism routes (including cultural routes), reorganization reforms within the Agency for Promotion and Support of Tourism and increase of their promotional budget (Kohl and Partner—Consulting Firm 2016). **Unrealized activities** include: a planned tourism development fund for local tourism products has not been created, lack of branding strategy, no continuous surveying of tourists has been conducted by the Agency for Promotion and Support of Tourism, a permanent working group within the Ministry of Culture of the Republic of North Macedonia for coordinating theme-based tourism activities (e.g. developing cultural sites and attractions) has not been created. From the analysis of the implemented/realized activities in the National Strategy for Tourism 2016–2021 it could be concluded that the unrealized/partially realized activities from the previous Strategy (2009–2013) mentioned above have not been taken into account and thus have not been implemented. Furthermore, additional activities could be observed that have also not been realized. Therefore, a significant discontinuity exists between the undertaken activities within the two past strategies for tourism development.

Regarding the National Strategy for Tourism Development that has been revised in 2011, its goals are systematized in priority groups: tourism products and services, accessibility and infrastructure, human resources, environment and cultural heritage, tourism marketing, economic development of tourism, investment policies, organization, tourism awareness (Fournier et al. 2011). These goals were examined in the sense of their inclusion in the local development plans of the municipality of Ohrid and their realization. Several goals could be identified that are related to the development of tourism products, cultural heritage protection and tourism marketing which have not yet been realized or are still in the process of implementation.

The following goals regarding **tourism products development** (related to cultural heritage) have been identified: increasing the number of accommodation capacities (accommodation in monasteries, preparing a register for these capacities and their promotion) has not been realized, developing wine tourism (signage and marketing activities) has not been realized, utilizing culture and handicrafts as tourism products is in the process of realization (calendar of events is being updated, cooperation of tour operators and organizers of events is fairly good and continuous, traditional crafts are already well developed in Ohrid but are insufficiently promoted), improving knowledge and awareness for cultural attractions is in the process of realization (such as training programs and seminars).

The following goal regarding **cultural heritage protection** has been identified: managing visits through better informing of visitors is in the process of realization (places for reception/informing and signage for trails, parking spaces, resting areas). However, it must be mentioned that some crucial goals within the priority of cultural heritage protection have fully been realized, such as: creating a general database for cultural heritage with identified values that need additional protection as well as improving the level of protection of current cultural sites.

Several goals related to **tourism marketing** (linked to cultural heritage) which have not been realized or are in the process of realization could be identified: developing a recognizable tourism brand has not been realized, identification of market segments and targeted consumers is in the process of realization, while improving the networking of tourist information centers has also not been realized.

It can be concluded that the municipality of Ohrid has put significant efforts in achieving several goals set in the National Strategy for Tourism Development. However, most of those goals are in the domain of cultural heritage protection, while there are still several crucial unrealized (or partially realized) goals within tourism product development and tourism marketing. The goals which are directly or indirectly linked to cultural heritage were examined. These identified and unrealized goals must be resolved in the near future in order to achieve sustainable cultural tourism development in the municipality of Ohrid.

From the national strategies mentioned above the *Strategy for Sustainable Local Economic Development of the* Municipality of Ohrid 2007–2018 was derived as a specific strategic document (Municipality of Ohrid 2007). Examined are those activities which have partially been realized (or have not been realized) and which are related to tourism and cultural values in the municipality. Activities that are **currently being realized** are: developing different types of tourism (including cultural tourism), protection and promotion of cultural-historical monuments and their use in tourism activities, maintaining the status of Ohrid as a World Heritage Site (not to be listed in the World Heritage List in Danger). On the other hand, identified are the following activities that have **not been realized**: defining absorption capacity and carrying capacities analyses (visitor management) as well as creating a tourism brand of Ohrid.

It can be concluded that the Ohrid region is trying to develop an appropriate selective type of tourism for a defined target market, thus creating a recognizable tourism brand. The protection of cultural-historical monuments and their inclusion in tourism activities is an ongoing process where limited success has been achieved. However, the Ohrid region is struggling in the efforts for maintaining the status of a UNESCO World Heritage Site. In the past two years there was a real possibility for the Ohrid region to be listed within the World Heritage List in Danger. There are several reasons for this, important ones being the absence of detailed management plans including tourism development and visitor management plans (absorption and carrying capacity analyses). Regarding the partially realized and unrealized activities in the previously mentioned strategic plans and programs, a detailed Strategy for Tourism Development in the Municipality of Ohrid has been prepared which addresses these issues and proposes concrete solutions for sustainable cultural tourism development.

4 Legal Protection Status of the Ohrid Region as a World Heritage Site

We have to examine the current national policies and legislature that are being implemented with implications on the cultural heritage of Ohrid and thus determine the tourism developmental directions of the region. These mechanisms aim to protect, conserve and enhance the cultural heritage and to reveal clear courses of action for developing sustainable tourism. The basic law that regulates the protection of the cultural heritage on the national level is the *Law on protection of cultural heritage* (Parliament of the Republic of Macedonia 2004) and its consecutive additions and changes (Official Gazette of the RNM No. 71/04, 115/07, 18/11, 148/11, 23/13, 137/13, 164/13, 38/14, 44/14, 199/14, 104/15, 154/15, 192/15, 39/16, 11/18, 20/19). The protection of the cultural heritage in the Ohrid region is mentioned in the category "Monument of culture", whereas the development of cultural tourism is mentioned in the sense of its protection and promotion. A more specific law is the *Law for proclaiming the Old City Core of Ohrid as a cultural heritage of particular importance* (Parliament of the Republic of Macedonia 2011) and its addition (Official Gazette of the RNM, 154/15). Within this law a Plan for integrated protection is envisaged to be created, where cultural tourism should be developed according to the principles of sustainable tourism. The most important law that should ensure a balance between protection and development is the *Law on the Management of the World Natural and Cultural*

Heritage in the Ohrid Region (Parliament of the Republic of Macedonia 2010). This is the core law of the World Heritage Site of Ohrid, because it includes not only the elements for managing (according to the UNESCO guidelines), but also the authorities with their rights and responsibilities related to the appropriate management activities. This law emphasizes the prevention of activities that could have negative impacts on the OUV of the Ohrid region. The concrete activities that could have such impacts are mentioned: large-scale projects in the public and private domains, intensification of urban development and unsustainable tourism development. The most specific strategic document that elaborates and regulates the protection of cultural (and natural) heritage and the development of tourism in the Ohrid region is the *Management Plan for the World Natural and Cultural Heritage in the Ohrid Region 2020–2029*. It mentions some of the main threats to the region's integrity: uncontrolled and uncoordinated urban development as well as pressures that arise from tourism development. The Management Plan's goals are determined by the main threats that have arisen from tourism development in the Ohrid region (Parliament of the Republic of N. Macedonia 2020). In this regard, a total of 13 threats have been identified, each with several planned measures:

- Inefficient management of the cultural and natural heritage in the Ohrid Region
- Impacts on the water quality of the Ohrid Lake watershed
- Unsustainable agriculture and forestry
- Unsustainable tourism
- Importing non-native fish species in the lake
- Loss and destruction of habitats
- Unsustainable fishing
- Landscape disturbances/destruction
- Climate changes
- Loss of urban-architectural values
- Loss of archaeological heritage caused by enhanced urbanization and built infrastructure
- Untimely and unprofessional conservation of protected values
- Low awareness for the significance of the natural and cultural heritage in the Ohrid Region.

As part of the identified threat regarding *Unsustainable tourism*, planned activities are: Controlling the lakeshore transformation due to mass tourism; Controlling the arranging/transformation of current beaches; Restrictions and controlling of water sports; Developing sustainable transport during the tourist season; Developing sustainable cultural and alternative tourism; Controlling overcrowding impacts in tourist sites/visitor management. However, tourism development is also included in a positive context in this strategic document. More specifically, it is mentioned that tourism could be regarded as a developmental opportunity, through respecting the protection status of the cultural and natural heritage. A main disadvantage of the Management Plan is that there is no clearly defined vision for a sustainable tourism product with any concrete tourism related goal. Therefore, the role of the current *Strategy for Tourism Development in the Municipality of Ohrid 2020–2025* is to fill the aforementioned gaps and deficiencies in this sense. The proposed sub-goals and measures related to the development of sustainable tourism (mentioned above) are incorporated in the Action Plan of the Strategy for tourism development. The Strategy and Action Plan were created by respecting defined sustainability criteria (Global Sustainable Tourism Council 2019). In this way it will be ensured that the basis is laid for the government's measures and actions to be in accordance with the Management Plan for the World Natural and Cultural Heritage of the Ohrid Region 2020–2029 and with the efforts to protect the region's OUV.

5 Existing Challenges and Identified Solutions for Cultural Tourism Development in the Function of Maintaining and Enhancing the World Heritage Site Status of the Ohrid Region

Besides the aforementioned gaps in implementing the set goals and activities in past development strategies, additional irregularities and weaknesses in the implementation of tourism development in the Ohrid region have been identified. These weaknesses are determined in the Management Plan for the World Natural and Cultural Heritage of the Ohrid Region 2020–2029, while the main goal of the newly adopted Strategy for Tourism Development in the Municipality of Ohrid 2020–2025 is to overcome those weaknesses (Andreeski et al. 2020). An important conclusion by the Reactive Monitoring Mission in 2017 was the need for a tourism strategy for the Ohrid region that would be OUV-based and sustainable, which since has been created (Joint World Heritage Centre/ICOMOS/IUCN 2017). One major deficiency and critique mentioned within the Management Plan is the creation of tourism development zones that could encourage mass tourism. Such zones were previously planned to be created in protected natural areas and would also have implications on the existing cultural heritage through the high influx of tourists. These tourism development zones have since been abandoned. According to the Management Plan, cultural heritage located in rural areas is planned to be enhanced through valorization and inclusion in tourism activities. However, the ways and methods for this to be achieved are not specified. Therefore, a new tourism zoning was conducted in the Strategy for Tourism Development in the Municipality of Ohrid. Furthermore, the Management Plan mentions the absence of a strategy for managing the cultural heritage in accordance with the principles of sustainable tourism. More specifically, archeological sites and cultural localities need to be included in newly created tourist offers. Another challenge that is mentioned in the Management Plan is the absence of maintenance of old buildings and awareness about their cultural and functional values. Some of these buildings could be adapted for accommodation of tourists. In this way tourists could experience "living culture" or "living heritage" which embodies both tradition and change through tangible and intangible elements (Robinson and Picard 2006). The remarks mentioned above have been taken into account while preparing the Strategy for Tourism Development in the Municipality of Ohrid and are included in the valorization process provided below (Table 1).

The main vision of the Management Plan states that the Ohrid region should represent an attractive destination where all the stakeholders will safeguard and recognize cultural values, interculturalism, traditional cultural events, ancient spirituality and identity. This is in accordance with the Operational Guidelines provided by the World Heritage Centre where stakeholders should have a clear understanding of what is inscribed on the list, and why a certain property has OUV (Badia 2018). Stakeholder involvement in management processes of heritage sites is essential, especially in visitor management activities. The vital ingredient to successful management of heritage is to have good relations between all stakeholders in the area (World Tourism Organization 2007): Stakeholders who provide the infrastructure (local, regional and national government); Stakeholders who manage the attractions (municipal government, religious groups, voluntary organizations/NGOs); Stakeholders who manage the supporting infrastructure services (tourist inspection, communal hygiene, health services, communication). Local communities should not only be involved in the decision making processes—together with other stakeholders, but should also be actively engaged in creating local tourism activities (Aas et al. 2005). The involvement of local communities is essential for improved planning and better balance between the needs of the community, heritage preservation and tourism (UNWTO 2018). Various stakeholders have been identified in the Ohrid region and are included as bearers of activities in the strategy's Action Plan. Regarding the local population, local and traditional handicrafts in Ohrid should be restored and reactivated, while recognizable characteristics of the region should be transformed into a tourism brand. Reactivating traditional handicrafts is a new way for the local community to maintain and value its culture through tourism (Grünewald 2002). This means that diversification of the current tourist offer in Ohrid is needed—oriented towards development of alternative types of tourism activities and combined with the implementation of visitor management plans for specific localities and values. Appropriate zoning of tourism areas in the Ohrid municipality and the Ohrid region is also evidently needed. This is essential in order for recognizable cultural attractions not to be endangered by mass tourism, while the promotion of less known cultural values has the aim to redirect current tourist flows. Such zoning was conducted in the Strategy for Tourism Development in the Municipality of Ohrid 2020–2025 (Andreeski et al. 2020). Examined in more detail are two zones located in the city of Ohrid: the *Old city core* and the *New part of Ohrid* with defined challenges and appropriate solutions for their mitigation. Although both of the aforementioned zones are located in the city of Ohrid, they must be regarded as separate territorial complexes because they face different challenges. Furthermore, almost all of the most visited cultural values are located in the old part of the city (Old city core) where most of the pressures on the cultural heritage exist. The system of protection of cultural heritage in the Old city core is different from the other part of Ohrid (New part). This directly influences the tourism development approaches for alleviating the identified challenges. In order to benefit from tourism, such frameworks need to be put in place that will mitigate the negative impacts, while optimizing the positive ones, based on the principles of sustainable tourism (Saarinen 2013). In that sense, we have identified the following challenges and solutions in Ohrid in order to determine the sustainable development directions of cultural tourism.

Challenges that the **Old city core** faces are:

- Traffic congestion
- Noise pollution
- Pressures on the cultural heritage
- Architectural endangerment (inappropriate construction of new buildings and illegal buildings)
- Low level of inclusion of cultural heritage in the hospitality offer
- Maintenance and improvement of infrastructure.

Regarding the challenges mentioned above, we have identified corresponding *solutions* that are incorporated also in the Strategy for Tourism Development. These are respectively:

- Conducting traffic regulation campaigns
- Conducting analysis for noise regulation with defining noise measurement points
- Restoration, conservation and protection of old houses and their adaptation into accommodation units

Table 1 Valorization of cultural complexes and attractions in the municipality of Ohrid

Value parameters	Uniqueness	Development benefits/assumptions
Urbanism and architecture		
Integration of the existing architecture in the surrounding landscape – Cultural landscape	**(UNESCO criterion iv)** Space with exceptional ambient values. Unique fusion of nature and urbanism—authentic landscape	Established image of the location of the old city core on the hill, from which its name derives ("o-hrid"—transl. "on-hill"). Possibilities for promotion activities and sightseeing Disturbances in the cultural landscape with new dominant buildings, interventions in the lakeside zone (promenades and terraces)
Content structure of architecture (urban units) *Old city complexes* – Ohrid-Old city core – Ohrid Bazaar	**(UNESCO criteria i, iv)** Well-preserved monumental ensemble that consists of old city architecture, complex of buildings, churches, monasteries, archeological remains of older buildings and structures, authentic urban composition Complex of shops, stores, cafes, restaurants as an economic center of the city that developed in the nineteenth–twentieth century, while nowadays it represents a fusion of traditional and modern elements	Multifunctional space in terms of tourism (accommodation in old-town type buildings; food in authentic restaurants with menus of local specialties; shops with items of traditional crafts—Ohrid pearls, wood carving items, icons; entertainment with traditional music; cultural heritage tailored for different tourists and visitors) The term "living heritage" can be attributed to this complex with a wide range of contents in the function of tourism The combination of tradition and contemporary needs in the Bazaar provides numerous entertainment and shopping activities that may be attractive to tourists. The characteristic pedestrian street offers opportunities for walks and sightseeing of the city's shopping center Serious interventions in the old part of the city with structures that do not follow the style and physiognomy of the older buildings, inappropriate extensions and interventions in the space, while in the Old Bazaar the renovated shops do not follow the old traditional shape and completely change the style and appearance of the street
Specific settlements – Settlements on the lakeshore (Peshtani, Trpejca) – Landscape settlements (Velestovo, Racha, Velgoshti, Elshani) – Complexes and settlements in rural areas (Kuratica, Rechica)	The settlements still include some of the traditions/customs; groups of rural architecture still exist in some of them. They represent an exceptional cultural space where buildings (old and new) are adapted to the terrain's dynamic morphology	Possibility for accommodation in traditional conditions with impressive viewpoints towards the city of Ohrid and the Ohrid Lake. There are restaurants with local specialties. In some of the settlements that are at some distance from the lake, opportunities exist for outdoor activities and exploring the values of the wider area Serious transformation of the rural concept of settlements by changing the contents of the rural economy, changing the traditional architecture, reducing the number of livestock, etc.
Cultural attractions		
Sacral architecture *Christian architecture* – Churches (St. Sophia, Holy Mother of God Perivleptos, St. John Theologian-Kaneo, St. Nicholas Bolnichki, Holy Mother of God Bolnichka, the small church of the Holy Unmercenaries, St. Demetrius, the small church of St. Kliment, St. Constantine and Helena, Holy Mother of God Chelnica, Holy Mother of God Zahumska, St.	**(UNESCO criteria iii, iv)** The complex of medieval Christian architecture possesses the highest monumental, aesthetic, historical and other cultural values. Some of them represent the highest architectural and fresco painting achievements of the Byzantine era. Their founders originate from the highest circles of the Ohrid Archbishopric and the Byzantine Empire. Their builders and painters belong to	This is the most important complex of cultural values for which Ohrid is inscribed in the UNESCO World Cultural Heritage List. This forms part of its brand, often referred to as "Jerusalem of the Balkans", a city of 365 churches. It should certainly be used in its tourism promotion. But what is important for tourists is the possibility to visit and see medieval masterpieces from the Byzantine cultural sphere in a relatively small area. Some

(continued)

Table 1 (continued)

Value parameters	Uniqueness	Development benefits/assumptions
Nicholas Gerakomija, Dormition of Mother of God-Kamensko, the great church of the Holy Unmercenaries, St. George, Holy Mother of God Peshtanska, St. Nicholas-Kosel, Holy Savior—Leskoec, Dormition of Mother of God-Velestovo – Monasteries (St. Panteleimon-Plaoshnik, St. Naum, St. Stefan—Pancir, St. Erasmus, St. Paraskeva-Velgoshti – Catholic Church *Islamic architecture* – Mosques (Haci Durgut-Krst Dzamija, Ali Pasha Mosque) – Tekkes and other buildings (Zeynel Abedin Pasha Tekke, Sinan Celebi Turbe)	the workshops of the most famous ones in the wider Byzantine cultural sphere of their time The churches that belong to the late Middle Ages and that were built during the "Byzantine-era Renaissance" have cultural values that link the great masters of the Middle Ages with their followers from the later centuries. Painting and woodcarving has flourished in the creation of ambitious works of art made by the most famous Mijak fellowships The monastery complexes, built in the most beautiful areas in the region are testimonies about their specific character as spiritual and educational centers Ohrid has the potential to become a center of interest for Islamic culture due to the existence of significant groups of Islamic buildings	of the events are held in these facilities. This allows tourists to experience the unique fusion of traditional and modern elements The old monasteries that are still in use today are the target of thousands of pilgrims who come during certain holidays. In addition, there is the possibility of providing longer tourist accommodation (of several days) This cultural heritage complex is well preserved and solidly presented, but lacks a better organization of visits and dispersion of visitors, because only some of the main attractions are currently being visited by most tourists
Profane and other architecture *Monuments* – Old city houses (Robevi, Uranija, Hristo Uzunov, Trimchevi, Krapchevi) – Buildings from the twentieth century (National Institution "Institute and Museum-Ohrid", Slavic Literacy, Gjurchinovi) – Gymnasium "St. Kliment Ohridski" – Clock Tower – Hammam – Sculptures and memorials	**(UNESCO criteria iii, iv)** An exceptional grouping of buildings that date from different periods and had different purposes The complex of buildings from the nineteenth–twentieth century presents the best examples of the architecture of the Balkan type of houses (Mediterranean type). There are also notable buildings that were used for administrative purposes	Much of the profane architectural complex could be used for tourism purposes as accommodation facilities. Their exterior and appropriately designed interior is particularly attractive. The more significant buildings have new functions and could attract tourists not only for their collections of artifacts and items, but also for their intrinsic value. Sculptures and memorials allow tourists to connect with important persons or events (of different religious, ethnic or other affiliation) The situation with the protection of these values is at a satisfactory level, but the inclusion of the cultural heritage corpus in tourism has to be improved. Some of them are not adapted for tourist visits at all. Furthermore, there is a lack of categorization and standardization of the facilities that have been adapted for tourist accommodation, while at the same time they need to retain their authentic architectural style
Archeological sites – Samuel's Fortress – Plaoshnik (early-christian basilica, polyconchous church) – Ancient Theater – Upper Gate – Early-christian basilica Studenchishta – Early-christian basilica St. Erasmus – Archeological site Manchevci – Ancient tomb "Karagjulevci"	**(UNESCO criteria i, iv)** Significant remains of buildings from antiquity and the Middle Ages located in the Old city core or in its surroundings Samuel's Fortress as part of the city's defensive rings marked the citadel and was a symbol of the city's strength Valuable mosaics are preserved in the early-christian buildings, which captivate visitors with their aesthetic qualities but also with the messages they convey	The complexes of buildings from the past, which are only partially preserved in the form of archeological remains, are a good basis for diversification of cultural tourism products. Some of them have received new purposes and uses and are easily accessible Archaeological heritage has attractive properties for tourism purposes, but is particularly susceptible to degradation and damage. Therefore its protection must be monitored and periodically be performed. Some archeological sites in Ohrid and its surroundings are not well protected and are not properly adapted for tourist visits
Museums – National Institution "Institute and Museum-Ohrid" – Icon Gallery – Museum on Water-Bay of bones – Museum of Slavic Literacy – Memorial House Grigor Prlichev	**(UNESCO criterion iii)** The collections contained in the Ohrid museums include items of global and universal significance. Archaeological collections are composed of artifacts from numerous sites belonging to a time span of	Museums play a significant role in all types of cultural tourism activities. Ohrid museums cover a wide range of themes. Some of them (Museum on Water) introduce new ways of presentation, thus gaining characteristics of visitor centers There is a noticeable discrepancy between the different museums,

(continued)

Table 1 (continued)

Value parameters	Uniqueness	Development benefits/assumptions
	about 5000 years (141 sites are registered, 16 of which are protected) The Icon Gallery is one of the most famous collections of Byzantine-era icons from the nineteenth–twentieth century, globally The Museum on Water is a unique reconstruction of a pillar settlement which is more than 3000 years old The two monasteries associated with the oldest Slavic literacy (Monasteries of St. Kliment and St. Naum) are known as the pillars of culture (Ohrid Literary School—St. Kliment's University). Together with the existence of the Ohrid Archbishopric, they made it possible numerous valuable written records to be part of museum collections (as originals or as reproductions)	seen in intensity of visits and their collections' cultural value. Thus, new and innovative tourist offers can be created
Events		
Religious events – Epiphany-Vodici – St. Kliment – St. Naum – St. Erasmus – St. Paraskeva – Dormition of Mother of God – Bajram	The Ohrid religious heritage associated with St. Kliment, St. Naum, St. Erasmus, as well as the fact that Ohrid is an archbishopric city, allows religious events of wider regional significance to be held	Religious events belong to a special type of cultural tourism offer that allows tourists to actively participate, thus providing them with unique experiences of local traditions and customs Events are short-lived in nature and their direct effect on tourism is limited, but they play a significant role in promoting the areas where they are being held
Cultural events – "Ohrid Summer"-Festival – Balkan Folklore Festival – Prlichevi sermons – Other events, concerts and parties	Some of the festivals (especially "Ohrid Summer"-Festival) include performers from all around the world. In this way they have gained international reputation through successfully developing their own brand	The possibility of utilizing large-scale events in tourism is seen through the prism of building a tourism brand in the areas related to organizing international cultural events More specific connections of the events with the tourist visits should be developed, while the length of stay within MICE-tourism should also be extended
Sports events – Ohrid Swimming Marathon – "Ohrid TrchaT" (transl. "Ohrid is running") – Sailing regattas	Traditional and newly created sports events that have prospects for further development	The event-related tourist offer in Ohrid has been expanded through the addition of new sports events (annually held), thus creating favorable conditions for increasing the level of visitor interest Most often such events are associated with significant costs that the organizers are not able to cover on their own, requiring additional financial support by other entities
Ethnographic heritage		
Traditional economic activities – Traditional woodcarving – "Ohrid Pearls" – Filigree – Icon painting – Manufacturing handmade paper – Shoemaking – Valavicharstvo – Winemaking *Regional culinary specialties* *Regional products* – Ohrid tomato-Jabuchar – Ohrid cherry – Ohrid trout (Salmo Letnica and Belvica) – Ohrid pearls	Handicrafts have a relatively long tradition of existence, while some of them (such as the manufacturing of Ohrid pearls) have already created their own brand For some, the future is more secure with the establishment of workshops and schools (woodcarving and icon painting), while handmade paper is a unique manufacturing that can not be found in the wider region Some of the culinary specialties are: Ohrid chomlek, traditional preparing grilled fish (Ohrid Trout, Carp), fish stew, fish-roe stew, stuffed carp, carp cooked in a clay pot, roasted plashici, gjomleze, compotes, kadaif, baklava, Ohrid cake	This type of heritage allows tourists to buy authentic products from the visited place. In that way, locals generate additional income, while enhancing the handicrafts' sustainability Through local/regional specialties tourists can taste the local cuisine in an attractive way. Some already are recognizable brands (such as the Ohrid trout) Most of the traditional handicrafts have secured their sustainability and their products are attractive to tourists. However, some of them need to adapt to the needs of modern tourists. Local/regional products are important from a promotional/marketing point of view

- Adopting a program for eliminating/demolishing illegal buildings and adaptation of protected buildings for tourism purposes
- Arranging, equipping and signaling of the most important cultural sites
- Adopting a plan for maintenance and improvement of infrastructure.

The identified *issues/challenges* in the **New part of Ohrid** are the following:

- Crowding and seasonal concentration
- Pollution
- Noise pollution
- Illegal buildings
- Non-registered accommodation
- Problems with tour guiding
- Insufficient parking spaces
- Discharge of motor oils and fuels from vessels in the lake, noise and accidents
- Mooring of boats and other vessels.

Respectively, the following *solutions* have been proposed to be included in the Strategy:

- Creating new locations for walks and stays
- Cleanup campaigns and projects for keeping hygiene
- Conducting analysis for noise regulation with defining noise measurement points
- Eliminating/demolishing illegal buildings
- Regulating accommodation through categorization and standardization, building new accommodation and accommodation expansion
- Raising the level of tour guiding through training activities
- Managing parking spaces and building multi-storey car-parks/underground garages
- Removing motor boats that run on fossil fuels
- Creating port facilities and marina.

The solutions to the identified challenges in the two tourism zones that are directly related to the cultural heritage in the city of Ohrid have been determined with the aim to protect the endangered OUV of the property listed as a WHS. These recommended solutions are also incorporated in the created Action Plan of the Strategy for Tourism Development in the Municipality of Ohrid 2020–2025 (Andreeski et al. 2020). The two proposed development zones are shown in Fig. 1.

However, it is needed to examine cultural values that are not only located in the city itself, but also in its surroundings. That is why an inventory-valorization process of the cultural values located in the municipality of Ohrid has been conducted, including several surrounding settlements. If cultural tourism is planned to be developed only in the city of Ohrid (New part and the Old city core) then the pressures on the most visited cultural values will inevitably increase. Therefore, additional 5 zones have been created besides the two already mentioned. These 5 development zones are situated in the municipality of Ohrid: Northeastern lakeshore; Former municipality of Kosel; Eastern lakeshore; Peri-Galichica; Galichica zone. All of the 7 zones are shown in Fig. 2.

The idea of such zoning is to alleviate the current pressures on the cultural values located in the city itself. This should be realized through the development of alternative tourist activities related to cultural tourism, such as gastronomic tourism, rural tourism and promoting the local ethnographic heritage. This will also create positive economic and social benefits for the local population in the smaller villages. A prevalent problem in the past decades is the emigration of the local population to the cities or abroad. The active inclusion of locals in the tourism offer should prevent such negative trends. These zones can furthermore be used as the basis for clustering activities in tourism.

Fig. 1 Map of the proposed tourism development zones—Old city core (left) and New part of Ohrid (right)

Fig. 2 Map of all the proposed development zones in the municipality of Ohrid (Old city core, New part of Ohrid, Northeastern lakeshore, Former municipality of Kosel, Eastern lakeshore zone, Peri-Galichica zone, Galichica zone)

Different attractions (natural and cultural) can then be interlinked based on their attractiveness and compatibility. Creating synergies between the different attractions is a very important element for sustaining their individual appeal and the appeal of the destination in general. This is particularly evident if the identified attractions have thematic compatibility and complementarity (Weidenfeld et al. 2016).

The municipality of Ohrid is the most important one within the region designated as a WHS. In that sense, the following pillars (for positioning the municipality of Ohrid as a recognizable tourism destination) have been created within the Strategy for Tourism Development:

- Cultural complexes and attractions
- Ohrid Lake and the natural surroundings
- Protection of the OUV of the region through sustainability
- Optimal inclusion of the local community.

These development pillars are elaborated in terms of identifying value parameters, analyzing their uniqueness, as well as developmental benefits/assumptions. In that sense, Table 1 examines the *Development pillar of cultural complexes and attractions* in the municipality of Ohrid which is most relevant to the topic of this paper. Included are settlements and cultural values located outside the city of Ohrid (but within the boundaries of the municipality), in accordance with the zoning mentioned above.

The value parameters in Table 1 represent the different cultural values and potentials (tangible and intangible) in the municipality of Ohrid, systematized by type. The values' uniqueness is determined by the type of UNESCO criteria they possess. These criteria were identified with the main purpose of protecting the OUV of the Ohrid region. This, on the other hand, determined the different development benefits/assumptions for each of the cultural values. The identified development benefits will help in the implementation of concrete and precise activities/measures that have practical implications for sustainable cultural tourism development. Furthermore, the identified attractions are the basis for further analysis of their complementarities and increasing their compatibility. This will also include assessing their spatial organization and their combination in defined clusters and/or tourism routes. This in turn will determine the movement of tourists and contribute to a more sustainable tourism development through the avoidance of overcrowding and exceeding the social and environmental carrying capacities of visitor attractions (Weidenfeld et al. 2016). Hence, a more diversified and more sustainable cultural tourism offer in the Ohrid region will be created which, on the other hand, will help to alleviate the current pressures that tourism causes on the most visited cultural attractions in the Ohrid region WHS. Several proposals for creating specific tourism brands are also provided with the analysis above. The identified development benefits/assumptions will transform Ohrid into a sustainable and recognizable destination.

The assumptions and expectations for Ohrid to become a world-renowned cultural tourism destination imply conducting a prediction analysis of various development scenarios. These development scenarios can be systematized into: *realistic*, *optimistic* and *pessimistic* scenarios. A model has been created about the prediction of overnight stays of foreign tourists in the municipality of Ohrid for 4 years (2020–2024). The statistical method of Time Series Analysis has been used for this prediction model (Andreeski et al.

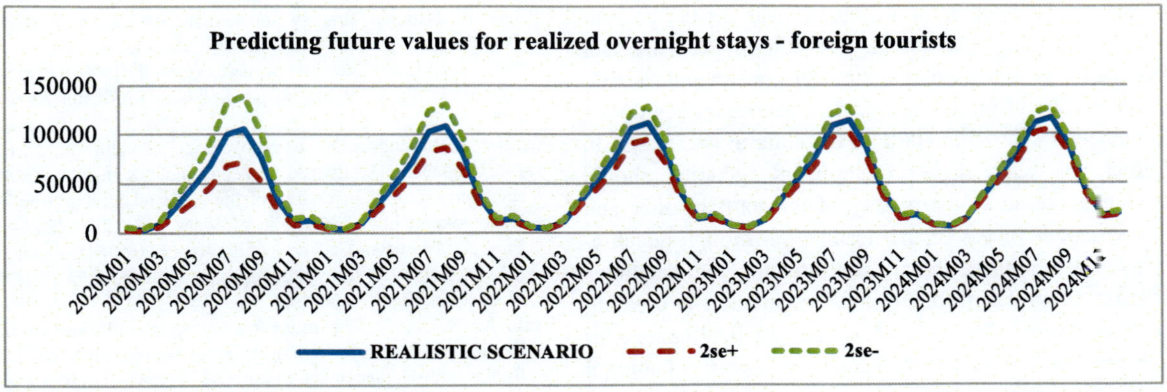

Fig. 3 Prediction model of future values for realized overnight stays of foreign tourists 2020–2024

2020). This prediction model includes a monthly-based analysis of the overnight stays of foreign tourists in the Ohrid region (Fig. 3).

The optimistic and pessimistic scenarios are represented by the calculated ±2 standard errors from the predicted realistic series. The predicted values derived from the analysis of overnight stays of foreign tourists in the past 5 years. These prediction scenarios can be accepted with some reservation, because they assume that the situation with the COVID-19 pandemic will stabilize in the near future. In Fig. 3 it can be seen that the number of foreign visitors' overnight stays is expected to rise, while the difference between the predicted optimistic and pessimistic scenarios from the realistic predicted series in the high season months (June, July and August) is significant and ranges between 20 and 30 thousand overnight stays. However, even if the optimistic scenario is realized, tourism development in Ohrid must shift away from quantity and move towards quality. This implies not attracting high numbers of tourists but tourists with a higher purchase power. Maintaining the status of a World Heritage Site of the Ohrid region will contribute to such a shift, given the fact that this designation already represents an important international tourism brand for the destination.

6 Conclusions

The Ohrid region is one of the few properties in the world that has been inscribed as a UNESCO World Cultural and Natural Site (mixed heritage status). Its socio-economic development is based mainly on tourism. However, urban and tourism development in the past two decades have created various pressures on its Outstanding Universal Value. These pressures endanger the WHS in terms that it could be listed in the World Heritage List in Danger. A Reactive Monitoring Mission (conducted in 2017) identified several issues and threats that must be eliminated in the near future. In this sense, it was identified that a new Strategy for Tourism Development in the municipality of Ohrid is needed which will be based on sustainable development in order to protect the OUV and the cultural values within the region. Therefore, the paper analyses the most important core strategic and legal documents related to tourism development and cultural heritage protection and usage. An important finding is that the national strategies for tourism development show no clear continuity of planned activities in the municipality of Ohrid related to developing cultural related local/regional tourism products as well as promotion/branding strategies. The analysis of the legal protection status of the Ohrid region showed that there is a general recognition of the importance of sustainable cultural tourism but no clear means for this to be realized. Therefore, the role of the new Strategy for Tourism Development in the Municipality of Ohrid 2020–2025 is to fill these gaps. Namely, this Strategy recognizes the need for diversification of the current cultural tourism offer, by appointing new uses of already attractive cultural attractions and valorizing less known cultural values, not only in the city of Ohrid but also its surroundings. This is needed because cultural heritage in the municipality of Ohrid (mostly the values located in the Old city core) is mainly endangered by the large influx of tourists. Identified are issues/challenges within the city of Ohrid that arise from tourism development. In this regard, proposed are appropriate solutions for their mitigation/elimination. Furthermore, a valorization analysis is conducted of the cultural values and attractions located in the city of Ohrid and the wider area within the municipality, by examining value parameters (inventory of cultural values), determining their uniqueness (according to defined UNESCO criteria) and development solutions and benefits. The identified solutions include proposals for sustainable tourist usage, branding/promotion as well as development of concrete tourism products. Maintaining the WHS status of the Ohrid region will create increased costs for the government's undertakings for solving the current problems and threats (such as increased spending for conservation activities, conducting visitor impact studies, decreasing/regulating the number of visitors in some of the cultural sites, etc.). In the long run, such higher government spending should ensure an increase in the quality of the tourism offer and higher competitiveness of the destination through developing sustainable tourist activities, redirecting tourism flows and active involvement of the local population in creating alternative tourist offers. Therefore, the practical benefits of this research are evident in the economic, social and cultural spheres. Furthermore, this paper's practical implication stems from the fact that the findings are incorporated in the newly adopted tourism strategy of the municipality of Ohrid. More precisely, it provides clear directions for cultural tourism development in the municipality that is based on the principles of sustainability. The findings provide proposals for alleviating some of the pressures on the cultural values and attractions in Ohrid, but also solutions for redirecting the current tourism flows towards different cultural localities and resources located outside the city core and within the boundaries of the municipality of Ohrid. These proposals and solutions have practical significance because they will be implemented through a detailed Action Plan within the adopted strategy. This in turn will lead the region of Ohrid to become a recognizable and sustainable tourism destination of global significance by maintaining its WHS status.

Acknowledgements Special thanks to our colleagues Professor Cvetko Andreeski and Professor Naume Marinoski from the Faculty of Tourism and Hospitality—Ohrid who were involved in the research activities for the preparation of the *Strategy for Tourism Development in the Municipality of Ohrid 2020–2025* from which some of the findings in this paper derived.

References

Aas, C., Ladkin, A., Fletcher, J.: Stakeholder collaboration and heritage management. Ann. Tourism Res. **32**(1), 28–48 (2005)

Alberts, H.C., Hazen, H.D.: Maintaining authenticity and integrity at cultural world heritage sites. Geograp. Rev. **100**(1), 56–73 (2010)

Andreeski, C., Marinoski, N., Korunovski, S., Risteski, M.: Strategy for tourism development in the municipality of Ohrid 2020–2025. Faculty of Tourism and Hospitality—Ohrid and Municipality of Ohrid, Ohrid (2020)

Badia, F.: The management plan for the world heritage sites as a tool of performance measurement and sustainability reporting: opportunities and limits in the Italian context. In: Makuzava, S. (ed.) Aspects of management planning for cultural world heritage sites, pp. 25–36. Springer International Publishing AG, Cham (2018)

Brida, J.G., Meleddu, M., Paulina, M.: The economic impacts of cultural tourism. In: Smith, M., Richards, G. (eds.) The Routledge handbook of cultural tourism, pp. 110–115. Routledge, New York (2013)

Flaneri, J.: National strategy for tourism development 2009–2013. Government of the Republic of Macedonia, Skopje (2009). Retrieved from http://www.merc.org.mk/Files/Write/Documents/03345/mk/ME_2009_Strategija-za-razvoj-na-turizmot-2009-2013.pdf

Fournier, J., Marinoski, N., Korunovski, S., Miladinoski, S., Risteski, M., Crnjanski, D., Efremova, A., et al.: Technical assistance for revision and implementation of the National Strategy for Tourism Development in the Republic of Macedonia 2009-2013—BERIS Project. Ministry of Economy of the Republic of Macedonia, Skopje (2011). Retrieved from http://eprints.uklo.edu.mk/6227/1/Planeth_Macedonia_Final_Report.pdf

Fyall, A., Rakic, T.: The future market for world heritage sites. In: Leask, A. (ed.) Fyall, Managing world heritage sites, pp. 159–177. Butterworth-Heinemann Elsevier Ltd., Burlington (2006)

García-Hernández, M., Calle-Vaquero, M., Yubero, C.: Cultural heritage and urban tourism: historic city centres under pressure. Sustainability **9**(8), 1–19 (2017)

Global Sustainable Tourism Council: GSTC destination criteria with performance indicators and SDGs. GSTC, Washington (2019)

Grünewald, R.A.: Tourism and cultural revival. Ann. Tourism Res. **29** (4), 1004–1021 (2002)

Jankuloski, B., Risteski, M., Marinoski, N., Cavdaroska, N., Geshoski, V.: Register of potentials for the development of rural tourism in the Southwest Planning Region. Prilep Region Enterprise Development Agency (PREDA PLUS)—Consulting Firm and Center for Development of the Southwest Planning Region, Struga (2016)

Joint World Heritage Centre/ICOMOS/IUCN.: Mission report of the reactive monitoring mission to the world heritage property of natural and cultural heritage of the Ohrid region. UNESCO World Heritage Convention, Krakow (2017). Retrieved from https://whc.unesco.org/en/documents/158740

Kohl and Partner—Consulting Firm.: National strategy for tourism 2016–2021—Republic of Macedonia. Ministry of Economy of the Republic of Macedonia, Skopje (2016). Retrieved from https://tourismmacedonia.gov.mk/wp-content/uploads/2019/04/Kohl-Partner_Tourism-Strategy-Macedonia_FINAL-REPORT_16-04_MK.pdf

Korunovski, S.: Cultural heritage. Faculty of Tourism and Hospitality - Ohrid, Ohrid (2011)

Korunovski, S., Marinoski, N., Andreeski, C., Nestoroska, I., Risteski, M.: Sub-strategy for development of cultural tourism in the Republic of N. Macedonia. Faculty of Tourism and Hospitality—Ohrid and Ministry of Economy of the Republic of N. Macedonia, Skopje (2019)

Leask, A., Fyall, A.: Managing world heritage sites. Butterworth-Heinemann Elsevier Ltd., Burlington (2006)

Loulanski, T., Loulanski, V.: The sustainable integration of cultural heritage and tourism: a meta-study. J. Sustain. Tourism **19**(7), 837–862 (2011)

Marinoski, N., Risteski, M.: Tourism potentials of the Southwest planning region. In: Conference Proceedings of The Second International Scientific Conference Challenges of Tourism and Business Logistics in the XXI Century–ISCTBL (pp. 289-296). Stip: Faculty of Tourism and Business logistics Goce Delcev University of Stip (2019)

Marinoski, N., Korunovski, S., Andreeski, C., Angelevska-Najdeska, K., Risteski, M.: Study on the tourism potentials in the Southwest planning region. Faculty of Tourism and Hospitality—Ohrid and Center for Development of the Southwest Planning Region, Ohrid (2013)

McKercher, B., Ho, P., Du Cross, H.: Relationship between tourism and cultural heritage management: evidence from Hong Kong. Tourism Manag. **26**(4), 539–548 (2005)

Municipality of Ohrid.: Strategy for sustainable local economic development of the municipality of Ohrid 2007–2018. Municipality of Ohrid, Ohrid (2007). Retrieved from https://ohrid.gov.mk/wp-content/uploads/2017/08/Strategija_Lokalen_Ekonomski_Razvoj.pdf

Parliament of the Republic of Macedonia: Law on protection of cultural heritage. Official Gazette of the Republic of N. Macedonia—No. 20/04, Skopje (2004)

Parliament of the Republic of Macedonia: Law on the management of the world natural and cultural heritage in the Ohrid region. Official Gazette of the Republic of Macedonia—No. 75/10, Skopje (2010)

Parliament of the Republic of Macedonia: Law for proclaiming the Old City Core of Ohrid as a cultural heritage of particular importance. Official Gazette of the Republic of Macedonia—No. 47/11, Skopje (2011)

Parliament of the Republic of N. Macedonia: Management plan for the world natural and cultural heritage in the Ohrid region 2020–2029 with action plan. Official Gazette of the Republic of N. Macedonia—No. 45/20, Skopje (2020)

Ringbeck, B.: The world heritage convention and its management concept. In: Makuzava, S. (ed.) Aspects of management planning for cultural world heritage sites, pp. 15–25. Springer International Publishing AG, Cham (2018)

Risteski, M.: Contemporary methods and approaches for tourist valorization. HORIZONS—Int. Sci. J. **27**, 79–89 (2020)

Risteski, M., Kocevski, J., Arnaudov, K.: Spatial planning and sustainable tourism as a basis for developing competitive tourist destinations. PROCEDIA—Soc. Behav. Sci. **44**, 375–386 (2012)

Risteski, M., Korunovski, S., Kocevski, J.: Analysis of potentials and attractions for the development of cultural tourism in the South-west planning region of the Republic of N. Macedonia. HORIZONS—Int. Sci. J. **26**, 7–17 (2020)

Robinson, M., Picard, D.: Tourism, Culture and Sustainable Development. UNESCO, Paris (2006). Retrieved from https://unesdoc.unesco.org/ark:/48223/pf0000147578

Saarinen, J.: Indigenous tourism and the challenge of sustainability. In: Smith, M., Richards, G. (eds.) The Routledge Handbook of Cultural Tourism, pp. 220–226. Routledge, New York (2013)

Sekulovska, M., Marinoski, N., Nestoroska, I., Risteski, M.: Macedonia. In: Vodenska, M. (ed.) Hospitality and Tourism in Transition in Central and Eastern Europe: A Comparative Analysis, pp. 261–295. Cambridge Scholars Publishing, Newcastle (2018)

Smith, K.M.: Issues in Cultural Tourism Studies. Routledge, London (2003)

Throsby, D.: The value of cultural heritage: What can economics tell us? In: Clark, K. (Ed.) Capturing the Public Value of Heritage—The Proceedings of the London Conference, pp. 40–44. English Heritage, London (2006)

Trajkov, A., Marinoski, N., Andreeski, C.: Analyzing the economic effects of tourism in the Municipality of Ohrid. Faculty of Tourism and Hospitality—Ohrid and Municipality of Ohrid, Ohrid (2019)

UNESCO World Heritage Centre: Operational Guidelines for the Implementation of the World Heritage Convention, January 10 (2008). Retrieved from UNESCO World Heritage Convention Web site: https://whc.unesco.org/archive/opguide08-en.pdf

UNESCO World Heritage Centre: Operational Guidelines for the Implementation of the World Heritage Convention, July 6 (2012). Retrieved from UNESCO World Heritage Convention Web site: https://whc.unesco.org/archive/opguide12-en.pdf

UNESCO World Heritage Committee: Convention Concerning the Protection of the World Cultural and Natural Heritage—3rd Session of the Committee. World Heritage Committee, Cairo, Luxor (1979). Retrieved from https://whc.unesco.org/archive/repcom79.htm

UNESCO World Heritage Committee: Convention Concerning the Protection of the World Cultural and Natural Heritage—4th Session of the Committee. World Heritage Committee, Paris (1980). Retrieved from https://whc.unesco.org/archive/repcom80.htm

UNESCO World Heritage Committee: Convention Concerning the Protection of the World Cultural and Natural Heritage—36th Session of the Committee. World Heritage Committee, St. Petersburg (2012). Retrieved from https://whc.unesco.org/en/sessions/36COM/documents/

UNESCO World Heritage Committee: Convention Concerning the Protection of the World Cultural and Natural Heritage—40th Session of the Committee. World Heritage Committee, Istanbul (2016). Retrieved from https://whc.unesco.org/archive/2016/whc16-40com-19-en.pdf

UNESCO World Heritage Committee: Convention Concerning the Protection of the World Cultural and Natural Heritage—43rd Session of the Committee. World Heritage Committee, Baku (2019). Retrieved from https://whc.unesco.org/archive/2019/whc19-43com-18-en.pdf

UNESCO World Heritage Convention: UNESCO world heritage list, December 5 (2021). Retrieved from World Heritage Convention Web site: https://whc.unesco.org/en/list/?type=mixed

UNWTO: Report on tourism and culture synergies. United Nations World Tourism Organization, Madrid (2018). Retrieved from https://doi.org/10.18111/9789284418978

Weidenfeld, A., Butler, R., Willimas, M.A.: Visitor Attractions and Events: Locations and linkages. Routledge, London (2016)

World Tourism Organization: A Practical Guide to Tourism Destination Management. UNWTO, Madrid (2007)

Dark Tourism Along the Silk Route: The Case of Thessaloniki

I. K. Mytaftsi and C. N. Tsironis

Abstract

Thessaloniki is the capital of the Central Macedonia and an important milestone in the Silk Road Routes. Historically speaking, it was a significant hub of merchandise, culture, philosophy sciences, religious ideas, language exchange between East and West. Thessaloniki is an urban destination which offers the possibility of associating the 15 UNESCO World Heritage Sites, which it hosts, with the Silk Road Routes for the development of new touristic experiences such as Dark Tourism. Dark Tourism as a segment of Cultural Tourism could operate as a key factor of local tourism development. The present paper presents the "silk" and the "dark" past of the city and it focuses on the potential development of a 'dark urban' route. It also explores the possibilities of a new polymorphic form of dark tourism in Thessaloniki, being a new distinctive destination of cultural heritage. The analysis concludes with a detailed dark route designed to accommodate a new, polymorphic, sustainable tourism strategy, which includes a dark past interconnecting the Silk Routes with places, sites and monuments as well as narrating stories which have been stigmatized by destruction and violence.

Keywords

Dark tourism • Dark urban route • Local development • Polymorphic tourism • Silk road routes • UNESCO world heritage sites

I. K. Mytaftsi (✉) · C. N. Tsironis
Department of Ethics and Sociology, Faculty of Theology, Aristotle University of Thessaloniki, Thessaloniki, Greece
e-mail: i.mitaftsi@gmail.com

C. N. Tsironis
e-mail: tsironis@theo.auth.gr

1 Introduction

Dark tourism is a type of special interest tourism with an "uneasy relationship with heritage tourism" (Light 2017). It has, in recent years, captured the interest of researchers, academics, but also of the business community engaged in the tourism sector. It is a modern form of tourism but not necessarily a new kind of activity. The goal of this paper is to bring to the foreground, the dynamic the city of Thessaloniki exhibits, in order to shed light to city's potential to become a dark tourism destination. More specifically, it indents to connect Thessaloniki with the Silk Road and Dark Tourism through the monuments of its religious and cultural heritage.

The city has been a historic hub along the Silk Road; it is home to fifteen UNESCO monuments; it possesses religious and cultural wealth; it has a "dark history". The following criteria were then posed in the form of relevant questions: which of the UNESCO monuments have a dark past; which are situated in the historic web of the city, and, finally, how many "death-incidents" (mass of deaths) occurred in them. Both the history of a place as well as its culture are poles of tourist attraction, important for local tourism development. The combination of all the above resulted in the design of a dark urban cultural route which may constitute a new tourism proposal by and for the city.

There is a deep and wide-ranging religious and cultural past and present in Thessaloniki. There is also a dark past, that includes historical events which have tainted the city and many places where deaths, torture and other atrocious acts have been committed, and recorded through the ages. Given this, Thessaloniki may become a new dark tourism destination in Europe.

The example set by Thessaloniki may function as a guide for other cities connected to the Silk Road, in order for them to relate their cultural heritage with their dark history and to be led to the design of similar dark routes. In the frame of this analysis an effort is made to connect the past with both

S. Kostopoulou et al. (eds.), *Cities' Vocabularies and the Sustainable Development of the Silkroads*, Advances in Science, Technology & Innovation, https://doi.org/10.1007/978-3-031-31027-0_29

the present and the future in order to revive paths, myth and stories along the Silk Road and for new tourist destinations to emerge. A listing of the tourism resources of Thessaloniki, but also of its dark spots, was carried out as part of this project.

2 Thessaloniki and The Silk Road: Yesterday and Today

Thessaloniki has been since the fourth century BC a key cosmopolitan hub for the exchange of merchandise, culture, philosophy, the sciences, religious ideas and language between the East and the West. It was also an important city in the Silk Routes network (Kostopoulou et al. 2016). Furthermore, the city is an urban destination which, by means of the fifteen UNESCO cultural heritage monuments it is home to, has the potential to be connected to the Silk Routes, in order for new tourism products to be developed, such as that of dark tourism. Dark tourism, being for many a sub-set of Heritage Tourism (Light 2017; Kunwar and Karki 2019), could function as a driver for local tourism development (Miller et al. 2017). In the present, post-covid era, where tourists-consumers turn to quality tourism products and experiences (Bowen 2022), special forms of tourism have the capacity to increase the popularity of an area and Silk Road constitutes a prime opportunity for highlighting new destinations and interconnecting historic routes as an added value for the countries involved. Experiential travel, the search for experiences and authenticity, which has recorded a huge increase in recent years (Cetin and Bilgihan 2016), serving also as the antipode to mass tourism, raises the interest on destinations along the Silk Road as part of unexplored travelling experiences.

This analysis focuses on the potential development of dark urban routes and aims to investigate the development potential of a new and polymorphic form of dark tourism in Thessaloniki, an exceptional cultural heritage destination. In this vein, it contributes to the currently underdeveloped research of the relationship between dark spots and World Heritage Sites (Kennell and Powell 2020). The analysis concludes by offering a planned in details dark route, designed to facilitate the new, multiform, and sustainable tourism product describing a dark past with places, sights and monuments, while narrating stories and events branded by death, destruction and violence, which have taken place across the entire history of the city.

The Silk Road is considered one of a major ancient commercial route in the world (Karluk and Karaman 2014). It was not a road, but many meeting trails and crossroads. In reality, it comprised of a labyrinthine network of commercial routes connecting the West with the East, empires and kingdoms, trading settlements, cities and sacred places (See: WTO 2018, UNWTO silk-road webpage). It crossed the entire continent of Asia, connected cities in China to India, Iran and Arabia, reaching the Mediterranean, Greece and Italy, via overland and sea routes. Nomads, artists, artisans, traders, businessmen, as well as conquerors travelled along the routes of silk (Karluk and Karaman 2014). These were the roads of a continuous circulation of commodities, such as spices, tea, salt, livestock, paper, but also luxurious goods like silk, porcelain, carpets, and precious stones. These were expensive products bought and sold by the higher economic classes (Williams 2014).

Besides being roads of commerce, they also, in reality, constituted an artery for the meeting of cultures, religions, ideas, beliefs, arts, sciences and knowledge. One could claim that the Silk Road constitutes a historic example of the interconnectivity of the then known world with silk being the most sought-after merchandise in Greece, Rome and Egypt (Mark 2014). The Silk Road established bonds of cultural dialogue and a spirit of the co-existence of people. Through this route and the revival of the Silk Road legacy important tourism destinations are shaped and developed (Kostopoulou et al. 2021).

Today, and specifically since 1993, several nations, headed by the World Tourism Organization, have made a coordinated effort to revive the Silk Road as an emerging tourist narrative which will bring new and old tourism destinations and routes to the attention of modern travellers, and connect countries with a rich cultural heritage under a unified international brand name.[1] Silk Road tourism is considered to be a special form of cultural tourism, and this renders it a branding & marketing tool for the development of new horizons and goals for tourism strategy (Kostopoulou et al. 2021). On the one hand, the new strategy adopted by China and entitled "One Road one Belt" (Aoyama 2016; Arase 2015) which includes the Silk Road, and, on the other hand, the development connecting two continents, Europe and Asia, have rekindled Europe's interest on the Silk Road and, thus, the programme also expanded to the West. This valuable raw material, silk, connecting continents, countries, cities and cultures, constitutes a treasure both in reality as well as metaphorically. While it can also form the reinforcing link between Greece and China, opening new silk roads.

It is important for cities which flourished in the past to be able to sustain their tourist activity today and, ideally, to be connected to one another with a silken thread and to constitute a long tourist route. This road of communication paved by silk is sustained and evolved by tourism. But also,

[1] The concept of a Silk Road tourism project was first raised at UNWTO's General Assembly in Indonesia in 1993. (Source: http://silkroad.unwto.org/sites/all/files/docpdf/bannerhistorywebsite.pdf.

with respect to regions which do not possess of some competitive tourism advantage, save for the imprint left on them by silk, they, too, can obtain an added value from it, and a boost to their tourism. Of course, today, in the digital era, cities and destinations are not only connected via the exchange of goods but also on account of their history, monuments, people and obviously through technology. Thus, in our days, the Silk Road is a network of destinations based on their tangible and intangible cultural heritage and geared towards the sustainable development of tourism. The revival of the Silk Road heritage will certainly mobilize new travellers and assist in the development of special forms of tourism, such as Religious and Cultural Tourism.

Besides being markets for products and services, the countries located on the Silk Road are potential tourism markets where local authorities must join forces with the tourism industry, always guided by the research-scientific work in progress on the Silk Road. Moreover, in the post-covid era, the modern motive for travellers revolves around the triptych: Tourist Routes—Sustainability—New Tourism Products. For sure, tourists' motivations should be examined in depth. As Farmaki (2013) states: "It is … pivotal that the conceptualisation of dark tourism incorporates understanding of the reasons advocating its development, the motivators for travelling to dark tourism sites and the preconditions surrounding the production and consumption of dark tourism. With the majority of literature focusing primarily on the supply of dark tourism, motivation for visiting dark tourism sites remains an under-researched area".

Tourist routes constitute an important tool for local economic development (Briedenhann and Wickens 2004). In recent years, a trend is noted for the design of an increasing number of new thematic tourist routes, aimed to attract visitors (Pattanaro and Pistocchi 2016). The idea of connecting isolated sights, which would not by themselves lure tourists, through a route and a narrative which will motivate visitors to consume this particular tourism product (Meyer 2004). These routes can share a theme and be a combination of natural and constructed sights which are completely accessible by mass transportation (Puczko and Ratz 2000). This common theme will connect them, promote them, and help bolster their popularity, their management as well as their preservation (Timothy and Boyd 2006; Stoddart and Rogerson 2009). Moreover, the thematic field of tourist routes varies. There are the traditional old Silk Road Routes, the Religious Routes, the Cultural Routes, as well as the Gastronomic, Wine-tasting, Nature-loving (bird watching) ones, as well as Thermalism Routes etc. The existing variety of touristic routes provides an opportunity of multifaced touristic products (Tsironis 2020) and especially a possible combination of dark tourism with other touristic perspectives (e.g. heritage, educational tourism etc.).

3 Dark Tourism

The engagement of the academic community with dark tourism begun expanding since the early days of the last century (Stone 2013) reflecting on the one hand, the potential of this particular form of tourism and, on the other hand, the increased interest shown by the late-modern society on new forms of tourism and specifically visits to places of death and destruction. It is considered by many to be part and sub-category of cultural tourism while it is increasingly recognised "as both an individual experience and a complex socio-cultural phenomenon" (Biran and Hyde 2013:191).

In the course of the academic study of dark tourism there have been attempts to offer a definition for this kind of tourism. Rojek (1993), introduced the concept of "Black Spots" and Blom (2000), talks of a kind of tourism which aims for artificial morbidity-related attraction and sudden death. Another approach refers to the visits to places of terrorist acts, mass deaths or the suppression of additionally human rights (Dalton 2014). Several other definitions have been put forward in the literature, such as "negative sightseeing" (Cannell 2013), "milking the macabre" (Dann 1998), "difficult heritage" (Logan and Reeves 2009), "tragic tourism" (Lippard 1999). Smith (1998) considered that the most attractive locations for tourists to visit are those relating to wars, the places where natural disasters have claimed human lives are also dark environments (Kang et al. 2012) and therefore dark places to visit. In other words, dark tourism focuses on places where humans have suffered torment, or even death, from natural or other disasters (Oren et al. 2021). Seaton (1996) describes travelling with motives focusing purely on death as "Thanatotourism". Following the survey of all definitions recorded in the literature, Yan et al. (2016) ascertained that the term used widely to describe this kind of tourism is Dark Tourism and should be attributed to Foley and Lennon (1996), who employed it to describe visits to places of death, violence and destruction. Thus, this frequently used term remains conceptually lacking (Jamal and Lelo 2010) and theoretically fragile (Stone and Sharpley 2008), since this field is very dynamic and to a degree "fluid", fact rendering any attempt to offer a "static" definition very precarious.

For dark tourism to be developed in a destination, said destination must feature a noteworthy cultural-historical heritage which has been stigmatized by atrocities, violence and death. The destination is evaluated by tourists, in order to be included in their itineraries, but it is also appraised by the agent, in order to be included in the latter's offer (Minić 2012). A dark spot-sight can be the actual location of a tragedy, or simple a spot commemorating a tragedy or death which took place elsewhere (Wight 2006). The actual place of death, violence and destruction is clearly more attractive for tourists since memory plays the key role. Thus, the authenticity of the place and the monument are key factors at the darker end of the spectrum (Stone 2006) and exercise a great attraction for dark tourism, just as they do in the monuments of cultural heritage (Powell et al. 2018). Visitors exhibit the trend of considering authentic those locations where the memorial ceremony takes place (Cohen 2011). While there are also tourists not familiar with the dark aspects of the place (Poria et al. 2004). Smith and Croy (2006) advocate that what motivates a dark tourist is the prevailing perception of a location as being a dark one (and not necessarily the characteristics of this location themselves).

There are four characteristics, according to Ashworth (2004) which make a place of death, a tourist attraction: (1) It is necessary for there to be a culprit and a victim, in order for the tourist to have the opportunity to identify with one of them; (2) the culprit must have been fully conscious of committing the crime, while the victim(s) must be innocent; (3) the atrocity must have entailed exceptional seriousness and (4) the event became widely publicised and people remember it.

4 Dark Tourism in Yesteryears and Today

For historiographers and social thinkers, visits to places of mourning, death or other adverse historical moments are a known phenomenon which has been recorded in the past (Casson 1994). According to Malinowski (1948), death is one of the oldest subjects humanity had to deal with. Travellers had always been attracted by death and, indeed, Kyle (2012) stresses that death had always been a spectacle. This assertion is demonstrated if one considers the visits, in the past, of people to places of public executions, fields of battles and cemeteries (Stone 2005). Despite the fact that dark tourism is considered a modern form of tourism (namely, exhibits the special characteristics of a tourism "product"), there are views advocating that travels and visits with a similar goal have also occurred in the past. Ever since antiquity, humans travelled to see or participate in "dark events", while Dale and Robinson (2011) suggest that dark tourism has been in practice since the eleventh century. Visits to fields of battle (such as the Waterloo) or Pompeii were a kind of tradition, with the latter considered as the greatest tourism destination of the Romantic period (Seaton 1996). What is clear is that visits to places with a dark nature existed as a phenomenon in the past, despite the fact that their form differed greatly from its present one and they were not organized. The intense presence of marketing in recent years led this to the implementation of this trend, as a new form of tourism. The relevant discussion has expanded to all directions.

Today, recreation for travellers has changed. But travel itself has also changed with respect to both its consumption and motives, which are almost always multi-faced (Božic et al. 2016; Griffin and Raj 2018). According to Skinner (2012), the search of a hedonistic experience, in dark places, is a result of post-modernism, where our ego is rejuvenated by the hardship of others. A turn has been observed in recent decades, away from mass tourism and towards alternative forms of tourism, or, better, special forms of tourism. Thus, dark tourism, makes its appearance increasingly intense, ushering visitors to consume death as a form of tourism which offers a unique experience. Scientific analysis focuses on this "meeting" of culture and death, in order to offer a better understanding of our tolerance for death (Korstanje 2017). The term dark tourism recognizes locations of cultural heritage or events connected to death and pain (Dimitrovski et al. 2017). Dark tourism is thus directly connected to cultural heritage tourism (Biran et al. 2011) and shares common fundamental characteristics with it. Visits to such places have educational, commemorating or even recreational goals (Stone 2005) and, thus, many papers on dark tourism have been published in the literature on cultural heritage tourism (Yankholmes and Akyeampong 2010). Just like cultural heritage tourists, dark tourists seek the answers to their questions in history, tragedy and death.

Sharpley and Stone (2009) suggest that an increase is noted, since the last century, both in the supply as well as the demand for dark tourism. It therefore, regards a new tourism product, the rise of which is attributable to the emergence of an increasing number of places where atrocities, torture and death have taken place (Biran and Hyde 2013). Globalization, the media, direct access to social media and new technologies, intense social concerns, were the causes for the wide spreading of a tragic event, leading the viewer directly and closely to death through the screen, photographs, TV series (Light 2017). The more recent examples of popular dark places include Ground Zero in New York City, the place where J.F. Kennedy was assassinated, Chernobyl and, of course, the infamous concentration camp Auschwitz.

In the year 2022 and following the pandemic brought by covid-19, Wuhan is, according to an article in The Telegraph (2020), one of the most popular dark destinations in the world. This is the city were patient zero of covid-19 was recorded and Wuhan is expected to attract tourists seeking adventure, who are daring and who believe that by visiting dangerous places they will set new trends in tourism. At any rate, it appears that the covid-19 pandemic will greatly affect the motivation and way of travelling in the field of cultural tourism (Tsironis et al. 2021). Moreover, the intense desire for social media and online fame in general will function as a motivating factor to visit Wuhan, as it was noted in the past with respect to Chernobyl and Ground Zero in New York (Yankovska and Hannam 2014; Stone and Grebenar 2021).

Virus covid-19 gave mankind direct experiences of the dying, something uncommon, even rare, in our days. This invisible and lethal enemy is responsible for the deaths of millions of people around the world. Dark tourism offers the potential to herald new places where visitors will commemorate the victims of covid-19. In order for the dead not to be forgotten, the establishment of a national covid-19 monument is discussed, which is also anticipated to become a pole of attraction for visitors-tourists (The Conversation 2020). Dark sanctuaries help to deal with pain or emotions related to disasters (Sather-Wagstaff 2016). Undoubtedly, the pandemic will constitute our common dark heritage. Demand based dark tourism will not fall after the pandemic. On the contrary, the coming years will be the years of revenge travel, where people following lockdowns (and almost vengefully) will travel much more. Dystopian theories make an appearance as well, describing a future for dark tourism (Wright 2019) in terms of inhumane activity, such as safari hunting of people, as if it were an expensive past-time for the extremely wealthy in 100 to 200 years from now, which may also be broadcast as a TV show.

5 Case Study: Thessaloniki as an Urban, Religion-Cultural and Dark Tourism Destination

5.1 Thessaloniki as an Urban Tourism Destination

The rise of tourism in cities recorded in recent years, has led researchers to the study of urban tourism (Maxim 2019; Wearing and Foley 2017). This trend for urban tourism is complemented by a post-modern culture prioritizing technology, culture and accessibility (Urry 2002). In order for a city to be classified as a tourist destination, it must meet key conditions (Berg et al. 1995). Principally, it must offer primary products which also constitute the principal reason for a tourist to visit the city. It must, namely, possess of the natural and historical assets and characteristics required and to offer tourist facilities such as museums, religion-cultural monuments, sports and conference facilities. In the second place, it must possess of the appropriate hospitality infrastructure, restaurants, info kiosks, which will facilitate tourists to stay at the destination and not just visit it on a daytrip. Thirdly, the city must be accessible and interconnected to surrounding tourist regions. Fourthly, it must be easy and safe for the tourist to cross the city or to wander in it and, finally, for the city must have a well-developed tourist image, as such is shaped by the existence of all of the above. Urban tourism exhibits minimal, almost zero, seasonality and offers the opportunity for the development of the cultural, religious and dark tourism (Richards 2007). It is, therefore, very important for cities to be able to offer innovative products and authentic experiences to tourists, in order for them to become poles of tourist attraction. Dark tourism is a form of tourism which can surely become a competitive advantage for a city (Stone and Sharpley 2008; Biran and Hyde 2013). Dark products are attractive and can constitute an additional motive, but also a reason for the increase of visits to an urban destination, especially if said destination also possesses of other tourism products, save for dark places (Crompton 1979).

Thessaloniki is the second largest city in Greece and is located in Northern region of the country. Due to its geographic location, it has functioned as an important hub for the exchange of goods and the communication of culture, philosophy, the sciences, religious ideas and language between the East and the West, just like the Silk Road was.

Based on the conditions posed by Berg et al (1995) it is an urban destination possessing a developed tourism image. More specifically:

A. It is home to more than thirty (30) museums as well as more than one hundred (100) monuments, spanning all the city's historical periods (Roman, Byzantine, Ottoman and up to the present day). This rich cultural heritage automatically renders the city a religious and cultural destination occupying a special place in the global tourism market. It exhibits a multi-cultural and multi-religious character, attributable, for the most part, to the various people who had lived in it across its entire history (Jews, people from Asia Minor Ottomans). Furthermore, Thessaloniki ranks amongst the well-known destinations for conferences, and it hosts major sports events, such as "Alexander the Great" International Marathon.
B. The hospitality infrastructure of the city is of the highest standards and based on the data provided by the Hellenic Chamber of Hotels for year 2018 (HCH 2019)[2] the city's hotel capacity is:
C. With respect to accessibility, Thessaloniki is connected to various European destinations by air, while domestic flights connect it to other cities, as well as islands, in Greece. It is the second largest commercial and passenger port in the country and is also connected by rail to all of Greece and the Balkans.
D. Finally, it is a city with a historic centre which the visitor can walk, enjoy and, thus, come to know the city better (Table 1).

[2] https://www.grhotels.gr/category/epicheirimatiki-enimerosi/statistika/plirotites-afikseis/.

Table 1 R.U. of Thessaloniki. *Source* Hellenic Chamber of Hotels 2019

Hotel capacity of the regional unit of Thessaloniki							
2018							
		5*	4*	3*	2*	1*	General total
Regional unit of Thessaloniki	Units	14	28	1	27	33	145
	Rooms	1854	2124	2301	950	778	8088
	Beds	3473	3876	4409	1789	1515	15,339

5.2 Thessaloniki as a Religion-Cultural Tourism Destination

Tourism researchers assert that cultural tourism is the fastest growing sector in the tourism industry (Richards 2000). It is a special form of tourism which is certainly less harmful than mass tourism (Richards 2011) and includes participation in festivals and other cultural events, sightseeing and visits to cultural and religious monuments, etc. (WTO 1985).

Thessaloniki, however, was not merely an urban destination. It is also a prime cultural destination. Thessaloniki is home to a wealth of monuments (over 100), while also being home to fifteen UNESCO World Heritage Sites. It is on account of this, that the city is also referred to as the UNESCO City. Thirteen of these monuments are Byzantine churches which are open to the public, while the other two formed part of the city's walls and also regard a Byzantine bath. Furthermore, Thessaloniki is home to infrastructure from the history of silk trade, while, on the academic front, the "European Interdisciplinary Silk Road Tourism Centre" operates in the Aristotle University of Thessaloniki since 2017.

Religion, silk, culture and modern tourism are interwoven in the city's history and present. In modernity, a visit to a religious monument integrates multiple religious, cultural, social and economic aspects and constitutes what we term as "Heritage" (Tsironis 2020). The major religious centres always constituted part of a land's identity, functioning as economic and political hubs between cities and empires, but also as the meeting points of cultural and artistic flows (Tsironis 2020). Visits to religious monuments constitute a visit to the most significant sights of a country and a culture. Silk, which historically connects the history of China, the people along the great trade routes in the East and the Byzantine Empire, was used in precious religious and also secular heirlooms from that era. In recent years, Thessaloniki, being the port of the Balkans, became a great hub, but also a trade centre of silk products (Lopez 1945; Yücekaya 2021). Today, this cultural heritage is connected with traditional forms of tourism, such as the religious, cultural and industrial tourism, but, more importantly, it can create new potential through the implementation of innovative tourism proposals. Dark tourism is one of the new tourism proposals which can be applied to the cities found along the silk roads. By virtue of its monuments, Thessaloniki can be connected to the Silk Road and develop new tourism products, such as dark tourism.

5.3 Thessaloniki as a Dark Tourism Destination

Thessaloniki has a dark past (Kornetis 2019). The city has witnessed many bloody, violent and catastrophic events since its foundation and there are even more spots and monuments signifying these unfortunate incidents. The historic centre of the city is home to monuments, places of memory, forgiveness and catharsis. Mazower (2007), refers to Thessaloniki as the "city of ghosts". The greatest slaughter in the history of the city happened in 390 AD, where more than 7500 civilians were slaughtered in the Hippodrome (the city's circus). Its modern history was marked by many political assassinations, such as those of the Consuls (the infamous Salonika Incident), of King George I, the assassination of journalist George Polk, of Yannis Zevgos, Grigorios Lamprakis, Tasos Tousis. The city of Thessaloniki and Eleftherias (Liberty) Square marked the beginning of the journey without return to Auschwitz concentration camp for 50,000 Jews. The city also features historic military cemeteries, remands of WWI and the Macedonian Front, which are also visitable. The city's walls, such as the Heptapyrgion and White Tower, which is also its landmark, were in the past amongst the toughest prisons, where torture, pain and death lurked in the shadows.

Thessaloniki has a plethora of dark spots, monuments and sights in general, which carry historical weight and if appropriately utilized could render the city a dark tourism destination. Unfortunately, literature shows that urban tourism destinations frequently do not identify and acknowledge their dark tourism products and do not promote them with consistency and effectiveness (Powell and Kennel 2016). Dark tourism, via its connection with religious and cultural tourism, could also constitute a tool for local tourism development. The unique religious, cultural, and monumental footprint of the city offers the potential and opportunity for a very significant development of this field.

6 Methodology

In order to design a dark route in Thessaloniki, historical and social data relating to the city were studied and recorded. On this basis the research team linked the city's cultural to its dark past. The objective was for this route to serve a new, multiform, and sustainable tourism experience describing a dark past with sights and monuments and to narrate stories and events which have been stigmatized by death, violence and destruction, while relating to the religious and cultural imprint of the city.

Account was taken of the fact that the following are necessary in order to design such a route[3]:

(a) The existence of material (monuments, archaeological sites, etc.) and intangible (festivals, etc.) cultural assets; (b) connection/accessibility infrastructure (trails, road network, bicycle lanes, etc.), and (c) collaboration & participation of public and private entities for the financing and promotion of the route.

Three stages are followed for the design of the route[4]:

(1) The establishment of the objective of the route; (2) the selection of its reference points (events, monuments, etc.) and (3) the marking of the spots on the route and on a printed map.

The UNESCO monuments and specifically those with a dark history have been selected as the thematic aspect of the route. Moreover, based on the space on which it will develop, it is an urban route which will include those UNESCO monuments located inside the city. The objective was for this route to constitute a new dark tourism option, for the residents of the city to also become aware of its dark history and, finally, for it to be an educational tool (Farmaki 2013; Kamber et al. 2016; Lennon and Teare 2017; Raine 2013). According to Cohen (2011), dark tourism can be used as an educational tool transmitting messages to society.

The UNESCO world heritage sites with a dark past were then selected, as the assets to be promoted and brought to the forefront. More specifically, these are the White Tower, Heptapyrgion and the Holy Church of the Virgin Mary Acheiropoietos. Both the White Tower as well as the Heptapyrgion belong to the city's fortifications and in the past served as prisons for those punished with long sentences, who were subjected to terrible torture and were executed inside their walls (Stefanou and Antoniadou 2021). The church, on the other hand, is also a monument where a civil dispute led to the slaughter of the Zealots rebellions (Smarnakis 2018). The Hippodrome Square was added, another landmark of the city, where no monument survives and thus its presence is intangible but was the location of the greatest slaughter in the city's history (7,500 dead), a shocking event with global impact which stigmatized the city, while it is situated on the axis of the route. Finally, another selection criterion was the death toll recorded in these spots, ranking them in the darkest spot of Stone's spectrum (2006).

Based on the information above, an urban dark route was designed, entitled: "***The UNESCO monuments in Thessaloniki: A dark route***" (Figs. 1). This route meets the criteria cited above. Thus, there are natural cultural, tangible and intangible assets. All stops along this particular route are monuments (tangible assets), while the surrounding area of the White Tower and Heptapyrgion, as well as the organization of Heptapyrgion Cultural Festival constitute its intangible assets. The stops on the route are connected by a road network and can be visited either by car, or mass transportation, on foot or bike. Finally, the tourism bodies in the city may collaborate for the promotion of these monuments and sights (guided tours, maps, cultural happening, participation in exhibitions and expos in Greece and abroad, etc.).

Using the google maps application, presented below is the route, which is a walking route spanning 3.5 km.

It is a route satisfactory in time and distance and in the course of which, the participant crosses the entire historical centre of Thessaloniki. It is a route constituting a synergy of the Greek-Orthodox tradition of the city, exemplified by its cultural monuments, as well as the many natural beauty sceneries the city possesses. But it is also a multi-level one with respect to the needs of travellers, as it serves their cultural, religious and educational pursuits. This route on the urban web of the city offers to the tourist-visitor a tour of the historical monuments in the city and its dark history, while also being entertaining. Parallel to this, the route incorporates other sights, and the tourist-visitor will be in touch with the city's rhythm, will smell the whiffs of local cuisine emanating from the taverns and will experience the authenticity this destination offers. This will automatically turn the traveller to a multifaceted tourist (Fig. 2), in a city with a multicultural profile.

7 Conclusion

Thessaloniki and its "dark" history can become a station along the great cultural route that is the "Silk Road". At the same time, it can also connect other cities, themselves stations along the Silk Road, the focus this time being not just silk, but the dark aspects of history, thus bringing a new tourism product, **the "dark trail" along the Silk Road**, to the forefront. Thessaloniki needs not devise of some history; the city is brimming with its very own, authentic dark

[3] International Scientific Committee on Cultural Routes (CIIC) of ICOMOS. (2008). The ICOMOS charter on Cultural Routes.

[4] International Scientific Committee on Cultural Routes (CIIC) of ICOMOS. (2003). CIIC ACTIVITIES, GENERAL SCOPE.

Fig. 1 The dark tourism urban route in Thessaloniki's UNESCO monuments (google map)

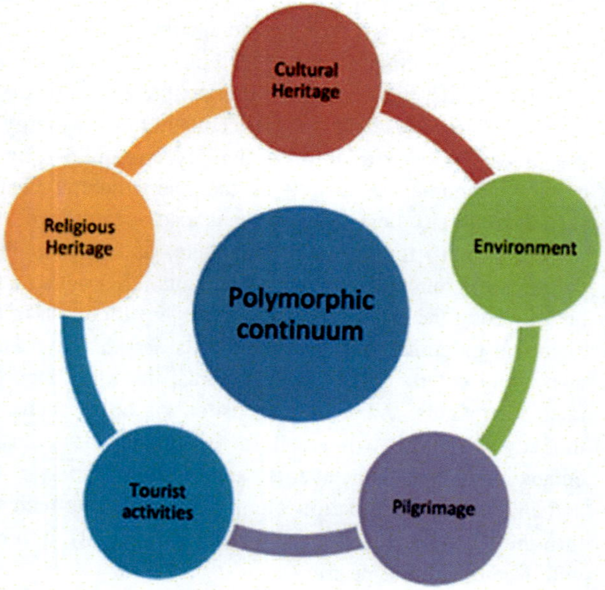

Fig. 2 The polymorphic touristic continuum (Tsironis 2020)

history. Nor does it need to invest in infrastructure, in order to shape spaces for dark events. It has plentiful of these.

In the present, post-covid era, where tourists-consumers turn to quality tourism products and experiences, special forms of tourism have the potential to increase the popularity of the region and the Silk Road is an excellent opportunity to highlight new destinations and to interconnect historic routes as an added value for the countries involved. The example set by Thessaloniki may function as a guide for other cities connected to the Silk Road, in order for them to relate their cultural heritage with their dark history and to be led to the design of similar dark routes.

References

Aoyama, R.: "One belt, one road": China's new global strategy. J. Contemp. East Asia Stud. **5**(2), 3–22 (2016)

Arase, D.: China's two silk roads initiative: what it means for Southeast Asia. Southeast Asian Aff. **2015**(1), 25–45 (2015)

Ashworth, G.J.: Tourism and the Heritage of Atrocity: Managing the Heritage of South African Apartheid for Entertainment, pp. 95–108. Cabi Publishing (2004)

Berg, L. Van den, Borg, J. Van der, Meer, J. van der: Urban Tourism; Performance and Strategies in Eight European Cities, Euricur Series, Ashgate, Aldershot (1995)

Biran, A., Poria, Y., Oren, G.: Sought experiences at (dark) heritage sites. Ann. Tour. Res. **38**(3), 820–841 (2011)

Biran, A., Hyde, K.F.: New perspectives on dark tourism. Inter. J. Cult., Tour. Hospitality Res. (2013)

Blom, T.: Morbid tourism-a postmodern market niche with an example from Althorp. Nor. Geogr. Tidsskr. **54**(1), 29–36 (2000)

Božic, S., Spasojević, B., Vujičić, M.D., Stamenkovic, I.: Exploring the motives for religious travel by applying the AHP method: a case study of Vujan Monastery (Serbia). Inter. J. Relig. Tour. Pilgrim. **4** (4), 33–47 (2016)

Bowen, D.: Contemporary Tourist Behaviour: Yourself and Others as Tourists. Cabi (2022)

Briedenhann, J., Wickens, E.: Tourism routes as a tool for the economic development of rural areas—vibrant hope or impossible dream? Tour. Manage. **25**(1), 71–79 (2004)

Buhalis, D., Amaranggana, A.: Smart tourism destinations. In: Information and Communication Technologies in Tourism, pp. 553–564. Springer, Cham (2013)

Cannell, D.M.: The Tourist: A New Theory of the Leisure Class. University of California Press (2013a)
Casson, L.: Travel in the Ancient World. JHU Press (1994)
Cetin, G., Bilgihan, A.: Components of cultural tourists' experiences in destinations. Curr. Issue Tour. **19**(2), 137–154 (2016)
Cohen, E.H.: Educational dark tourism at an in populo site: the Holocaust Museum in Jerusalem. Ann. Tour. Res. **38**(1), 193–209 (2011)
Crompton, J.L.: Motivations for pleasure vacation. Ann. Tour. Res. **6** (4), 408–424 (1979)
Dale, C., & Robinson, N.: Dark tourism. In: Robinson, P., Heitmann, S., Dieke, P. (eds), Research Themes for Tourism, Vol. 38, pp. 193–209 (2011)
Dalton, D.: Dark Tourism and Crime. Routledge (2014)
Dann, G.M.: The Dark Side of Tourism. CIRET-International Center for Research and Studies in Tourism (1998)
Dimitrovski, D., Senić, V., Marić, D., Marinković, V.: Commemorative events at destination memorials–a dark (heritage) tourism context. Int. J. Herit. Stud. **23**(8), 695–708 (2017)
Farmaki, A.: Dark tourism revisited: a supply/demand conceptualisation. Inter. J. Cult., Tour. Hospit. Res. **7**(3), 281–292 (2013)
Foley, M., Lennon, J.J.: JFK and dark tourism: a fascination with assassination. Int. J. Herit. Stud. **2**(4), 198–211 (1996)
Griffin, K., Raj, R.: The importance of religious tourism and pilgrimage: reflecting on definitions, motives and data. The Inte. J. Religious Tour. Pilgrimage **5**(3), 2–9 (2018)
Jamal, T., Lelo, L.: Exploring the conceptual and analytical framing of dark tourism: From darkness to intentionality. In: Tourist Experience, pp. 49–62. Routledge (2010)
Kamber, M., Karafotias, T., Tsitoura, T.: Dark heritage tourism and the Sarajevo siege. J. Tour. Cult. Chang. **14**(3), 255–269 (2016)
Kang, E.J., Scott, N., Lee, T.J., Ballantyne, R.: Benefits of visiting a 'dark tourism'site: The case of the Jeju April 3rd Peace Park, Korea. Tour. Manage. **33**(2), 257–265 (2012)
Karluk, S.R., Karaman, S.C.: Bridging civilizations from Asia to Europe: The silk road. Chinese Business Rev. **13**(12), 730–739 (2014)
Kennell, J., Powell, R.: Dark tourism and World Heritage Sites: a Delphi study of stakeholder perceptions of the development of dark tourism products. J. Herit. Tour. **16**(4), 367–381 (2020)
Kornetis, K.: Public history and the revival of repressed sephardic heritage in Thessaloniki. In: Retelling the Past in Contemporary Greek Literature, Film, and Popular Cult*ure*, 29 (2019)
Korstanje, M.E.: Towards new horizons in dark tourism studies. Gazing at death, 1 (2017)
Kostopoulou, S., Toufengopoulou, A., Kyriakou, D., Malisiova, S., Sofianou, E., Xanthopoulou-Tsitsoni, V.: The Western Silk Road in Greece. Aristotle University of Thessaloniki, Silk Road Programme, 2019-09 (2016). Available on line: https://webunwto.s3.eu-west-1.amazonaws.com/2019-09/aristotleuniversitynationalswotanalysis-westernsilkroadgreece.pdf. Retrieved: 18.05.2022
Kostopoulou, S., Sofianou, P.K., Tsiokanos, K.: Silk road heritage branding and polycentric tourism development. Sustainability **13** (4), 1893 (2021)
Kunwar, R.R., Karki, N.: Dark tourism: Understanding the concept and recognizing the values. J. APF Comm. Staff College **2**(1), 42–59 (2019)
Kyle, D.G.: Spectacles of death in ancient Rome. Routledge (2012)
Lennon, J.J., Teare, R.: Dark tourism–visitation, understanding and education; a reconciliation of theory and practice? In: Worldwide Hospitality and Tourism Themes (2017)
Light, D.: Progress in dark tourism and thanatourism research: an uneasy relationship with heritage tourism. Tour. Manage. **61**, 275–301 (2017)
Lippard, L.R.: On the Beaten Track: Tourism, Art and Place. New Press (1999)
Logan, W., Reeves, K.: Places of pain and shame, pp. 1–19. Routledge, Dealing with Difficult Heritage. Nueva York (2009)
Lopez, R.S.: Silk industry in the Byzantine Empire. Speculum **20**(1), 1–42 (1945)
Malinowski, B.: Magic, Science and Religion: And Other Essays, by Bronislaw Malinowski. Selected, and with an Introduction by Robert Redfield. Beacon Press, Boston (1948)
Mark, J.J.: Silk road. In: Ancient History Encyclopedia, 28 Mach (2014)
Maxim, C.: Challenges faced by world tourism cities–London's perspective. Curr. Issue Tour. **22**(9), 1006–1024 (2019)
Mazower, M.: Salonica, City of Ghosts: Christians, Muslims and Jews 1430–1950. Vintage (2007)
Meyer, D.: Tourism routes and gateways: key issues for the development of tourism routes and gateways and their potential for pro-poor tourism. In: ODI discussion paper (2004)
Miller, D.S., Gonzalez, C., Hutter, M.: Phoenix tourism within dark tourism: Rebirth, rebuilding and rebranding of tourist destinations following disasters. In: Worldwide Hospitality and Tourism Themes (2017)
Minić, N.: Development of "dark" tourism in the contemporary society. Зборник радова Географског института" Јован Цвијић" САНУ **62**(3), 81–103 (2012)
Oren, G., Shani, A., Poria, Y.: Dialectical emotions in a dark heritage site: a study at the Auschwitz Death Camp. Tour. Manage. **82**, 104194 (2021)
Pattanaro, G., Pistocchi, F.: Linking destinations through sustainable cultural routes. Symphonya. Emerg. Issues Manag. **1**, 83–96 (2016)
Poria, Y., Butler, R., Airey, D.: The meanings of heritage sites for tourists: the case of Massada. Tour. Anal. **9**(1–2), 15–22 (2004)
Poria, Y., Ashworth, G.: Heritage tourism—Current resource for conflict. Ann. Tour. Res. **36**(3), 522–525 (2009)
Powell, R., Kennell, J.: Dark cities? Developing a methodology for researching dark tourism in European cities. In: Tourism and Culture in the Age of Innovation, pp. 303–319. Springer, Cham (2016)
Powell, R., Kennell, J., Barton, C.: Dark Cities: A dark tourism index for Europe's tourism cities, based on the analysis of DMO websites. Inter. J. Tourism Cities (2018)
Puczko, L., Ratz, T.: Tourist and resident perceptions of the physical impacts of tourism at Lake Balaton, Hungary: Issues for sustainable tourism management. J. Sustain. Tour. **8**(6), 458–478 (2000)
Raine, R.: A dark tourist spectrum. Inter. J. Cult., Tour. Hospit. Res. **7** (3), 242–256 (2013)
Richards, G.: Cultural tourism: challenges for management and marketing. In: Trends in Outdoor Recreation, Leisure and Tourism, 187–196 (2000)
Richards, G. (ed.).: Cultural tourism: Global and local perspectives. Psychology Press (2007)
Richards, G.: Creativity and tourism: the state of the art. Ann. Tour. Res. **38**(4), 1225–1253 (2011)
Rojek, C.: Fatal attractions. In: Ways of Escape, pp. 135–172. Palgrave Macmillan, London (1993)
Sather-Wagstaff, J.: Heritage that hurts: tourists in the memoryscapes of September 11. Routledge (2016)
Seaton, A.V.: Guided by the dark: from thanatopsis to thanatourism. Int. J. Herit. Stud. **2**(4), 234–244 (1996)
Sharpley, R., Stone, P.R. (Eds.): The Darker Side of Travel: The Theory and Practice of Dark Tourism. Channel view publications (2009)
Skinner, J. (Ed.).: Writing the Dark Side of Travel. Berghahn Books (2012)

Smarnakis, Y.: Thessaloniki during the Zealots' revolt (1342–1350): power, political violence and the transformation of the urban space. Scand. J. Byzantine Modern Greek Stud. **4**, 119–148 (2018)

Smith, V.: War and tourism: an American ethnography. Ann. Tour. Res. **25**, 202–227 (1998)

Smith, N., Croy, W.G.: Presentation of dark tourism: Te Wairoa, the buried village. In: Taking Tourism to the Limits, pp. 223–238. Routledge (2006)

Stefanou, E., Antoniadou, I.: Eptapyrgio, a modern prison inside a world heritage monument. In: Contested Antiquity: Archaeological Heritage and Social Conflict in Modern Greece and Cyprus, p 272 (2021)

Stoddart, H., Rogerson, C.: Tourism routes, local economic promotion and pro-poor development The case of the crocodile ramble. Africa Insight **38**(4), 10–26 (2009)

Stone, P.: Review: dark tourism—cashing in on tragedy?" Paper presented at the Tourism Society Seminar Event, 17 October, London (2005). Available at: www.dark-tourism.org.uk (Retrieved 06 September 2013)

Stone, P.R.: A dark tourism spectrum: towards a typology of death and macabre related tourist sites, attractions, and exhibitions. Tourism: An Int. Interdis. J. **54**(2), 145–160 (2006)

Stone, P.: Dark tourism scholarship: a critical review. Int. J. Cult., Tour. Hospit. Res. (2013b)

Stone, P., Grebenar, A.: Making tragic places: dark tourism, kitsch and the commodification of atrocity. J. Tour. Cult. Chang., 1–18. https://doi.org/10.1080/14766825.2021.1960852

Stone, P., Sharpley, R.: Consuming dark tourism: a thanatological perspective. Ann. Tour. Res. **35**(2), 574–595 (2008)

The Conversation: Dark tourism memorial sites will help us heal from the trauma of coronavirus, 24 June (2020). Available: https://theconversation.com/dark-tourism-memorial-sites-will-help-us-heal-from-the-trauma-of-coronavirus-139164. Retrieved: 18.05.2022

The Telegraph: Wuhan to become popular 'dark tourism' destination, June 1 (2020). Available:https://www.telegraph.co.uk/travel/comment/wuhan-dark-tourism-coronavirus/?fbclid=IwAR1Bg0Cm6-gsNsFD7R2Bdq-Dd2p6MWLhxruzbe5834-iorq5qMapnztTDZQ. Retrieved: 18.05.2022

Timothy, D.J., Boyd, S.W.: Heritage tourism in the 21st century: valued traditions and new perspectives. J. Herit. Tour. **1**(1), 1–16 (2006)

Tsironis, C.N.: Religious routes and destinations: a polymorphic social approach. Sociología a Spoločnosť/sociology and Society **5**(2), 56–70 (2020)

Tsironis, C.N., Sylaiou, S., Stergiou, E.: Risk, faith and religious tourism in Second Modernity: visits to Mount Athos in the COVID-19 era. J. Heritage Tour., 1–17 (2021)

UNWTO, Silk road. Available online: https://www.unwto.org/silk-road Retrieved: 18.05.2022

Urry, J.: The Tourist Gaze, 2nd ed. Sage, London (2002)

Wearing, S.L., Foley, C.: Understanding the tourist experience of cities. Ann. Tour. Res. **65**, 97–107 (2017)

Wight, A.C.: Philosophical and methodological praxes in dark tourism: controversy, contention and the evolving paradigm. J. Vacat. Mark. **12**(2), 119–129 (2006)

Williams, T.: The Silk Roads: An ICOMOS Thematic Study. ICOMOS (2014)

Wright, D.W.M.: Sport hunting and tourism in the twenty-second century: humans as the ultimate trophy. Foresight (2009)

World Tourism Organization. Secretary-General: The State's Role in Protecting and Promoting Culture as a Factor of Tourism Development and the Proper Use and Exploitation of the National Cultural Heritage of Sites and Monuments for Tourism: Report of the Secretary-General on the General Programme of Work for the Period 1984–1985 (1985)

World Tourism Organization: Western Silk Road Roadmap. UNWTO, Madrid (2018)

Yan, B.J., Zhang, J., Zhang, H.L., Lu, S.J., Guo, Y.R.: Investigating the motivation–experience relationship in a dark tourism space: A case study of the Beichuan earthquake relics, China. Tour. Manage. **53**, 108–121 (2016)

Yankholmes, A.K., Akyeampong, O.A.: Tourists' perceptions of heritage tourism development in Danish-Osu, Ghana. Int. J. Tour. Res. **12**(5), 603–616 (2010)

Yankovska, G., Hannam, K.: Dark and toxic tourism in the Chernobyl exclusion zone. Curr. Issue Tour. **17**(10), 929–939 (2014)

Yücekaya, H.: Thessaloniki Sericulture. Iğdır Üniversitesi Sosyal Bilimler Dergisi **26**, 754–768 (2021)

Author Index

S. Kostopoulou et al. (eds.), *Cities' Vocabularies and the Sustainable Development of the Silkroads*, Advances in Science, Technology & Innovation, https://doi.org/10.1007/978-3-031-31027-0